AF454337

Biodeterioration of Cultural Property and Researches on Conservation of Heritage Buildings

Biodeterioration of Cultural Property and Researches on Conservation of Heritage Buildings

— A Festschrift in honour of
Padma Shri Dr. O.P. Agrawal –

— Editors —

Arun Arya

Virendra Nath

Mandana Barkeshli

Abduraheem K.

2023

Daya Publishing House®
A Division of

Astral International Pvt. Ltd.
New Delhi – 110 002

Published by   :   **Daya Publishing House®**
A Division of
Astral International Pvt. Ltd.
– ISO 9001:2015 Certified Company –
4736/23, Ansari Road, Darya Ganj,
New Delhi-110 002
Ph. 011-43549197, 23278134
E-mail: info@astralint.com
Website: www.astralint.com

Late Padma Shri Dr. O.P. Agrawal

(22-10-1931- 15-8-2021)

Dedicated in the loving Memory of

Padma Shri Dr. O.P. Agrawal

(Pioneer in the Field of Conservation Science in India)

आचार्य देवव्रत
राज्यपाल, गुजरात
गांधीनगर-३८२०२१

'2 3 MAY 2022

MESSAGE

It gives me immense pleasure that in this year of Azadi Ka Amrut Mahotsav, the International Council of Biodeterioration of Cultural Property (ICBCP) is publishing a volume in the honour of Padma Shri Dr. O. P. Agrawal, founder Director of NRLC, Lucknow. He was also one of the founders of INTACH, and founder President of ICBCP. Based at Lucknow he helped to restore and enrich the cultural heritage in India and abroad and also trained the experts in the field of conservation. The present volume entitled "Biodeterioration of Cultural Property and Researches on Conservation of Heritage Buildings," published by ICBCP is an apt tribute to his services to the society.

Beyond the Indus valley civilization and the existence of large port city of Lothal in Gujarat, as well as the seats of higher learning Nalanda and Taxsilla, beautiful Taj Mahal and temples of Jagannath Puri and Sun Temple – Konark are not the only prominent Heritage sites of the county. Places like Statue of Unity, Rani ki Vav in Patan, Champaner monument and Rann of Katchh and Gir lion sanctuary, are tourist attractions and important heritage sites of Gujarat. Due to biodeterioration, pollution and urbanization, many of the important heritage sites are under threat of damage. It is in this view the valuable opinion of experts is needed to conserve the tangible and intangible heritage.

I am sure, the chapters will not only provide knowledge about historical monuments but will infuse the love for our traditional cultural heritage and respect for Mother Nature. I thank the members of ICBCP for talking a right step at right time and presenting the volume for the benefit of readers and expert alike.

(Acharya Devvrat)

Message

Dr. Virendra Nath

India has vast diversity and a rich cultural heritage. A large number of monuments and historical buildings spread all over the country are facing challenges of destruction due to changing climate and biodeterioration. To safeguard the cultural heritage Central government passed The Ancient Monuments and Archaeological Sites and Remains Act, 1958, which was amended in 2010 to strengthen its penal provisions, to prevent encroachments and illegal construction close to the monuments. Archaeological Survey of India is doing a great job to protect the damage. We need to analyse the beauty and importance of 40 UNESCO World Heritage sites recognized in the country and historical significance of these and many more monuments. Conservation experts should come forward in help to restore, preserve these valuable assets of human civilization by identifying the causes, improve the stability and cohesion of the archival material with due consideration to the aesthetic affect. In the name of progress and era of commercialization our oral cultural and social traditions should not erode and craft skills should not be lost. We have to imbibe the new zeal and enthusiasm among youngsters of the country to have reverence for ancient heritage and love for the Mother Nature.

International Council of Biodeterioration of Cultural Property (ICBCP) and INTACH both at Lucknow, were initiated by Padma Shri Dr. O.P. Agrawal. These organizations helped to promote the scientific research in the field of biodeterioration, particularly related to care and conservation of Art and valuable Cultural heritage. In a tropical country like India, growth of microorganisms and insects on the materials of cultural property is formidable. Even hard and durable substances such as stones and marbles are attacked and damaged by bacteria, algae, lichens, bryophytes and higher plants. Paper manuscripts books textiles and woollen objects are destroyed by insects of various types very rapidly. The control is not an easy task and no solution of a permanent type is available to control them. ICBCP is registered under registration society act 21, 1861 at Indian Conservation Institute at Lucknow. One of the mandates of ICBCP is to hold conferences in different countries and to bring out publications. It has organized 5 International conferences so far. They were held at Lucknow in India, Japan, Thailand, Iran, and at Australia. National conferences were held at Delhi, Kanpur, Aligarh and Vadodara. Through these International as well as National conferences the mission for the study of Biodeterioration of Cultural property is under progress. Council has organised Padma Shri O.P. Agrawal Memorial lecture in his honour on 7th February 2022, which was delivered by Prof. Arun Arya and another in the fond memory of its General Secretary late Dr. Shashi Dhawan, on Museum's day, 18th May 2022, which was delivered by Dr. Mamta Mishra, Director INTACH Lucknow.

It gives me immense pleasure that ICBCP is publishing a volume entitled **Biodeterioration of Cultural Property and Researches on Conservation of Heritage Buildings,** in his honour. I really thank the contributors for their articles on varied subjects related to conservation. This work will provide an excellent review on various aspects to ponder upon tangible and intangible heritage, and will generate enthusiasm among experts working in the field to try the newer technologies available. It will be equally useful for academicians and common citizen.

I convey my good wishes to the editorial team and especially to Prof. Arun Arya for taking initiative and compiling the book.

Dr. Virendra Nath
President ICBCP
Senior Scientist, NBRI
Lucknow

Preface

Conservation Science is an interdisciplinary approach, complex in nature, global in character, comprising both theoretical and technological advances, and adopts the modern concept of integrated conservation. This concept aims to satisfy the dual purpose of preserving and disseminating knowledge about cultural heritage in an integrated way, in close connection with socioeconomic and cultural development at micro and macro levels. Out of this developed the concepts of collaborative conservation and participatory conservation having in addition to the conservator and the renowned specialists in the field like curator, restorer, *etc.*, should include representatives from the pure sciences (Geology and Mineralogy, Archaeology, Chemistry, Biology, Microbiology, Applied Science, Environmental Science, Molecular biology), from the fields of technology and art history and even the members (artists and local natives) of communities from regions with tangible heritage value.

Human development in different parts of the world has proceeded in many ways. Each society found their own ways to live life happily, to educate, to dress, to govern itself, to communicate with world, entertain and to generate wealth. With rapid changes in technology, globalization, use of alternative sources of energy, internet for information and communication, we are now living in a changed society. Humankind's achievements are proceeding exponentially, we are in a process to create small Sun like sources of energy, linked the voice of man across the globe already, transforming species by gene transfer it is impossible to predict its future. Corona like pandemic outbreaks have warned the human beings that we are still a fragile creature on this planet Earth. We have to ponder in our past look how the societies flourished during Vedic, Harrapan, and Egyptian civilizations and vanished ultimately. Our cultural

Heritage can help us to learn the lessons from past and we can preserve our culture and Good Values for future generations.

International Council of Biodeterioration of Cultural Property (ICBCP) and INTACH both at Lucknow, were initiated by Padma Shri Dr. O.P. Agrawal Ji. These are devoted towards promotion of scientific research in the field of biodeterioration of materials. Particularly related to care and prevention of Art and cultural heritage objects. In a tropical country like India, growth of microorganisms and insects on the materials of cultural property is formidable. Even hard and durable substances such as stones and marbles are attacked and damaged by bacteria, algae and lichens. Paper manuscripts, books, textiles and woollen objects are destroyed by insects of various types very rapidly. The control is not an easy task and no solution of a permanent type is available to control them. ICBCP is registered under registration society act 21, 1861 at Indian Conservation Institute at Lucknow. One of the mandates of ICBCP is to hold conferences in different countries and to bring out publications. It has organized 5 International conferences so far. They were held at Lucknow in India, Japan, Thailand, Iran, and at Australia. National conferences were held at Delhi, Kanpur, Aligarh and Vadodara. Through these International as well as National conferences the mission for the scientific study of Biodeterioration of Materials and Cultural property has gained momentum. On the sad demise of its founder respected Agrawal Ji on 15th August 2021, a decision was taken to publish a volume in his beloved memory. The present volume is outcome of that. Council also organised Padma Shri O.P. Agrawal Memorial lecture in his honour on 7th February 2022, which was delivered by Prof. Arun Arya and another in the fond memory of its General Secretary late Dr. Shashi Dhawan, on Museum's day 18th May, 2022, which was delivered by Dr. Mamta Mishra, Director INTACH.

Not only cultural loss of valuables, the biodeterioration has direct link to our economy as it is related to survival of UNESCO heritage monuments of International or of National importance which attract lot of tourists across the world. Protecting old monuments and statues as well as paintings, cave murals, and artefact from microbial damage is very essential. All round efforts are needed and scientific inputs are required to restore the damaged objects and protect the old ones. Contribution of our ancestors is very immense in evolving a civilized society with different cultures and social values but common needs of food, water and recreation by music and games. In this age of electronic revolution and booming social media it is very necessary that we should protect the basic human values of faith, sincerity and willingness to live together in harmony with Nature. The knowledge of folklore and religion can be utilized to bridge the difference, among communities and dwell a path of happy and satisfied life. The new policies on the approach to cultural heritage consider the safeguarding and inclusion of cultural heritage assets within a global system of values, the development of cultural tourism as a way of guaranteeing the right of access to culture and the integration of active participation of the population in cultural heritage conservation policy. Even if the concept of integrated conservation is relatively modern, the attempts to attract members of the public/community to the activities aimed at preserving cultural heritage have a longer history. The role of community in the cultural heritage conservation process (preservation, restoration, recovery and hoarding), which imply the concepts of collaborative and participatory conservation, started in 1964 with the Venice Charter.

The invited articles presented in the Book are divided into three separate sections. A.) Biodeterioration of Cultural Objects B.) Biodeterioration of Heritage Monuments, and C.) Hindu Festivals: Promoting Harmony and Heritage. There are total 26 chapters and the last two include conservation of intangible heritage, two festivals celebrated in the country mainly by the Hindu community, at the same time equally popular in other nations as well. We heartily thank Acharya Devvrat, Hon. Governor of Gujarat for sending his good wishes for the present volume.

We would like to thank late Dr. Shashi Dhawan from NRLC and Secretary, ICBCP, Prof. R.N. Mehta in the field of Archaeology and Ancient Indian Culture, late Shri Hira Lal Shah who contributed immensely in development of Children's museum at Amreli, late Dr. I.K. Bhatnagar of National Museum Institute, New Delhi, Dr. Hideo Arai from Japan, Dr. Vinod Daniel, from Australia, late Dr. S.M. Nair, former Director National Museum of Natural History, Past heads of Dept. of Museology of M.S. University of Baroda, late Shri V.L. Devkar, Prof. R.T. Parikh, Dr. S.K. Bhowmik, Former-Director Baroda Museum and Picture Gallery, late Shri Madhav Gandhi associated with CSMVS, Mumbai, Dr. R.B. Srivastava, former Deputy Director DRDO, and many others for their services in the field of Conservation of cultural heritage and Museology in the country.

Editorial team members thank their parental organizations for giving permission and support to produce this work for Technical and research use and public awareness in the field of conservation. Assistance received from NRLC and INTACH is gratefully acknowledged. We thank team of Astral International for nicely publishing the volume.

Prof. Arun Arya

Dr. Virendra Nath

Dr. Mandana Barkeshli

Prof. Abduraheem K.

Contents

Part II: Biodeterioration of Heritage Monuments

Part III: Hindu Festivals: Promoting Harmony and Heritage

The Editors

Prof. Arun Arya, worked as Head, Department of Botany and Environmental Studies in The Maharaja Sayajirao University of Baroda, Vadodara INDIA. He is a Botanist, Phytopathologist, Expert on Biodiversity and Environment, Philatelist and a Popular Science writer. He retired from active services in 2018. He is Vice President of International Council for Biodeterioration of Cultural Property (ICBCP), and Joint Secretary of Society for Clean Environment (SOCLEEN), Vadodara. His work on deterioration of Egyptian Mummy brought him to International scenario. He served as honorary President of Indian Association for Air Pollution Control, he is Fellow of Indian Botanical Society, Indian Phytopathological Society and Fellow of Gujarat Science Academy. He served as Head of the Department of Botany during 2008 – 2013. Dr. Arya is a recipient of Young Scientist award by DST in 1986, and Prof. V. Puri Medal award by IBS. His areas of specialization include fungal taxonomy, Aerobiology, Biodeterioration and Endomycorrhizal research. He visited Japan, Quebec-Canada, Italy, Morocco and Singapore to present his research findings. He had an honour to co-chair the session on Eco regions in 12ᵗʰ World Forestry Congress held at Canada in 2003. This visit to Quebec city, Canada was funded by World Food Organization. He has published more than 150 research papers and 16 books. One of his books on Mushrooms is published by Springer, Singapore, and another entitled Management of Fungal Plant Pathogens, is published by CABI, U.K. in 2010.

Dr. Virendra Nath is President of International Council for Biodeterioration of Cultural Property (ICBCP), he was Former Chief Scientist (Sci. 'G'), CSIR-Emeritus Scientist. He served as Head Bryology group at National Botanical Research Institute, Lucknow. Trained under the able guidance of world renowned Prof. Ram Udar, F.N.A. he is providing his expertise in the field of Bryology for the last 45 years. His expertise includes Taxonomy (Monographic and Floristics), Palynological, Conservation and Bioprospection studies on Bryophytes. He has discovered 14 new species to the Science and 15 new records for India. In the year 1986 he was deputed to visit Poland under CSIR-PAS exchange programme and in 1988 to Royal Botanic Gardens, Kew London. He attended world Conference of Bryology held at Malaysia in 2007. He is recipient of Prof. Ram Udar Medal by Associations of Plant Taxonomists of India in the year 2012. He is Fellow of Association of Angiosperm Taxonomy (F.A.P.T.), Indian Botanical Society (F.B.S), Society of Ethnobotanists (F.E.S), Palaeobotanical Society (F.Pb.S.), The Society of Plant Reproductive Biologists (F.S.P.R.B.). He is Regional Secretary for South Asian Region of International Association of Bryologists (IAB) and contact person for India. He has published more than 160 research papers and 6 books/proceedings.

Prof. (Dr.) Mandana Barkeshli is a Conservation Scientist, associated with National Museum Institute of History of Art, Conservation and Museology, New Delhi, India,. She is currently Head of Research and Post Graduate school of De Institute of Creative Arts and Design, at University College Sedaya International (UCSI), Kuala Lumpur, Malaysia. She is also visiting professor and honorary principle fellow in Faculty of Arts of University of Melbourne, Australia. She has held both senior academic and museum positions, including faculty member of Art University at Tehran and Isfahan, International Islamic University, Malaysia and the first Head Curator of the Islamic Arts Museum Malaysia. Besides her academic and professional work, Prof. Barkeshli is also a keen textile design artist.

Dr. Abduraheem K. is serving as Senior Professor and Chairman, Dept.of Museology, Aligarh Muslim University, Aligharh, U.P. He is Secretary and Ex- Vice President, Indian National Committee for International Council of Museums (ICOM-Paris), General Secretary, International Council for the Biodeterioration of Cultural Property. His areas of specialization are Museology, Conservation and Preservation, Museum Architecture,

Control of Biodeterioration, Museum Research, Documentation *etc.* Awardee of ECONS National Educational Summit, New Delhi, Member of Oxford Round Table Conference, London. Fellow of Indian Social Science Congress. He is a recipient of ARE Egyptian scholarship. Prof. Abduraheem presented 65 research papers in National and International Conferences and has 60 research papers published in Journals of International repute to his credit. He renders his services as expert committee member, Board of studies member, Selection committee member for staff, member of Jury of various institution like M.S. University of Baroda, BHU- Varanasi, Cochin University, National Museum Institute New Delhi, National. Research Laboratory for Conservation (NRLC), Lucknow, INTACH Lucknow, National Museum of Natural History New Delhi, AITC, New Delhi *etc.* He successfully organized 7th International Conference on Biodeterioration of Cultural property of ICBCP in Feb. 2013 at AMU, Aligarh.

Pioneer in the Field of Conservation Science: Padma Shri Dr. O.P. Agrawal – as I knew him

Dr. Neeta Das

For conservation architects, and all architects in general, getting rid of termites is a near impossibility. Termites secretly enter our buildings, find the dark and damp corners in the wooden cupboards, make a large colony, and spread far and wide into other rooms before you find them! Sitting in the shaded verandah of Dr. O.P Agrawal's Lucknow house, sipping a cup of hot tea with savories and sweets, lovingly put out by Madam, Mrs. Usha Agrawal, discussing issues such as these with him, I learnt all the tricks of my trade. In his cool, calm way he would explain with the logic of a scientist, that rather than wasting so much effort and money on expensive chemical anti-termite solutions, one simply needs to understand the insect, its habitat, and food habits. Changing these conditions will force the colony of the termites to perish or move elsewhere! In other meetings he would explain the whole problem of deterioration of materials and how they want to go back to the original state from which they were extracted, how iron rusts, and lime becomes limestone, and while it was impossible to stop these natural processes, the conservation methods could slow them down. When others talked of complicated science, Dr. Agrawal came up with simple and logical solutions in the field of material conservation, and this was in my understanding his greatest strength.

I had heard a lot about Dr. Agrawal and had a great desire to meet him, but kept putting it off with the notion that he would be difficult to approach and like other famous people, very high handed. I still remember first seeing him at a conference in the State Museum in Lucknow, dressed in his usual half sleeve shirt, and loose trousers. When I approached, he looked around in his calm and composed matter, and started talking as if he had known me all his life. Years rolled by and I found myself more and more by his side, learning on the way he conducted his classes, his conversational technique of lecturing, while connecting with his audience through eye contact and interactions. One came out after his lecture refreshed and educated, never bored and intimidated.

**Dr. O.P. Agrawal Ji Receiving Prestigious Honor
Padma Shri for his Contribution to the Field of Conservation from
Smt. Pratibha Patil Honorable President of India.**

He always emphasized that all research in labs should be followed by practical training or workshops to check for their practicality and for disseminating the latest information to others in the field. Dr. Agrawal never believed in keeping any useful information secret or for his personal growth. He had perfected a simple solution to organize, conduct, and document training sessions. In turn he also trained trainers, encouraging capacity building and skill training for the nascent field of material conservation, many of whom were women.

Dr. Agarwal graduated from the University of Allahabad with M.Sc. in Chemistry, and applied his knowledge in the field of material conservation. He joined the conservation team of Archeological Survey of India in 1952 and moved to the National Museum soon after. In 1959 he went to Rome on a fellowship to master the technique of transferring wall paintings on paper and placing them back after restoration.

Dr. O.P. Agrawal during a Hands on Workshop and Conducting a Seminar at the ICI, Lucknow.

He established the National Research Laboratory for Conservation of Cultural Property (NRLC), in Lucknow. His wife and partner, Mrs. Usha Agrawal, often talked about these times and how much labor Dr. Agrawal Ji has put in to get the paper work done together and get the requisite permissions, while she took care of her children and other family members, theirs being a joint family. But she was quick to add, that those were the best years of their life. Dr. Agrawal did not stop here and continued to set up conservation labs in Colombo, Jakarta, Tehran, and Bangkok. After his retirement he joined the Indian Trust for Art and Heritage Culture, and continued to create more labs in all parts of India under the aegis of Indian Conservation Institute (ICI). Most of the work done by INTACH is through these centers, or by those trained in these centers.

It is impossible to account for all the work done by Dr. Agrawal. He successfully completed many projects on material conservation, conducted useful research, wrote and edited innumerable books and articles, received many grants, awards, fellowships, and accolades personally and on behalf of his institution. But what strikes out as admirable is that he also touched the lives of many professionals, inspiring them to continue the work that he was so passionate about. His wife mentions in a newspaper article published soon after his passing away that, 'his brain was a mirror of his actual heart. He spoke of his projects and methods as someone talking about a beloved daughter and expected the same respect and sensitivity from others.' One such person he inspired most was late Dr. Shashi Dhawan, who continued to follow in his footsteps, as Senior Scientific Officer, NRLC, and even after retirement as a member of ICBCP. She was equally passionate and dedicated about issues related to material conservation, and despite her worrisome family conditions, with an invalid brother and an ailing mother, and later her own ill health. She did not leave the side of Dr. Agrawal or his partner, Mrs. Usha Agrawal. They could often be seen discussing conservation, family problems, and world affairs over a cup of tea, a tradition that the Agrawal couple continued till the end.

Dr. Agrawal's career spanned over sixty years after which he received the highest civilian award, the Padma Shri in 2011, from none other than the President of India, for his pioneering work in the field of Material Conservation. But this was also the day that Dr. Agrawal suffered a stroke and remained bed ridden since then for the remaining ten years of his life. One would imagine a man as dynamic, creative, and passionate about his work to break down after such a tragedy, but his inner strength and character, a satisfaction of a life well lived and used, and good family values instilled in him by his Jain parents held him, and his family, together. But the real credit goes to his wife and partner for lovingly and patiently caring for him and handling his work, occasionally taking a quiet break to play solitaire on the computer, her favorite pastime. All his family members, friends, and students like myself, visited them as time permitted, and came back intellectually enriched and warmly loved. It was inspiring to see the legend that even though he gradually lost physical strength and mobility, but his mind remained bright and clear till the end. Only on rare occasions did he seem disturbed by his physical condition, and was always calm and peaceful in his forced retirement, as he was during the peak of his career. Such lives are rare and I am indeed blessed to have been a part of it.

Dr. Neeta Das,
3A Prathama, 320 Chak Garia,
Kolkata – 700 094, West Bengal, India
www.neetadas.com

Life and Works of Padma Shri Dr. O.P. Agrawal

Dr. Virendra Nath and Prof. Arun Arya

Padma Shri late Dr. O.P. Agrawal is well-known for his contribution in the field of conservation of cultural heritage, not only in India, but internationally. After completing his Master's Degree in Chemistry from the University of Allahabad, he preferred to join the Conservation Branch of the Archaeological Survey of India in 1952. He went to Rome in 1959 on a one year fellowship of the Italian Government to study at the famous institution namely the Central Institute of Restoration, Rome. While in Rome, he particularly mastered the technique of transfer of wall paintings from wall and placing them back

Dr. O.P. Agrawal: Pioneer in the Field of Scientific Conservation.

on the walls after repair. The services that he had rendered to archaeology, museums, scientific conservation of art heritage, are unmatched. He had conducted hundreds of workshops and training programmes and his students numbering in hundreds are to be found almost in all countries. He has helped in the establishment of conservation laboratories not only in India, but throughout the world. He was the first expert to use the techniques in India, to save priceless wall-paintings in Chamba, in Kullu, in Sikkim, in Rashtrapati Bhawan and many other places. He was the founder and former Director of the National Research Laboratory for Conservation of Cultural Property (NRLC) established in 1976. He was responsible for the establishment of this great institution from the concept to the finish. One of his book, 'Preservation of Art Objects and Library Materials' has been translated into Hindi, Tamil, Bangla and Marathi. Another of his book 'Care and Preservation of Museum Objects' is translated in Singhalese and Persian. Till 1976 he was with the National Museum, New Delhi and became its Chief Chemist and Head in 1972.

**President Hon'ble Smt. Pratibha Patil Ji Honouring
Dr. O.P. Agrawal Ji with Highest Civilian award Padma Shri in 2011.**

Under his leadership, this institution attained a great height. He served as the Director General of the INTACH, Indian Conservation Institute, Lucknow under which he set up ten conservation centres throughout India and several projects and programmes were successfully implemented. To him his work was his definition of patriotism. He was of the opinion that Cultural legacy of the country should be conserved for posterity, and did everything possible for it. INTACH has been described by the International Council of Museums as the largest network of conservation centres in the world. Besides, he had prepared reports on the basis of which a large number of conservation institutions like the Regional Conservation Laboratories in several states of India like Kerala, Rajasthan, Punjab, U.P. Madhya Pradesh and several others were set up. He also prepared reports for the setting up of the conservation laboratories in several Asian countries like National Museum – Colombo Sri Lanka, Central Museum – Jakarta Indonesia, SPAFA – Bangkok, Department of Fine Arts -Bangkok, National Research Centre for Conservation – Tehran, Iran and several others.

Mrs. U. Agrawal and Dr. O.P. Agrawal with Shri S.P. Gupta and Dr. Nigam.

Dr. Nigam and Mr. and Mrs. Agrawal.

Prof. O.P. Agrawal Working in his Office.

A Group Photograph taken in Rashtrapati Bhawan after Conferring him Civilian Award Padma Shri by President of India.

Dr. Agrawal has been connected with several national and international professional associations and organizations. He was the President of the Museums Association of India (M.A.I.) for two terms. He is the founder and former Secretary of the Indian Association for Study of Conservation of Cultural Property, New Delhi. Later, he became its President and occupied that position for a number of years. He was the Vice President of the International Centre for the Study of Preservation and Restoration of Cultural Property at Rome (ICCROM) founded by UNESCO. He was a member of the Executive Council of

International Institute of Conservation, London. He was the founder President of the ICBCP. He refused to take the U.N. assignment to work as Director of International Centre for Study of the Preservation of cultural property around his retirement only to make him available to set up country's first National Research laboratory of Cultural Property.

Dr. Agrawal was a prolific writer and 38 books are in his credit, out of which several are manuals on the subject of conservation. He has written about 200 articles and papers on conservation, archaeology and museology. On account of his life-time service to conservation, he was conferred by the International Council of Museum, Brazil, the title 'Personality of Conservation 1989'. The International Centre for Conservation Rome, founded by UNESCO conferred upon him the 'ICCROM Award – 1993'. The Kanpur University conferred upon him D.Litt. (HonorisCausa) in 1994. He was specially felicitated by the Governor of Uttar Pradesh in 1993. A Felicitation Volume entitled 'Recent Trends in Conservation' was published by his admirers in 1995. A few years back Honorary Fellowship of the International Institute for Conservation, London has been conferred upon Dr. Agrawal. There are only 18 scientists throughout the world on whom this honour has ever been bestowed. Lucknow Management Association (LMA) honoured him for his life-long work in the service of the community. In 2008, he received INTACH-SATTE Award for his contribution to the promotion of tourism through conservation, *etc.* For his tireless services in the field of conservation, he was conferred with prestigious civilian award Padma Shri in 2011.

With his Wife during a visit to Rome, Italy

Some of his notable conservation projects were, Restoration of Canvas Paintings and Ceiling paintings in Rashtrapati Bhavan, New Delhi, wall-paintings in various heritage buildings in Santiniketan, Wall-paintings of

Moti Mahal, Gwalior, conservation of wall-painting sites, at least 40 spread all over India, conservation of a large number of art objects, housed in museums, libraries, temples, churches, palaces, and so on. Contributions of Dr. Shashi Dhawan, first as a student and later as his close associate in the field of conservation are praiseworthy. His wife Mrs. Usha Agrawal also decided to share Dr. Agrawal's load and joined INTACH in 1984. She accompanied her on many restorative trails. While she has been witness to every turning point of his life, Usha Ji tells about memories of the project at Tulja Bhawani temple of Kullu, where an ancient wall painting was to be extracted and brought to the National museum, Delhi. Citing an example Mrs. Agrawal tells " when representatives from the Manglorean church approached him, the team asked if he will take the responsibility if the outcome of his restoration work was negative. He immediately stated that please see me like a doctor of the painting.... I will treat he will cure. The team went back. But almost a decade later they came back and thanked him, the church is a masterpiece today."

With his Daughter Ms. Poonam Agrawal.

"When I began my career, there were only 5-6 conservators across the country. Today the number has increased over 2000, but this is too small going by the volume of heritage India has," was told by Dr. Agrawal when Founder's of Lucknow Society, met him and discussed the conservation aspects in Lucknow. Lucknow Society honoured him for being one of the pillars of conservation, who has dedicated more than 50 years for the preservation and promotion of our heritage and culture. Lucknow is really proud of him.

Dr. Agrawal was really a noble soul, his two grand-daughters, Nimisha and Kritika tells that he believed in doing the work with full commitment... pour your heart into it and strive for excellence...... success will follow on its own.... This was the great lesson he taught and this is all about his legacy, we still cherish.

**Dr. Agrawal was Honoured by Members of Lucknow Society.
In Picture Shri Avnish Awasthi Offering a Shawl.**

Ph.D. Theses Supervised by Dr. O.P. Agrawal

Sl.No.	Name	Topic	Year
1.	Ms. Rashmi Pathak	Rajasthani Wall Paintings	1993
2.	Ms. Mandana Barkeshli	Scientific Study of the Materials of Miniature painting of India and Persia	1997
3.	Ms. Azadeh Taheri Karimi	Consolidation of limestone with new material: A case study of Sepahasalar Mosque in Iran	2009
4.	Ms. Mehrnaz Azadi Boyaghchi	Finding an appropriate solution to tackle the problem of acidity in paintings	2010
5.	Ms. Tanushree Gupta	Scientific studies on Conservation issues of Acrylic paintings	2016

Part – I

Biodeterioration of Cultural Objects

हमारी सांस्कृतिक धरोहर

मनुष्य एक दिन
अपनों से दूर चला जाता है
टूट जाते हैं रिश्ते सभी
छूट जाता है सगे संबंधियों का साथ
रहती है तो बस उसकी कहानी
सदियों से चली आ रही है यह परंपरा
हमारी भाषा, हमारा व्यवहार, हमारे संस्कार
हमारे त्योहार और कला से प्यार
बस यही बनाती है मानव की अनमोल धरोहर
जो मानव संस्कृति को
विकास की एक नई दिशा प्रदान करती है।
बचाएं हम सांस्कृतिक विरासत को
हमारी यह धरोहर जीवन को
और गति दे मधुरता , और प्यार दे
मानव संस्कृत के विकास का अध्ययन
जो सिंधु घाटी से शुरू होकर
मिश्र और मेसोपोटामिया से होता हुआ
परमाणु युग की ओर बढ़ रहा है
हम सहिष्णु बनें
विकासवान तो हों पर मानव मूल्यों को भी न भूलें।
हमारी संस्कृति हो वसुधैव कुटुंबकम की
हम प्रकृति का करें संरक्षण
जैव विविधता को दें प्रधानता
चारों ओर फैले सुरभि हमारी
सलामत रहे सांस्कृतिक धरोहर
और सभ्यता की अप्रतिम दास्ताँ हमारी ॥

Chapter 1

Sizing Materials Used in Persian Manuscripts and their Behaviour against the Mould Fungus *Aspergillus flavus*

Mandana Barkeshli and Ina Stephan[1]

Conservation Scientist, Honorary Principal Fellow,
The Grimwade Centre for Cultural Materials Conservation, Faculty of Arts,
Level 3, Thomas Cherry Building (Bldg #201),
The University of Melbourne, Victoria 3010 Australia
e-mail: mandana.barkeshli@gmail.com, bmandana@unimelb.edu.au
[1]BAM Federal Institute for Materials Research and Testing,
Division 4.1 "Biodeterioration and Reference Organisms",
Unter den Eichen 87, 12205 Berlin, Germany
e-mail: ina.stephan@bam.de

ABSTRACT

Our earlier study (Barkeshli 2003; 2015) based on historical and scientific analysis showed that Iranians used a considerable range of materials in the sizing process. Unlike many nations who used limited sizing materials to improve the mechanical strength and to smoothen the paper surface, Iranians have used various materials for sizing process from Taimurid (15th century) to Safawid (16th century) and Qajar (19th century) periods. Further to our earlier study on the Sizing Materials used in Persian manuscripts and miniature paintings, scientific analysis was carried out to investigate their behaviour against the mould fungus Aspergillus flavus. In the first stage fourteen different sizing materials that was identified from Persian historical recipes were reconstructed. In the second stage fungicidal property of each sizing samples against Aspergillus flavus fungus were identified that will be presented in this paper.

Keywords: Sizing, Materials, Persian manuscripts, Deterioration, Fungus, Aspergillus flavus.

1.1 Introduction

Paper sizing (*āhārdādan*) is a process of preparing the surface of the paper to make it suitable for writing, illuminating, or painting on. After a sheet of paper has been formed and dried, the cellulose fibre it contains can continue to absorb water unless it has been 'sized', *i.e.*, impregnated with a substance like starch, glue, or wax to prevent such penetration. There are different techniques available for sizing paper, depending on requirements, such as soaking or applying one or several layers of sizing material on the paper surface with the help of a soft cotton ball or brush.

Certain scientific investigations have revealed valuable information on materials used in the sizing process. According to Wulff (1976)and based on chemical investigation, the Persian papermakers at Samarqand made an important contribution to papermaking by introducing a new method in sizing paper to make it more suitable for writing on with ink and a reed pen. Wheat starch and later gum tragacanth (*katira*) or emerusus (*seriš*) were used as sizing substances. Our earlier investigations (Barkeshli 2003; 2015) based on historical and scientific analysis showed that a number of different sizing materials were prescribed in Persian historical treatises mainly from the late Seljuk and early Ilkhanidto Timurid, Safawid and Qajar periods.

In a number of Persian historical treatises, it has been advised to apply sizing materials to make fragile paper strong enough, reduce the fluffiness of paper fibres and make the surface of the paper smooth enough to write on. For example, in his treatises *Ādāb al-Mašq*, *Rasm al-Ḵatt* and *Savād al-Ḵatt*, Soltān Ahāmad Majnun Rafiqi Heravi advised the use of soft, smooth and even paper to write or draw on.

[1]In his treatise *Ādāb al-Mašq*, writes in a poetic way:

وز عشق هوای مشق داری	ای طرفه پسر که عشق داری
بریان و لطیف و صاف و هموار	رو کاغذ طرفه ای بدست آر

peerless boy that has love in your heart

and from this love you desire to write

bring a fresh new paper

which is crisp, thin, pure, and even.

---

1. Soltān Ahmad Majnun Rafiq Heravi, *Ādāb al-Mašq*, 217.

In his other treatises *Rasm -al Katt*, writes the following couplets poetically:[2]

اگر خواهی تو ای گنج معانی	که وصف کاغذ نیکو بدانی
کمالش آن بود کآید پدیدار	سفید و نرم و بریان، صاف و هموار

Oh you who are treasure of mystical meanings,

if you want to know the qualification of fine paper;

it's perfection becomes the one

that is white, soft, crisp, pure, and even

In *Savâd al-5amm* he also advises the same as above in prose:[3]

بدان که کاغذ خوب آنست که سفید و نرم و بریان و صاف و هموار باشد...

Know that fine paper is one which is white, soft, crisp, pure, and even….

Different techniques are described in Persian historical recipes based on the sizing materials and the type of paper that is meant to be achieved. Two general techniques are advised for sizing paper either by application on the surface or dipped and submerged in the sizing solution. The first method is usually recommended for thicker substances like rice starch, whereas the second method is advised for more thinner substances like mucilage.

In most historical recipes, after the techniques of sizing are described, recommendations are made to dry the paper in the shade or under sunlight, either on a flat surface or hung on a rope depending on the paper size, followed by burnishing to smoothen the surface and to polish the paper.

In Persian, burnishing paper is called *mohreh zadan* or *mohreh kardan* and burnishing tools is known as *mohreh*. A number of burnishing materials were employed according to the recipes, such as agate stone (*'aqiq*), jade (*yašm*), ivory (*'āj*), glass (*zejāj, abgine*), crystal (*bolur*) and onyx (*jiz'*). Some of these materials, their characteristics and their medical benefits have been described in earlier sources for example by Biruni[4], Jamali-ye Yazdi[5], and Teflisi[6]. Neishāburi[7] in *Jawāhirnāmeh-e Nezāmi* has written different chapters in detailon agate stone (*'aqiq*), onyx (*jiz'*) and crystal (*bolur*) and has described their differences such

---

2. Abid, *Rasm-al Katt*, 166.

3. Abid, *Savād al-ḵaṭṭ*, 194.

4. Abu Raihān Biruni, "*Al-Saydaneh fi al-Tebb*", 735 (*'aqiq*), 1008 (*yašm*).

5. Jamali-ye Yazdi, *Farrokhnameh*, 169 (*jiz'* and *'aqiq*).

6. Teflisi, 388 (*'aqiq* and *bolur*), 389 (*yašm*).

7. Neishāburi, M. A. J. "*Jawāhirnāmeh-e Nezāmi*", 200-204 (*'aqiq*), 205- 209 (*jiz'*), 209-214 (*bolur*).

as between glass (*zejāj*) and crystal (*bolur*)[8]. In his book Neishāburi also advises on burnishing tool and highlights that the best stone to burnish gold on paper or woven in textile is onyx (*jiz'*).The recipe reads as follows[9]:

و خاصیت مهین او آن است که زر هایی که بر کاغذ نهاده باشند بدان مهره زنند تا برّاق شود. و همچنین زر حلّ کرده را چون بدان چیزی نوشته باشند تا بدان مهره نزنند براق نشود. و جامه های زر کشیده هم بدان مهره زنند تا روشن و هموار شود، و این خاصیت مختص است به جزع و هیچ سنگ را این خاصیت نیست.

It's most important property (onyx stone) is that it is used for burnishing gold sprinkled on papers to make it glossy. Furthermore, if gold ink is used for writing, it wouldn't become glossy if it is not burnished with onyx stone. To make smooth and shiny, the clothing made of gold thread is burnished and this would be obtained only with onyx stone as other stones do not have this property.

Sometimes even a person's bare hands were used to smoothen the surface of paper. A hard and smooth base made of flint (*čaqmāq*) was also employed, and a wooden board was used as the base for burnishing and sizing the paper. For example, Bābā Shāh-e Isfahanīin *Adāb al-Mašq* after advising to dye paper with henna, saffron and drop of ink, mentions to size paper with starch; he has a recipe on the process of burnishing that reads as follows:[10]

باید که ساز آهار از نشاسته کنند و در وقت آهار کردن آن قدر دست بمالند که در جسم کاغذ نفوذ کند، و در وقت مهره کردن کاغذ را نم کنند تا از گرمی مهره نسوزد. و خوبی مهره آن است که چنان روشن شود که عکس روی در وی نماید.

Starch should be made for sizing and after sizing, the paper should be rubbed till it penetrates thoroughly to the body of the paper. At the time of burnishing, the paper needs to be moistened so that the heat that is produced by the burnisher (mohreh), doesn't burn the paper. The good thing about burnishing is that the paper becomes shiny in a way that one can see his face on it.

To make fragile paper strong enough, reduce the fluffiness of paper fibres and make the surface of the paper smooth enough to write on, Mohammad Bokāri also recommends the techniques for sizing paper in his work *Favāyed al-Ḳotut* in details as follows:[11]

و دیگر طریق مهره کشیدن آن است که اوّلاً تختهٔ مهره پاکیزه می باید و شیشهٔ او پاکیزه و بی گره و بی رگ می باید. و مهره را اعتدال باید کشید نه سخت و نه سست. و باید که رخی و خطّی در کاغذ نیائند و اگر بائند نماند. و از هر طرفی که مهره کشیدی با آن طرف دیگر را بر بالای او باید کشیدن تا هموار و به یک منوال آید.

---

8. Abid, 210.

9. Abid, 208.

10. Bābā Shāh-e Isfahanī, *Adāb al-Mašq*, 157.

11. Mohammad Ibn-e Dust Mohammad Bokāri, *Favāyed al-Ḳotuṭ*, 375.

....and the technique for burnishing is that the wooden board should be clean, and the glass used for burnishing should be clean without any peaks and valleys. Burnishing tool should be applied moderately, not strong, nor weak and it shouldn't make any trace of line or cleavage on paper. To make paper even and homogenous, burnishing process should be done on both sides.

که رخ رخ بر او نه بنماید   مهرهٔ کاغذ آنچنان باید

Burnishing process should be done in a way

that it does not cause cleavages to the paper

و دیگر بعد از این که مُهره کشیدی، باید که آن کاغذ را در تَه تَختهٔ سنگی مانی چندان روز که گذشت بعد از آن گرفته به او کتابت کنی که بسیار خوش می شود. و اگر خواهی که الحال به همان کاغذ تازه مُهره نویسی، باید که در آن کاغذ دم اندازی. همچنین که به حکم کاغذ مذکور می گردد و نیز خوش قلم می شود.

Moreover, after burnishing the paper should be kept under a flat stone for a few days before writing on it and in this way the writing will come out nice. However, if you desired to write immediately after burnishing, the paper needs to be aired. In this way the paper will also become suitable to write, and calligraphy will be nice on it.

An anonymous author in his treatises, *Resāleh dar Bayān-e Ḵaṭṭ va Morakkab va Kāḡad va Sāḵtan-e Ranghā* (19th century) has given the same tips for burnishing process as mentioned above in poetry form. The recipe reads as follows:[12]

که رح رخ بر و نه بنماید   مهرهٔ کاغذ آنچنان باید

زور بازو ولی نه سخت و نه سست   تَختهٔ مهره پاک باید شست

Burnishing process should be done in a way

that it does not cause cleavages to the paper

Burnishing board should be washed clean

it needs force of arm but not too strong nor weak

1.2 Research Background

Our previous study on Persian sizing materials revealed that Persians have used a considerable range of materials in the sizing process of paper manuscripts. It was found that from 15th to 19th century based on historical references from Taimurid to Safawid and Qajar periods sizing materials can generally be categorised as vegetable and animal base. Vegetable base sizes include starches (wheat and rice), gums (gum Arabic and gum tragacanth, eremurus), plant mucilage (rice, fleawort, cucumber, melon, marshmallow, and myrtle), fruit juices (sweet melon, grape syrup), sugar (rock sugar) and animal base sizes include fish glue.

---

12. Anonymous, *Resāleh dar Bayān-e Ḵaṭṭ va Morakkab va Kāḡad va Sāḵtan-e Ranghā*, 534-535.

In our previous study, in the next stage of the investigation, the large collection of sizing materials, which were prepared on the basis of the historical recipes were compared with the spectra of variety of original historical paper manuscript samples belonging to 16th to 19th Century from Museum and private collection. For identification of the sizing materials used in the original samples, laboratory analysis was carried out using a staining method and Fourier transform infrared spectrometry (FTIR). Out of nine Persian miniature paintings and illuminated manuscripts examined, one of the sizes used was starch, seven were cucumber-seed mucilage and one was a mixture of tragacanth and cucumber seed (Barkeshli 2003)

Our previous research also showed that each sizing material has its own particular properties and gives a certain effect to the paper (Barkeshli 2015). We found that starch paste gives more body to paper and makes it firmer. In contrast, plant mucilage – especially from cucumber seed – gave the paper a lighter body and made it softer than the effect obtained using starch paste. Fruit and sugar syrups gave more shine to the surface of the paper we examined and made it stiffer compared to plant mucilage. As for vegetable and animal glue, they both added a lustre to the paper, but animal glue made the paper stiffer than vegetable glue did.

In terms of drying, the sizing material sometimes took a while to dry, depending on the duration of dipping and the number of layers of it that had been applied. The drying process was prolonged if the dipping or application process was repeated during the sizing procedure. In our observation we found that cucumber-seed mucilage was what dried the fastest and paper sized with it was also easier to burnish than with other sizing materials (Barkeshli 2015).

1.3 Research Methodology

Further study was carried out to identify the behaviour of traditional sizing materials against biological deterioration and their fungicidal property. In the first stage the data that was collected from historical recipes reviewed and additional research was carried out on Persian historical treatises from 9th to 19th century to explore more information on sizing materials and techniques for sample preparation based on Iranian historical recipes. In the second stage fourteen different sizing materials which were identified from historical recipes were reconstructed based on Persian historical recipes. In the third stage a microbiological test was conducted to explore the medieval sizing material against the mould fungus *Aspergillus flavus* at BAM Federal Institute for Materials Research and Testing in Division 4.1 Biodeterioration and Reference Organisms.

1.4 Materials and Techniques

All sizing ingredients were prepared from the materials obtained from traditional herb stores of Tehran, Iran. Sizing solutions and pastes were prepared based on historical recipes including starches, vegetable gums, animal glue, mucilage, fruit juices, and sugar syrup. Handmade grape syrup was prepared at home, and ready-made ones were purchased from traditional herbal stores at the Bazaar from Boroujerd in Lorestan provenance. In general, the main ratio of 1:10 was selected for soaking or diluting the ingredients in water, where no mention of ratios was available in the recipes, unless it was not applicable due to the nature or consistency of the material (Figures 1.1-1.8). Table 1.1 showing fifteen sizing materials that is found in the historical recipes and Table 1.2 shows the summary of the process of sizing preparation along with the recipes and their ratio used.

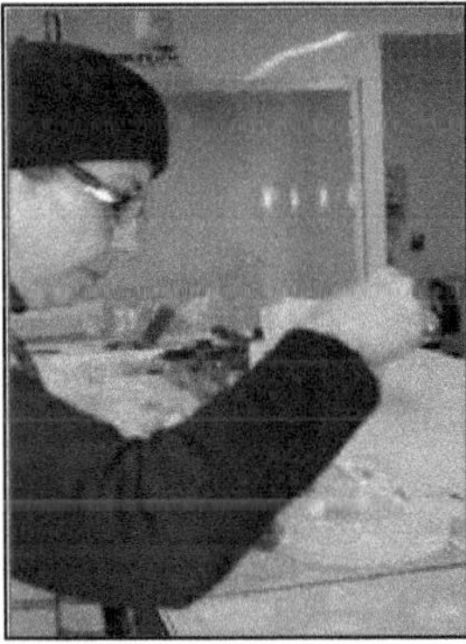

Figure 1.1: Preparation of Paper Sizing According to Persian Historical Recipes.

Figure 1.2: Prepared Sizes kept in a Jar with Label for Paper Sizings and for the Tests.

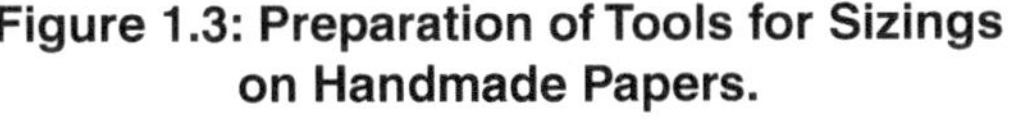

Figure 1.3: Preparation of Tools for Sizings on Handmade Papers.

Figure 1.4: Paper Sizing According to the Historical Recipes: Application Technique.

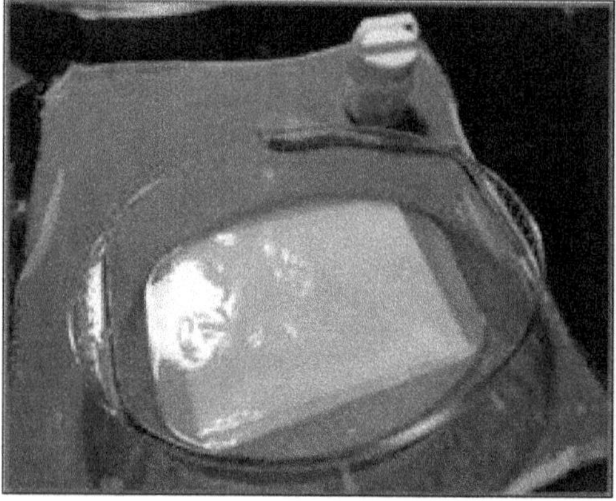

Figure 1.5: Paper Sizing according to the Historical Recipe: Dipping Technique.

Figure 1.6: Drying the Papers after the Sizings.

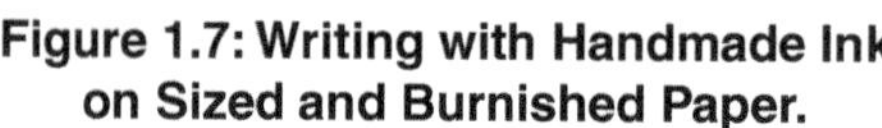

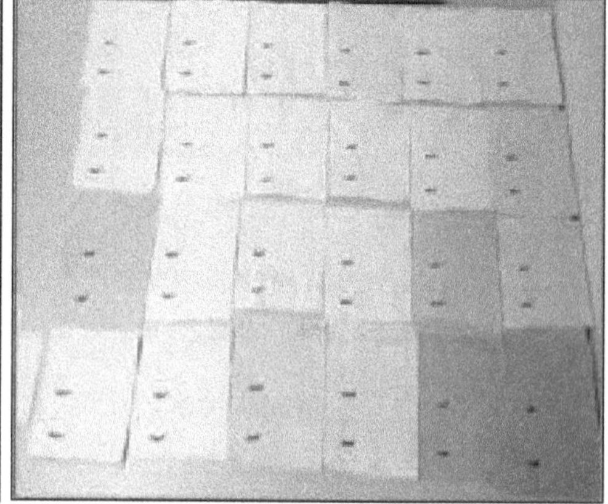

Figure 1.7: Writing with Handmade Ink on Sized and Burnished Paper.

Figure 1.8: Collection of Sized Paper According to the Historical Recipes for Tests.

Table 1.1: Showing Fifteen Sizing Materials that are Found in the Historical Recipes

The Category of Sizings Based on Historical Analysis		
Sl.No.	*Sizing Category*		*Sizing material*
1	Vegetable Based Size	Starches	Rice starch (*neshasteh-e berenj*)
2			Wheat starch (*neshasteh-e gandom*)
3		Gums	Gum arabic *(samg-e arabi)*
4			Gum tragacanth *(katira)*
5			Asphodel *(seris)*
6			Rice mucilage *(lo¿ab-e berenj)*
7			Fleawort seed *(espaghol, esfarze, qetòuna)*
8			Cucumber seeds *(tok1m-e k1iar)*

The Category of Sizings Based on Historical Analysis		
Sl.No.	*Sizing Category*		*Sizing material*
9	Vegetable Based Size	Gums	Melon seeds *(ab-e tok1m-e khiarein)*
10			Marshmallow mucilage *(lo¿ab-e k1etòmi)*
11			Myrtle seeds *(tok1m-e mord)*
12		Fruit Juices	Juice of a sweet melon *(k1arboze)*
13			Grape syrup *(sirey-e angur)*
14		Sugar syrup	Egyptian rock-sugar solution (*ab-e nabat-e mesòri*)
15	Animal Based Size	Animal Glue	Fish glue *(sirisum-e mahi)*

Table 1.2: Shows the Summary of the Process of Sizing Preparation along with the Recipes and their Ratio Used

Sizing Recipes				
Sizing Category		*Sizing Material*	*Ratio*	*Recipe*
Vegetable Base Size	Starches	Wheat Starch	10g/100 ml	Cold water added and left over night. It is then boiled and mixed with wooden stick for 20 minutes until paste is made
		Rice Starch	10g/100 ml	Cold water added and left over night. It is then boiled and mixed with wooden stick for 20 minutes until paste is made
	Gums	Gum Arabic	10g/100 ml	Hot water added and left over night.
		Gum Tragacanth	10g/100 ml	Hot water added and left over night.
		Espaghol	10g/100 ml	Cold water added and left over night.
	Plant Mucilage	Rice Mucilage	1 cup	Rice is boiled for half an hour and the mucilage is taken from the top
		Fleawort seed mucilage	10g/100 ml	Water added and left over night. It is then filtered 4 times
		Cucumber seed Fresh	10g/100 ml (Fresh)	Water added and left over night and filtered
			20g/100 ml (Fresh)	
		Cucumber seed Dry	5g/50ml (Dry)	
		Hock melon seed	10g/100 ml	Water added and left over night and filtered
		Honey Dew Seed	10g/100ml	Water added and left over night and filtered
			20g/100 ml	
		Marshmallow Seed	2g/20 ml	Water added and left over night and filtered
		Marshmallow Flower		

Sizing Recipes				
Sizing Category		Sizing Material	Ratio	Recipe
Vegetable Base Size	Fruit Juice	Rock melon juice		Juice made in blender, then filtered
		Honey dew melon juice		Juice made in blender, then filtered
		Grape syrup Home Made		Grapes are blended, filtered and boiled for 30 minutes
		Grape Syrup Bazar		Grape syrup from Iran traditionally made
	Sugar Syrup	Sugar syrup	10g/100ml	Boiled for 20 minutes
Animal Base Size	Animal Glue	Fish glue		

As for test samples, Indian handmade paper from Avani Trading Sdn Bhd, Taman Sri Batu Caves, Kuala Lumpur, Malaysia was purchased and twenty five paper sizing samples were prepared using both sizing techniques, application (A) and dipping (D) based on historical recipes and where it was applicable. Table 1.3 shows twenty six coated paper test samples indicating the sizing techniques (A or D) including control sample. For reference each sample had been given a number to be referred to in the results and discussion section.

Table 1.3: Showing Twenty Six Coated Paper Test Samples Indicating the Sizing Techniques (A = Application, D= dipping) including Control Sample

Sizing Test Samples					
Sl.No.	Sizing Category		Sizing Material	Sizing Technique	
				A	D
1	Vegetable Base Size	Starches	Wheat Starch	A	
2			Rice Starch	A	
3		Gums	Gum Arabic		D
4			Gum Tragacanth	A	
5			Espaghol		D
6		Plant Mucilage	Rice Mucilage	A	
7					D
8			Fleawort seed mucilage		D
9			Cucumber seed Fresh		D
10			Cucumber seed Dry		D
11			Rock melon seed		D
12			Honey Dew Seed		D
13			Marshmallow Seed		D

		Sizing Test Samples		
Sl.No.	Sizing Category		Sizing Material	Sizing Technique
				A	D

Sl.No.	Sizing Category		Sizing Material	A	D
14	Vegetable Base Size	Plant Mucilage	Marshmallow Flower	A	
15					D
16		Fruit Juice	Rock melon juice	A	
17					D
18			Honey dew melon juice		D
19			Grape syrup Home Made (H)	A	
20			Grape Syrup Bazar (B)	A	
21					D
22		Sugar Syrup	Sugar syrup	A	
23					D
24	Animal Base Size	Animal Glue	Fish glue	A	
25					D
26			Control	-	-

Note: Sizing Techniques A = Application, D = Dipping.

The laboratory test concerning mould growth was conducted at BAM Federal Institute for Materials Research under the team of Lab Testing Division 4.1 "Biodeterioration and Reference Organisms".

Fungal tests were conducted on the sizing samples in two glass vessels (40 x 39 x 30 cm) at 30 ± 1 °C with relative humidity (RH) of ≥ 90 per cent using *Aspergillus flavus* ATCC 9643. Twenty six different coated papers, two pieces of each type, delivered in one container. One piece of each type was inoculated with a solution containing spores of the fungus mentioned above in a glass vessel. In addition, the second half of the test specimens were incubated as delivered under the same conditions, but in a separate glass vessel. These test specimens were not inoculated with *A. flavus*. To achieve RH ≥ 90 per cent each vessel had about 2 cm of water at the bottom and a stainless steel mesh to separate the test specimens from the water surface. During incubation the glass vessels were closed with a lid. As a result, the humidity in the glass vessel rose to above 90 per cent.

To test the viability of the spores of the test fungus, a Petri-dish containing nutrient-agar was inoculated with the spore-suspensions of *Aspergillus flavus*. After a 2-week incubation the test specimens were visually examined and rated at a magnification of up to 50-fold. The test specimens were classified into three different groups considering the intensity of the mould growth (Table 1.4). Within each group an attempt was made to rank them by naked eye based on the level of mould growth (Figures 1.9-1.16).

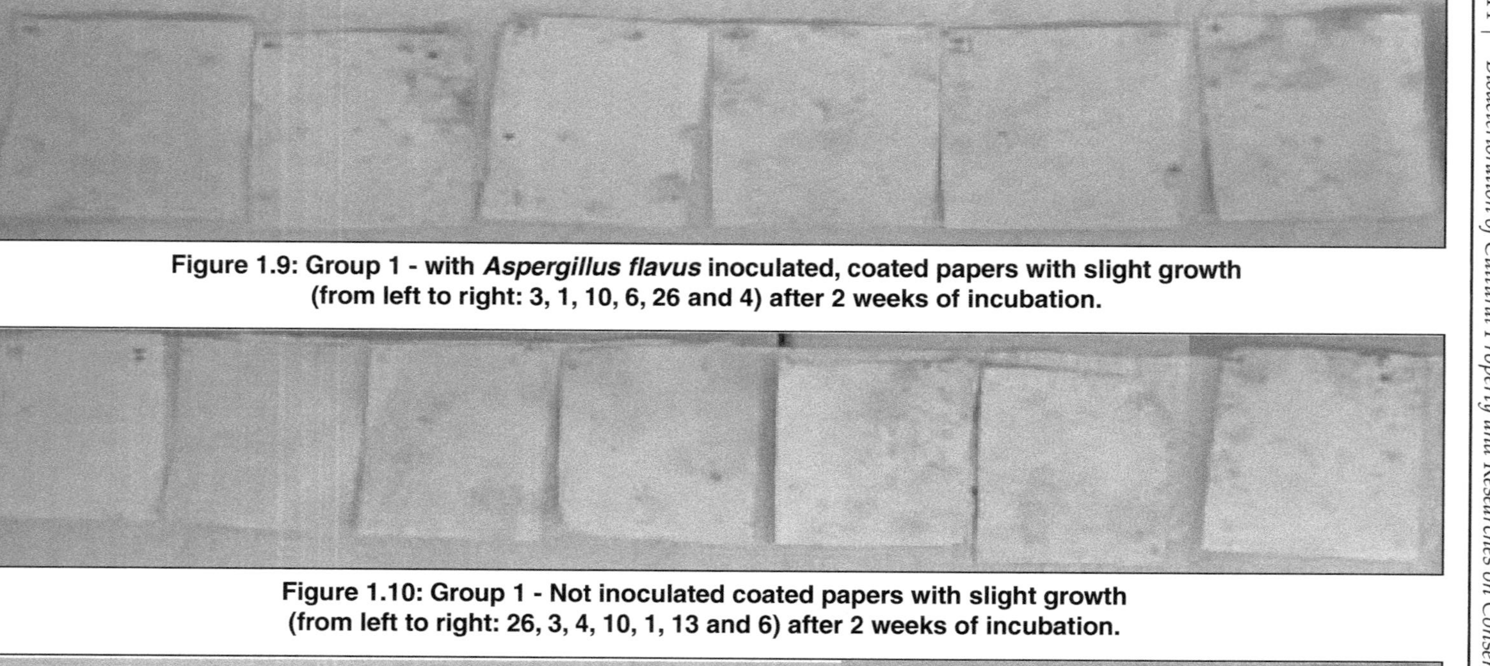

Figure 1.9: Group 1 - with *Aspergillus flavus* inoculated, coated papers with slight growth (from left to right: 3, 1, 10, 6, 26 and 4) after 2 weeks of incubation.

Figure 1.10: Group 1 - Not inoculated coated papers with slight growth (from left to right: 26, 3, 4, 10, 1, 13 and 6) after 2 weeks of incubation.

Figure 1.11: Group 2 - with *Aspergillus flavus* inoculated, coated papers with moderate growth (from left to right: 13, 14, 8, 22, 7 and 11) after 2 weeks of incubation.

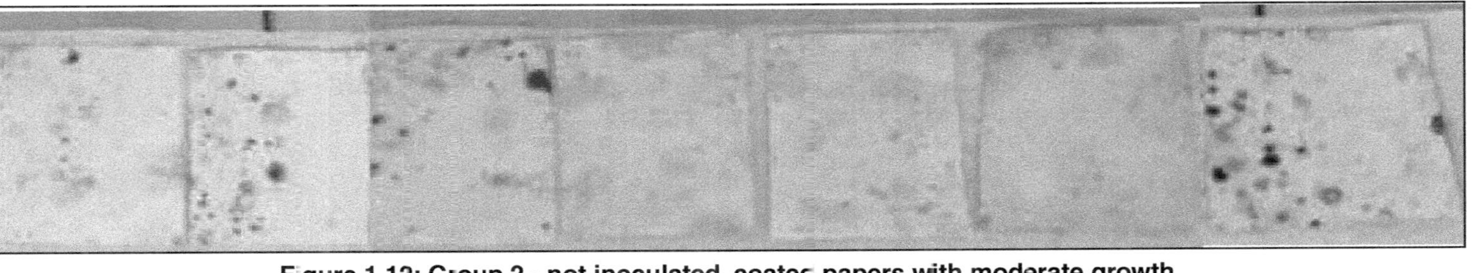

Figure 1.12: Group 2 - not inoculated, coated papers with moderate growth (from left to right: 22, 23, 14, 7, 8 and 19) after 2 weeks of incubation.

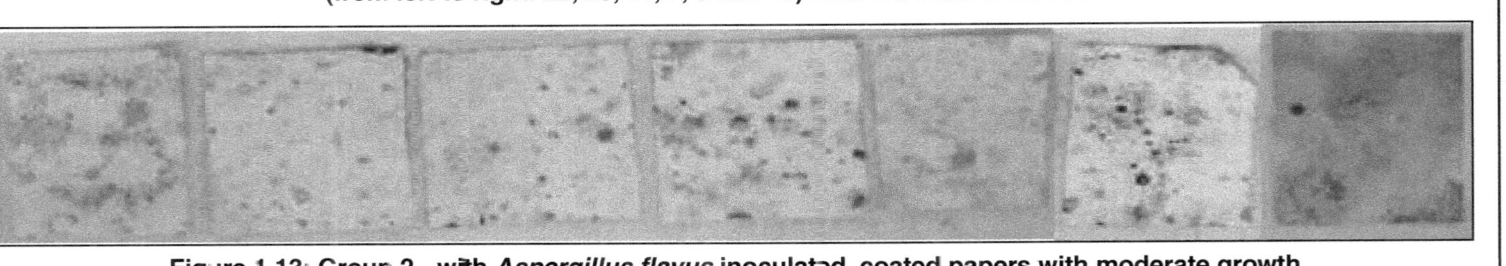

Figure 1.13: Group 2 - with *Aspergillus flavus* inoculated, coated papers with moderate growth (from left to right: 12, 21, 20, 23, 2, 21 and 5) after 2 weeks of incubation.

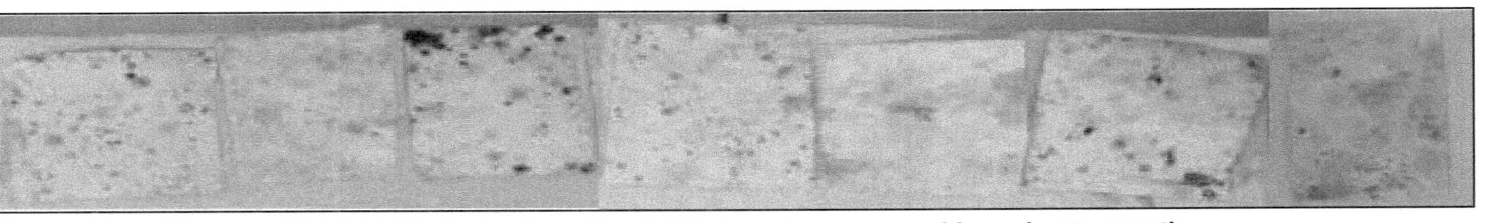

Figure 1.14: Group 2 - not inoculated, coated papers with moderate growth (from left to right: 22, 23, 14, 7, 8, 19 and 5) after 2 weeks of incubation.

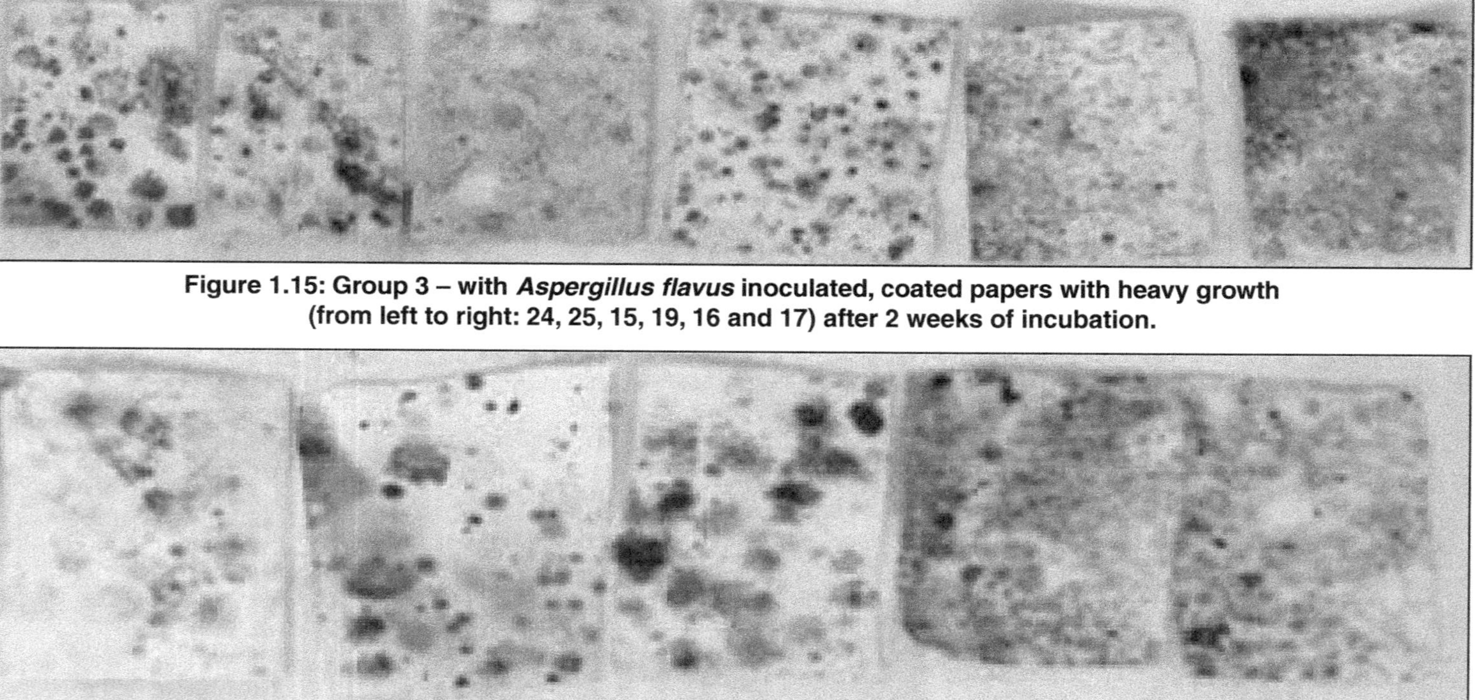

Figure 1.15: Group 3 – with *Aspergillus flavus* inoculated, coated papers with heavy growth (from left to right: 24, 25, 15, 19, 16 and 17) after 2 weeks of incubation.

Figure 1.16: Group 3 - not inoculated, coated papers with heavy growth (from left to right: 25, 15, 24, 17 and 16) after 2 weeks of incubation.

Table 1.4: Showing the Test Specimens Classified into Three different Groups Considering the Intensity of the Mould Growth

Group Ranking	
Group 1	*Aspergillus flavus* inoculated, coated papers with slight growth after 2 weeks of incubation
Group 2	*Aspergillus flavus* inoculated, coated papers with moderate growth after 2 weeks of incubation
Group 3	*Aspergillus flavus* inoculated, coated papers with heavy growth 2 weeks of incubation

1.5 Results and Discussion

The Petri-dish inoculated with *Aspergillus flavus* was overgrown after one week. This demonstrated that the test organism used was vigorous and that the test conditions were favourable for its growth. Furthermore, the coated papers that were incubated as delivered without inoculation with *A. flavus* also showed the growth of mould. This indicated that they were already contaminated with fungal spores, which germinated and grew under the test conditions (Figures 1.10, 1.12, 1.14 and 1.16).

As the result our investigation showed that Wheat Starch-A, Gum Arabic-D, Gum Tragacanth-A, Rice Mucilage, Cucumber Seed Dry-D sizing's showed slight mould growth. Whereas, Rice Starch-A, Espaghol, Rice Mucilage-D, Fleawort Seed-D, Cucumber Seed Fresh-D, Rock Melon Seed-D, Honey Dew Seed-D, Marshmallow Seed-D, Marshmallow Flower-A, Honey Dew-D, Grape Syrup-B-A, Grape Syrup-B-D, Rock Sugar-A, and Rock Sugar-D sizing's showed moderate mould growth. And Marshmallow Flower –D, Rock Melon-A, Rock Melon-D, Grape Syrup H-A, Fish Glue-A, and Fish Glue-D sizing's showed heavy mould growth.

Table 1.5: Shows the Rating of Fungal Colonisation on Test Specimens

Rating of Fungal Colonisation on Test Specimens			
Rating Group		Number for Identification of the Coated Paper	
Group	Characteri-zation	Inoculated with *Aspergillus flavus*	Not Inoculated with *Aspergillus flavus*
1	Slight growth	1, 3, 4, 6, 10, 26	1, 3, 4, 6, 10, 13, 26
		Wheat Starch-A (1), Gum Arabic-D (3), Gum Tragacanth-A (4), Rice Mucilage-A (6), Cucumber Seed Dry-D (10), Control (26)	Wheat Starch-A (1), Gum Arabic-D (3), Gum Tragacanth-A (4), Rice Mucilage-A (6), Cucumber Seed Dry-D (10), Marshmallow Seed-D (13), Control (26)
2	Moderate growth	2, 5, 7, 8, 9, 11, 12, 13, 14, 18, 20, 21, 22, 23	2, 5, 7, 8, 9, 11, 12, 14, 18, 19, 20, 21, 22, 23

Rating of Fungal Colonisation on Test Specimens			
Rating Group	Number for Identification of the Coated Paper		
Group	Characteri-zation	Inoculated with Aspergillus flavus	Not Inoculated with Aspergillus flavus
		Rice Starch-A (2), Espaghol (5), Rice Mucilage-D (7), Fleeworth Seed-D (8), Cucumber Seed Fresh-D (9), Rock Melon Seed-D (11), Honey Dew Seed-D (12), Marshmallow Seed-D (13), Marshmallow Flower-A (14), Honey Dew-D (18), Grape Syrup-B-A (20), Grape Syrup-B-D (21), Rock Sugar-A (22), Rock Sugar-D (23)	Rice Starch-A (2), Espaghol (5), Rice Mucilage-D (7), Fleeworth Seed-D (8), Cucumber Seed Fresh-D (9), Rock Melon Seed-D (11), Honey Dew Seed-D (12), Marshmallow Flower-A (14), Honey Dew-D (18), Grape Syrup Home Made (19), Grape Syrup-B-A (20), Grape Syrup-B-D (21), Rock Sugar-A (22), Rock Sugar-D (23)
3	Heavy growth	15, 16, 17, 19, 24, 25	15, 16, 17, 24, 25
		Marshmallow Flower –D (15), Rock Melon-A (16), Rock Melon-D (17), Grape Syrup H-A (19), Fish Glue-A (24), Fish Glue-D (25)	Marshmallow Flower –D (15), Rock Melon-A (16), Rock Melon-D (17), Fish Glue-A (24), Fish Glue-D (25)

With the exception of paper 13 (Marshmallow Seed-D) and 19 (Grape Syrup-H), papers that had been inoculated with *A. flavus* were in the same rating group as the non-inoculated test samples. For paper 13 (Marshmallow Seed-D) and 19 (Grape Syrup-H), inoculation led to a higher rating group. Ranking inside a rating group was difficult and could only be achieved to a minor extent (Figures 1.9-1.16).

1.6 Conclusion

The case study presented here focused on the sizing materials used in medieval Persia and early modern Iran. Present study was based on historical recipes which showed that Iranians used a considerable range of materials in the sizing process. The amount of growth of *Aspergillus lavus* differed depending on the sizing material used. In our earlier investigations, we have found a number of paper manuscripts that were sized with cucumber seed mucilage. The reason for this choice may have been to reduce the biological attack on the paper as our study showed little growth of *A. flavus* on paper sized with cucumber seed mucilage. However, more research and data on identification of sizing materials used on original manuscripts is necessary to find out the reasons behind the use of certain materials and their effect on physical, and chemical deterioration such as changes in mechanical strength, colour and pH as well as biological growth.

Acknowledgements

This paper and the research associated with it would not have been possible without the exceptional support of Federal Institute for Materials Research and Testing (BAM). I would like to thank Thomas Dimke for the technical work in the microbiological laboratory at the Lab Testing Division 4.1 "Biodeterioration and Reference Organisms". I also would like to thank Prof. Dr. Michael Friedrich the director of Centre for the Study of Manuscript Cultures (CSMC) at University of Hamburg for his generous support to conduct part of this research at their Scientific laboratory through Petra Kappert Fellowship grants award.

REFERENCES

Abu Raihān Biruni (2004) *"Al-Saydaneh fi al-Tebb"*, 440 A.H./1048 A.D. Translated by B. Mozaffarzadeh, 1383/2004. Tehran: Farhangestane-e Zaban va Adabiyat-e Farsi. [in Persian]

Anonymous (1596) 1005 A.H./1596 A.D. *Majmu'at al-sanaye'*, Manuscript in Central Library of Tehran University, Registration No. 3875

Anonymous (1993) *Resāleh dar Bayān-e Kāḡaḏ, Morakkab va Ḥall-e Alvān* also known as Resāleh dar Bayān-e Kāḡaḏ, Morakkab-e Alvan va Khat-e Ouhal, Mid 9th A.H./15th A.D., In *Ketab-Arayi Dar Tamaddun-I Eslami*,. ed. N. Mayel Heravi (1372/1993), Mashhad: Islamic Research Centre of Astan-e Quds-e Razavi. 57-67. In [Persian].

Anonymous (1993) *Resāleh dar Bayān-e Ḵaṭṭ va Morakkab va Kāḡaḏ va Sāḵtan-e Ranghā*, 13th Hijra/19th A.D., In *Ketab-Arayi Dar Tamaddun-I Eslami*, ed. N. Mayel Heravi (1993), 533-542, Mashhad: Islamic Research Centre of Astan-e Quds-e Razavi. [In Persian]

Bābā Shāh-e Isfahanī (1993) *Adāb al-Mašq* (Manners of writing) (Mid 10th A.H./16th C.E.), In *Ketab-Arayi Dar Tamaddun-I Eslami*, ed. by N. Mayel Heravi (1993), 147-157, Mashhad: Islamic Research Centre of Astan-e Quds-e Razavi. In [Persian]

Barkeshli M (2003) Historical and Scientific Analysis on Sizing Material Used in Iranian Manuscripts and Miniature Paintings', The Book and Paper Group Annual, 22: 9–16.

Barkeshli M. (2015) Manuscripts of Mystical Persian Literature : Material Technology and Science, Journal of The Manuscript Cultures, Centre for the Study of Material Culture, vol. 8 : 187 – 215, Hamburg, Germany, ISSN No 1867-9617.

Hossein Aqili Rostamdari (1993), "Ḵaṭṭva Morakkab", 978 A.H./1571 A.D.). In *Ketab-Arayi Dar Tamaddun-I Eslami*, ed. by N. Mayel Heravi (1372 1993). Mashhad: Islamic Research Centre of Astan-e Quds-e Razavi. 323-342. [In Persian]

Jamali-ye Yazdi A M (2007) *Farrokhnameh*, 580 A.H./1184 A.D, edited by I. Afshar,), Amirkabir Publication, Tehran, 1386/2007. [in Persian]. Also partly in Mayel Heravi, N. *Ketab-Arayi Dar Tamaddun-I Eslami*, Mashhad: Islamic Research Centre of Astan-e Quds-e Razavi, 1372/1993, 1046-1048.

Mohammad Ibn-e Dust Mohammad Bokāri (976 A.H./1568 A.D.- ?), *Favāyed al-kotut* (Advantages of Scripts) (995 A.H./1587 A.D.), In *Ketab-Arayi Dar Tamaddun-I Eslami*, edited by N. Mayel Heravi (1372/1993), Mashhad: Islamic Research Centre of Astan-e Quds-e Razavi. 357-456. [In Persian]

Neishāburi M A J (2004) *"Jawāhirnāmeh-e Nezāmi"*, 592 A.H./1196 C.E. Edited by. I. Afshar and M. R. Daryagasht 1383/2004. Tehran: Miras-e Maktub Press. [in Persian]

Soltān Aḥmad Majnun Rafiqi, *Rasm -al Katt,* (909 A.H./1503 C.E. or 940 A.H./1533 CE), In *Ketab-Arayi Dar Tamaddun-I Eslami*, ed. by N. Mayel Heravi (1993), 161-181, Mashhad: Islamic Research Centre of Astan-e Quds-e Razavi, In [Persian]

Soltān Aḥmad Majnun Rafiqi, *Savād al-katt,* (after 930 A.H./1523 CE), In *Ketab-Arayi Dar Tamaddun-I Eslami*, edited by N. Mayel Heravi (1993), 185-206, Mashhad: Islamic Research Centre of Astan-e Quds-e Razavi. In [Persian]

Soltān Aḥmad Majnun Rafīqī Heravī), *Ādāb al-Mašq*, (Mid 10th A.H./mid 16th C.E.), In *Ketab-Arayi Dar Tamaddun-I Eslami*, edited by N. Mayel Heravi (1993), 209-236, Mashhad: Islamic Research Centre of Astan-e Quds-e Razavi. In [Persian]

Soltan Ali Mašhadi (1993) (841 A.H./1437 A.D.-926 A.H./1520), *Ṣerāt al-Ṣotur* (920 A.H./1514 A.D.), In *Ketab-Arayi Dar Tamaddun-I Eslami*, edited by N. Mayel Heravi (1993), 71-83, Mashhad: Islamic Research Centre of Astan-e Quds-e Razavi. In [Persian]

Teflisi H (1956) (515-600 A.H./1121-1203 A.D.), *Bayān al-Ṣenā'at*, 12th Century, In *Farhang-ī Irān-Zamīn*, ed. by I. Afshar (1336/1956) Vol. V,. Tehran: Anjoman-e Asar-e Melli-ye Iran Press, 298–457.

Wulff H E (1976) The Traditional Crafts of Persia. Their Development Technology and Influence of Eastern and Western Civilizations, (2nd ed. Cambridge, Mass. M.I.T. Press).

Chapter 2

Effects of Environmental Factors on Air-borne Fungi in Bangladesh National Museum in Relation to Deterioration of Museum Specimens during Pandemic of COVID-19

Shikha Noor Munshi

Department of Natural History, Bangladesh National Museum, Dhaka 1000, Bangladesh
e-mail: shikhanmunshi@yahoo.com

ABSTRACT

Bangladesh is a humid tropical country. The indoor environment in museums is influenced by several factors, such as environmental conditions (e.g. relative humidity, temperature and rainfall), gaseous and particulate matter pollution and microorganisms. The present work has been completed in two parts. In the first part, an attempt was made to investigate the reduction in the concentration of various air-borne fungi associated with the deteriorating museum specimens of Bangladesh National Museum (BNM) during pandemic of COVID-19 induced after lockdown period in the Dhaka city.

In this circumstance, the second part of the study dealt with the secondary data of the environmental factors such as relative humidity, temperature and rainfall for a period of 9 months before lockdown period during 1989, 1990 and 1991 on air-borne fungi in relation to *deterioration of museum specimens of BNM and correlated with the present result. The isolates recovered from the air showed that the environmental factors always regulated the fungal growth and considerable variations with the previous secondary data.*

Keywords: Environmental factors, Air-borne Fungi, Bangladesh, National Museum, Deterioration, Specimens, COVID-19

2.1 Introduction

Environmental factors play an important role on the works of arts and antiquities. It has a direct effect on museum specimens and indoor air-borne microorganisms around the museum specimens. These factors influence the deterioration of museum specimens in two ways. Firstly, they affect on or within the materials that result from free radical and ionic reactions. The first kind includes auto-oxidation reaction, usually initiated by thermal or photochemical energy input from the environment. Secondly, they affect on physiological process involved in the growth and development of the saprophytic microorganisms, which develop on museum specimens (Mills and White 1987).

Nowadays, the megacity Dhaka is one of the top polluted hotspots of air pollution for its distinct natures like high pollution density, unfit vehicle movement that emits industrial emission; dust in and around the city. According to estimation, the total population of Dhaka city stands more than 15 million, with a density of 33,878 per square kilometer, which is the highest figure in the World (Demographia 2020).

A worldwide pandemic of COVID-19 has forced the Government of Bangladesh to implement a lockdown during April 2019 to May 2020 by restricting people's movement, shutdown industries and motor vehicles; closing markets, public places. This type of strict measurers caused an outcome the reduction of city areas were recording massive reductions in a range of pollutants associated with internal combustion engines. The BNM has collaborated with the data to visualize the changes with especial relation to indoor airborne fungi of the Natural History Galleries of Bangladesh National Museum. The indoor air quality before and after lockdown period, the concentration of indoor airborne micro flora and the environmental parameters have been analyzed. The presence of viable indoor airborne micro flora after lockdown period during 2021 was considered against the comparison of the environmental parameters. The time series environmental data was based during 2019- 2021 and collaborated with secondary time series data during 1989, 1990 and 1991 to evaluate the prevailing environmental conditions before and after lockdown period respectively. The meteorological data of the two time series were obtained from the Climatic Division of Bangladesh Meteorological Department of Dhaka. Bangladesh National Museum (BNM) possesses multifarious type artifacts of historical and social significance. It plays a unique role for balancing preservation and conservation of cultural heritage. A large number of materials of organic and inorganic nature are preserved in BNM. Artifacts are exposed to various chemical and biochemical reactions owing to physical and biological agents. Physical agents as a humid tropical environmental conditions are responsible for deterioration of organic materials (Plenderleith and Werner 1974).

Coremans (1968) described physico-chemical aspects and observed that temperature, relative humidity and rainfall are important environmental factors for a tropical country. He reported that the tropical environmental condition was very much conducive for a vast range of microorganisms to grow on organic or inorganic materials and causing their deterioration. Among the recent publications, the works of Dhawan *et al.* (1991), Nigam and Pathak (1991) testified that they did some experimental investigation on environmental factors in India.

Bangladesh has a special humid, tropical geographical location and this type of environmental condition is very much conducive for a vast range of microorganisms to grow on organic or inorganic materials. When relative humidity is 60 per cent - 90 per cent, air-borne microorganisms rapidly attack organic materials. Nevertheless, no systematic and scientific study was made so far on the effects of environmental factors on deterioration of museum specimens with special relation to indoor air-borne microorganisms in Pandemic of COVID-19. The present paper deals with the occurrence of indoor air-borne fungi in the BNM Gallery and the prevailing environmental factors of the Dhaka city during 2019, 2020 and 2021 after lockdown period.

2.2 Materials and Methods

The present work has been completed in two parts. In the first part, an attempt has been made to study the prevailing environmental factors, *viz.*, relative humidity (RH), temperature and rainfall of the Dhaka city during 2019,-2021 to evaluate the prevailing environmental conditions after lockdown period. The meteorological data of the Dhaka city were obtained from the Climatic Division of Bangladesh Meteorological Department of Dhaka.

Table 2.1: The Site of Galleries and their Respective Environmental Conditions

Name of the Site	Environmental Conditions
The Natural History Gallery (first 10 galleries) of BNM, Dhaka.	The indoor air circulation of the Natural History Gallery of BNM was confined. The prevailing humidity and temperature inside the gallery was more ($12^{\circ}C$ - $13^{\circ}C$) than that of the outside, presence of diffused florescent light, absence of sunlight and window.

An attempt has been made also to determine the identity of the microorganisms, *viz.*, bacteria and particularly indoor air-borne fungi in relation to the deteriorating museum specimens of the BNM Gallery after lockdown period. These were isolated from the indoor air randomly around the first 10 galleries of the Natural History of BNM. The experimental work was carried out in the winter month of December 2021.

2.2.1 Collection of the Samples

A total of 30 plates were examined randomly employing 'Moist Filter Paper Exposure' Method. A total of 30 sterilized Petri plates were collected from the Natural History Gallery (first 10 galleries) of BNM. Amongst them 3 plates from each Gallery.

2.2.2 Screening of the Samples for the Presence of Viable Air–borne Microorganisms

As most of the fresh slides did not show good growth or sporulating structures of the microorganisms. These were subjected to moist chamber culture. Moist chamber culture often encourages good growth and sporulation of various groups of fungi.

(i) *Moist cotton buds culture*: A total of 50 moist cotton buds were sterilized in test tubes. A sterilized cotton bud was gently rubbed over glass or exhibit of the BNM gallery where microbial growth was observed. A total of 10 sterilized Petri plates from each gallery were taken randomly. The cotton buds were placed on one plates from each gallery and taken to the laboratory for incubation.

(ii) *Moist chamber culture*: A moist chamber culture was made by placing 2- layers of moist filter paper on the bottom of the Petri plate, which was covered with its upper lid and steam sterilized in an autoclave under 15 lbs. pressure at 120°C for 10 minutes. In each Petri plate, a small portion of a collected swabbed cotton bud/exposed filter paper was incubated under room conditions. For each gallery a total of two moist sterilized filter paper Petri plates (90 mm dia.) were placed 0.76m above the floor and exposed for one min. All exposed plates were then incubated for five days at room condition. After five days of incubation, plates were observed and isolates were counted.

After a suitable incubation period (5days), sporulating cultures of fungi were produced in profusion, whereas bacterial growth was not detectable with naked eye. The moist chamber culture samples were then used for microscopic examination. A total of 30 moist filter paper samples, thus gathered following above two methods were carried to the laboratory. After five days of incubation, isolates were examined with a microscope (Nikon Microphot, Japan) for detecting the presence of viable cultivable microorganisms and incidence of the individual genus was calculated.

2.2.3 Microscopic Examination

A series of six slides for each culture sample were prepared for microscopic examination. The preparations of the slides were as follows:

A small portion from an incubated sample was placed in lactophenol mixed with aniline blue (whenever necessary), mounted with cover glass and examined under an oil immersion lens of a light microscope. Even after incubation in moist chamber, it was found difficult to detect the presence of bacteria in culture samples. The number of each species of fungi was calculated and the incidence of the individual genus was calculated.

The second part of the present work dealt with the previous data of the environmental factors such as RH, temperature and rainfall for a period of 9 months during 1989, 1990 and1991 recorded before lockdown period, and the indoor air-borne fungi in relation to deterioration of museum specimens of BNM Gallery as secondary data. These previous recorded data were correlated with the present data obtained after lockdown period during 2019-2021.

2.3 Results and Discussion

A. Effect of Environmental Factors

The monthly relative humidity (RH), average temperature and rainfall of the Dhaka city after lockdown period during 2019, 2020 and 2021 are presented in the Figures 2.1–2.3 respectively. The data obtained show a significant correlation between the environmental factors and deterioration of museum specimens. Plenderleith and Werner (1974) recommended that limits of atmospheric RH between 50 per cent - 65 per cent at temperatures of 16°C - 25°C were good for preservation of museum specimens.

B) Effect of Relative Humidity

The Figure 2.1 shows that the RH in percentage of the Dhaka city. The data show the RH of the Dhaka city was always above the limits of Plenderleith and Werner (1974). The condition was liable for deterioration of museum specimens. The monthly average RH value of the Dhaka city was above than 60 per cent. The highest RH (85 per cent) in July 2020 and 2021; the lowest RH (57 per cent) was observed in March 2019, 2020 and 2021 in the Dhaka city. The RH plays a significant role in deterioration of cultural property because it has influence on moisture contents around the museum specimens, as well as fungal sporulation and spore distribution in air. Again the RH over 60 per cent favors the growth of microorganisms that destroy organic materials. All of these factors have direct effects on museum specimens (Plenderleith and Werner 1974; Dhawan *et al.*1991).

C) Effect of Temperature

The monthly average temperature in degree Celsius of the Dhaka city during

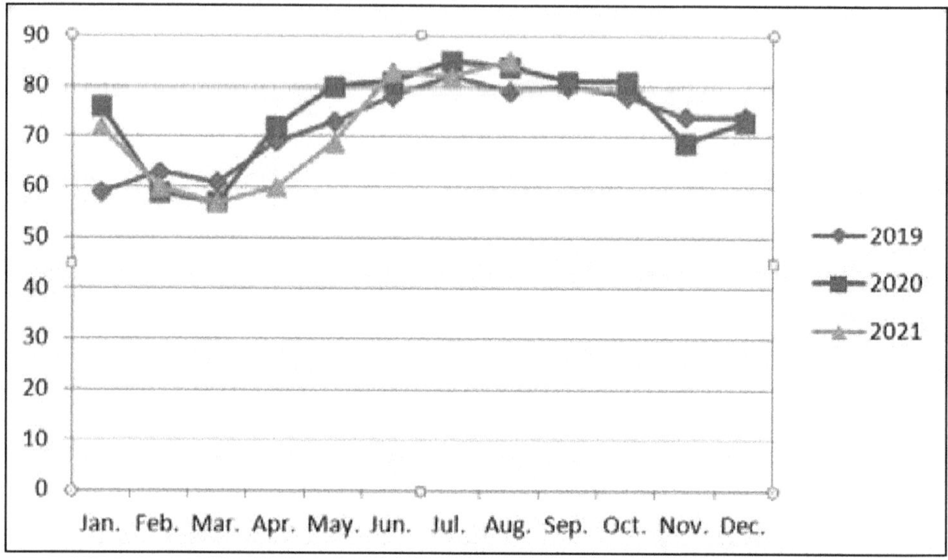

Figure 2.1: Monthly Relative Humidity in Percentage during 2019, 2020 and 2021.

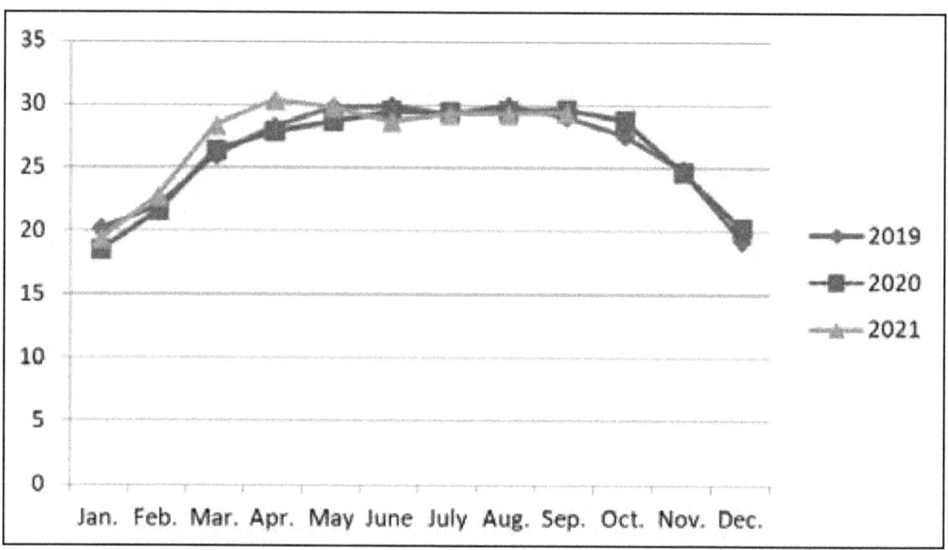

Figure 2.2: Monthly Average Temperature in °Celsius during 2019, 2020 and 2021.

2019, 2020 and 2021 are presented in the Figure 2.2 The data indicate that the ranges of the temperature were more than the limits of Plenderleith and Werner (1974), *i.e.*, above 19°C, (February to November). In addition, the fluctuation of high temperature in daytime and subsequent cooling at night along with its high RH, the overall environmental conditions were deleterious for museum specimens. In case of winter (December to January) during the three years the temperature was always 19°C. The overall environmental condition of winter was less detrimental than that of the other two seasons, *i.e.*, summer and the rainy season.

The chemical composition of the museum specimens may gradually alter through exposure to oxygen, humidity, light and other factors in the environment. The molecules of organic materials are oxidized, decomposed and polymerized. High temperature accelerates the decomposition. The higher the temperatures, the more rapidly museum specimens will age. Temperatures accelerate photochemical reactions also. The Figure 2.2 indicates that the prevailing temperature ranges of the Dhaka city was very much conducive for the growth of fungal organisms and hence harmful for museum specimens. The temperature and humidity inside the gallery of BNM vary with outside temperature and humidity of Dhaka City by +2°C - +3°C.

D) Effect of Rainfall

The Figure 2.3 contains the data of monthly rainfall of the Dhaka city during 2019, 2020 and 2021. The obtained data show that highest monthly rainfall was observed in Dhaka city (546 mm) in June 2021 and no rainfall was recorded in January and February 2021.

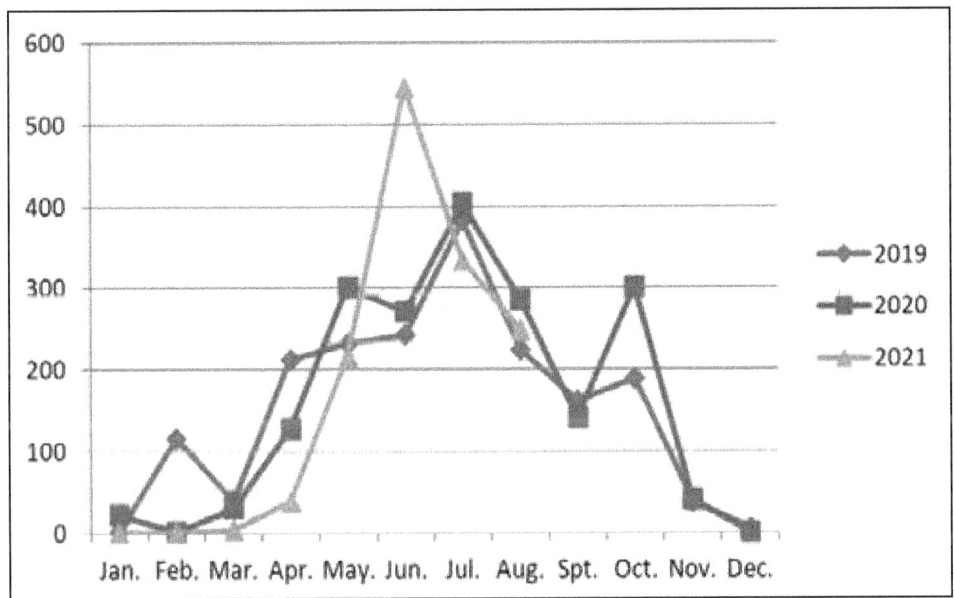

Figure 2.3: Monthly Rainfall in Millimeter during 2019, 2020 and 2021.

E) Effect of Air-borne Fungi

The indoor air-borne fungal isolates after lockdown period recovered from the 'Moist Filter Paper Exposure' Petri plate listed in the Table 2.1. Indoor air samples were collected in two sequential repetitions on each 'Moist Filter Paper Exposure' Petri plate for a single screening method only.

Table 2.2: List of the Isolated Air-borne Fungi from the Natural History Gallery of BNM during December (Winter) in 2021

Name of the Isolates	No. of Isolates	
Aspergillus flavus Link.	53	
A. fumigatus Fresenius	39	126
A. niger van Teighem	34	
Penicillium chrysogenum Thom	29	55
Penicillium. spp.	26	
Sterile dark mycelia	25	-
Total number of isolates	**206**	-

A total of 206 air-borne fungal isolates mostly the members of Fungi Imperfecti belonging to two genera, *viz.*, *Aspergillus* and *Penicillium* were recorded. The list of the air-borne fungal isolates from the BNM Gallery is presented in the Table 2.2

The members of *Aspergillus* and *Penicillium* (181) were the most common fungi. The members of *Aspergillus* (126) showed the maximum percentage incidence and the second frequent genus was *Penicillium* (55). The genus *Aspergillus* was represented by three species, *viz.*, *A. flavus*, *A. fumigatus* and *A. niger*. The genus *Penicillium*, was represented by *P. chrysogenum* and one unknown *Penicillium* sp.

A good number of air-borne fungal isolates having white and colored (black, green, yellow, brown and chocolate) mycelium were observed on the incubated filter paper, the determination of the identity of these fungi was not possible because of their non-sporulating nature on the 'Moist Filter Paper Exposure' Petri plates.

The second part of the present work dealt with the review of the secondary data of environmental factors (such as relative humidity, temperature and rainfall) on air-borne fungi before lockdown period of 9 months during 1989, 1990 and 1991 represented in Figure 2.4. The secondary data obtained during 1989, 1990 and 1991 was correlated with the present environmental factors of the Dhaka city after lockdown period during 2019, 2020 and 2021 (Figures 2.1–2.3) respectively.

In addition, an overview of the indoor air-borne mycoflora recovered from the 'Culture Plate Exposure' on Czapek Dox medium during 1989, 1990 and 1991 listed in the Figure 2.4. A total of 4,475 air-borne fungal isolates were recorded over a period of 9 months during 1989, 1990 and1991 in three different seasons before lockdown period and compared with the indoor air-borne fungi obtained employing 'Moist Filter Paper Exposure' Method during 2021 after lockdown period (Table 2.2). In both circumstances the isolates recovered from the indoor air of the BNM Gallery showed that the environmental factors always regulated the fungal growth and considerable variations with the previous secondary data and the present records.

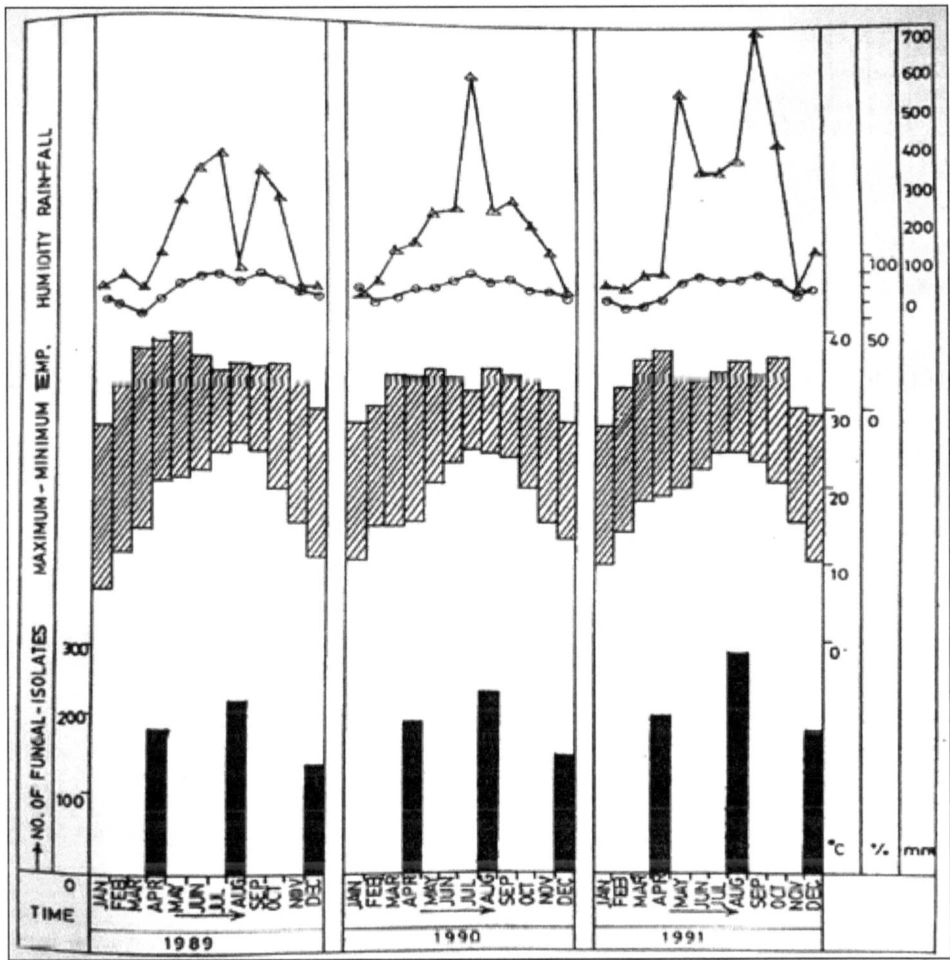

***Figure 2.4: Indoor Aero-mycoflora Recovered from the 'Culture Plate Exposure' on Czapek Dox medium during 1989, 1990 and 1991 in BNM and Min. and Max. Temp in °C and Humidity/Monthly Rainfall in mm.**
***Part of the Ph.D. thesis by Munshi (2002)**

The presence of a few number of air-borne fungi during 2021 after lockdown period was considered as a significant indication for their presence and association with the deteriorated of museum specimens in Bangladesh. The presence of different air-borne fungal spores determines the seat of their origin so that some effective measures might be evolved to control the growth of fungi right at their origin (Plenderlieth and Werner 1974, Dhawan *et al.,* 1991). This would eventually results in awareness against microbial deterioration of our national heritage. Lockdowns in Bangladesh along with other countries have resulted in significant improvement in environmental quality. The increased concern of the virus spread in confined spaces due to meteorological factors has sequentially fostered the need to improve indoor air quality.

Acknowledgements

The author expresses her gratitude to the authority, Bangladesh National Museum, Dhaka, for providing necessary facilities for carrying out the present work.

REFERENCES

Coremans P (1968) Climate and Microclimate. Museums and Monuments: The Conservation of Cultural property. Published by UNESCO, Rome, Italy. pp. 27-39.

Demographia (2020) World Urban Areas reports. June 26th edition

Dhawan S, Pathak N, Garg K L, Misra A (1991) Effect of Temperature on Some Fungal Isolates of Ajanta Wall Paintings: Biodeterioration of Cultural Property. Proc. of the International Conference, 1989. Lucknow, India. Macmillan India Ltd. Lucknow. pp. 339-352.

Mills J S, White R (1987) The Organic Chemistry of Museum Objects. Butterworths, London. pp. 1 - 159.

Munshi S N (2002) Microbial Deterioration of Museum Specimens in Bangladesh and their possible control- Ph.D. thesis under the supervision of Prof. A. Z.M. Nowsher Ali Khan and Prof. M. R. Khan, Dept. of Botany, Univ. of Dhaka.

Nigam R K, Pathak N C (1991) Environmental Mycology of a Tannery and a Textile Unit at Kanpur: Allergical aspects. Biodeterioration of Cultural Property (Proc. of the International Conference, 1989. Lucknow, India). Macmillan India Ltd. pp. 375-386.

Plenderlieth H J, Werner A E A (1974) The Conservation of Antiquities and Works of Art. Oxford University Press, London. pp. 1-394.

Chapter 3

Glimpses of Egyptian Civilization and Conservation of Mummies

Arun Arya

Department of Environmental Studies, Faculty of Science, The Maharaja Sayajirao University of Baroda, Vadodara – 390 002, Gujarat, India
e-mail: sarojarun10arya@rediffmail.com

ABSTRACT

Ancient Egyptian, Chinese and South American civilizations witnessed the mummified remains of human and animals. The bodies if buried in hot sand of desert, cold ice may be preserved. But during Egyptian civilization the art of mummification was practiced and human bodies are preserved with utmost care and are known as mummies. These are excavated and are displayed in different museums of the world. Since these are organic in nature and about 3000 years old, which are subjected to biodeterioration needs our immediate attention.

The origin of Mesopotamia first civilization dates back so far that there is no known evidence of any other civilized society before them. The timeline is held to be from around 3300 BC to 750 BC. Mesopotamia is generally credited as being the first place where civilized society truly began to take shape. It was somewhere around 8000 BC that people developed the idea of agriculture and domestication of animals. People had already been creating art well before the Mesopotamians, but this was part of human culture, not human civilization. It was the Mesopotamian civilization that refined this, adding to and formalizing all these systems. They prospered in the regions of modern-day Iraq, then known as Babylonia, Sumer, and the Assyria Highlands. A sophisticated and technologically advanced urban culture was seen in the Indus Valley civilization from 2600 BC to around 1900 BC making its capital the first urban center in the region. The great accuracy in measuring length, mass, and time, was achieved during the civilization. Based on artifacts found in excavations, it is evident that the culture was rich in arts and crafts as well.

Ancient Egypt is one of the oldest and culturally rich civilizations on the banks of the Nile. It is known for its prodigious culture, its pharaohs, the enduring pyramids, and the Sphinx. Ancient Egypt reached its pinnacle during the New Kingdom, when pharaohs like Ramesses the Great ruled with such authority that another contemporary civilization, the Nubians, also came under Egyptian rule. Mummies in London, Turin, and Klagenfurt and other well-mummified corpses of this group are discussed. A masked female mummy in Klagenfurt (Austria) was perhaps not originally associated with the coffin, all certainly belong to the same workshop and date range. Seven such rare objects are preserved in Indian museums, which arouse curiosity about ancient life on this planet and attract attention of a common visitor in museums. An extensive review is presented about the process of mummification and recent researches related to mummies and their conservation. The science of Egyptology has developed with studies related to such precious objects their conservation and their burial process in the pyramids of Egypt. Problems of deterioration associated with mummies placed in Vadodara, Lucknow, Kolkata and Jaipur are discussed.

Keywords: Egyptian Civilization, Conservation, Mummy, Pyramids, Fungi, Baroda museum, Emericella nivea

3.1 Introduction

In 1959 the governments of the United Arab Republic (U.A.R.; now Egypt and Syria) and Sudan turned to UNESCO for help in salvaging the ancient sites at Nubia. The sites were threatened with destruction by the great lake due to construction of new dam at Aswan. On an appeal from UNESCO the largest archaeological rescue operation in history took place. And a movement to protect the Cultural Heritage started. The Nubia Museum in Aswan became a reality after fifteen years and opened its doors in November 1997. It literally mobilized worldwide attention and fostered willingness and commitments to safeguarding heritage as an objective of international co-operation.

Ancient Egypt is one of the oldest and culturally rich civilizations on the banks of the Nile. Other major civilizations include Indus valley civilization present in Harappa and Mohenjo-Daro, the Sumerian Civilization (4500 BC to 1900 BC) or Mesopotamian civilization, and Ancient Maya Civilization (1000 BC to AD 1520) in Mexico. Egyptian civilization is known for its prodigious culture, its pharaohs, the enduring pyramids, and the Sphinx. Ancient Egypt reached its pinnacle during the New Kingdom, when pharaohs like Ramesses the Great ruled with such authority that another contemporary civilization, the Nubians, also came under Egyptian rule. Mummies in London, Turin, and Klagenfurt. Other well-mummified corpses of this group (excerebrated, eviscerated, and with the arms extended) include a male mummy in London, wrapped in a bead-net-painted shroud (British Museum U.K.; Dawson and Gray 1968; Riggs 2005); a similarly decorated boy's mummy in Turin Italy (Egyptian Museum; De Lorenzi and Grilletto 1989); and a masked female mummy in Landes Museum, Klagenfurt Austria. Although this mummy was perhaps not originally associated with the coffin, all certainly belong to the same workshop and date range (Riggs 2005). The Science of Egyptian mummification, excavation and its conservation and other studies related to that period has evolved in to a new branch referred as Egyptology. Designing a correct conservation treatment process requires an interdisciplinary work that involves professionals from various fields, in order to combine and incorporate several approaches and data to reach an overview, as complete as possible, of the artifact. In particular, for archaeological objects, besides notes on the specific excavation activities, very few documented information are available on the artistic techniques and conservation history of the mummy and other buried objects.

Mummification was not only limited to Egypt. In modern-day people of Papua New Guinea still mummify the deceased. Eight mummified bodies were recovered from Greenland. These mummies dated 1460 A.D. were preserved by freeze drying method (Lynnerup 2016). Beyond that, funeral homes in west often embalm dead bodies to slow down the decomposition and allow time for ceremonies to take place. Few anatomical medical laboratories also preserve the bodies. Recently (2/2/22) Archbishop Desmond Tutu was cremated in Cape Town by an ecofriendly way called Aquamation, or alkaline hydrolysis, is actually a cremation by water and not fire. In it, for three to four hours, the body of a deceased is immersed in a mixture of water and a strong alkali, such as potassium hydroxide. This is done in a pressurized metal cylinder and heated to around 150 ° C. The process liquefies everything except for the bones. They are then dried with the help of an oven and reduced to white dust. Finally, it is placed in an urn and handed over to the grieving relatives.

The mummified remains provide the research and educational information to the scientific community as well as the general public, and hence are valuable components of natural history collections. To look to their ongoing worth in the museums of the future, we must look to the field of conservation for approaches to preserving this material without compromising the specimens' integrity for research. A case of mishandling occurred with a mummy that was shipped from Arica, Chile to Lincoln, Nebraska in the late 1800s. Although the mummy appears to be in good shape on the outside, CT-scans of the mummy showed mechanical breakage within. This is especially true of the neck region where a crack has formed. These problems are probably due to the posture of the mummy on exhibit, in storage, and during shipment which put subtle pressures on the neck. The mummy had also been unwrapped prior to exhibition, losing any support from the bundle (Meier *et al.*, 1998). Keeping a mummy bundle intact, however, presents many different kinds of conservation concerns. A typical South American mummy bundle contains cloth, ceramic, leather, bone, shell, metal, fibre and wood artefacts besides the mummy. The process of decay can take many years, but the end result is the loss of an important specimen. If you look closely at a mummy on display you can often see the clues to decomposition. Tiny holes in the skin and a brownish dust on the floor of the case are strong indicators of insect activity (Meier 2001).

Sometimes mixed in with the dust are the carcasses of the insects that produced the debris. Mould is usually inactive on a dried specimen such as mummified skin, but the mould spores are still viable. When the environmental conditions are right they can start to damage the skin and spread to other organic materials in the area. The presence of an active mould growth is a strong indication that seriously high relative humidity is occurring (Strang and Dawson 1991).

3.1.1 Civilizations in Past

Indus valley civilization or Harappa civilization, is the earliest known civilization in Indian subcontinent, which appeared about 2500–1700 BCE, though the southern sites may have lasted later into the 2nd Millennium BCE. Among the world's three earliest civilizations–the other two are those of Mesopotamia and Egypt. Indus civilization was the most extensive. Mohenjo-Daro is a world of Sindhi means "Mound of Dead" which is present in present Pakistan and was discovered in 1920s. Fired bricks were used in structures excavated. The Indus civilization apparently evolved from the villages of neighbors or predecessors, using the Mesopotamian model of irrigated agriculture with sufficient skill to reap the advantages of the spacious and fertile Indus valley river. Annual floods occurred in the region that simultaneously fertilized and destroyed the crops. Having obtained a secure foothold on the plain and mastered its more immediate problems, the new civilization, doubtless with a well-nourished and increasing population, would find expansion along the flanks of the great waterways an inevitable sequel. The civilization subsisted primarily by farming, and supplemented by an appreciable but often dealing with trade. Wheat, six-rowed barley, field peas, mustard, sesame, and dates were grown. Some of the earliest known traces of cotton are seen. Domesticated animals included dogs and cats, humped and short-horned cattle, domestic fowl, and possibly pigs, camels, and buffalo. The Asian elephant probably was also domesticated, and its ivory tusks were freely used (Sharma 1975).

The Sumerians were responsible for the first system of writing, cuneiform; the earliest known codes of law; the development of the city-state; the invention of the potter's wheel, the sailboat, and the seed plow; and the creation of literary, musical, and architectural forms that influenced all of Western civilization. This cultural heritage was adopted by the Sumerians and the Amorites, a western Semitic tribe that had conquered all of Mesopotamia by about 1900 BCE. Under the rule of the Amorites, which lasted until about 1600 BCE, Babylon became the political and commercial center of the Tigris-Euphrates area, and Babylonia became a great empire.

Civilization in Shang's started about 1600 BCE and has identified the dynasty's end as being 1046 BCE. The latter part of the Shang dynasty, from the reign of the Panging emperor onward (*c.* 1300 BCE), has also been called the Yin dynasty. Shang China was centered in the North China plain and extended as far north as modern Shandong and Hebei provinces and westward through present-day Henan. The king appointed local governors, and there was an established class of nobles as well as the masses, whose chief labor was in agriculture. The king issued pronouncements as to when to plant crops, and the society had a highly developed calendar system with a 360-day year having 12 months of 30 days each. It was during the Shang that Chinese writing began to develop, and the symbol for "moon" was–as it has remained–that also for "month." The calendar took cognizance of both lunar and solar cycles. The musical instruments were developed. Legend traces the origin of pipes of bamboo earlier, even before the mythical Xia. The architects of the Shang period built houses of timber over rammed-earth floors, with walls of wattle and daub and roofs of thatch. There was a three-legged *li* for cooking, and upon it could be fitted a bronze *zeng*, a bowl with a pierced bottom to function as a steamer–together called a *yan*. Serving bowls were often stemmed, and pouring vessels, such as the *gu*, had long spouts. Those vessels were often richly decorated. Pottery was shaped on a potter's wheel. And it included dishes and bowls in a white glaze for ceremonial and ritual use. Ceremonial weapons of jade were made. Other funerary art ran a gamut in size from tiny objects of jade or carved bone and ivory. Numerous records and ceremonial inscriptions and family or clan names exist, carved into or brushed onto bone or tortoise shells. Three kinds of characters were used–pictographs, ideograms, and phonograms–and those records are the earliest known writings in China (Anonymous 1990).

3.1.2 Worship of the Mummified Mermaid to get Rid of Corona

In Japanese mythology mermaids are associated with immortality. The Japanese folklorist have a myth that when a fisherman from Wakasa Province caught a fish and the daughter of a man who forgot to throw the evil fish away "lived to 800 years old." Ningyo – a mermaid was described with shining golden scales and a monkey's mouth, offering longevity to those who ate of its flesh. But the act of hunting and catching a *ningyo* was believed to bring storms and bad luck. Margaritott (2022) reported that according to chief priest, Kozen Kuida from the Enjuin temple in Japan the mummified mermaid was put on display some 40 years ago and is now kept in a safe place. Alarmingly, this object was worshipped recently with a hope that it would help in containing Corona virus in Japan.

3.1.3 Ancient Practice of Mummification in Egypt

The present knowledge of the mummification techniques is based on early period, not very detailed records available. It can be further upgraded only by investigating the mummies still extant in museums: roughly 150 corpses are now datable as Roman. The majority of these are in all likelihood made for wealthy people (Borg 1998). Many of these mummies represent a predominantly urban class. The painted mummy portraits were, for instance, not in use at Kharga or Dakhla. The observation that the various kinds of mummy and coffin decoration–as well as the manifold types of portrait and mask (Figure 3b)–reflect distinctive local workshops has long become the general consensus (Parlasca and Seemann 1999; Riggs 2005). By striking contrast, embalming techniques are often described as typically Roman without taking comparable variations into account.

Even though similar features in mummification methods have been observed in a variety of locations, differences from one cemetery to the next, even within the same oasis, occur. In order to distinguish between local, inter-regional, social, and chronological changes, the results from the various sites need further comparative study with the inclusion of mummies in museums. For example, while the use of natron mixtures as the main dehydrating substance is still attested, whether among remnants of embalming material (Dunand and Lichtenberg 2008), or put in linen bags into the body cavities and occasionally left there by forgetful embalmers (Macke 2002), its complete absence in the Roman mummies from Dakhla remains exceptional (Aufderheide 2005). Hence, the absence of natron might be due to local availability, but this issue can only be revised on the basis of a much larger set of data. Although interdisciplinary cooperation on excavations in Egypt is well established and highly effective, mummy research in museums has been conducted rather intermittently and according to different standards (Raven and Taconis 2005). Nevertheless, mummy research is ongoing worldwide, backed by steadily advancing technologies. As appeared in scholarly journals (Christensen 1969; Mininberg 2001; Lynnerup 2007). Egyptian mummy was usually wrapped in linen bandage (Figure 3.1).

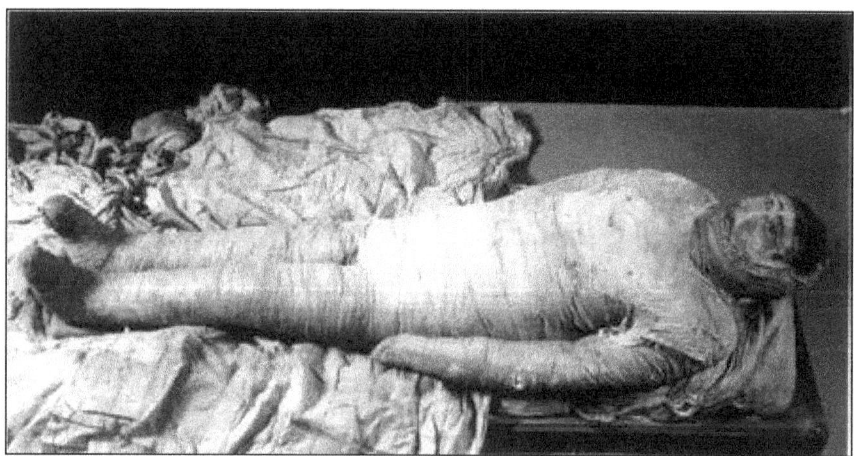

Figure 3.1: Wrapped in Linen Bandage an Egyptian Mummy Placed in Turin Museum, Italy.

3.2 Egyptian Civilization: Salient Features

The use of textile like linen was wrapping material for mummies with objects like bronze and stone statuettes, making them crossing the boundary between earthly life and divine. Most Egyptians were probably descended from settlers who moved to the Nile valley in prehistoric times, with population increase coming through natural fertility. Most people lived in villages and towns in the Nile valley and delta. Dwellings were normally built of mud brick and have long since disappeared beneath the rising water table or beneath modern town sites, thereby obliterating evidence for settlement patterns. Nearly all of the people were engaged in agriculture. They were expert in making bread (Figure 3.4 a) and articles like basket and toy making. The spices like onion, garlic, black pepper were are found in collection. Broadly speaking, the style of textiles follows the same lines as the stonework. The use of color was an essential part of the decorative nature of Coptic art in general and may be judged from the fourth century onwards, mainly in the cemeteries of Actinopolis and Akhmim (James 1979).

The native Egyptian breed of sheep became extinct in the 2nd millennium BCE and was replaced by an Asiatic breed. Sheep were primarily a source of meat; their wool was rarely used. Goats were more numerous than sheep. Pigs, ducks and geese, and other migratory birds were hunted and trapped. Desert game, principally of various species of antelope and ibex, were hunted by the elite; it was a royal privilege to hunt lions and wild cattle. In addition, the Egyptians had a great interest in, and knowledge of, most species of mammals, birds, reptiles, and fish in their environment.

Just as the Egyptians optimized agricultural production with simple means, their crafts and techniques, many of which originally came from Asia, were raised to extraordinary levels of perfection. The Egyptians' most striking technical achievement include, massive stone building, also exploited the potential of a centralized state to mobilize a huge labor force, which was made available by efficient agricultural practices. The construction of the great pyramids of the 4th dynasty (*c.* 2575–*c.* 2465 BCE) has yet to be fully explained and would be a major challenge to this day.

Figure 3.2.a depicts stone and granite sculptures exhibited in a Gallery in Turin Museum of Mummies. Figure 3.2d shows illustration from an Egyptian Book of the Dead, (*c.* 1275 BCE), showing the jackal- headed god of the dead, Anubis, weighing the soul of the scribe, Ani. The Hieroglyph writing can also be seen (Figures 3.3d and 3.5). Ramesses III was able to foster trade and engage in extensive building work. Great mortuary temple is situated at Medinet Habu (James 1979). Many pre-dynastic burials contain small female figurines of ivory and other materials thought to be primitive fertility or mother-goddess. In early times most Gods were worshipped in the form of animals and inanimate objects (James 1979).

3.2.1 Pyramids in Egypt

Tombs of early Egyptian kings were bench-shaped mounds called *mastabas*. Around 2780 BCE, King Djoser's architect, Imhotep, built the first pyramid by placing six mastabas. This Step Pyramid stands on the west bank of the Nile River at Sakkara near Memphis. Like later pyramids, it contains various rooms and passages, including the burial chamber of the king. The transition from the Step Pyramid to a true, smooth-sided pyramid took placed during the reign of King Snefru, founder of the Fourth Dynasty (2680–2560 BCE). At Medum, a step pyramid was built, then filled in with stone, and covered with a limestone casing. About halfway up, however, the angle of incline decreases from over 51 degrees to about 43 degrees, and the sides rise less steeply, causing it to be known as the Bent Pyramid. The largest and most famous of all the pyramids, the Great Pyramid at Giza, was built by Snefru's son, Khufu, known also as Cheops, the later Greek form of his name. The pyramid's base covered over 13 acres and its sides rose at an angle of 51 degrees 52 minutes and were over 755 feet long. It originally stood over 481 feet high; today it is 450 feet high. Scientists estimate that its stone blocks average over two tons a piece, with the largest weighing as much as fifteen tons each. Two other major pyramids were built at Giza, for Khufu's son, King Khafre (Chephren), and a successor of Khafre, Menkaure (Mycerinus). Also located at Giza is the famous Sphinx, a massive statue of a lion with a human head, carved during the time of Khafre.

Pyramids did not stand alone but were part of a group of buildings which included temples, chapels, other tombs, and massive walls (Figure 3.3c). Remnants of funerary boats have also been excavated; the best preserved is at Giza. It is thought that the king's body was brought by boat up the Nile to the pyramid site and probably mummified in the Valley Temple before being placed in the pyramid for burial. There has been speculation about pyramid construction. Egyptians had copper tools such as chisels, drills, and saws that may have been used to cut the relatively soft stone. The hard granite, used for burial chamber walls and some of the exterior casing, would have posed a more difficult problem. Most of the stone for the Giza pyramids was quarried on the Giza plateau itself. Some of the limestone casing was brought from Tura, across the Nile, and a few of the rooms were cased with granite from Aswan. Marks of the quarry workers are found on several of the stone blocks giving names of the work gangs such as "craftman-gang". Part-time crews of labourers probably supplemented the year-round masons and other skilled workers.

3.2.2 Hieroglyph: Language Used in Ancient Egypt

A vital innovation was the introduction of vernacular forms into the written language. This led in later decades to the appearance of current verbal forms in monumental inscriptions. The vernacular form of the New Kingdom, which is now known as Late Egyptian, appears fully developed in letters of the later 19th and 20th dynasties. Hieroglyph, meaning "sacred carving," is a Greek translation of the Egyptian phrase "the god's words," which was used at the time of the early Greek contacts with Egypt to distinguish the older hieroglyphs from the present hand writing (Figure 3.5 Egyptian hieroglyphic writing).

Egyptian hieroglyphic writing was composed entirely of pictures, though the object depicted cannot be identified in every instance (Figure 5). The earliest examples that can be read show the hieroglyphs used as actual writing, that is, with phonetic values, and not as picture writing such as that of the Eskimos or American Indians. It apparently arose in the late predynastic period (before 2925 BC). Hieroglyphic writing followed four basic principles. First, a hieroglyph could be used in an almost purely pictorial way. The sign of a man with his hand to his mouth might stand for the word "eat." Similarly, the word "sun" would be represented by a large circle with a smaller circle in its center. Second, a hieroglyph might represent or imply another word suggested by the picture. The sign for "sun" could as easily serve as the sign for "day" or as the name of the sun God Re. The sign for "eat" could also represent the more conceptual word "silent" by suggesting the covering of the mouth. Third, the signs also served as representatives of words that shared consonants in the same order. Thus the Egyptian words for "man" and "be bright," both spelled with the same consonants, *hg*, could be rendered by the same hieroglyph. Fourth, the hieroglyphs stood for individual or combinations of consonants.

Writing was a major instrument in the centralization of the Egyptian state and its self-presentation. The two basic types of writing–hieroglyphs, which were used for monuments and display, and the cursive form known as hieratic–were invented at much the same time in late pre-dynastic Egypt (*c.* 3000 BCE). Writing was chiefly used for administration, and until about 2650 BCE no continuous texts are preserved; the only extant literary texts written before the early Middle Kingdom (*c.* 1950 BCE) seem to have been lists of important traditional information and possibly medical treatises.

A fabric face mask is exhibited in the British Museum, inspired by the Rosetta stone. This re-usable cotton mask has outer and a bamboo inner lining, making it an eco-friendly option (Figure 3.2). Soldiers in Napoleon's army discovered the Rosetta stone in 1799 while digging the foundations of an addition to a fort near the town of el-Rashid (Rosetta). Exhibited in the British Museum since 1802, the Rosetta stone was the key to unlocking the ability to read and translate ancient Egyptian hieroglyphics. The Stone contains the same decree in three different scripts, and scholars were able to use the Greek and Egyptian Demotic sections to decipher the ancient hieroglyphs. The text contains a decree passed by priests in 196 BC to honor King Ptolemy V for his services to Egypt.

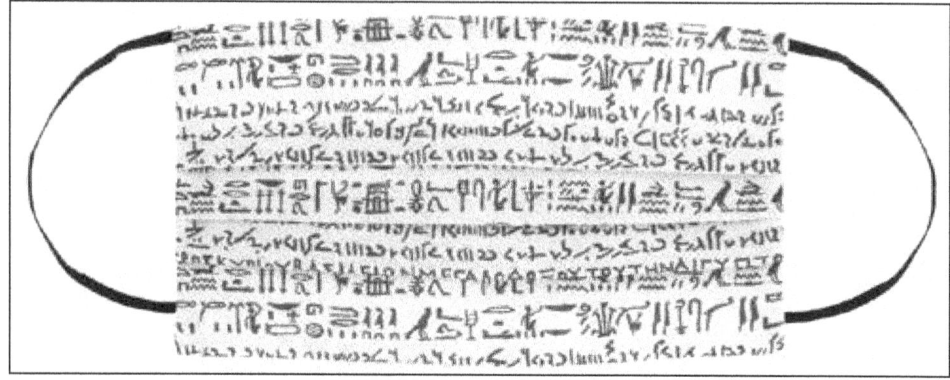

Figure 3.2: A Fabric Face Mask Exhibited in the British Museum, Inspired by the Rosetta Stone.

Egyptian literary and writings include poetry, lyrics and hymns. The Nile and many gods were worshipped. The hymn to the Nile (Papyrus Sallier II, Papyrus Anastasi VII and Papyrus Chester Beatty V, 10182,10222,10685) was composed in 1250BC.

Hail O Nile, who issues fourth from the earth

Who comes to give life to the people of Egypt.

Secret of movement, a darkness I the daylight.

Praised by his followers whose fields your water.

Created by Re to give life to all ho thirst. (James 1979).

They represented the sun-go, Amen-Re, as the universal creator:

Creator are you, fashioner of your own limbs,

One who brings into being, himself unborn,

Unique in his qualities, traversing eternity

Upon roads with millions under his guidance.

3.2.3 Egyptian Mummies: Researches in Museums

The Egyptian practice of embalming made its way to Rome and to the Roman Empire in conjunction with other 'exotic' Egyptianizing trends from the first century to the first half of the 4th century CE. This phenomenon is documented by both the inscriptional and the archaeological evidence (Counts 1996; Budischovsky 2004). The present state of fieldwork should not obscure the fact that some of the most important Graeco-Roman cemeteries, such as Hawara, Abusir el-Meleq, el-Hibeh, Hermopolis, Akhmim (Panopolis), and Antinoopolis,

Figure 3.3a–d

(a) Stone and granite sculptures exhibited in a Gallery in Turin Museum of Mummies; (b) Face mask of a mummy; (c) A picture of Pyramids in Egypt; (d) Illustration from an Egyptian Book of the dead, *c.* 1275 BCE, showing the jackal-headed god of the dead, Anubis, weighing the soul of the scribe, Ani.

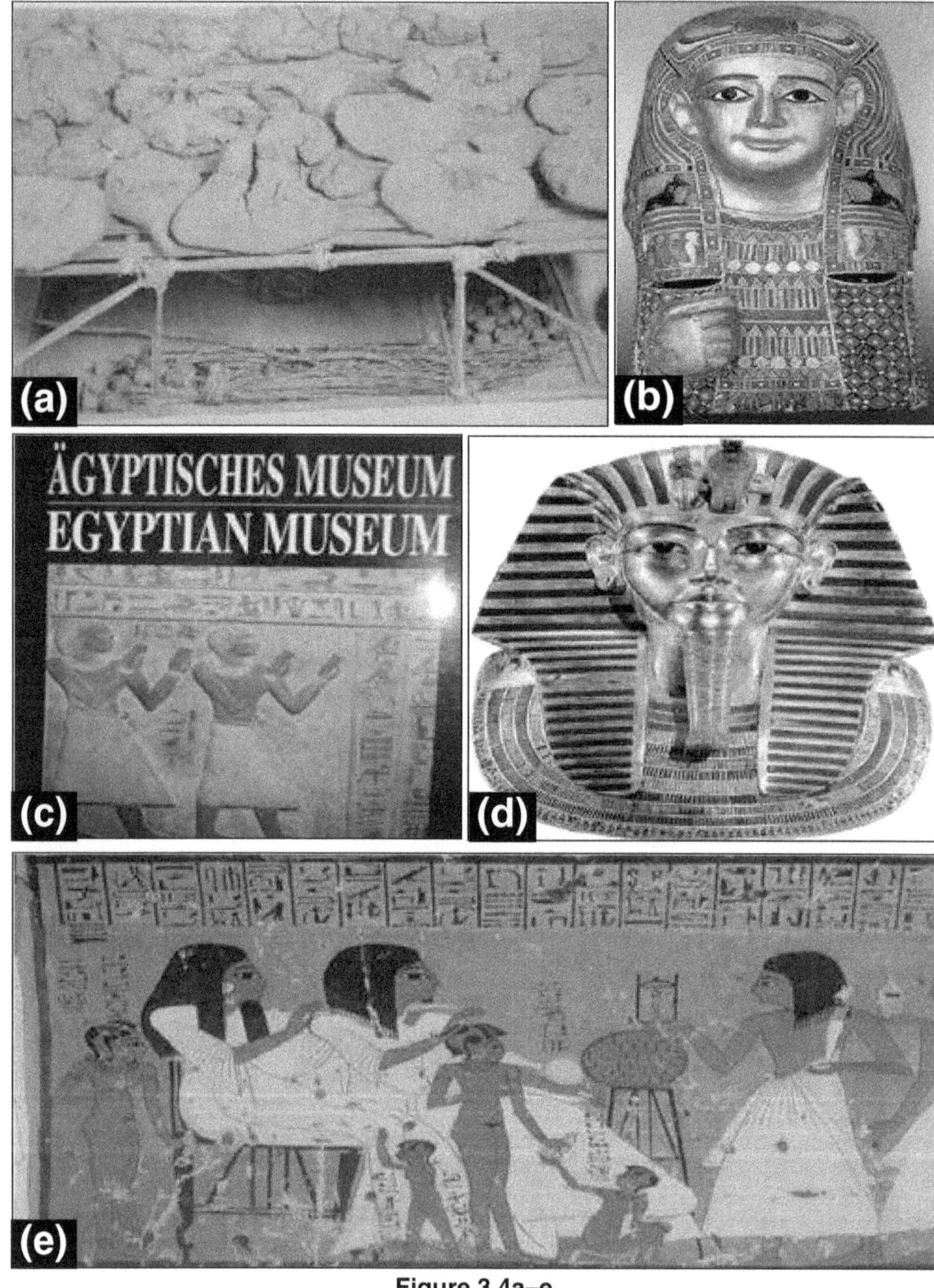

Figure 3.4a–e

(a) Picture showing wheat bread and garlic used as food in ancient Egypt; (b) A mask of the mummy of a girl; (c) A catalogue of Egyptian Museum Ancient Egypt sphynx; (d) The mask of Tutankhamun: A gold mask of the 18th-dynasty ancient Egyptian Pharaoh Tutankhamun (reigned 1334–1325 BC); (e) Mural of women with children, from the tomb of Khai–Inherkha, Thebes. (Photo by DEA/G Dagli Orti/De Agostini).

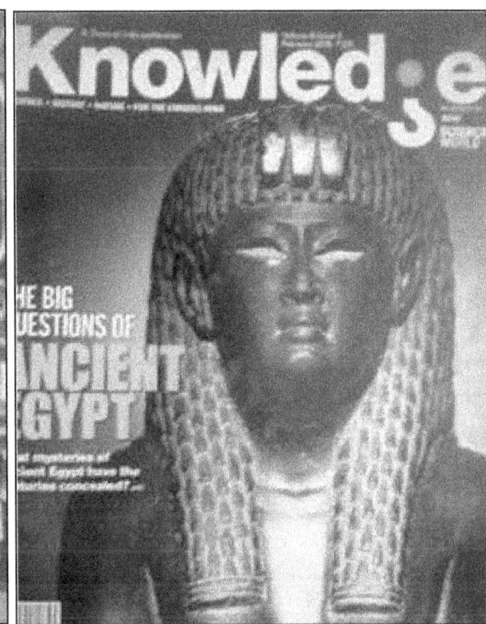

Figure 3.5: Egyptian Hieroglyphic Writing; Sacred Carving by the Greeks (c 1950 BC).

Figure 3.6: The Basalt Statue of Cleopatra. Placed in St. Petersburg's Hermitage Museum (Courtesy BBC Knowledge Magazine, Mumbai, 2016).

were pillaged and excavated from the nineteenth century onwards, and are in the main lost forever. Hence, our knowledge of the mummification techniques at these places is based on early, not very detailed records. It can be upgraded only by investigating the mummies still extant in museums: roughly 150 corpses are now datable as Roman (plus an unknown number of unproven specimens). The majority are expensively decorated and in all likelihood made for wealthy people (Borg 1998).

At Alexandria mummified bodies are only sporadically preserved. These were probably never became as important as inhumation or as the adopted Greek custom of cremation (Schmidt 2003). Nevertheless, Strabo's description of the necropolis in the western part of the city mentioning embalming institutions (ταρέχcßα) cannot be ignored (20 BCE). The reference to the embalming of Antony and Cleopatra in Dio Cassius (second–third century CE) also documents Egyptian funerary practices in Alexandria for the end of the Ptolemaic era (Counts 1996; Grimm 1997).

The exceptional status of mummification there during Roman times became apparent when a mummified body was found at Gabbari (excerebrated, with a gilded face and wrapped up in many layers of linen fabric): it was placed among inhumations with only the bones preserved in pit burial (Boës *et al.*,

2002). However, Egyptian burial practices used in parallel with hellenistic ones (corpses wrapped in clothes or textiles, without a 'mummiform' shape) have also been observed at Antinoopolis (Calament 2005). The observation that the various kinds of mummy and coffin decoration–as well as the manifold types of portrait and mask–reflect distinctive local workshops has long become the general consensus (Parlasca and Seemann 1999; Riggs 2005). By striking contrast, embalming techniques are often described as 'typically Roman' without taking comparable variations into account.

For example, while the use of natron mixtures as the main dehydrating substance is still attested, whether among remnants of embalming material (Dunand and Lichtenberg 2008), or put in linen bags into the body cavities and occasionally left there by forgetful embalmers (Macke 2002), its complete absence in the Roman mummies from Dakhla remains exceptional (Aufderheide 2005). Hence, the absence of natron might be due to local availability, but this issue can only be revised on the basis of a much larger set of data.

The pioneering studies from the 1960s onwards are still useful (Dawson and Gray 1968; Gray and Slow 1968; Cockburn and Cockburn 1980; David 1986), but partially outdated because of the advances in radiology over the past twenty-five years, especially CT scanning, and because of changes in Egyptological knowledge and interpretation. Meanwhile, some museums have updated the studies on their human and animal remains from Egypt, above all the Rijksmuseum van Oudheden in Leiden, in a path-breaking collaboration with the Academic Medical Centre of the University of Amsterdam (Raven and Taconis 2005). Other collections have published interdisciplinary mummy projects including Graeco-Roman mummies, either as monographs (De Lorenzi and Grilletto1989; Küffer *et al.*, 2007). Nevertheless, mummy research is ongoing worldwide, backed by steadily advancing technologies.

3.2.4 The Queen: Tutankhamun

Akhenaton had six daughters by Nefertiti and possibly a son, perhaps by a secondary wife Kiya. Either Nefertiti or the widow of Tutankhamun called on the king Suppiluliumas to supply a consort because she could find none in Egypt; a prince was sent, but he was murdered as he reached Egypt. Thus, Egypt never had a diplomatic marriage in which a foreign man was received into the country. After the brief rule of Smenkhkare possibly a son of Akhenaton, Tutankhaten, a nine-year-old child, succeeded and was married to the much older Ankhesenpaaten, Akhenaton's third daughter. Around his third regnal year, the king moved his capital to Memphis, abandoned the Aton cult, and changed his and the queen's names to Tutankhamun and Ankhesenamen. In an inscription recording Tutankhamun's actions for the gods, the Amarna period is described as one of misery and of the withdrawal of the gods from Egypt.

Just as Akhenaton had adapted and transformed the religious thinking that was current in his time, the reaction to the religion of Amarna was influenced by the rejected doctrine. In the new doctrine, all gods were in essence three: Amon, Re, and Ptah (to whom Seth was later added), and in some ultimate sense they too were one. The earliest evidence of this triad is on a trumpet of Tutankhamun and is related to the naming of the three chief army divisions after these gods; religious life and secular life were not separate. This concentration on a small number of essential deities may possibly be related to the piety of the succeeding Ramesside period, because both viewed the cosmos as being thoroughly permeated with the divine. Tutankhamun's modern fame comes from the discovery of his rich burial in the Valley of the Kings. His tomb was superior in quality to the fragments known from other royal burials.

3.2.5 Tutankhamun Funerary Mask

Gold funerary mask of King Tutankhamun (Figure 3.4d), can be seen Thebes, Egypt, 14th century BCE in the Egyptian headrest; in the form of the god Shu with two crouching lions, from the tomb of Tutankhamun, c. 1340 BCE; in the collection of the Egyptian Museum, Cairo.

Gilding the Skin: Gold, the 'Flesh' of the Gods From ancient times, gold was considered the color of the life-giving sun-god and of gods in general. As Egyptian funerary belief entailed the deceased's transition into the divine sphere, mummies could be equipped with gold colored (for kings, solid gold) face masks and coffins. The gilding of the skin itself, mainly on the face, chest, arms, and legs, or even a complete gilding of the body with very thin gold leaf, became common only in the Roman period; its earlier antecedents are unclear (Dunand and Lichtenberg 1998). For Antinoopolis, gilding was reportedly done with thin plates around 4 cm long, the longest one up to 10 cm. The most spectacular mummy of this type is housed in the Museum of Dunkirk (Zimmer 1993; Calament 2005). Zimmer lists twenty-eight complete mummies with skin gilding, twenty-five heads, and several parts of bodies (Aubert and Cortopassi 1998). In addition to the distinctive feature of the extended arm position, gilding can thus be a dating criterion for Roman period mummies, used, for instance, to date three unwrapped children's mummies in the British Museum (Dawson and Gray 1968; Zimmer 1993). There are similarly ornamented mummies of children in Frankfurt (Schultz *et al.*, 1995). Besides, or as an alternative to, gilding the skin, mummies could be equipped with gold leaf in the form of small objects placed into the mouth or directly on the skin, in order to protect and to revive vital parts of the body, such as the eyes, lips, tongue, breasts, navel, genitals, and the nails of fingers and toes (Andrews 1994).

3.3 Need for Mummification

Preserving the dead body was a widespread practice in Egyptian civilization. Not only human beings but along with them pets were preserved to give them company even after death. In Indian Mythology also a description is given that with Dharmraj Yudhisthira eldest of the Pandwas, a dog was also following till his death and he was suddenly stopped by Yamraj (the God of death) that the pet is not permitted to enter in heaven. This practice followed some 3000 or more years ago and has given various vital clues as to how the people were looking at that time and when they died what was their health status. A large numbers of studies give us the features of social life prevailing at that time.

Egyptians loved life and believed in immortality. They thought that the human body was needed even after death. Thus preserving the body in same form was the basic idea. They believed that the body housed, one soul or spirit. If the body is destroyed, the spirit could be lost. The tomb preparation was crucial ritual in Egyptian society. This process began long before a person's death, and involved the storage of articles that will be required afterlife, such as furniture, clothing, food and valuable ornaments *etc.* Mummification was practiced around 2600BC and only Pharaohs were initially entitled to the process. In 2000BC commons were also granted access to the after world. A seminal anthropological theory developed through companions between the mortuary rituals of various traditional cultures. Death was conceived as being outside of nature than part of it (Day 2016). Mummification developed to preserve the body.

3.4 Mechanism of Mummification

The question of whether mummies from Roman Egypt attest a decline or a late heyday of mummification techniques can be answered firmly in favor of the latter. The different but generally high standards of embalming, using traditional as well as innovative Egyptian craftsmanship including excerebration, evisceration, and large quantities of embalming resins and linen (Figure 3.1), were dependent on two factors: cost and local practices (Counts 1996). The use of extensive linen wrappings, mummy decoration (masks, shrouds, portraits, *etc.*), amulets, jewelry, and other burial goods, and the gilding of the skin, often correspond to high-quality mummification. Mummification was a 70 day process. Based on social class three different methods were practiced. However, the extremely high expense of a first-class burial in Roman Egypt was beyond almost everyone's reach (Sarkar 2021).

Elite Class

- ☆ The brain was drawn out through the nostrils using a crooked piece of iron.
- ☆ Contents of belly were taken out and the interior was cleaned using palm-wine.
- ☆ The belly was filled with pure myrrh, cassia and other spice and then sewn together again.
- ☆ The corpse was covered in Natron (NaCl) for 70 days
- ☆ The body was washed and rolled up in fine linen and covered with gum. And finally returned to family members.

Middle Class

☆ The syringes filled with oil from cedar tree were injected into the abdomen, dissolving the bowels and interior organs.

☆ The body was covered with natron treatment for 70 days, after which the cedar oil was cleaned out and the body was left as skin and bones.

☆ And finally the body in unwrapped form was returned to family members.

Poor Class

☆ The body was cleansed with oil enema.

☆ The body was covered with natron treatment for 70 days, and finally returned to the family members.

3.5 Mummy: Recent Researches

Knowledge accumulated by Invasive techniques has helped to analyze the mummies by radiological techniques in South Korea. CT imaging was done by Kim *et al.* (2016) and preservation of internal organs was studied. Autopsy of a tuberculoid Chachapoya-Inca mummy was done in Canada (Nelson *et al.*, 2016). The diagnosis of tuberculosis in human mummy from Pre Columbian times in the Andes continues to be the object of debate. It was found in late Pre Columbian times. Nelson *et al.* (2016) found TB in a 40 years old lady of Chachapoya Inca mummy (CMA 218), on chest, neck and pelvis showed calcified nods.

A few months after the discovery of x-ray in 1895, the Physicist Walter Koening conducted the first radiographic investigation of an ancient child mummy of unknown archaeological provenience from Senckenberg Museum Frankfurt (Inv. No. As 18). The mummy was reexamined recently. Radiocarbon analysis yielded an age between 378 and 235cal BC- the beginning of Ptolemaic period in Egypt. Dental status revealed a child aged between 4 and 5 years. Remains of desiccated brain are preserved in skull base and within the cervical spine. Intestines are located inside the chest and abdomen cavity. The liver seems to be enlarged. Analysis of thorax revealed a ***pectus excavatum*** deformity. Both femora show a longitudinal cleft in the ventral part of the diaphysis. Harris lines are indicators for metabolic stress during growth are visible on long bones of the lower limb. The recent CT scanning revealed comprehensive anthropological results about the first-ever X-rayed human mummy (Zesch 2016).

Preliminary results of Canopic jar project were revealed by Bouwman (1916). All Canopic jars were studied with x-ray, CT scan and histological and molecular studied were performed. The chemicals used in embalming were studied. The metagenomic High Throughput Sequencing data has shown a wealth of bacterial data. Some evidence of dietary or ritual plant remains was also present in DNA.

3.6 Mummy: Paleopathology

Human remains from tomb K93.12 in the ancient Egypt were analyzed (Losch *et al.*, 2016), which is situated opposite to modern city of Luxor in Upper Egypt on the western bank of river Nile. In Total 79 individuals could be partly reconstructed and investigated in detail. The male predominance was observed. Percentage of young children (>6years) and adults in age of (20-40 years. Dental diseases, degenerative diseases and a possible case of ankylosing spondylitis. Additionally, 13 mummies of an intrusive waste pit could be attributed to three different groups belonging to earlier time period.

Atherosclerosis is thought to be a disease related to modern lifestyles. Over one third of 300 mummies examined had definite or probable atherosclerosis. It was common in all of the preindustrial populations studied, including pre-agriculture hunter gatherers and was seen in Bronze age in Europe, and dating back to 5000 years ago (Thompson *et al.*, 2016).

Researchers during a routine DNA test on a male Egyptian have made an astonishing discovery after finding a 23cm iron orthopedic screw inside his knee. The pin is held in place by organic resin, similar to modern bone cement. Not only were the researchers astonished that the pin is ancient, but the highly advanced design had the visiting surgeons in awe. "The pin is made with some of the same designs we use today to get good stabilization of the bone," said Dr. Richard Jackson, an orthopedic surgeon from Brigham Young University. "I have to give the ancients a lot of credit for what they have done," added Dr. Wilfred Griggs, who led the team of scientists conducting DNA research on the mummy when they made this incredible find (Paschall 2015). Pathogenic *Mycobacterium tuberculosis* complex organisms (MTBC) primarily cause pulmonary tuberculosis (PTB); however, MTBC are also capable of causing disease in extra pulmonary (EP) organs, which pose a significant threat to human health worldwide. Extra pulmonary tuberculosis (EPTB) accounts for about 20–30 per cent of all active TB cases and affects mainly children and adults with compromised immune systems (Gopalaswamy *et al.*, 2021). Tuberculosis is a bacterial disease predominantly located in lungs, which can disseminate to other areas, including kidney and bones. Latent tuberculosis is present in one third of the world's population but less than 10 per cent of infected develop active disease. Lymph nodal involvement is common. Six out of 7 mummies investigated showed pulmonary TB (Piombino-Mascali *et al.*, 2011). Based on skeletal observations, it was concluded that during first centuries of the first millennium, the population increased and agriculture intensified, which likely caused the disease to spread.

Abnormalities in immune system leading to osteo-immunologies were found in mummies by Barkhordarian and Chiappelli (2011). The temporomandibular joint (TMJ) articulates the mandible with the maxilla. Temporomandibular joint disorder (TMD) are dysfunctions of this joint. Auriculotemporal (AT) nerve, a branch of mandibular nerve plays a critical role in TMD (Demerjian 2011).It was further observed that AT nerve can lead to significant neurologic and neuro-muscular disorders, including Cervical Dystonia (Torticolli), Blepharospam, Strabismus, gait or balance disorders and Parkinson's disease.

During the Roman period, the Fayum became the granary of Egypt and Rome. The lake continued to shrink during the Roman Period (35 BC- AD 385) and declined from -7 meters in the 2nd century AD to -17 or lower in the 3rd century BC (Davoli 2012). The whole canal system continued to be used, the agricultural productivity especially of wheat and barley was further optimized. Settlements formed with periods of high population density. The most reasonable guess for the overall population density in the Roman Period is about 200–300 people per km squared (Scheidel 2001). For ancient Fayum's populations, we hypothesize that reclamation of lands for cultivation and the constant contact between malaria's vectors and human beings might have determined an increased risk of acquiring malaria. Furthermore, dense crowding of people may have been a significant factor also for the manifestation and spread of human tuberculosis and the risk of contracting concomitant infections might have been, therefore, augmented. High frequencies of malaria and tuberculosis co-infections (4/16) (25 per cent) were identified in pre-adolescents (2 cases) and young adults (2 cases) which appeared to be, as it occurs in modern populations, the most affected age-classes (Lalremruata *et al.*, 2013).

3.7 Problems of Biodeterioration

Embalming techniques, as well as funerary rituals and burial practices, underwent major changes, surveys include (Dunand and Lichtenberg 1998; Taylor 1996). It is well known that the methods of mummification developed from experimental stages to various types, which are characteristic of their respective eras. There has also been common consensus that, after a peak during the 21st dynasty, the techniques gradually declined throughout the late period. This was mainly owing to increasing production and negligent workmanship. Especially for the Graeco-Roman era, when the emphasis seemed to be placed on the decoration and equipment of a mummy rather than on the body itself, embalmers and their workshops became discredited in the Egyptologist's eyes (Smith and Dawson 1924; Bataille 1952; Gray and Slow 1968).

Deterioration in mummies is caused by several factors, including environmental conditions, physical damage, or damage caused by previous inept conservation attempts (David 1986, David and David 1995; David 2001). Conditions, which existed prior to excavation particularly flooding of the burial site, can cause deterioration. However, poor display or storage conditions in the museum environment are often responsible. High relative humidity, inadequate air movement, and darkness provide an almost ideal environment for the cultivation of fungal spores (Hino *et al.*, 1982), microbiological deterioration, and insect attack.

This physical type of damage can happen at the archaeological site or in the museum, and in the past was often the result of inadequate or careless handling by early archaeologists or investigators. In addition to human mishandling, mummies are sometimes subjected to rodent or insect attack (Curry 1979). It is necessary to store the mummies in a suitable environment, ideally with a relative humidity, of 40-55 per cent and a constant temperature of 10-15° C. And, if the damage cannot be eradicated thorough environmental control alone, it is necessary to apply other methods to arrest the damage.

3.7.1 Biodeterioration of Mummy in Vadodara

The mummy of a boy is placed in Egypto-Babylonian Gallery of the Baroda museum and Picture Gallery, in a glass chamber of the size of 214 cm × 91.5 cm × 81.5 cm (Figures 3.7a, b). It was purchased by Maharaja Sayajirao Gaekwad III from New York in 1895 for 175 dollars. It is kept a wooden dome of height 116 cm at the top of the glass chamber. The size of the mummy kept in the museum is 154 cm × 35 cm × 17 cm (Figure 3.7a). It has a coffin box made up of wood obtained from Sycamore tree (*Ficus sycomorus*), belonging to the family Moraceae. This mummy belongs to late new empire of the Alexandrian period. At that time, Ptollemy II and his wife, his full sister were the co-rulers. Although oldest known mummies date back to 2500 BC, the period of this mummy has been ascertained to be 230 BC by an Egyptian mummy expert, Mr. Nasry Iskander.

In order to solve the curiosity of the visitors, the linen bandage was partly removed from the feet of the mummy by the museum curator fifty years ago. The fingers and toes of legs are thus exposed to air. An unusual white growth was observed in 1998 on the first toe of the right leg. After six months, an almost similar but smaller spot was noticed on the third finger (Figure 7a). Now the deterioration has started in this toe also. To confirm the involvement of any fungus, the isolations were made from these spots in 1999. The fungal growth was completely removed with the help of 90 per cent ethanol after collecting the samples. Similar spots reappeared after one month. The same fungus was re-isolated on 5 May 2001. On the basis of morphological characters and color of the colony, the fungus has been identified as *Emericella nivea* Wiley and Simmons. Agharkar Research Institute, Pune, confirmed the identity of the fungus. Colonies of *E. nivea* on Czapek-Dox's agar were white becoming creamy with age (Figure 3.7d). Colonies were slow-growing on PDA slants. Conidiophores measure 600 µm to 1000 µm long, heads with white globose vesicle 20 µm to 30 µm in dia (Figure 3.7e). Phialides in two series, primary sterigmata 5–8 µm × 2.5–3 µm, secondary sterigmata 5– 7 µm × 2–2.5 µm. Conidia colourless 2.2–3.3 µm in size, globose, smooth. Old culture tubes containing PDA slants showed the presence of perfect state. Rounded black cleistothecia were seen in culture. These were 350– 400 µm in dia (Figures 3.4c). Asci were evanescent, ascospores smooth-walled with a pulley, 4.4–5.5 µm in dia (Figure 3.7e). Presence of *E. nivea* in indoor air and on the mummy kept at the Baroda Museum is new to the

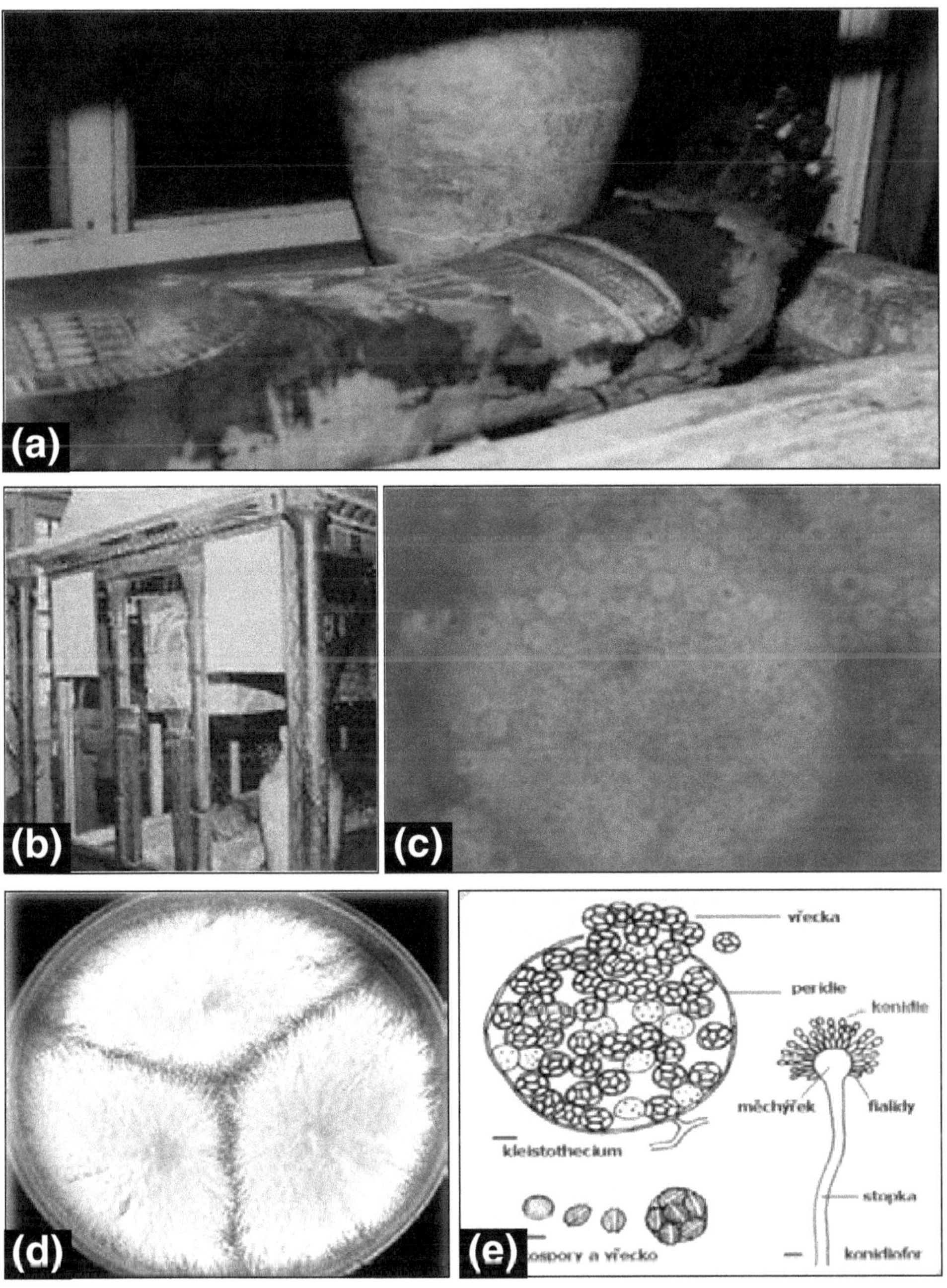

Figure 3.7a–e

(a) Picture of Egyptian mummy placed in Baroda museum. Note the white spots in fingers of right leg; (b) Wooden show case housing mummy placed in Egyptobabylonian gallery; (c) Microphotograph of fungus *Emericella.* Note the thick walled Hulle cells of Cliestotheium; (d) Culture of *Emericella* in Petridish; (e) The picture showing Cliestotheium and conidia of asexual stage of fungus.

Science. Earlier the anamorph of this fungus, *Aspergillus niveus* Blochwitz was isolated from air in northern Italy at 1700 m (Thom and Raper 1945). Studies have shown the presence of another species of *Emericella, E. nidulans* (Eldam) Vuilll., in the indoor air of different galleries of the Baroda museum.

Fungi are major contributors to the deterioration of cultural property in tropical countries like India (Sarbhoy 1994). During this study members belonging to Zygomycotina group were also recorded from the mummy chamber and Egypto-Babylonian Gallery (Arya *et al.,* 2001). Since the coffin box was infested with insects, the mummy was transferred to a glass plate. Lopez-Martinez *et al.* (2007) isolated 469 fungal colonies from 12 mummies deteriorated with fungi. Biological attack was observed in cartonnage in Egypt. Among the isolated fungi, *Penicillium, Cladosporium* and *Aspergillus* were identified (Mohmoud *et al.,* 2021).

3.7.2 Biodeterioration of Mummy in state Museum, Lucknow

According to Fatmi and Zarrin (2014) state museum of Lucknow purchased mummy of a 13 year old girl in 1952 from a UK national J.J.E. Porter. Examination depicts that it is the finest type of mummy belonging to some royal family of dynasty 22- 25. A painted wooden cover (Sarcophagi) having decoration on it, covers the mummy. Since the mummy has been purchased by museum it was displayed in a traditional air tight showcase (Figure 3.8). And from time to time remedial action has been taken by museum conservationist. To maintain humidity in the showcase (2.4 X 1.5 m area) 15 kg of Silica gel is spread on the floor of the showcase. To avoid any interaction of chemical with mummy Silica gel is placed at a distance. Mummy is raised through an

Figure 3.8: Mummy Displayed in State Museum at Lucknow.

Aluminum stand. Scientist from National Research laboratory for Conservation of cultural property (NRLC), Lucknow, when visited the mummy. On their recommendation the mummy has been shifted to a new air tight showcase. However, the environmental factors like temperature, light, pollution are gradually affecting the mummy. It is showing signs of unwrapping. Further consultation from a mummy expert from Egypt or England is required to take to immediate steps for restoration and conservation of this unique heritage object. Apart from the mummy there is painted and gilded wooden coffin (165X41cm) without any mummy belonging to Sema-periest Patosiris hailing probably from province Akhmim. It was purchased in 1952 from Spinkson, U.K. London originally coming from Marquis of Duffering and Ava collection.

3.7.3 Biodeterioration of Mummy in Kolkata and Jaipur

The Figure 3.9a and c shows the mummy place in national Museum at Kolkata. The head of mummy in Kolkata museum is exposed and damage can be seen in the mummy. Better preservation is needed for this mummy.

The mummy was first brought to India in 1883 for an exhibition in Jaipur as a gift by Brughsch Bey, then curator at the Museum of Cairo, to Sawai Ishwar Singh, the ruler of Jaipur at the time. The Egyptian mummy named Tutu was from a family of priests who were devotees of Kehm in Egypt's Penopolis. Tutu was a teenaged girl belonging to the Ptolemaic dynasty, the last dynasty of ancient Egypt that lasted from early 300 BC to 30 BC. The mummy was kept in the museum's basement when heavy rains lashed Jaipur. Rakesh Chholak, superintendent at the Albert Hall Museum, told the media, "As soon as we were informed that the basement was getting waterlogged, in 2020, we rescued the mummy after breaking the glass-box. We brought it upstairs." Diya Singh, a BJP MP from Rajsamand and member of Jaipur's royal family raised her concern about safety and conservation of the mummy and other artefacts in the Albert Hall Museum of Jaipur (http://timesofindia.indiatimes.com/articleshow/78265402) (Figures 3.9b,d).

3.7.4 Problems of Biodeterioration in Certain Other Mummies

A) Mummies in London, Liverpool, Turin, and Klagenfurt

To date, only one group of very well-mummified corpses has been considered as being of exceptional quality for Roman times. These mummies–males, females, and one child–were bandaged with the limbs extended and wrapped separately before the outer bandaging took place. Any outer wrappings are now lost, and the original methods of burial remain unknown, since the mummies were purchased around the 1840s from the Egyptian antiquities trade. The corpses received careful treatment, including excerebration and evisceration; occasionally some visceral packages were returned into the body cavities. This group, of a very similar, integral type, is represented by eight anonymous examples, some of them with a painted face-wrapping: one in the

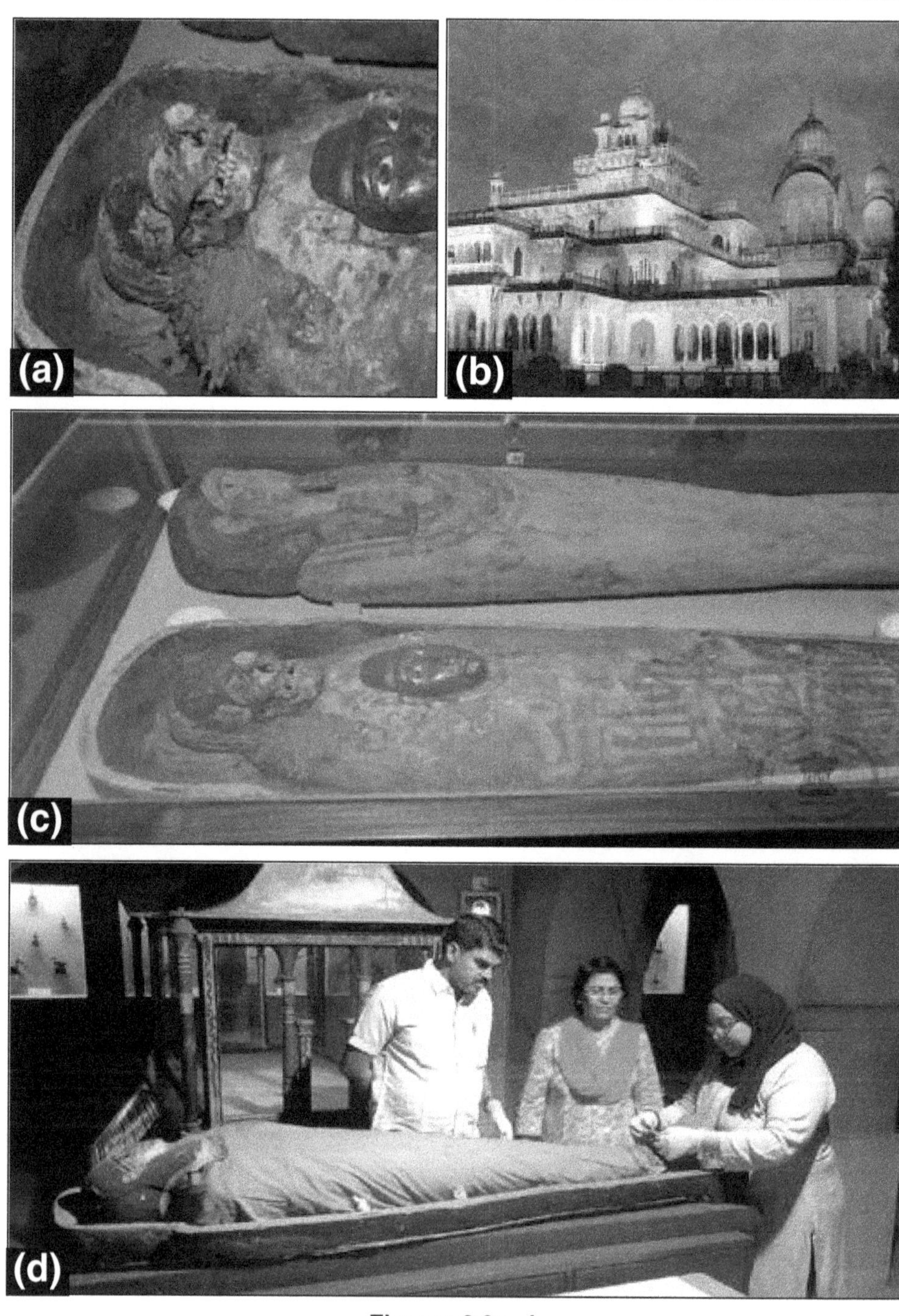

Figures 3.9a–d

(a and c) The mummy placed in National Museum at Kolkata. Exposed mouth and teeth can be seen in (a and b) Prince Albert museum of Jaipur; (d) The removal of upper half of coffin box of mummy placed in Jaipur museum.

British Museum (Dawson and Gray 1968); two in Liverpool (Gray and Slow 1968).

B) A New Method of Sampling in Brazil

dosReis and deSouza (2016) working in National Museum Rio de Janeiro, Brazil developed a new method of microbes collection using a tube and vacuum cleaner. They collected 12 fungi in 18 samples. The method was useful to trap fungi trapped in cloths and holes. The method can also be used in other sampling of pulverized material from archaeological contexts.

C) Studies on "Three Sisters" Group of Mummies Displayed in the Museo Egizio

Oliva (2016) conserved the "Three Sisters" group of mummies, dated to the Late Period (22nd -25th Dynasty) and displayed in the Museo Egizio Turin. The mummies of Tamit, Tapeni and Renpetnefret were acquired with the Drovetti collection in 1824. Their provenance is not recorded but according to textual and stylistic criteria they were probably discovered in the Theban necropolis. The mummies are preserved inside their coffins which, based on iconographic studies date from the 25th dynasty. They are called the 'Three Sisters' because, as is stated in the inscription, all three are the daughters of the Amun priest Ankhkhonsu and of the lady Neskhonsu. Conservation followed the radiological and chemical analyses for the identification of fibers and dyestuffs. The skeletons were in a good condition, showing similar embalming techniques, even if each mummies presented differences in their external construction (two of them had an outer shrouds dyed red, one had traces of a bead-net). All three mummies were in a state of advanced degradation, caused in part from natural decay of the organic materials, and in part due to being on display for a long time, but the most significant damage was due to ancient looting, when Tamit had been partly destroyed by looters in their search for amulets and jewelry. Tamit is displayed in a wide showcase with the other two sisters at the first floor of the renewed Museo Egizio in Turin.

The mummy of Tamit was moved onto a transparent board of Plexiglass in order to have full access to the back of the body and to proceed with the usual graphic and digital documentation. Textiles and shrouds identified on the mummy were studied, and for each of them a record sheet was created, containing the following information: fiber, torsion, weaving reduction, selvedge, starting and finishing borders, fringe. Tamit had been wrapped in one external unbleached shroud, folded back over the head and feet, leaving the center of the back uncovered, and knotted, with the final part of the straps under the neck. It seemed as if she had been wrapped in at least seven different textiles, of which two probably belong to wider shrouds. They showed decorative bands, one alternating with blue and the other with red wefts, with long and twisted fringes.

Dust deposits from the exposed sections of wrappings and human remains, which were heavily darkened and dehydrated was removed. Vacuum cleaning of the entire surface, with the help of soft brush, and the gentle action of vulcanized sponges was performed to remove the dirt from the surface. The major challenge was to re-construct the original shape of Tamit's head wrappings to what it had been prior to looting, as that act had resulted in the cutting off and removal of the bandages in the head area and in a visible loss of mechanical tension in the textiles. The bandages, still *in situ*, were first re-shaped with cold steam and then breaks and cuts in them were stitched locally on nylon net. The cushion was made from polyester wadding covered with dyed linen stitched onto itself, so as to avoid the use of any glues that could contaminate the original fabrics.

Figure 3.10: Egyptian Mummies kept in Turin Museum, Italy.

D) CT Scanning of Amenhotep I Mummy in Cairo

The perfectly wrapped mummified body of the Egyptian pharaoh Amenhotep I was digitally "unwrapped" with high-tech scanners (Saleem 2021).With the use of three-dimensional computer topography (CT) scanning technology, hitherto unknown details about his appearance and the lavish, unique jewelry have been observed. Egyptologists knew from decoded hieroglyphics that the mummy had been unwrapped once in the 11th century BCE – more than four centuries after his original mummification and burial. She said: "This fact that Amenhotep I's mummy had never been unwrapped

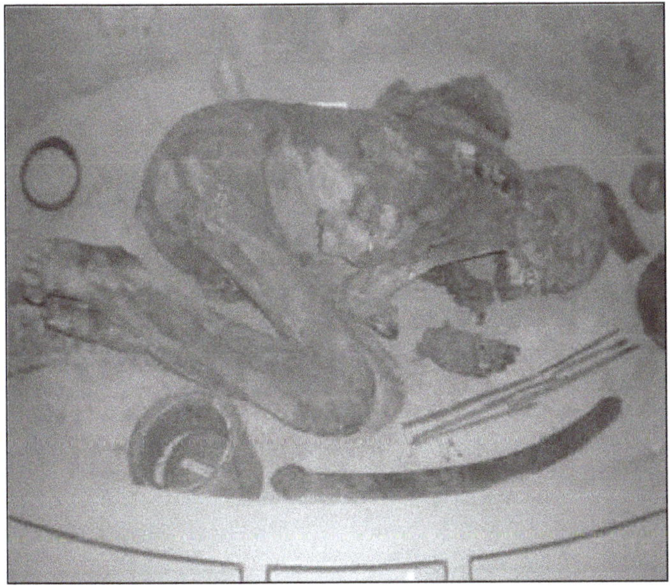

Figure 3.11: A Deteriorated Mummy kept without any Linen Bandage, and with Associated Articles.

in modern times gave us a unique opportunity: not just to study how he had originally been mummified and buried, but also how he had been treated and reburied twice, centuries after his death, by High Priests of Amun."By digitally unwrapping of the mummy and 'peeling off' its virtual layers – the facemask, the bandages, and the mummy itself – we could study this well-preserved pharaoh in unprecedented detail. "We show that Amenhotep I was approximately 35 years old when he died. He was approximately 169cm tall, circumcised, and had good teeth. Within his wrappings, he wore 30 amulets and a unique golden girdle with gold beads.

"Amenhotep I seems to have physically resembled his father: he had a narrow chin, a small narrow nose, curly hair, and mildly protruding upper teeth." He was the second pharaoh of Egypt's 18th dynasty after his father Ahmose I, who had expelled the invading Hyksos and reunited Egypt (Figures 3.12a,b,c). After his death, he and his mother Ahmose-Nefertari were worshipped as gods. His entrails had been removed by the first mummifier, but not his brain or heart. "We show that at least for Amenhotep I, the priests of the 21st dynasty lovingly repaired the injuries inflicted by the tomb robbers, restored his mummy to its former glory, and preserved the magnificent jewelry and amulets in place."

3.8 Preventing Deterioration

Ideally before any mummy is considered for exhibition, a thorough exam, including CT scans, needs to be conducted to determine the stability

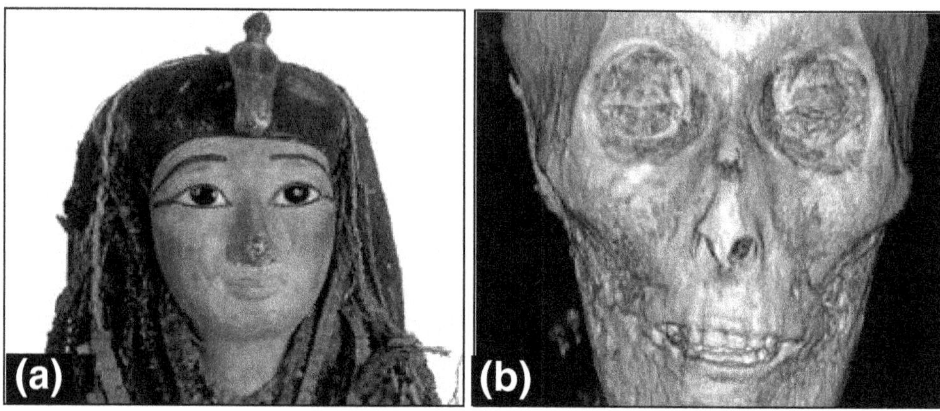

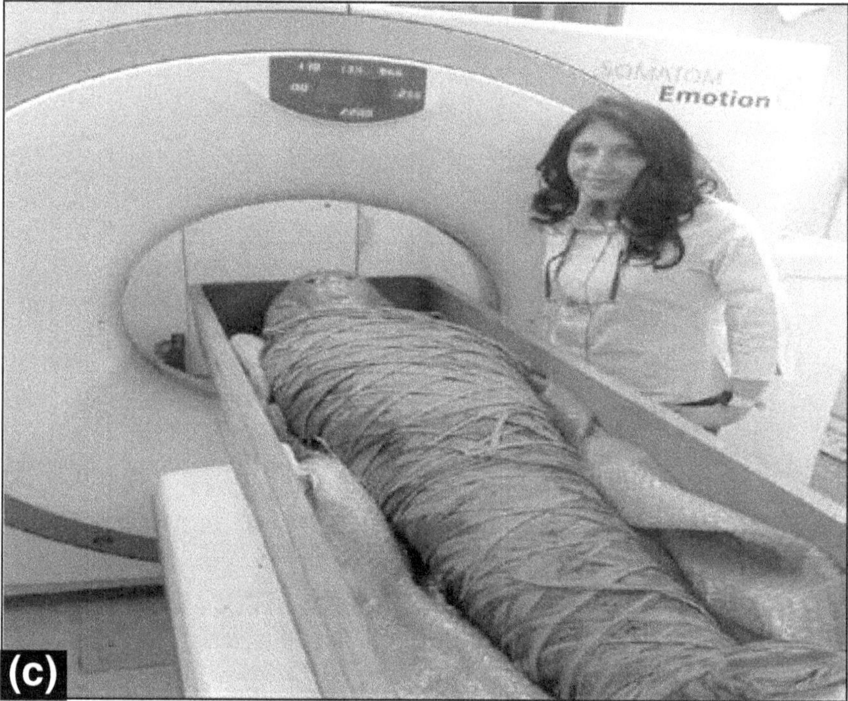

Figure 3.12a–c

(a) Mummy of Amenhotep was discovered in Southern Egypt in 1881; (b) Picture showing scanned image of mummy; (c) Use of advanced X-ray technology and computerized tomography scanning to catch a glimpse of King Amenhotep I's mummified body. (Photo issued by Sahar Saleem of the Amenhotep sarcophagus. Photograph: Dr Sahar Saleem/University of Cairo).

and conservation needs of the mummy. If insect damage or other kinds of decomposition are found, a treatment needs to be started right away to stop any further degradation. Pesticides can be used, but aren't highly recommended due to our inadequate understanding about interaction of chemicals with a mummy (Williams and Cato 1995; Agnew and Levin 1996).

Freezing is an option for pest control with mummified material. Another successful means of treating specimen decomposition in collections, that is relatively safe for both organic and inorganic materials with no side effects, is the anoxic (no oxygen) environment. Nitrogen, a relatively inexpensive gas, is used to replace the oxygen in the specimen's enclosure. The Getty Conservation Institute's case design is being used in Cairo to stop the oxidation and microbial deterioration of the royal mummies that are on display. An oxygen scavenger is used in conjunction to the nitrogen to ensure an oxygen-free environment. Provide a micro environment that discourages the perpetrators of deterioration from reaching your mummy, would be recommended. A microenvironment case meets these criteria: 1.) it is sealed as much as possible from the exterior environment, 2.) it is constructed on the interior from inert, stable materials, and 3.) it is fitted with a material to buffer rapid changes in relative humidity. A common buffering material is silica gel.

Inert Gas Control method was developed for displaying the Egyptian mummies in the Turin collection, this method is only effective if the mummy shows no existing signs of deterioration (Raphel 1994). It is also expensive to install and requires continuous monitoring. (i) The mummy is placed inside the fumigation chamber. Air is then evacuated from the chamber by vacuum pressure and chemical sterilization is done with carbon dioxide, argon or nitrogen (Valentín 1994). (ii) The mummy is placed inside a tightly sealed polythene bag into which the fumigant is introduced. However, this method is less effective than earlier one. Radiation can bring about physical and chemical changes in the cells of microorganisms (Van der Molen *et al.*, 1980), and the use of x-rays, gamma rays and electron beam radiation has been employed, the most famous instance was the treatment of the mummy of Ramesses II in Grenoble, France (Belyakova 1960). Florian (1986) suggested use of freezing process. We need to strive to have a mummy be as valuable the day it comes off exhibit as the day it went on.

3.9 Mummy: Restoration

Gonzalez *et al.* (2016) found parchment and vellum more suitable for restoration of Andean mummy in Spain than Japanese paper for missing parts. Oliva *et al.* (2016) helped in conservation of a 20-30 years old lady mummy from Deir-el-Medina, where skull and thoracic bones were reconstructed and mummy was restored. In the conservation of a mummy belonging to the Museo Egizio in Turin (Inv. Suppl. 5227 XXV Dynasty), the outer shroud that was dyed red was covered with thick dust deposits that had altered its original color to a grayish brown and had weakened the structure, already damaged by historical looting. The shroud had already been removed from its original position on the body, as was ascertained from previous documentation, and showed several gaps and cuts on its surface (Oliva *et al.*, 2006).Following the cleaning of a mummy, the subsequent challenge posed to the conservator is providing fragile and degraded wrappings and human remains with adequate support. The choice of methods are determined not only by the weakness of the artefact, but also by its three-dimensional shape and future "museum life". In mummy conservation, the main problem to solve is surface consolidation, keeping the fragments in place without producing any chemical and visual alteration in the fibers. Most of the time, there is no access from the back of the textiles, so it has to be consolidated from the front or displayed side. Thus, any new material that is used to support the old has to be as transparent as possible, easy to manipulate and to dye (in order to match the original color of the wrappings), with possibly non fraying-edges and enough elasticity to follow the mummy's shape (Wills *et al.*, 2014).

3.10 Conclusions

Mummies and other preserved bodies are a unique and irreplaceable resource for the study of ancient cultural practices such as methods of mummification, for the identification of the occurrence and patterns of disease, and for genetic and population studies. It is therefore essential that conservation treatments be developed, which preserve the bodies in an effective way, while not destroying their value as a source of biological and biomedical evidence. The methods of embalming, or treating the dead body, that the ancient Egyptians used is called mummification. Using special processes, the Egyptians removed all moisture from the body, leaving only a dried form that would not easily decay. It was important in their religion to preserve the dead body in as life-like a manner as possible. So successful were they that today we can view the mummified body of an Egyptian and have a good idea of what he or she looked like in life, 3,000 years ago. Mummification was practiced throughout most of early Egyptian history. The earliest mummies from prehistoric times probably were accidental or by chance, during 2600 BCE, the 4th and 5th Dynasties, Egyptians probably began to mummify the dead intentionally. The recent researches in the field and deterioration of mummies in Indian museums and their conservation aspects are discussed.

REFERENCES

Andrews C (1994) Amulets of Ancient Egypt. London: British Museum Press; Austin: University of Texas Press.

Anonymous (1990) India: Prehistoric and Protohistoric periods. Pub. Div. New Delhi 52pp.

Arya A, Shah AR, Sadasivan S (2001) Indoor aeromycoflora of Baroda museum and deterioration of Egyptian mummy. Current Science, 8(7): 793-700

Aubert M-F, Cortopassi R (1998) Portraits de l'Égypte romaine. Paris: Réunion des Musées Nationaux

Aufderheide A C (2005) The Scientific Study of Mummies. Cambridge: Cambridge Uni. Press.

Barkhordarian A, Chiappelli F (2011) Fundamental osteoimmunology: from stem cells to Bone-immune metabolism. 7th World Congress on Mummy studies. California 12-16 June. 2011

Bataille A (1952) Les Memnonia. Recherches de papyrologie et d'epigraphie grecques sur la necropole de la Thebes d'Egypte aux epoques hellenistique et romaine. (RAPH 23) Cairo.

Belyakova L A (1960) Gamma irradiation as a means of disinfection of books against spores of Mould Fungi. Microbiology (Moscow) 29: 762765.

Borg B E (1998) Der zierlichste Anblick der Welt Mainz: von Zabern

Boës É, Georges P, Alix G (2002) 'Des momies dans les ossuaires de la Nécropolis d'Alexandrie: Quand l'éternité a une fin', in A. Charron (ed.), La mort n'est pas une fin. Arles: Musée de l'Arles Antique, 68

Bouwman A (2016) Preliminary results from the Canopic Jar project. 9th World congress of mummy studies, Lima Peru. Symposium Autopsy and Metagenomics 135pp.

Budischovsky M C (2004) Temoignages de devotion isiaque et traces culturelles le long du limes danubiens. In : L. Bricault (ed.) Isis en Occident Leiden: Brill, 171-191.

Calament F (2005) La révélation d'Antinoé par Albert Gayet: Histoire, archéologie, muséographie, 2 vols. Cairo: Institut Français d'Archéologie Orientale

Christensen O E (1969) 'Un examen radiologique des momies égyptiennes des muséesdanois', La Semaine des Hôpiteaux de Paris 45/28 (14 juin), 1990–8.

Cockburn A, Cockburn E (1980) Mummies, Disease and Ancient Cultures. Cambridge University Press, Cambridge.

Counts D B (1996) Regum Externorum consuetudine: The nature and foundation of embalming in Rome. Classical Antiquity 15: 189-202

Curry A (1979) The Insects Associated with the Manchester Mummies. In The Manchester Mummy Project, edited by A. R. David, pp. 113118.: Manchester Museum, Manchester.

Day J (2016) His flesh is Slain: fear of decay as the origin of Egyptian mummification. 9th World congress of mummy studies, Lima Peru. Symposium Free Topics 113pp.

David A E (1986) Conservation of mummified egyptian remains. In Science in Egyptology: proceedings of the Science in Egyptology Symposia, edited by A. R. David, pp. 8790. Manchester University Press, Manchester.

David A R, David A E (1995) Preservation of human mummified specimens. In: The Care and Conservation of Palaeontological Material, edited by C. Collins, Butterworth Heinemann, Oxford pp. 7388.

David A R (2001) Benefits and disadvantages of some conservation treatments for egyptian mummies. Chungará (Arica) v.33 n.1 Arica ene. 2001. http: // dx.doi.org/10.4067/S0717-73562001000100020

Davoli P (2012) The Archaeology of Fayum, in: C Rigg (ed), The Oxford Handbook of Roman Egypt. Oxford Hanbooks Online, 153-169. https: // doi.org/10.1093/oxfordhb/9780199571451.001.0001

Dawson W R, Gray P H K (1968) Catalogue of Egyptian Antqiuities in the British Museum: Mummies and Human Remains BMP, London, 1968

De Lorenzi E, Grilletto R (1989) *Le mummie del Museo Egizio di Torino, N.13001-13026: Indagine antroporadiologica,* La Goliardica, Milano.

Demerjian G (2011) Systemic correlates of temporomandibular joint disorders. 7th World Congress on Mummy studies. California 12-16 June. 2011

Dunand F, Lichtenberg R (1998) Les momies et la mort en Égypte. Paris: Errance

Dunand F, Lichtenberg R (2008) 'Dix ans d'exploration des nécropoles d'el-Deir (Oasis de Kharga): Unpremier bilan', Chronique d'Égypte 83: 258–76

Fatmi Al S, Zarrin A (2014) Egyptian mummies in Indian museums, their biodeteriorating factors, conservation measures; special reference to state museum Lucknow. In: Biodeterioration of Cultural Property-7 Dhawan S. abduraheem K. and Nath V (eds). Pub. By Sukriti Nikunj, Lucknow.95-104

Gonzalez M, Begerock A M, Valls A, Castillo C, Prieto P (2016) An old material for an innovative intervention: the restoration of the Andean mummy of Museu Dader (Banyoles, Spain). 9th World Congress on mummy studies, 10-13 Aug. Lima, Peru. 209pp.

Gopalaswamy R, Dusthackeer V N A, Kannayan S, Subbian S (2021) Extrapulmonary Tuberculosis–An Update on the Diagnosis, Treatment and Drug Resistance*J. Respir.* 2021, *1*(2), 141-164; https://doi.org/10.3390/jor1020015

Grimm G (1997) Verbrannte Pharaonen? Antike Welt 28(3): 233–234

Gray PHK, Slow D (1968) Egyptian Mummies in the City of Liverpool Museums. Liverpool: Liverpool Corporation.

Gray H K, Slow D (1968) Egyptian Mummies in the City of Liverpool Museums. Liverpool.

Hino H, Ammitzb Jll T, M Jller R, Asboe-Hansen G (1982) The Ultrastructure of Bacterial Spores in the Skin of an Egyptian Mummy. Acta Pathologica et

Microbiologica Scandinavica, Section B, Microbiology (Denmark) 90 (1): 2124.

http: //timesofindia.indiatimes.com/articleshow/78265402.cms?utm_source=contentofinterest and utm_medium=text and utm_campaign=cppst

James T G H (1979) An introduction to ancient Egypt. Pub. by British Museum London, pp. 286.

Kim M J, Yoo D S, Shin D H (2016) Invasive technique in accumulation of knowledge for a better non-invasive studies on mummies. 9th World Congress on mummy studies, 10-13 Aug. Lima, Peru. 2016

Küffer A, Frank R, Thomas B, Susanne D (2007) Unter dem Schutz der Himmelsgöttin: Ägyptische Särge, Mumienund Masken in der Schweiz. Zurich: Chronos. 1: 223. ISBN : 978-3-03-400854-9

Lalremruata A, Ball M, Bianucci, Welte B, Nerlich A G, Kun JFJ, Pusch C M (2013) Molecular Identification of Falciparum Malaria and Human Tuberculosis Co-Infections in Mummies from the Fayum Depression (Lower Egypt). Plos One. April 2, 2013, https: //doi.org/10.1371

López-Martínez R, Hernández-Hernández F, Millán-Chiu B E, Manzano-Gayosso P, Méndez-Tovar L J (2007) "Effectiveness of imazalil to control the effect of fungal deterioration on mummies at the Mexico City Museum 'El Carmen'" (article in Spanish), Revista Iberoamericana de Micología 24: 283-288

Losch S, Rummel U, Zink A (2016) Mummification remains from tomb K93.12 at Dra'Abu el-Naga, Thebes, Egypt. 9th World Congress on mummy studies, 10-13 Aug. Lima,Peru. 2016.112 pp.

Lynnerup N (2007) Mummies. Yearbook of Physical Anthropology. Pub. by Wiley InterScience. 50: 162

Lynnerup N (2016) Experiences from the Greenland mummies. 9th World Congress on mummy studies, 10-13 Aug. 2016, Lima, Peru. 82pp.

Macke A (2002) 'Les momies de la Vallée des Reines', in F. Dunand and R. Lichtenberg(eds), Momies d'Égypte et d'ailleurs. Monaco: Rocher, 78–91.

Mahmoud H M, Hassan A R A, Soliman M A AM, Roshdy N S, Ahmed MA, Fattah M A (2021) Saving the mummy's shell: An interdisciplinary approach for analysis and restoration of cartonnage mummy case from el lahun excavations, middle Egypt. Mediterranean Archaeology and Archaeometry 21(1): 185-203. DOI: 10.5281/zenodo.4394072

Margaritoff M (2022) This Centuries-Old 'Mermaid' Mummy Housed In A Japanese Temple Is Finally Being Studied By Scientists. https: //allthatsinteresting.com/asakuchi-mermaid-mummy

Meier D (2001) Mummies on display: conservation considerations. Chungará (Africa) v.33 n.1 Aricaene. 2001http://dx.doi.org/10.4067/S0717-73562001000100013

Meier D, Reinhard K, Stavis J (1998) Role of Radiography in Museum Conservation Assessment: a Case Example from Arica, Chile.

Mininberg D T (2001) 'The Museum's Mummies: An Inside View', Neurosurgery49/1: 192–9.

Nelson A, Gullien S, Conlogue C, Gonzaler R, Zadori P, Bravo A (2016)Autopsy of a tuberculoid Chachapoya-Inca mummy: implications for differential diagnosis. 9th World congress of mummy studies, Lima Peru. Symposium Radiography 121pp

Oliva C (2016) The Conservation of Egyptian mummies in Italy. La restauration des momies égyptiennes en Italie. Archives de l'humanité : les restes humains patrimonialisés. 44: 122-126. https://doi.org/10.4000/techne.1205

Oliva C, Marocchetti F E, Doneux K, Curti A, Janot F (2006) The mummies of Kha and Merit: Embalming Ritual and Restauration Work, in Rabino Massa E.(ed.), *Proceedings V World Congress on Mummy Studies, Journal of Biological Research*, n. 1, 2005, Rubettino, Catanzaro. p. 243-247

Oliva C, Cadot L, Boano R, Boria M (2016) A lady from Deirel-Medina: a case study of multidisciplinary conservation program. 9th World congress of mummy studies, Lima Peru. 94pp.

Parlasca K, Seemann H (1999) Augenblicke: Mumienporträts und ÄgyptischeGrabkunst aus Römischer Zeit. Munich: Klinkhardt and Biermann.

Paschall J (2015) Was surgery performed on Egyptians? Prosthetic pin in 3000-year-old mummy discovered. Express News Paper. 17: 46, Thu, Jul 9, 2015

Piombino-Mascali, Jankauskas R, Tamosiunas A, Valancius R, Gill-Frerking H, Spigelman M, Panzer S (2015) Evidence of probable tuberculosis in Luthianian mummies. Homo- Journal of Comparative human biology. https://doi.org/10.1016/j.jchb.2015.01.004

Piombino-Mascali D, Jankauskas R, Tamosiunas A, Valancius R, Gill- Frerking H, Spigelman M, Panzer S (2011) Evidence of probable tuberculosis in Lithuanian mummies. Homo- Journal of Comparartive Human biology. Xx: 1-12

Raphael T (1994) An Insect Pest Control Procedure: The Freezing Process. Conserv O Gram 3 (6). National Park Service, Washington D.C.

Raven M J, Taconis W K (2005) Egyptian Mummies: Radiological Atlas of the Collections in the National Museum of Antiquities in Leiden. Turnhout: Brepols.

dosReis R F A, deSouza S M (2016)Sampling of microorganisms from Egyptian mummies of the National museum of Rio de Janeiro, Brazil: a successful adaptation of the vacuum cleaner model. 9th World congress of mummy studies, Lima Peru. 177pp.

Riggs C (2005) Beautiful burial in Roman Egypt. OUP, Oxford, 2005Riggs C. (2012) The Oxford Handbook of Roman Egypt. ISBN-13: 9780199571451 Pub Oxford Handbooks Online: Nov-12 Subject: Archaeology, Archaeology of the Near East, Egyptian Archaeology, Ritual and Religion DOI: 10.1093/oxfordhb/9780199571451.001.0001

Saleem S (2021) Ancient Egyptian mummy digitally 'unwrapped' for first time. The Irish Times. Tuesday Dec. 28. 2021

Sarbhoy AK (1994) In: Advances in Mycology and Aerobiology (ed. Talde, U. K.), Today and Tomorrow's Printers and Publishers, New Delhi, 1994 pp. 161–169.

Sarkar D (2021) The Mummification Process: How Ancient Egyptians Preserved Bodies for the Afterlife. Secrets behind the ancient Egyptian mummification process, revealed. Newsletter. Mar 10, 2021.

Schultz M, Gessler-Löhr B, Kollath J (1995) 'The First Evidence of Microfilariasis in an Old Egyptian Mummy', in Actas del I Congreso Internacional de Estudios sobre Momias 1992/Proceedings of the I World Congress on Mummy Studies 1992. Santa Cruz de Tenerife: Museo Arqueologico y Etnografico de Tenerife, 317–20.

Sharma R N (1975) Society and culture in India. Pub by Rajhans Prakashan, Meerut. 210pp

Scheidel W (2001) Death on the Nile: disease and demography of Roman Egypt. Leiden, The Netherlands: Brill. 82–99.

Smith G E, Dawson W R (1924) Egyptian mummies, London, G. Allen and Unwin, Ltd. 189pp

Strang T J K, Dawson J E (1991) Controlling Museum Fungal Problems. Canadian Conservation Institute Technical Bulletin 12.

Taylor J H (1996) Unwrapping a Mummy: The Life, Death, and Embalming of Horemkenesi (Egyptian Bookshelf) – March 1, 1996. Pub by University of Texas Press; 111pp

Thom C, Raper K (1945) A manual of the Aspergilli. Baltimore : The Williams and Wilkins Company, 373pp.

Thompson R, Allam A, Lombard G (2016) Evidence of atherosclerosis in 300 human mummies from around the world. 9th World congress of mummy studies, Lima Peru. Symposium Palaeocardiology 139pp.

Valentin N (1994) Tratamientos no Tóxicos de Desinsectación con Gases Inertes. *APOYO* Newsletter 5 (2): 5-6.

Van der Molen J R *et al. (*1980) Growth Control of Algae and Cyanobacteria on Historical Monuments by a Mobile UV Unit (muvu). Studies in Conservation 25: 7177.

Williams S L, Cato P S (1995) Interaction of Research, Management, and Conservation for Serving the Long term Interests of Natural History Collections. Collection Forum 11 (1): 16-27.

Wills B, Ward C, Sáiz Gómez V (2014) "Conservation of Human remains from Archaeological Context", in Fletcher A, Antoine D, Hill J. D. (ed.) *Regarding the Dead: Human Remains in the British Museum,* The British Museum, London, p. 49-73.

Zesch S (2016) From first to latest imaging technology- Revisiting the first mummy investigated with X-rays in 1896 by using dual-energy computed tomography. Symposium 14- From autopsy to diagnostic imaging. 9th World congress of mummy studies, Lima Peru. Symposium Autopsy and Metagenomics 131pp

Zimmer T (1993) 'Momies dorées: Matériaux pour servir à l'établissement d'un corpus', Acta Antiqua Academiae Scientiarum Hungaricae 34: 1–38

Chapter 4

Conservation of Panjis and Manuscripts of Mithila Heritage, Darbhanga, Bihar

Mamta Mishra

INTACH Conservation Institute, B-42, Nirala Nagar, Lucknow – 226020, Uttar Pradesh, India
e-mail: intachicilko86@gmail.com

ABSTRACT

Panji or Panji Prabandh are extensive genealogical records maintained among the Maithil Brahmins of the Mithila region similar to the Hindu genealogy registers at Haridwar. The Panji (Chronicle) system and the Panjikars (Chroniclers) have a great influence on the social and religious life of the Maithils. To save the Mithila heritage of Panjis and Manuscripts that were in neglected and dilapidated condition and needing attention for conservation, INTACH had taken up a pilot project on conservation with the team of INTACH Conservation Institute, Lucknow in 2021-2022. A temporary laboratory was established at various sites to conserve the valuable collection belonging to individual panjikars, custodians, museums and universities at Darbhanga, Bihar. The paper reports about the challenges faced in conservation of such traditional but old Panjis and manuscripts and awareness about conservation was generated through campaigns.

Keywords: Conservation, Panjis, Manuscripts, Mithila heritage, Palm leaves, INTACH

4.1 Introduction

On the basis of survey, INTACH submitted a preliminary detailed report on the condition of manuscripts, panjis on paper and palm leaf to the authority of Mithila Sanskrit Post Graduate Studies and Research Institutes, Darbhanga and Maharajadhiraja Kameshwar Singh Sanskrit University, Darbhanga. After consistent efforts and hard work of over one year of the Convenor of INTACH Patna Chapter with the Government of Bihar, no action was taken for the approval. Finally Chairman, INTACH took an initiative to save the valuable manuscript collection as a Pilot project with free of cost as a sample of conservation and this will be further approved by the Government of Bihar. Conservation on manuscripts was started with the establishment of a temporary laboratory and inauguration on 27-11-2021 at Shri Maharajadhiraj Laxmishwar Singh Museum, Darbhanga.

4.1.1 About Panjis and Panjikars

Panji or Panji Prabandh are extensive genealogical records maintained among the Maithil Brahmins of the Mithila region similar to the Hindu genealogy registers at Haridwar.

Figure 4.1: Manuscripts Damaged by Physical and Biological Decay like Torn and Loss Edges and Fungus.

Panjikars or record keepers are the persons authorized to maintain genealogical records that are said to be more than 700 years old and are dying a natural death. The Panjis have enormous value when arranging marriages, as they ensure that incestuous relationships do not occur, delineating the last seven generations from paternal side and six generations from the maternal side of prospective bride and groom. The Maithil Brahmin delegates assembled in a conference to deliberate upon new marriage alliances duly checked with the respective panjikars at a place near *Madhubani* of India called as *Saurath*, the conference itself was called *Saurath Sabha*. Due to progressive loss of panjis, panjikars are taking up modern professions and increasing cosmopolitan behaviour, the practice of fixation of marriage by consulting Panjis is dying. There have been cases reported of sale of panjis to foreign agencies. The recent Saurath Sabha are all deserted. Increasing, people are looking forward to more modern methods of match making like internet, rather than century old palm leaves and paper. The genealogical records are under the threat of being lost forever, as much because Saurath is losing preference as wedding negotiations rendezvous. The panji system was launched in 14th century during the reign of Maharaja Harisimhadeva the last ruler of Karnat dynasty of Mithila Harsimhadev did not quite invent this system; it was already in practice in less organized ways since ancient time.

4.1.2 Types of Manuscripts at Mithila Heritage

Valuable and rare collection belong to Mithila Sanskrit Post Graduate Studies and Research Institute, Darbhanga having 12462 manuscripts in *Bengali, Devanagiri and Mithilakshari script* that contain the cultural history of Mithila. It was established in 1951 by the Govt. of Bihar to promote advanced studies and research in Sanskrit learning, to bring together modern scholars with their technique of research and investigations, to publish works of permanent value to scholars.

Another collection belong to Maharajadhiraja Kameshwar Singh Sanskrit University, consisting 5000 manuscripts including 150 palm leaf manuscripts. These collection are in *Devnagari* and *Mithilaksheri* written on poems, prose, dramas and religious topics. These manuscripts are having the cultural history of Mithila and every citizen of the locality is proud of it. It is thus believed that while some are 300-400 years old.

Individual collection of manuscripts and panjis are of 1896 period, requiring immediate action for the treatment. A special conservation laboratory was established at Maharajadhiraj Laxmishwar Singh Museum, Darbhanga, Bihar for the conservation of the collection of individual owners.

4.2 Conservation Problems in Manuscripts and Palm Leaves

The deterioration of a collection of manuscripts on paper or palm leaves was due to a combination of physical, biological and chemical factors worked together to cause damage over a period of time. Inappropriate lighting, improper temperature, relative humidity levels; excessive dust, dirt, insects, mould, mildew, fungus infestation, accidents, incorrect handling, inappropriate repair *etc.* were also some of the common problems.

A) Physical Factors

The main cause of physical damage was indirect effect to the collection by light, heat, moisture and mishandling. Light, heat and moisture cause photo chemical or oxidative change in pages, while mishandling or neglect may cause mechanical damage.

Damages generally noticed were deposition of dust and dirt; presence of stains due to bad adhesive used for repair; tears, loss of areas and broken edges in the folios; unscientific old repairs using improper material like thick paper had been used to repair the tears due to which brown stains developed which were difficult to remove. Also, old repairs converted the paper to yellow or brown colour; folios were badly stuck as a solid board due to moisture; binding has become weak with broken stitches, the cardboards used in base and end sheets has become highly acidic, brittle and weak along with the spine of manuscripts.

Figure 4.2: Discoloration of the Folios, Torn Edges and Presence of Acidity in the Manuscripts.

B) Biological Factors

Biological agents like fungi, insect cause damage to paper. Most of the collection is infected by insects like cockroaches, silver fish, book worms, rodents and termites. Damage by insects and micro-organisms has been noticed in form of holes made by book borers and presence of fungi. Fungal infection has left blackish and greenish stains on the surface of paper.

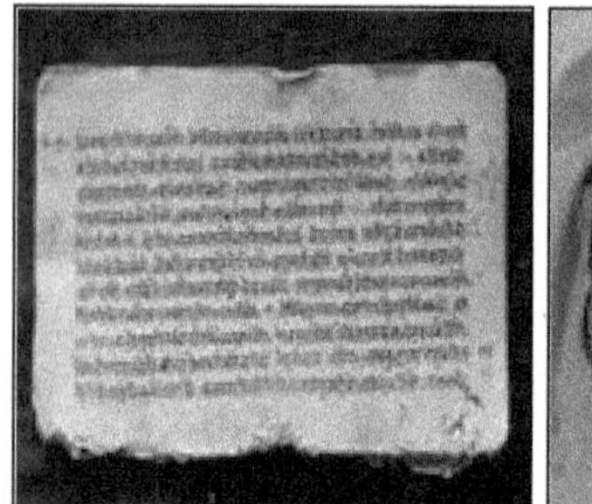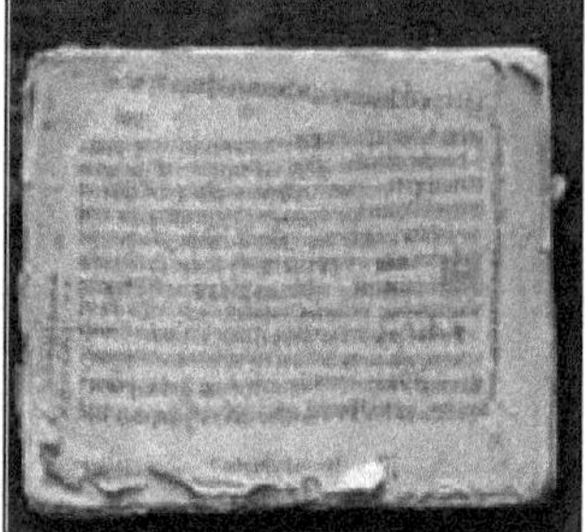

Figure 4.3: Condition of Manuscripts Showing Discoloration of the Folio, Stuck and Presence of Acidity.

C) Chemical Factors

Heat, light and moisture caused by chemical damage like photochemical change, oxidation and hydrolysis in paper due to chemicals present in paper or in atmosphere as acidity and lignin of paper at the time of manufacturing. There was discolouration of paper due to ageing and charring of paper, wherever iron gall ink is used for the writing purpose.

Figure 4.4: Condition of Palm Leaf and Paper Manuscripts in Alarming Stage of Deterioration.

4.3 Curative Conservation at the Site

A team of Conservators started the conservation work from 27.11.2021 and completed on 8.4.2022 at the temporary laboratories established at various Institutions like: -

☆ Mithila Sanskrit Post Graduate Studies and Research Institute, Darbhanga, Bihar

☆ Maharajadhiraja Kameshwar Singh Sanskrit University, Darbhanga, Bihar.

☆ Maharajadhiraj Laxmishwar Singh Museum Darbhanga (M.D.L.S.M), Bihar.

4.3.1 Establishment of a Temporary Conservation Laboratory at M.D.L.S.M., Darbhanga

Figure 4.5: Receiving of the Manuscripts from the Authority and a View of Conservation at the Laboratory.

Figure 4.6: Conservation Work at the Site by the Team of ICI, Lucknow.

4.3.2 Curative Conservation Carried by the Team on Manuscripts/Panjis in the following Steps

☆ **Documentation:** A detailed condition report of each manuscript/panji was prepared. Photography was done in three phases *i.e.* before, during and after treatment. It consists of photography of the whole object and parts of object which was damaged or were important like the signature.

☆ **Examination:** Physical, chemical and biological examination of the object was done to know the actual condition of the object and decided a conservation procedure accordingly. The examination process involved- test of solubility of inks/pigments, test for presence of acidity by measuring the pH value by pH meter flat head electrode, test for checking the fibre strength.

☆ **Conservation:** Based on the observations noted during the examination, the solvents, chemicals and materials were chosen judiciously and a suitable line of treatment was decided. According to that manuscript was conserved *viz.* fungicide treatment to kill fungus, spores, *etc.* removal of binding and separation of folios especially the stuck folio or mechanically fixing of fugitive inks, cleaning to remove superficial dust and accretions; removed ingrained dust, depositions; aqueous or non-aqueous deacidification method, depending on solubility of ink or colour, to remove acidity; removal of previous restoration if they are actively damaging the object like lens tissue lining, mending and finally binding if it was originally presented.

Figure 5.7: Manuscript Showing Acidity, Loss of Areas and same after Treatment and Finally Wrapped in Starch Free Red Cotton Cloth.

Total conservation of 62 manuscripts/panjis having 4381 folios was carried out at different sites.

4.3.3 A Glimpse of Restoration Work by the Conservators

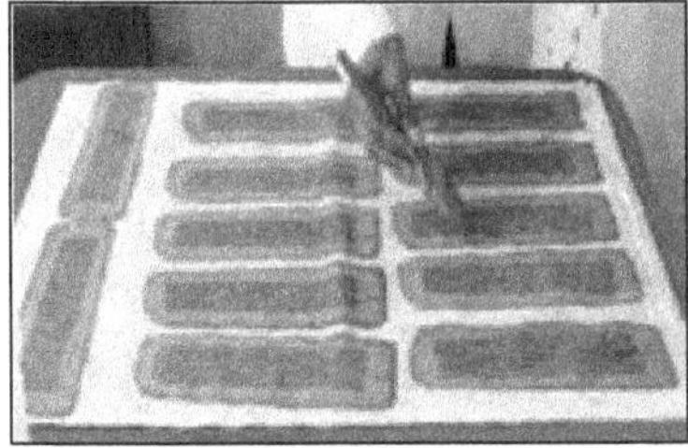

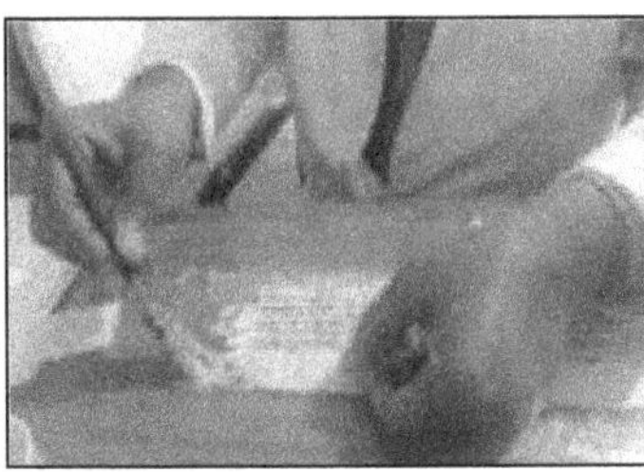

Figure 4.8: Solvent Cleaning of the Manuscript by Brush.

Figure 4.9: Repairing of the Manuscript.

Figure 4.10: Lining of the Manuscript by the Conservator.

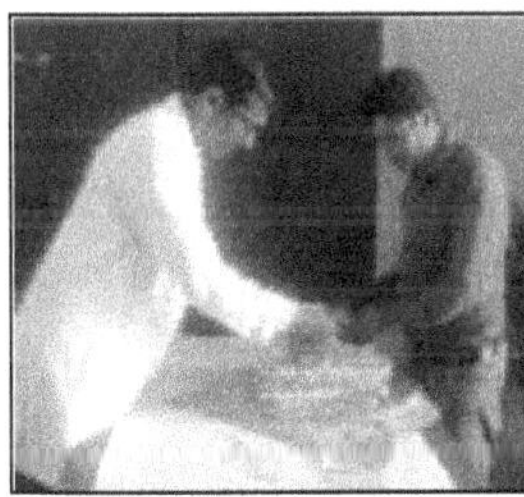

Figure 4.11: Conservation Treatment at Temporary Laboratory Established at Maharajadhiraja Kameshwar Singh Sanskrit University, Darbhanga.

Figure 4.12: Stuck Manuscripts before and after Conservation Treatment.

4.4 Preventive Conservation Carried at the Site

Library materials need preventive conservation throughout their life. Therefore, it can be said that preventive conservation is a continuous process which include all indirect actions that do not disturb the structure and materials of the object being preserved *i.e.* the appearance of the object is not modified under preventive conservation. As per the selected manuscripts that do not require curative treatment, preventive conservation treatment was provided on 232 manuscripts having 7464 folios which was carried by dry cleaning, serial checking, providing acid free alkaline board on up and bottom of the manuscripts, wrapping in handmade paper and finally providing starch free red cotton cloth (*VESHTAN*) for protecting them from micro-growth at the library collection of Mithila Sanskrit Post Graduate Studies and Research Institute, Darbhanga and Kameshwar Singh Darbhanga Sanskrit University, Darbhanga, Bihar.

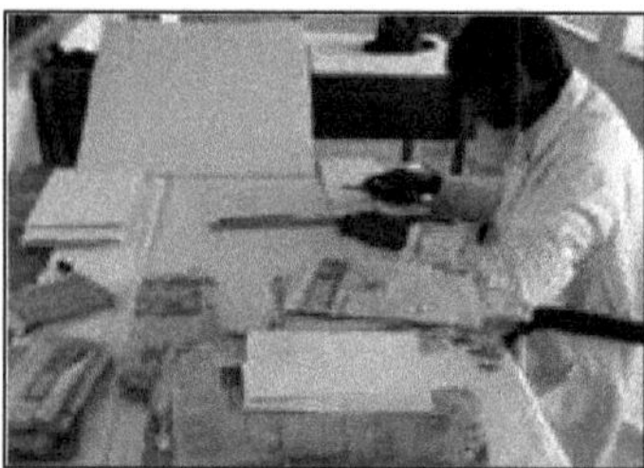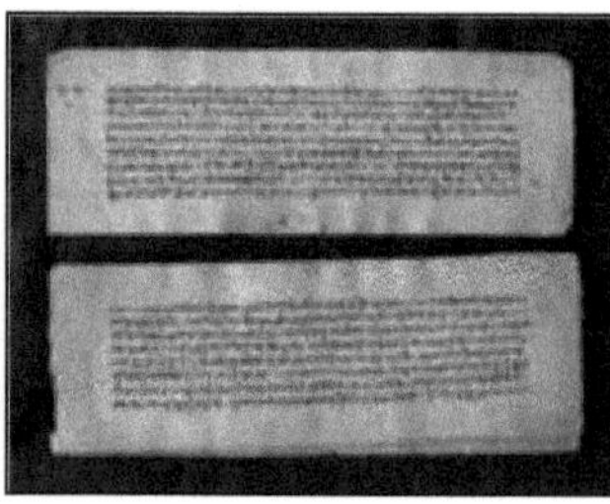

Figure 4.13: Measurement and Cutting the Acid Free Mount Board as per the Size of the Manuscripts, Cleaning of each Folio and Wrapping in Starch Free Red Cotton Cloth as a Preventive Measure.

4.5 Awareness Campaign about Conservation

Two Awareness campaigns were organized at Madhubani district on 27th March 2022. The objective of the Campaigns was to aware the causes of deterioration and prevent the damage by care and maintenance which was carried along with practical exercises so that the panjikars or custodians recognize the damage and care their collection themselves from biological, chemical and mechanical decay. Total 55 participants along with 45 online participants attended the awareness campaign at **Mithila Lalit Museum, Saurath, Madhubani** and **Yadunath Public Library, Paitaghat, Lalganj**.

Figure 4.14: Awareness Campaign on Preventive Conservation of Mithila Heritage of Manuscripts at Mithila Lalit Museum, Saurath, Madhubani.

Figure 4.15: Awareness Campaign Held at Yadunath Public Library, Paitaghat, Lalganj, Madhubani.

4.6 Outcome of the Project

With the efforts of Chairman INTACH, the valuable panjis and manuscripts were conserved to survive for further research scholars, custodians and panjikars. Also, the panjikars were encouraged to donate their panjis and manuscripts to the museum so that they may be saved for future generations. Awareness campaigns were successfully conducted on *"Preventive conservation of manuscripts/panjis"* for the panjikars so that they know how to keep the collection safe for further generation. All these efforts helped to preserve and carry forward the legacy of panjis and manuscripts which is the pride of Mithila.

Acknowledgements

Conservation was a joint effort by the team. Author thanks and acknowledges the assistance received from Maj. Gen. (Retd.) L.K. Gupta, Chairman INTACH, INTACH Bihar and Darbhanga State Chapter, Conservation team and the Institutions for the efforts. The coordination and facilities provided at site for establishing a temporary laboratory *i.e.* Maharajadhiraj Laxmishwar Singh Museum, Darbhanga, Mithila P.G. Studies and Research Institute, Darbhanga, Maharajadhiraj Kameshwar Singh Darbhanga Sanskrit University Darbhanga are gratefully acknowledged.

REFERENCES

Jha Pt R. Maithil Brâhamano ki Pañji Vyavasthâ (Hindi), published by Granthâlaya, Darbhangâ.

Karna B K (1973) Panji System in Maithil Karna Kayastha: A Sociological Evaluation.

Thakur G (2009) Genome Mapping- 450 AD to 2009 AD- Mithilak Panji Prabandh, pub. by Shruti Publication, 2009. Delhi ISBN No.978-81-907729-6-9

Varma B B (1973) Maithili Karna Kayasthak Panjik Sarvekshan. Madhepura: p. 212. OCLC 20044508.

Chapter 5

Microbial Deterioration of Miniature Paintings, Paper, Books, Herbarium Sheets, and Palm Leaf Manuscript

N.R. Shah[1] and Arun Arya[2]

[1]*Department of Museology, Faculty of Fine Arts,*
[2]*Department of Environmental Studies, Faculty of Science,*
The Maharaja Sayajirao University of Baroda, Vadodara – 390002, Gujarat, India
e-mail: sarojarun10arya@rediffmail.com

ABSTRACT

Scientists have studied aeromycoflora of cities like Bhopal, Jabalpur, Lucknow and Vadodara, etc. Numerous fungi from different vegetable and fruit markets of Vadodara were identified. Study of aeromycoflora of Agra city resulted into occurrence of 102 fungal spp. In the indoor survey of hospitals in Agra, maximum concentration of fungal spores was obtained during rainy season. A large number of fungi are reported to damage the valuable cellulosic material. Pigments produced by fungi can cause brown or black spots on paper and cloth objects.

Museums are centres of non-formal education, a good museum attracts, entertain, arouses curiosity, leads to questioning and this promotes learning. Study on damage to anthropogenic objects like Miniature paintings, palm leaf manuscript and damage to French paintings was undertaken. Enzymes cellodextrins and cellobiose break down cellulose. Hydrolyses of cellulose occurs which leads to degradation of paper. Damaged Miniature paintings in Baroda Museum showed the presence of C. gangligerum, A. versicolor, A. terreus, Penicillium sp. and Streptomyces griseflavus. Palm leaf manuscript showed the occurrence of Nigrospora. Association of fungi was observed from deteriorating herbarium sheets placed in Baro Herbarium of the botany department of the M.S. university of Baroda. Percentage loss of different papers by fungi was calculated. Use of certain botanicals in form of Vacha (Acorus calamus) rhizome powder, methanolic extracts of Adenocalymma alliaceum, Hyptis suaveolens and Cymbopogon oil was found effective to minimize the damage. The chapter discusses how mycoflora present in indoor air can damage the paper objects placed in Baroda Museum and Picture Gallery, museum of Botany Department of the M. S. University of Baroda and Maharaja Fatehsingh Museum, Vadodara. Mechanisms of deterioration of valuable cellulosic objects due to fungal organisms are described and certain control methods are suggested.

Keywords: Biopollutants, Fungi, Aspergillus, Paper, Deterioration, Miniature paintings, Palm leaf manuscripts.

5.1 Introduction

Sculptures, paintings, manuscripts and historical monuments present all over the country constitute the rich cultural wealth of our country. Everyone knows about Taj Mahal, one of the seven wonders of the world, but our country has still older monuments than this. Earlier rocks and metallic plates were used for writing and conveying the message of rulers. The use of leather, parchment and Bhojpatra (*Betula utilis* D. Don) was made in India. The bark from the tree was used for writing Sanskrit scriptures and texts. *Cyperus papyrus* L. or *Cyperus* grass was the plant growing along river Nile which yielded the present day source of paper from its peduncle (Figure 5.5). Paper is a cellulosic material obtained from a large number of plants and even from cloth. Discoloration and damage to paper material may be caused by bacteria, fungi and insects like silverfish (*Lepisma* Figures 5.7g, 5.13b).

In today's competitive world, we have to preserve the monuments as well as museum materials and showcase the valuable heritage to the next generation as the contributions or achievements of our ancestors (Figures 5.1a,b). Damp surfaces are readily colonized by microbes settling from the air. Fungi have a

Figure 5.1: Baroda Museum and Picture Gallery.
(a) Building of the Museum made in Indo saracenic style of Architecture;
(b) Inner view showing show cases in the gallery.

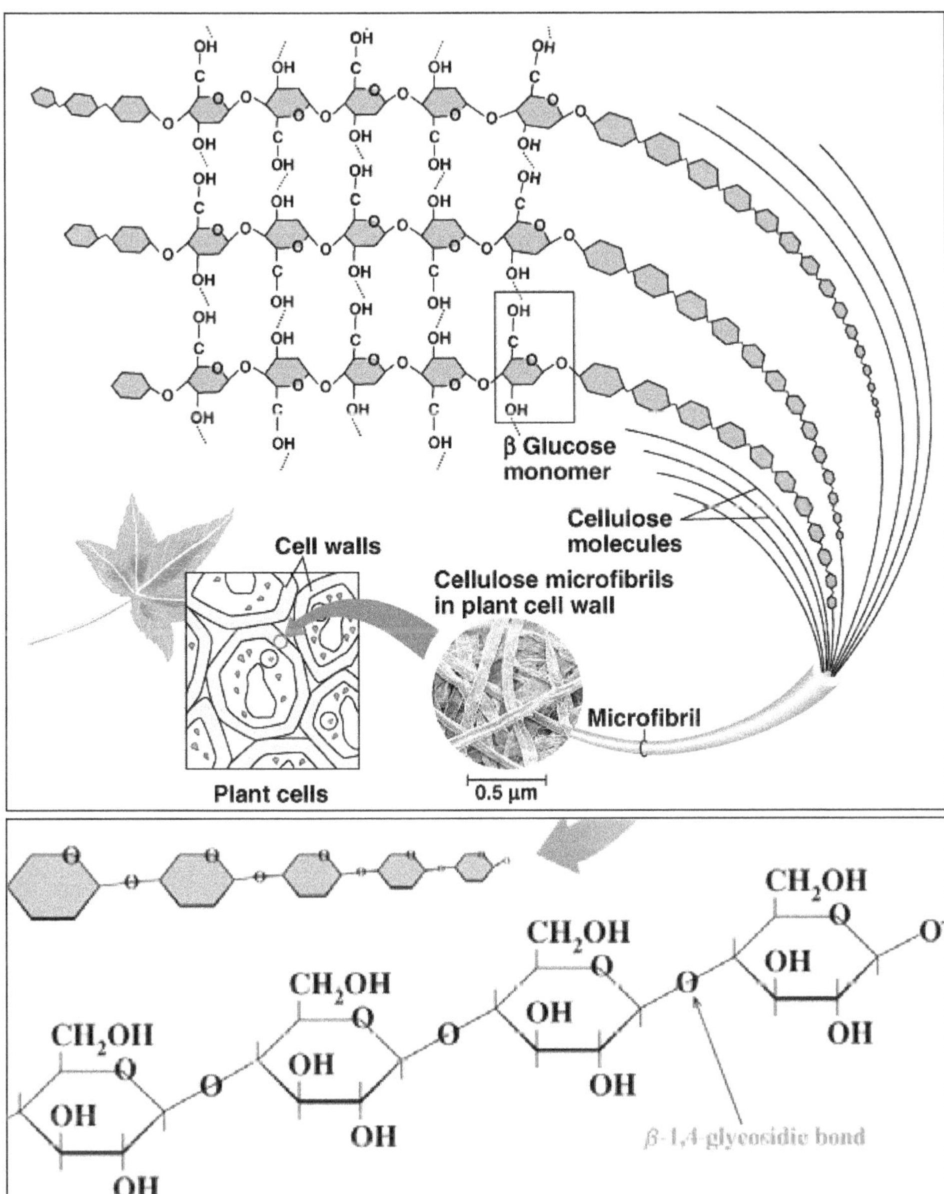

Figure 5.2: Cellulose Fibers: Structural Formula.

Cellulose

☆ Is a polysaccharide of glucose units in unbranched chains.

☆ Has β-1,4-glycosidic bonds.

☆ Cannot be digested by humans because humans cannot break down β-1,4-glycosidic bonds.

definite role in deterioration of cultural heritage (Sterflinger 2020) Cellulose degrading fungi like *Chaetomium* may degrade paper and cloth. Fungal agents like *Alternaria* and *Curvularia* produce black pigments and discolor the objects. The fungal organisms may show the fast growth at increase temperature. They show increased production of enzymes at higher temperature. The fungi which grow at higher temperature are termed thermophilic fungi.

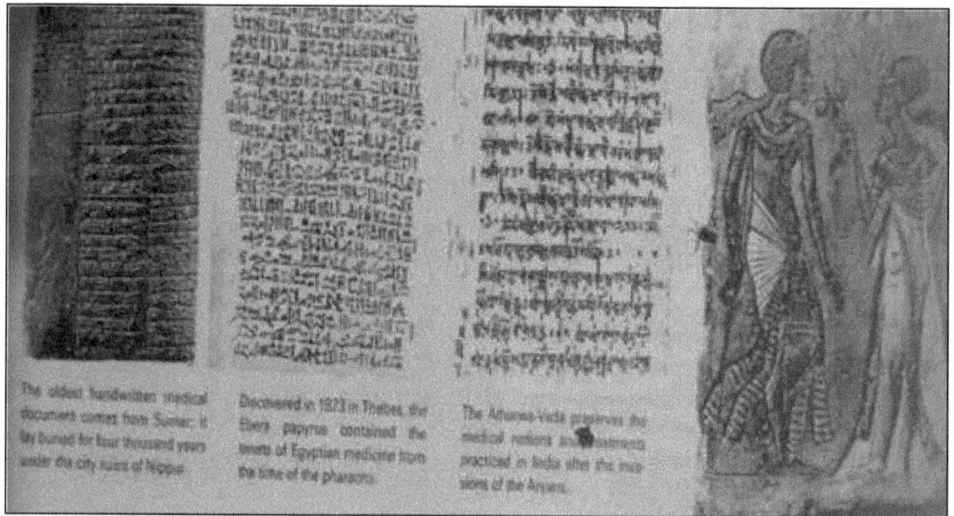

Figure 5.3: Page of a Book Showing Old Hand Written Documents.

Figure 5.4: An Old Egyptian Painting on Cloth.

Figure 5.5: *Cyperus papyrus* a Grass.

Why is 500-year old paper often in better condition than paper from 50 years ago? In other words, what makes some papers deteriorate rapidly and other papers deteriorate slowly? The rate and severity of deterioration result from internal and external factors: most importantly, the composition of the paper and the conditions under which the paper is stored. Paper is made of cellulose – a repeating chain of glucose molecules (Figure 5.2) – derived from plant cell walls. One measure of paper quality is how long the cellulose chains, and subsequently the paper fibres are. Long-fibered paper is stronger and more flexible and durable than short-fibered paper. In the presence of moisture, acids from the environment (*e.g.*, air pollution, poor-quality enclosures), or from within the paper (*e.g.*, manufacturing process), repeatedly cut the glucose chains into shorter lengths. This acid hydrolysis reaction produces more acids, and cause degradation.

Pliny made mention of the writings of Numa, who lived about 670 BC, whereas, Joshi (1989) and Sindall (1919) have stated that documents written on *Papyrus* have been found in Egypt and are as old as 3000 to 2000 B.C. Earlier Joshi (1947) reported use of parchment and tender skin of animals for writing purpose by Persians. Before the mid-19th century, western paper was made from cotton and linen clothing rags and by a process that largely preserved the long fibers of the raw material. While fibers may shorten with age, rag papers tend to remain strong and durable, especially if they have been stored properly in conditions not overly warm or humid. Starting in the mid-19th century, wood replaced rags as the raw material for paper manufacture. Wood is processed into paper by mechanical or chemical pulping, which produces paper with shorter (compared with rag paper) fibers. Mechanical pulping produces paper with the shortest fiber length and does not remove lignin from the wood, which promotes acid hydrolysis. Newspapers are printed on mechanically pulped paper. Chemical pulping removes lignin and does not cut up the cellulose chains as thoroughly as mechanical pulping, yielding a comparatively stronger paper, but which is still not as durable as rag paper.

Wood pulp paper used before the 1980s also tends to be acidic from alum-rosin sizing (added to the paper to reduce absorbency and minimize blurring of inks), which, in the presence of moisture, generates sulfuric acid. Acids also form in paper by the absorption of pollutants – mainly sulfur and nitrogen oxides. Book leaves that are more brown and brittle along the edges than in the center clearly illustrate this absorption of pollutants from the air. Research by the Library of Congress has demonstrated that cellulose itself generates acids as it ages, including formic, acetic, lactic, and oxalic acids. Measurable quantities of these acids were observed to form under ambient conditions within weeks of the paper's manufacture. Moreover, paper does not readily release these acids due to strong intermolecular bonding. This explains why pH neutral papers become increasingly acidic as they age. Acids form in alkaline paper as well, but can be neutralized by the alkaline reserve. Besides acid hydrolysis, paper is susceptible to photolytic (damage by light) and oxidative degradation. The role of oxidative degradation is limited as compared to acid hydrolysis, except in the presence of nitrogen oxide pollutants. Generally speaking, good quality paper stored in good conditions (cooler temperatures; 30-40 per cent relative humidity) are able to last a long time – even hundreds of years.

5.2 The World's art: Under Attack of Microbes

India is known for its rich, diversified, cultural properties kept in World heritage sites, National monuments, State monuments UNESCO, ICOMOS, ICCROM, ICOM *etc.* Gujarat is famous for Somnath and Akshardham temples, historical monuments, artistic sculptures, Adalaj vav (step well located in Gandhinagar) *etc.* There is need for conservation and benefit of these requires special strategies to find out the deterioration particularly of organic objects and reduce the losses.

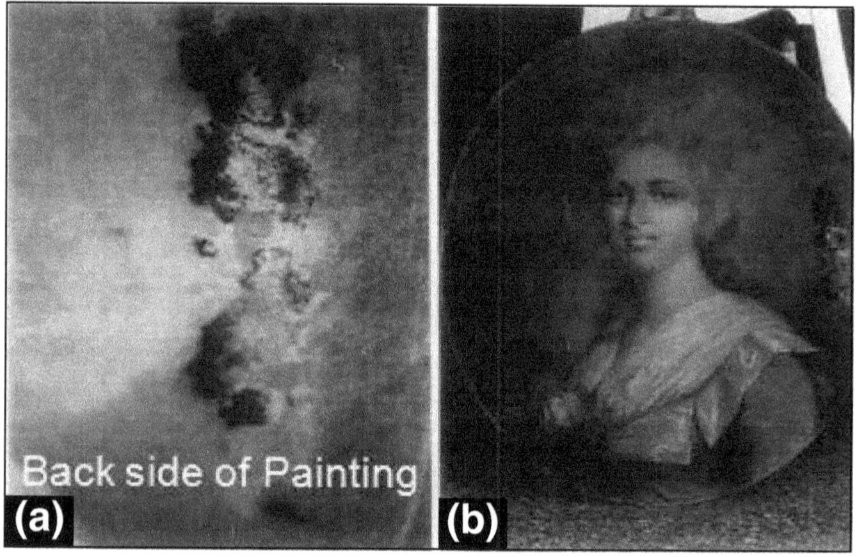

**Figure 5.6: Fugal Damage to Pastel Colour Painting
(a) Back, (b) Front Side of French Painting.**

The bacterial and fungal colonies are a nuisance because their dark pigments leave stains like ink drippings along the walls on monuments, cloth and paper objects. The moist environment attracts other microbes that damage and weaken it. "That humidity really sets up the conditions where everything can get a toehold," says Paula DePriest, Deputy Director of the Smithsonian's Museum (Baggaley 2017). Fugal damage to pastel color painting made by a French artist on hard board and placed in Fatehsingh Museum Baroda can be seen due to species of Aspergilli (Figure 5.6). Arya and Arya (2007) reported numerous fungi from different vegetable and fruit markets of Baroda. Study of aeromycoflora of Agra city resulted into occurrence of 102 fungal types. In the indoor survey of hospitals in Agra, maximum concentration of fungal spores was obtained during rainy season (Kulshrestha and Chauhan 2001). A large number of fungi are reported to damage the valuable cellulosic material. Pigments produced by fungi cause brown or black spots on paper and cloth objects (Arya 2006).

Biodeterioration phenomena represent a complex of physical and chemical alteration processes in various materials, such as those constituting the objects that represent our cultural heritage (Allsopp *et al.*, 2004). Association of *Aspergillus* and perithecia like structure forming genera *Ascotricha* was reported by Singh (2014) from deteriorating pape manuscripts in Vadodara. The *Ascotricha* and *Chaetomium*, two common cellulose degrading fungi are described in detail by Chivers (1915).The biodegradation of paper is conditioned by several variables such as the materials from which cellulose is obtained, the manufacturing processes employed, the occurrence of other affecting substances such as lignin or metallic compounds, and by the environmental conditions in which papers are stored. In this study, biodeterioration of paper was artificially induced in order to evaluate the role of a range of chemical and physical variables on damage caused by cellulolytic fungi. A variable pressure SEM instrument was used to characterize paper samples with different fibre origins, and alterations obtained *in vitro*. Two fungal strains, *Aspergillus terreus* Thom. and *Chaetomium globosum* Kunze, which are cellulolytic species frequently associated with paper spoilage, were used to produce stains with characteristics close to those observable on art objects made from paper. The stains obtained on the different samples of paper were compared at both low and high magnification, in order to visualize the macro and microscopic characteristics of paper fibres, inorganic constituents, impurities, and the deteriorating agents related to the spoiled areas. During this survey it was observed that single paper characteristics can strongly influence the intensity and the results of the fungal action. For example, the activity of a fungal strain on paper grades containing fibres of the same origin, but with different sizing, led to the formation of profoundly different stains and alterations. Moreover fungal structures, analysed by low vacuum SEM, in areas on paper corresponding to the stains appeared in different physiological states suggesting an important effect of paper constituents on fungal growth and their sporulating ability. According to Nair (1972) fungi that can grow on any substrata and play an important role in the deterioration of museum materials.

5.3 Paper: Production of a Unique Material

The principal component of paper is cellulose, a polymer of β-D-glucose. Hydrogen bonding between cellulose chains sticks them together to form fibrils, which further associate to form fibers, the basis of paper. In the digital age of the 21st century, our reliance on paper is rapidly declining. Smartphone or tablet screens and electronic paper displays like those of Amazon's Kindle are becoming the norm for everything from cinema tickets to best-selling books. But our history is written on paper and we face an ever-growing urgency to preserve paper-based artefacts before they are lost forever. About one third of the paper items in large libraries are too brittle to handle, with another third in need of attention over the coming century. Papers are of different types, the news print paper is rough, writing paper is white and bright, tissue paper is soft. When Dr. Arya visited Quebec City in Canada; it was told by an industry that 75 per cent paper is produced from recycled material. The pulping of wood, bleaching, sizing, pressing and drying are the main steps in manufacturing. The paper is produced in long sheets, then cut and rolled and sent for market use. In the 1980s, paper manufacturers began adding alkaline buffers to wood pulp papers intended for lasting use and today this is common practice. Alkaline buffers retard or prevent acid hydrolysis by neutralizing acids that attack the cellulose chains. Alkaline wood pulp papers stored under good conditions are long lasting. Since the 1990s, books published in the U.S. that conform to ANSI/NISO paper permanence standards (*e.g.* ANSI/NISO Z39.48 - 1992) are likely to be printed on chemically purified wood pulp alkaline paper. The magnitude of losses and poor quality of fiber depends on the type of pulp and length of storage period (Gray 1959). Paper mill storage yard should be well drained. Fungi will not grow in wood if the moisture content is lower than 20 per cent (Prescott and Dunn 2011).

Laboratory filter paper consists of almost pure cellulose, it is poor for writing and weak, when wet. Whatman's No. 1, 42 and 41 are differentiated on the basis of their pore size. Chromatographic paper is yet another costly paper used for identification of compounds in labs. Additives are included to strengthen the interaction between the fibres. Traditional additives include gelatine and aluminium sulphate, which strengthen the paper and prevent ink from blurring. Paper in Europe was originally made from cellulose sourced from linen and cotton rags. This made strong paper structures, owing to the long cellulose chains. However, most paper in our hands today is made from cellulose extracted from wood pulp. Cotton and linen sourced cellulose is now usually reserved for special purposes such as banknotes and artists' materials. While wood is a much more readily available source, the resulting paper has shorter cellulose chains (with a degree of polymerisation around 600–1000) and a weaker structure.

There are two principal chemical degradation pathways of paper: acid-catalysed hydrolysis and oxidation. The pH of 1 g of a piece of paper in 50 cm^3 of water gives a measure of its acidity. Early work on paper conservation chemistry found that old paper that was alkaline was much stronger than acidic paper. Paper can become acidic either by absorbing pollutants such as SO_2 and NO_x, or during manufacture. Traditionally, aluminium sulphate (or 'paper maker's alum') was added to harden, or 'size', the paper. It provides initial strength, but is a source of acidity. The β-acetal oxygen bridge joining the glucose molecules of cellulose together is susceptible to acid hydrolysis. This breaks the chains and weakens the fibres. Paper that decomposes this way becomes hard and brittle, and disintegrates easily. It is lignin that is the main cause of photo-yellowing of paper. It contains several chromophores with conjugated aromatic rings and carbonyl groups that absorb in the near UV spectrum (300–400 nm). When these chromophores absorb light they can decompose into yellow-coloured ketones and quinones, turning the paper yellow.

Understanding the degradation pathways reveals preservation methods. From an elementary chemical point of view, if acid is causing decomposition, the solution is to de-acidify it. This involves washing paper in a bath of mild alkali such as calcium hydroxide, calcium bi carbonate or magnesium bi carbonate. One method examined by the US Library of Congress was using diethylzinc gas to neutralise the acid and leave an alkaline deposit of zinc oxide. While this appeared to be perfect as it involved no solvent and worked well in principle, it was abandoned owing to the pyrophoric nature of diethylzinc and the cost of establishing a plant. There are existing commercial products that use microparticles – such as one called Bookkeeper that contains microscopic MgO particles, which form alkaline $Mg(OH)_2$ on application. However, nanoparticles of calcium and magnesium hydroxide can penetrate the paper structure more easily, resulting in more complete deacidification. While the deacidification washes may also remove some yellowing, more persistent discolouring can be removed by bleaching. The most common reagent is hydrogen peroxide (0.5–3 per cent), since hypochlorites can damage the cellulose fibres. The active bleaching agent is probably the peroxide anion, ^-OOH, which is formed by dissociation of peroxide in water. Conservation science is understandably a conservative area that is slow to change, given the nature of the materials involved. One area of current research showing promise is using dispersions of metal hydroxide nanoparticles to neutralise acidity.

$$H_2O_2 + H_2O \rightleftharpoons H_3O^+ + {}^-OOH$$

This reaction only produces an appreciable amount of peroxide anion in the pH range of 8–10, and while slow, is considered among the most effective bleaching strategies.

The gall ink used for printing was formed by reacting gallic acid (derived from tannins extracted from gall-nuts) with iron (II) sulphate. The presence of excess Fe (II) ions can catalyse the oxidation of cellulose through the production of hydrogen peroxide according to the Fenton mechanism. This can lead to significant destruction of the paper along the lines of the ink so much, so that the paper can fall out of the structure leaving a lace-like pattern. This decomposition can be stopped by introducing chelating agents – preferably phytate (inositol hexakisphosphate) – to complex the Fe (II) in the ink. The paper is washed in a solution of calcium phytate, which chelates any iron (II) present, blocking the Fenton pathway.

5.4 Microbial Damage to Paper and Books

Fungal contamination is a major cause of deterioration of libraries, archives and museum materials (Florian 1997). The presence of biological agents is also a potential health risk because some species are highly pathogenic or are known to cause allergies and asthma and exert toxic effects (Lacey 1991; Mishra *et al.*, 1992; Singh and Singh 1999). Mould and fungi spores are always present in the air and will start to grow wherever conditions are favourable. Filamentous fungi are organisms with an important role in the degradation of organic waste, such as wood and paper (Chadeganipour *et al.*, 2013). Humidity, Insects, termites and rodents can also damage the books. The fungal ability to produce extracellular enzymes is well established. They can produce hydrolytic enzymes such as cellulase, xylanase, pectinase, *etc.* Also, they spoil valuable documents mechanically, chemically and aesthetically because they form hyphae, produce and excrete pigments and organic acids (Micheluz *et al.*, 2015). Paper is susceptible to a larger number of colonizing microorganisms (80 genera of fungi and 59 genera of bacteria) (Pinzari and Gutarowska 2021). Isolated fungi from paper and books may be spp. of *Chaetomium, Myxotridium, Eidamella, Aspergillus, Acrostalagmus, Spicaria, Torula, Cephalosporium, Stachybotrys, Dematium, Cladosporium, Stemphillium, Alternaria, Stysanus* and *Fusarium* (See 1919). Foxing of paper and colour change in paper was reported by Sartory *et al.* (1935). They reported occurrence of fungi like *Aspergillus, Cladosporium, Fusarium* and *Monilia*. Verona (1938) reported association of *Cephalosporium, Coniosporium* and *Phoma* with books and paper.

In the case of fungi, culture-dependent methods are widely used in the laboratory, and results indicate that 58 per cent of fungal species can be identified using culture methods, 26 per cent using molecular methods and 15 per cent using both methods (Gutarowska 2016). So far, only a few genera inhabiting book materials have been identified using both methods (culture-dependent and molecular), with just 18 genera of fungi, on paper that have been identified using both approaches. Microbial community analysis using the NGS (next-generation sequencing) method makes it possible to detect higher biodiversity than the culture-dependent approach yields (Kraková *et al.*, 2018) and some early culture-independent approaches based on cloning (Michaelsen *et al.*, 2006). Rakotonirainy *et al.* (2003) developed ATP bioluminescent method to detect presence of viable fungi. They found presence of minimum1000 spores per ml and it did not differentiate bacteria and fungi. Process is expensive but simple and quick than culture method.

Some fungal species are able to live at low water activities for that are classified as xerophilic fungi; they are perfectly adapted to indoor environments and thrive in dusty environments, lack of ventilation or water retention by hygroscopic materials for these materials with a very low water activity can be colonized by xerophilic species (Khan and Karuppayil 2012). They can be found in the indoor air of archives, libraries and museums where much paper exists. Dust is a good source for these fungi to feed and grow; these conditions intensify fungal contamination (Borrego *et al.*, 2010). Fungal degradation and the documentary materials deterioration is a worldwide problem that causes great damage to especially paper documents stored in the archives, libraries and museums (Rakotonirainy *et al.*, 2007; Mesquita *et al.*, 2009). Each document or book evaluated represents one microbial ecosystem composed by a fungal community which is formed by one or more fungal species. For this reason, different genera could be detected. *Aspergillus, Penicillium* and *Cladosporium* were the predominant genera and were classified ecologically as abundant, frequent and common, respectively. Figure 5.7 shows damage caused by insect and fungi on books and paper. The structural nature of ascospores as progenitors of future growth allows the fungi to survive severe conditions, and ascospores are consequently harder to inactivate than the vegetative hyphae. In unfavourable conditions resting spores have low water content and their metabolism is inactivated but it is reversible (Deacon 2005).

Discoloration (Figures 5.7a, b), consolidation of paper (Figure 5.7c), and foxing (Figure 5.7d) (Arai 2000; 2014; Aryanak 2005; Borregoa *et al.*, 2010) was observed in most cases. An example of slow-forming damage that gradually makes itself apparent is foxing (Choi 2007; Modica *et al.*, 2019). The formation of foxing stains on paper can be caused by fungi as a result of limited water availability. Absolute tonophilic fungi germinate on paper in the presence of very low water availability (aw < 0.80) and release cello-oligosaccharides, aminobutyric acid and amino acids. These compounds condensate in a spontaneous chemical reaction (Maillard condensation) to form brown-coloured compounds known as "melanoidins" (Arai 1987; 2000), which in many cases are responsible for foxing stains. When free water is scarce, the microbiological attack on the materials takes place differently. It is slower, and there are only a few or single extremophile species that can germinate and grow. Calcium carbonate generally gives opacity and bulking characteristics. As well as filler calcium carbonate is used to buffer acidity due to several causes (*i.e.* pollution, acid inks, *etc.*) and it is also used in permanent papermaking to provide an alkaline reserve (Gallo *et al.*, 1998). Acidic environments promote biodeterioration by means of fungal growth mainly deuteromycetes (*A. terreus*), while alkaline environments can stimulate specific groups of ascomycetes such as *C. globosum*, whose growth is optimal at the pH interval 7.1–10.4 (Domsch *et al.*, 1980). The positive effect of neutral to alkaline reaction in paper on its spoilage by *C. globosum* was confirmed by this study. According to SEM observations the neutral to alkaline pH also promoted the maturation of the perithecia and the production of ascospores (Pinzari *et al.*, 2006).

Aspergillus and *Penicillium* fungal species are among those most frequently associated with the production of organic acids and other metabolites, which can react with various components of paper. The presence of starch, gelatin, rosin and other glues, or that of salts and mineral compounds can exert an influence on fungal metabolism and determine, for example, the kind of stains produced. The fungal pigments that affect paper have been associated to polyketide quinones, carotenoids and other compounds whose synthesis or colour can depend on the

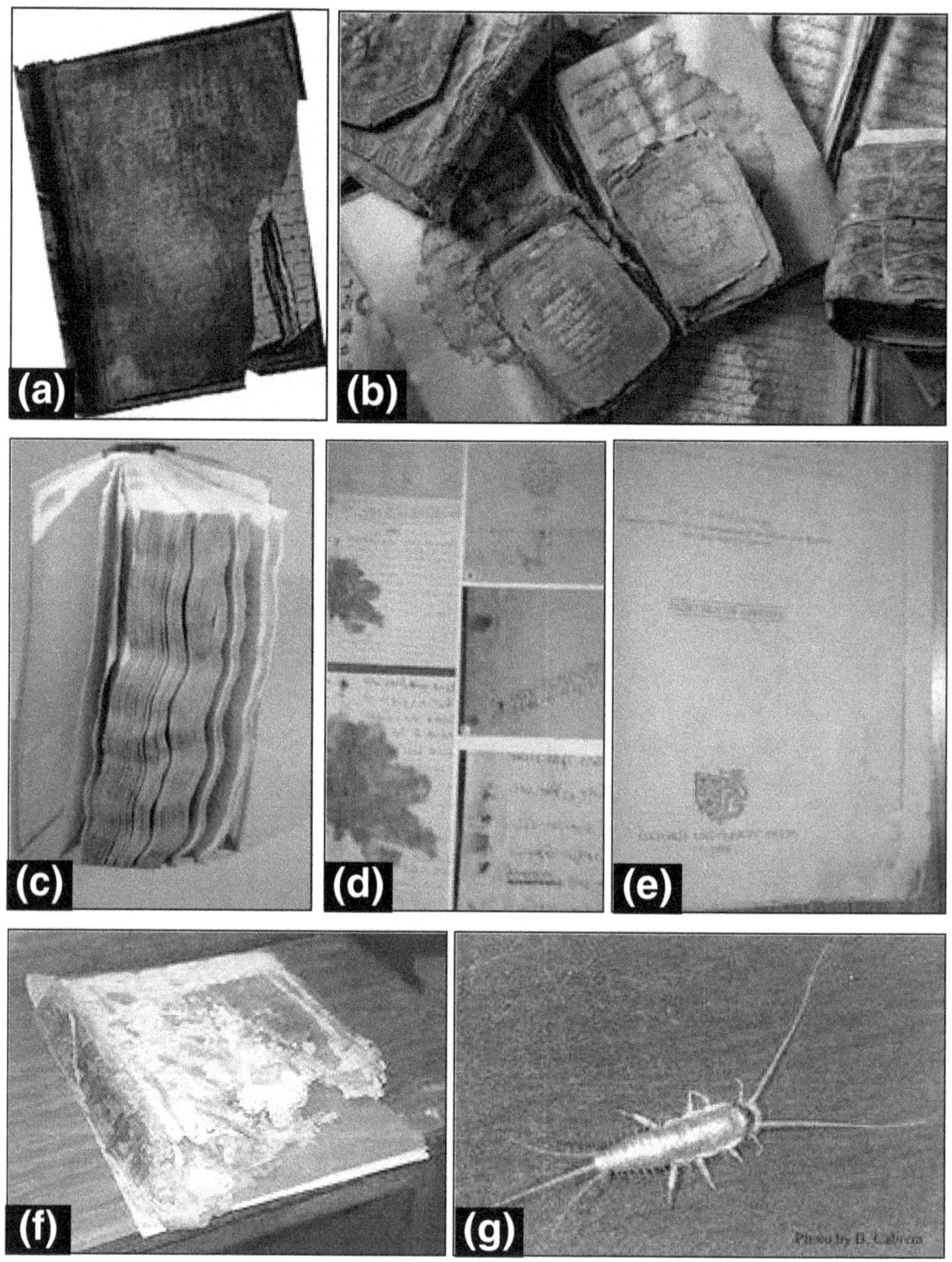

Figures 5.7a–g

(a) Book deteriorated by insect attack; (b) Old books became brown in colour; (c) Books with consolidated leaves; (d) Foxing or browning of court stamp paper; (e) Damage to book by humidity; (f) Book rotten by Fungi; (g) Silver fish.

availability of nitrogen, the prevailing pH and the presence of other limiting nutrients and enzyme cofactors such as metals and cations (Fe, K, Ca, Mg, Mn) (Melo *et al.*, 2019). However, in order for fungi to be able to produce abundant enzymes and synthesize complex metabolites, there must be no limiting factors, and therefore water must be available together with the organic matter.

Metabolomics and proteomics provide information on biological mechanisms and potential biomarkers in samples. Recently, metabolomic analysis based on the AuNPET SALDI-ToF-MS method was applied in the study of foxing stains marring some 19th-century papers. This technique enabled the authors (Szulc *et al.*, 2018) to demonstrate the occurrence in the stained areas of several metabolic pathways, including sugar degradation, amino acid and protein metabolism, ubiquinone and other terpenoid-quinone biosynthesis, 2-methyl-6-phytylquinol and delta-, gamma-, beta-tocopherols (responsible for the yellowish-brown colour of foxing spots) and 3-hydroxy-L-kynurenine (a fluorescent, yellow compound). These pigments can all contribute to the mechanism underlying the appearance of foxing caused by microorganisms (Szulc *et al.*, 2018).

Eurotium halophilicum is a xerophilic fungus, with high tolerance to water stress; it was first isolated from dry food and indoor dust in association with other xerophilic fungal species and dust mites (Montanari *et al.*, 2012). Since the first reports of it as a book contaminant, it has been identified in dozens of libraries and archives in Italy as the main, if not the sole fungus flourishing on books and archival materials (Montanari *et al.*, 2012; Polo *et al.*, 2017). It has also been linked to the appearance of foxing-like stains on materials (Sclocchi *et al.*, 2016). The minimum water activity value observed for its successful germination and growth is 0.675 (Christensen *et al.*, 1959). It has repeatedly been found in association with the covers of books and other surfaces within museums, libraries and archives, even when the overall environmental conditions are in line with those recommended for the conservation of the materials (relative humidity 50 to 60 per cent, and a temperature 20 to 22°C). The niches preferred by this fungus are characterized by infrequent ventilation and water condensation events after a drop in temperature or night/day thermic cycles. Another xerophilic fungus, *Aspergillus penicillioides*, has often been isolated from books in association with *E. halophilicum*, probably due to its similar ecology. Other fungi have frequently been isolated from *E. halophilicum* mycelium: *Aspergillus creber, A. protuberus, Penicillium chrysogenum* and *P. brevicompactum* (Micheluz *et al.*, 2015). These species grow on the dead mycelium of *E. halophilicum*. It has been observed that their presence is associated with historic infections of the fungus. As long as the mycelium is alive, the fungus remains the only organism on materials, and therefore represents the primary colonizer. Once established and mature, its mycelium serves as a substrate for other xerophilic organisms. Dead structures of the fungus *E. halophilicum* have also been found in foxing spots on paper and other materials. Metal oxalates have often been documented in association with fungal hyphae (*i.e.* calcium oxalates) (Pinzari *et al.*, 2010). These findings are consistent with the hypothesis that absolute tonophilic fungi germinate on paper, releasing organic acids, oligosaccharides and amino acids. The damage that this fungal species causes to materials consists of pale brown to dull grey stains and other forms of decolouration (Figure 5.9b). When decomposition occurs, it is limited to small areas and consists of a sort of superficial erosion. This fungus also produces volatile organic compounds (Micheluz *et al.*, 2016) and represents a potential hazard for workers who manage infected materials. GC-MS analysis showed that *E. halophilicum* produces at least 20 different volatile compounds, with acetone and 2-butanone being the main products. A total of eight secondary metabolites were detected through LC/MS-MS, *e.g.* deoxybrevianamid E, neoechinulin A and tryprostatin B (Micheluz *et al.*, 2015).

Aranyanak (2005) working in National Museum, Thailand reported 62 fungi causing stains on paper in different libraries.

Table 5.1: Different Stains Produced in Paper and Causal Fungal Agents Isolated

Sl.No.	Stains Produced	Fungi Isolated	Comments
1.	Black pigment/dark	*Cladosporium, Alternaria, Curvularia, Drechslera, mycelia sterilia*	The coloured mycelium penetrates the fibres. Removal of stains difficult
2.	Brown- black spots	*Asperillus niger, Chaetomium* spp. *Memnoniella echinata, Nigrospora Periconia, Stachybotrys atra, Torula herbarium.*	Stains are cause by mycelium. Which produces melanin. Stains can be removed.
3.	White spots	*Eurotium halophilicum*	Produce superficial colonies
4.	Pink- orange fluffy growth	*Monilia*	Removal of stains difficult
5.	Violet	*Alternaria*	Violet colour is water soluble
6.	Orange red stain	*Monascus* sp.	Stains are water soluble
7.	Yellow, Green spots	*Aspergillus fumigatus, A. flavus, Penicillium citrinum, Trichoderma*	Spots are produced by fungal metabolites. These can be cleaned by moist cotton swabs

Keraton Kasepuhan Cirebon has collection of five old manuscripts of European origin from 19th century, and the former library of Faculty of Humanities University, Indonesia has collection of four old manuscripts of European paper origin from 19–20th centuries. The objectives of this study were to isolate and morphologically characterize fungi from nine old manuscripts of European origin. All manuscripts showed brown and black spots, and fungal spores on the surfaces. Twenty fungal isolates were able to grow on the old paper strips indicating that the old manuscripts are liable to fungal degradation.

Morphological characterization described the fungal isolates in the genera of *Aspergillus, Penicillium,* and *Eurotium.* Three fungal isolates could not be described and were grouped as Mycelia sterilia. Basic local alignment search tool (BLAST)-based identification and phylogeny-based identification using the data set of internal transcribed spacer (ITS) sequences is required to determine the species identities of the fungal strains (Oetari *et al.,* 2018).

The samplings were conducted in 2011 at library of Keraton Kasepuhan Cirebon, and March 2011 at former library of Faculty of Humanities UI. Samples of old manuscripts of European origin were obtained from there (five manuscripts from the 19th century), and the former Library of Faculty of Humanities Universitas Indonesia (4 MSS from 19–20th centuries). Fungal growth on the surface of manuscripts was detected by the presence of spores, mycelia and brown spots. Adhesive tape method and sterile cotton swabs were used to obtain samples suitable for further fungal culturing and identification. Swab was wiped along the damaged margins of the outer and inner pages of the manuscripts (Oetari *et al.,* 2018). Based on the morphological characteristics, twenty fungal isolates were described to the genus level as follows: *Aspergillus* (9 isolates), *Penicillium* (7 isolates), and *Eurotium* (1 isolate). Three isolates did not show structures to be identified to the genus level and were assigned as Mycelia sterilia. *Aspergillus* and *Penicillium* were anamorphic fungi since the asexual spores were only found. Molecular study was required to determine whether these isolates were members of the same genera as mentioned above, or they were members from different genera. Pinzari and Montanari (2011) and Sterflinger and Pinar (2013) reported that typical fungal infections in libraries, colonizing documents made of paper, are caused by xerophilic fungi of the genera *Aspergillus, Penicillium, Paecilomyces,* and *Cladosporium,* and slow-growing Ascomycetes.

5.5 Biodeterioration of Historic Paper: Preliminary FTIR and Microbiological Analyses

According to Zotti *et al.* (2008) the paper is subjected to numerous biodeterioration processes, which may cause the irreversible degradation of important documents and works of art. Many chemical and physical factors can affect these processes and their behavior, and fungi seem to play a key role in biodeterioration. This study is mainly aimed at verifying the presence of fungi in biodeteriorated 18th century etchings, and characterizing the paper surface by means of Fourier transform infrared (FTIR) spectroscopy and fluorescence under UV radiation. The laboratory tests highlight the presence of fungal entities from all the samples investigated. Specifically, 14 species were identified; three of them were never isolated from paper until now. Furthermore, the data gathered do not confirm the theory according to which there is a correspondence between fluorescence of the stains under UV radiation and the vitality of micro fungi.

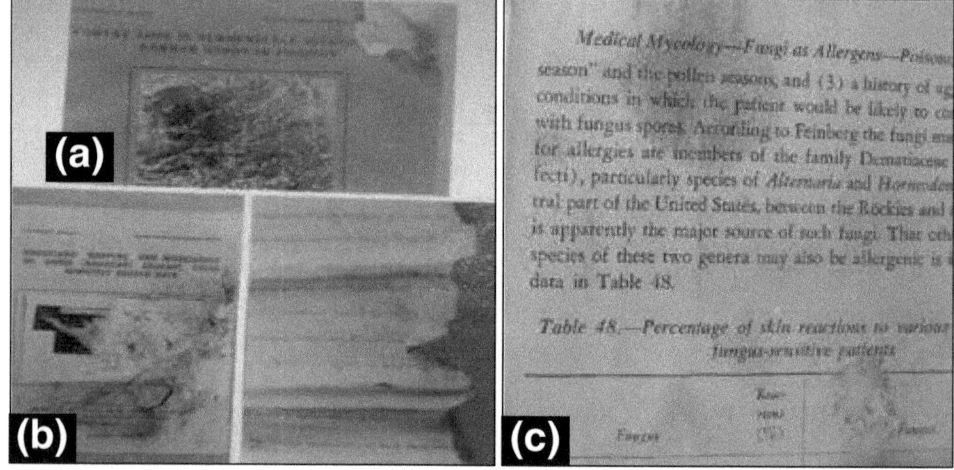

Figure 5.8a–c

(a): A project report damaged by fungi; (b) Stains produced by fungal growth; (c) Blackening of a page in book caused by *Aspergillus niger.*

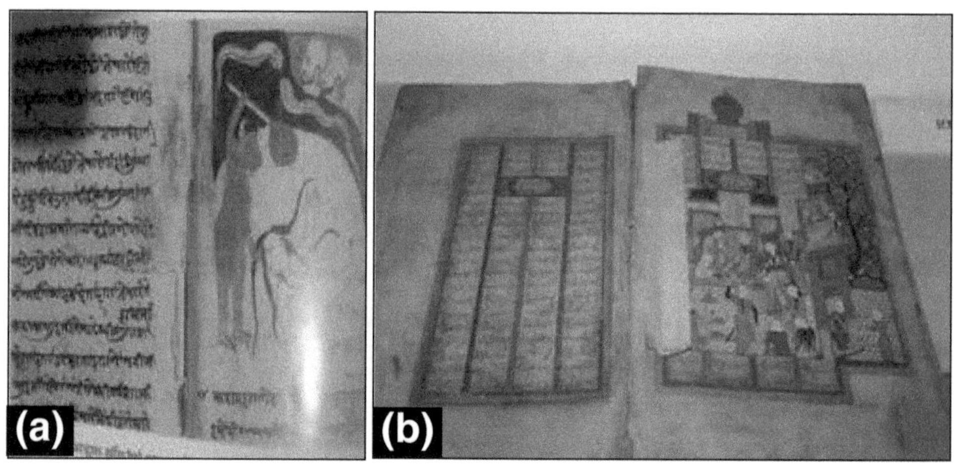

Figure 5.9: Damage Caused in Pictorial Books.

(A) Book placed in Dept. of Museology; (b) A pictorial book placed in national Museum Kolkata.

The decay of cotton, Whatman paper, and chemical pulp caused by *Trichoderma harzianum* and *Paecilomyces variotii* was studied. The structural changes of the paper were evaluated by Infrared Spectroscopy (FTIR) and Scanning Electron Microscope (SEM). The SEM results show differences in hyphae colonization and paper decay patterns between studied species under the current study; *P. variotii* caused an eroded structure in the cotton (cavity forming), whereas the initial *T. harzianum* colonization produced rupture and erosion (soft-rot decay type II) for the three types of paper, the gaps were elongated with sharp pointed ends, which consisted either of individual cavities or in chains. Moreover, FTIR results confirmed that there a relationship could be observed between fungal decay and crystalline cellulose content because the intensity of peaks at 1335 and 1111 cm^{-1} significantly decreased due to the fungal decay. Furthermore, the intensity of O–H stretching absorption slightly decreased, and this may be attributed to hydrolysis of cellulose molecules (Figure 5.10, Hassan and Mansour 2018). The Whatman paper, with cotton linter generally supported better fungal growth than rag paper or the Mezzofino paper made from cellulose and wood pulp (Michaelsen *et al.*, 2013).

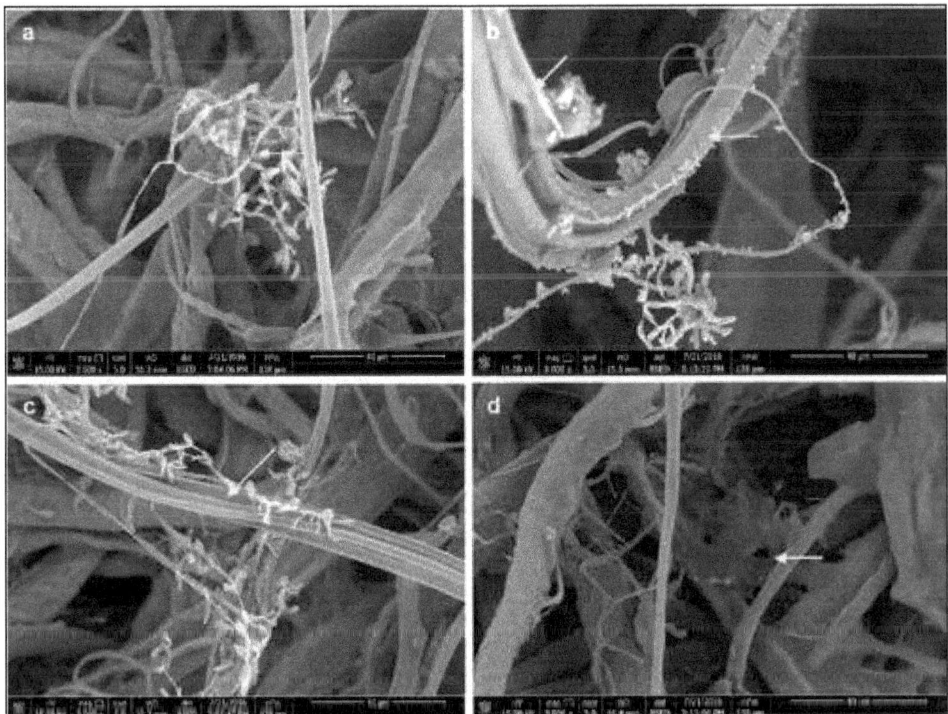

Figure 5.10: SEM Picture of Colonization of Cellulose Fibers by Fungal Hyphae and Conidia (Hassan and Mansour 2018).

5.6 Microbial Damage to Herbarium Specimens

In the herbarium collected plaints are identified by experts, pressed, and then carefully mounted to archival paper in such a way that all major morphological characteristics are visible. The mounted plants are labelled with their proper scientific names, and the name of the collector. This is useful in identification of unknown plants by comparison. Observation were done on certain sheets

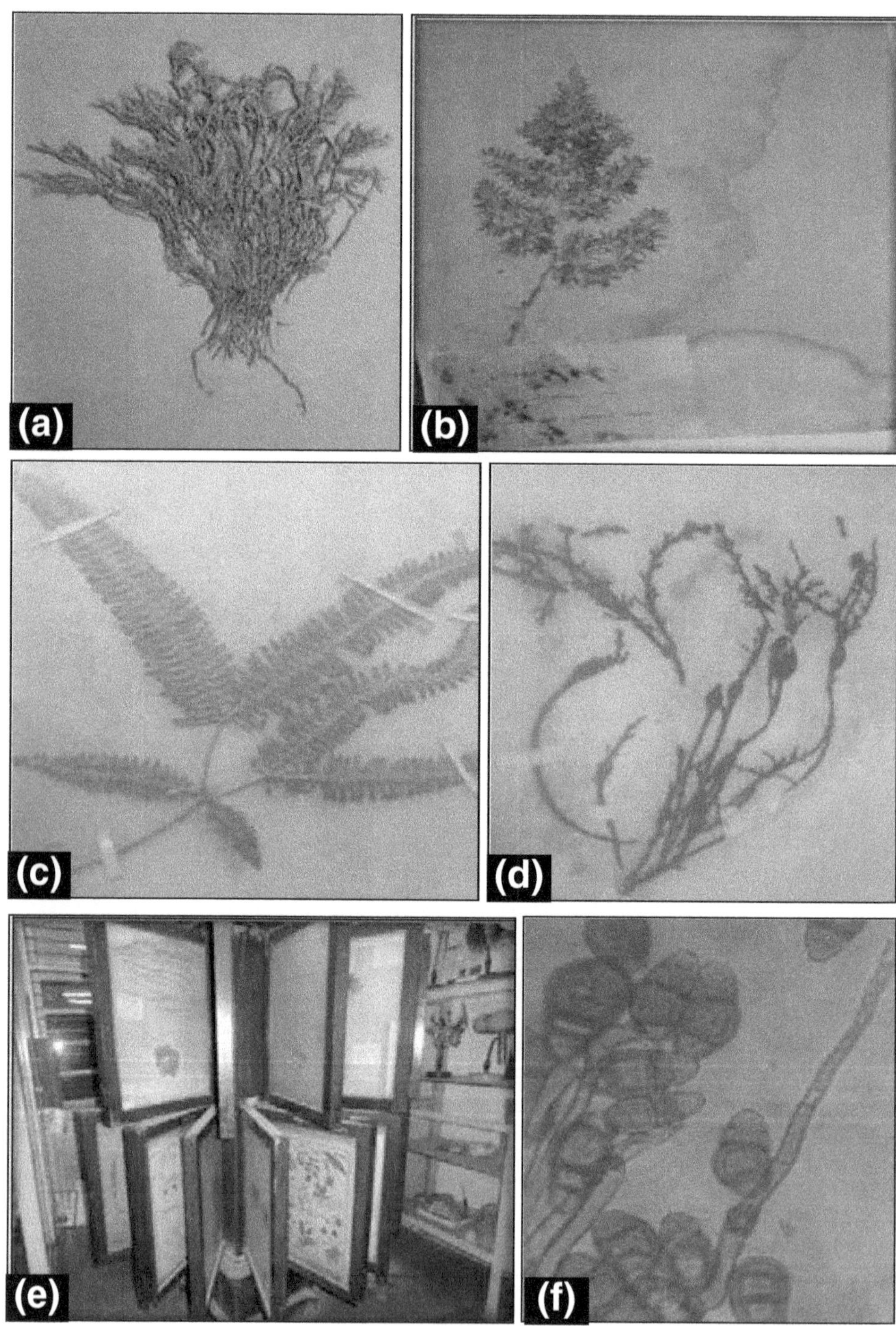

Figure 5.11: Herbarium Sheet of (a) *Selaginella lepidophylla;* (b) *Lygodium* 1908; (c) *Pteridium aquilinium;* (d) Sea weed *Ascophyllum nodosum;* (e) Display board of herbarium sheets; (f) Conidia of fungus *Curvularia.*

of **Baroherbarium,** a collection of dried specimens of the Botany department of The M.S. University of Baroda.

A herbarium sheet is a thick white paper of the size 42x29cm and plants are attached with a thread or gum. The plants or herbarium specimens are treated with 0.1 per cent solution of $HgCl_2$ to avoid any fungal infestation, then dried by pressing. A sheet having a frond of fern *Lygodium* was procured from the Cryptogamic Plant collection from Pusa, and showed the occurrence of *Curvularia pallescens* Boedijn.

This sheet ia as old as of the year 1908. Another sheet of fern *Gleichenia glauca* Sw. and *Selaginella lepidophylla* (Hook. And Grev.) Spring, showed the presence of *A. niger* and *Cladosporium* in it.

A herbarium sheet shows a brown alga belonging to *Fucaceae* called *Ascophyllum nodosum* (L.) Le Jolis. It has long tough and leathery fronds, irregularly dichotomously branched alga with large, egg-shaped air bladders set in series at regular intervals along the fronds and not stalked. The fronds can reach 2 m in length and are attached to rocks by hold fast. The fronds which were olive-green, in colour were turned to black and showed the presence of *Aspergillus niger* and *Chaetomium globosum*. A collection of herbarium was placed in Baroda Museum but due to floods in 1942, most of the specimens were sent to Baroda College. Now it is termed as Baro Herbarium.

5.7 Damage to Palm Leaf Manuscripts by Fungi

Earlier palm leaves were used as writing material. They were preserved in special decorted wooden boxes. The maximum life of these leaves is approximately 500 years. The engraving was done on the leaves and then graphite was filled to darken the written script. From an Orrian Palm leaf manuscript, isolations were made in Jan- Feb. 2004. Black coloured Dematiaceous Hyphomycetes fungus *Nigrospora sphaerica* Penz. was isolated from black coloured spots of the manuscript (Figure 5.12a). The conidia are dark black in colour and this is a leaf blight or infecting fungus producing dark pigment during growth, thus damaging the beauty of this rare museum object (Figure 5.12b).

5.8 Microbial Potential to Deteriorate Miniature Paintings

The fungi most commonly recovered from indoor air were *Aspergillus niger*, and *Penicillium citrinum*, the percentage occurrence of these two fungi was more in samples collected in 2018. Literature has suggested the association between damp environment, microbial exposure, and higher prevalence of respiratory symptoms and diseases. The study began by evaluating the airborne fungal concentrations at urban and suburban areas of a typical metropolitan city in southern Taiwan for the estimation of related health risks. A group of representative homes, based on the housing characteristics questionnaires completed earlier, were selected from two parts of the city; urban and suburban.

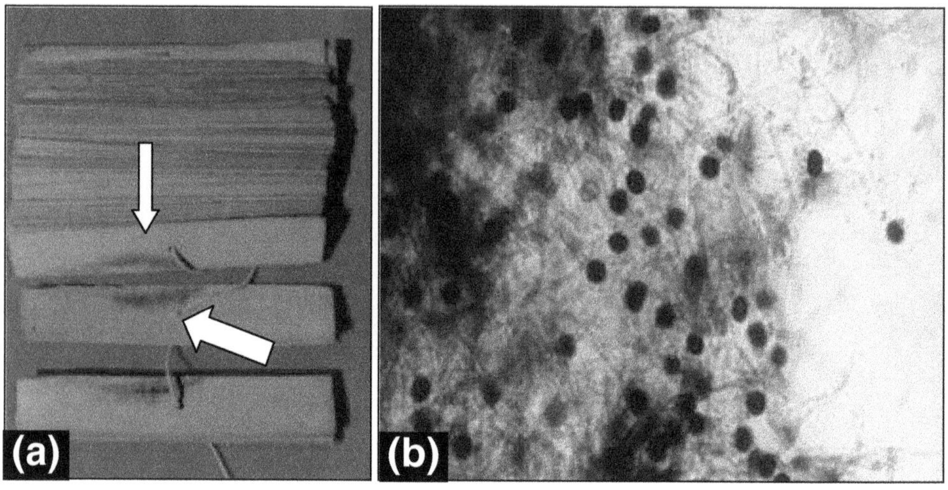

Figure 5.12a-b

(a) Blackening of Orrian Palm leaf manuscript; (b) Fungus *Nigrospora sphaerica* Penz.

Burkard sampler (BURKARD, Rickmansworth, England) was used to collect airborne fungi onto agar plates with malt-extract. The species belonging to the first group are common both in the soil and in the air, they generally develop in moderate and humid climate where organic substances are available. Hyphomycetes include species with colourless or brightly coloured colonies (hyaline Hyphomycetes) and species producing dark brown, green-black, or black colonies (Dematiaceous Hyphomycetes). Hyphomycetes secrete organic acids and can actively dissolve carbonates, they often produce different kinds of pigments, but only a few species are melanin producers (Sterflinger, 2000).

Hand made paintings drawn on paper sheets by artists and coloured with natural vegetable colours were produced in large numbers popularly known as miniature paintings. Isolations were made from a beautiful Rajasthani Miniature painting placed in Baroda Museum, which showed the presence of *Chaetomium gangligerum, Aspergillus versicolor, A. terreus, Penicillium* sp. and an actinomycetes member *Streptomyces* sp (Figure 5.13a). Another damaged Miniature painting showed the presence of Ascomycetes member cellulose degrading fungus *Chaetomium,* and damage to laccate red border was caused by silver fish (*Lepisma*) (Figure 5.13b).

Damage to paper (sulfate pulp, cotton half-stuff, and flax half-stuff) caused by the *Aspergillus niger, A. sclerotiorum,* and *Penicillium chrysogenum* is investigated by Raman spectroscopy, Fourier-transform infrared spectroscopy, and scanning electron microscopy. Its application allows one to identify the initial stages of damage from a decrease in the degree of crystallinity of the

Figures 5.13a-b

(a) A beautiful Rajasthani Miniature painting placed in Baroda Museum showed the presence of *Chaetomium gangligerum, Aspergillus versicolor, A. terreus, Penicillium* sp. and *Streptomyces* sp.; (b) A damaged Miniature painting showed the presence of, fungus *Chaetomium,* and damage to red border caused by silver fish *Lepisma.*

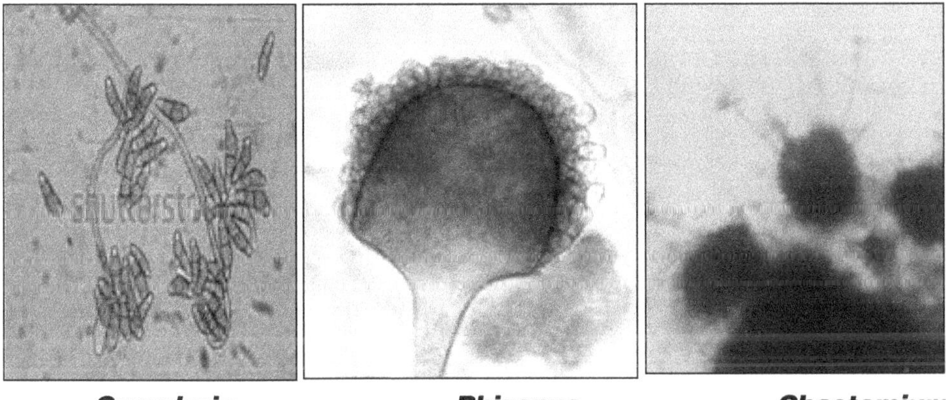

Figure 5.14: Photo Micrographs of Fungal Organisms Affecting Paper and other Objects.

cellulose contained in paper. The absorption band near 900 cm⁻¹ is used as an indicator of early stages of damage. An increase in the amide II peak at 1550 cm⁻¹ and spectral changes in the region of valence vibrations of the C–H bonds (2800–3000 cm⁻¹) are observed in the case of heavier damage. The obtained data indicate that the vibrational spectroscopy techniques are promising in the study of damage of archival document.

Using 1 per cent α-amino isobutyic acid solution in water-ethanol mixture the fungal growth caused by *Chaetomium globosum* was inhibited (Bhardwaj and Bhatnagar 2002). Earlier Elliot and Watkinson (1989) suggested its use on *Chaetomium* causing wood decay.

5.9 Percentage Loss of Eight different Papers

Percentage loss of eight different papers is recorded in Table 5.2 given below. In 25 ml of Asthana and Hawker's modified medium 'A' 0.5 g of different papers was added. After 25 days fungal mat and paper mass was determined and percentage loss was calculated (Shah and Arya 2000). It is evident that maximum percentage loss was observed in Japanese tissue paper by *Chochliobolus verruculosus* and *Aspergillus ustus*. In card board by *Penicillium citrinum* and brown paper by *Chaetomium cymbeforme* a cellulose degrading member of Chaetomiales group in Ascomycotina.

Table 5.2: Percentage Loss of Eight different Papers by Four different Fungi

Sl.No.	Type of Paper	Percentage Loss			
		C. verruculosus	P. citrinum	A. ustus	C. cymbiforme
1.	Whatmans filter paper	15	11.4	9.5	12
2.	Common	14.2	48	2.5	14.3
3.	Brown	37	24	20.6	35.6
4.	News Paper	29	16	18.8	24.4
5.	Old Manuscript	36	14.4	15.8	28.4
6.	Card board	44.4	51.4	25	22
7.	Japanese Tissue paper	49	28	49	22
8.	Nepalese Tissue paper	48	12	20	10

Source: Shah and Arya (2000).

Out of four fungi three caused negligible loss of Nepalese tissue paper and hence it can be safely used for restoration work. Verma and Johri (1992) studied cellulose decomposition of fungi and Ricelli (1999) studied the fungal growth on paper samples caused by sp. of *Aspergillus, Chaetomium, Penicillium* and *Stachybotrys* stored in high relative humidity.

5.10 Paper and Use of Adhesives

In paper conservation practice, adhesives are used for several purposes, such as mending tears and gaps, or paper consolidation. The criteria to choose one or another adhesive should be based on the knowledge of the properties and stability of those adhesives. However, the several different adhesives available on the market still lack enough information to help the process of a rational decision-making. In the present work, five adhesives currently used in the paper conservation field (starch paste, unsupported *Archibond*, carboxymethylcellulose, hydroxyl-propylcellulose and methylcellulose) were analysed for their chemical stability and fungal bioreceptivity (the ability of a material to be colonized by fungi). Bioreceptivity of products used in conservation and restoration is a still poorly explored subject, despite its great relevance for the preservation of objects. The chemical and physical properties of the adhesives, before and after moist heat artificial ageing, were analysed by thermogravimetry, capillary viscometry, measurement of water absorption capacity, colorimetery, and pH measurement.

Fungal bioreceptivity of the adhesives was tested on two different substrates (paper and glass) against three fungal species: *Aspergillus niger, Aureobasidium pullulans* and *Penicillium pinophilum*. Along 56 days of incubation, the colonization area on the adhesives was measured through digital photo analysis.Starch paste was the most bioreceptive adhesive, but on other hand was also the most stable adhesive to artificial ageing, regarding colour alteration, degree of polymerization and pH.

Carboxymethyl cellulose and *Archibond* showed chemical deterioration with ageing. Nevertheless, these two adhesives presented only scarce bio receptivity to the tested fungi. Methylcellulose and hydroxypropyl cellulose showed the best relationship between higher chemical stability with artificial ageing and lower fungal bio receptivity.

5.11 Management Strategies

The good news is we can ward off microbial infestations by storing artefacts in conditions that don't encourage spores to settle and germinate. This means keeping things cool and dry. "In an indoor environment, like a museum or a library...you can control the temperature, humidity, and light," Palla says. Conservators also use UV-C fans to filter spores from the air and zap them with ultraviolet light.

Collections that are open to the public can't be chilled very much. "It's not very practical in museum settings, where you actually want people to see objects without being cold," DePriest says. One exception is Ötzi the Iceman, a natural mummy who froze in the Italian Alps 5,300 years ago. He's kept in a cold cell in the South Tyrol Museum of Archaeology in Italy that emulates the glacier he was discovered in.

In Oriental institute of the M.S. University Godavach (*Acorus calamus*) rhizome is used in cupboards in clean white cloth (Figure 5.16a). The oil was good for polishing the palm leaf manuscript (Figure 5.12). Eucalyptus leaves may be effective due to the presence of Cineole in them Arya (1988). It has been observed that the activity of extracts may be lost due to autoclaving (Lokesha *et*

Figures 5.16a–e

(a) Small packets made of white cloth containing rhizome powder of; (b) *Acorus calamus*, which worked as antifungal and insect repellent; (c) *Cymbopogan;* (d) *Adenocalymma alliaceum*; (e) *Datura metal* used as eco-friendly alternatives.

al., 1986). Methanolic extract of *Adenocalymma alliaceum* or garlic wine showed wide spectrum of antifungal activity (Sharma and Srivastava 2014). Fumigation by Bactygas in museums or library (Florian 2000; Gatenby and Townley 2003). Gray and Martin (1947) reported use of fungicide Dowicide G at a level of 0.7 per cent incraft paper. The leaching in paper can be prevented by asphalt coating 3 per cent pentachloro phenol. To prevent the fungal invasion in libraries and books fumigation is recommended (Gray 1959).

Three different conservation treatments, namely freeze-drying, gamma rays, and ethylene oxide fumigation, were compared and monitored by Michaelsen *et al.* (2013) to assess their short- (one month) and long-term (one year) effectiveness to inhibit fungal growth. Denaturing gradient gel electrophoresis technique was used to study the structure of the microbial communities colonizing artworks (Gonzalez and Saiz-Jimenez, 2004).

After the inoculation with fungi possessing cellulose hydrolysis ability – *Chaetomium globosum*, *Trichoderma viride*, and *Cladosporium cladosporioides* – as single strains or as a mixture, different quality paper samples were treated and screened for fungal viability by culture-dependent and -independent techniques. Results derived from both strategies were contradictory. Both gamma irradiation and EtO fumigation showed full efficacy as disinfecting agents when evaluated with cultivation techniques. However, when using molecular analyses, the application of gamma rays showed a short-term reduction in DNA recovery and DNA fragmentation; the latter phenomenon was also observed in a minor degree in samples treated with freeze-drying. When RNA was used as an indicator of long-term fungal viability, differences in the RNA recovery from samples treated with freeze-drying or gamma rays could be observed in samples inoculated with the mixed culture. Only the treatment with ethylene oxide proved negative for both DNA and RNA recovery.

5.12 Conclusion

The study of fungal population or community dynamics on cellulosic objects, before and after restoration, needs more investigations and these outcomes could make an important contribution to the science of Conservation in Cultural Heritage. SEM studies showed colonization of cellulose fibres by fungal hyphae. Table 5.2 clearly showed that degradation of all the 8 different papers occurred by test fungi. *In vitro* by dry weight method loss results were observed for *Cochliobolus verruculosus*, *Aspergillus ustus*, *Penicillium citrinum* and by *Chaetomium cymbeforme*. Maximum percentage reduction was observed in case of cardboard due to *P. citrinum*. Certain preventive and control methods using plant products like rhizome powder of *Acorus calamus* and leaf extract of *Eucalyptus*, *Adenocalymma alliaceum* and *Cymbopogon citratus* are suggested.

REFERENCES

Allsopp D, Seal K, Gaylarde C (2004) Introduction to biodeterioration. Cambridge University Press, Cambridge, pp 1–10

Arai H (1987) Microbiological studies on the conservation of paper and related cultural properties. Part 5. Physiological and morphological characteristics of fungi isolated from foxing, formation mechanisms and countermeasures. Sci Conserv 26: 43–52

Arai H (2000) Foxing caused by fungi: twenty–five years of study. Int Biodeter Biodegr 46: 181–188. doi.org/10.1016/S0964-8305(00)00063-9

Arai H (2014) Microbiological problems in biodeterioration of cultural property: Forty years of study.In: Biodeterioration of cultural property-7. (eds) S. Dhawan, Abduraheem K. and V. Nath. Pub. by ICBCP, Lucknow.1-10pp.

Aranyanak C (2005) Fungal stains on paper In: Biodeterioration of Cultural Property-4. (eds) O.P. Agrawal and S. Dhawan. Pub. by ICBCP, Lucknow.40-50pp.

Arya A. (1988) Control of *Phomopsis* fruit rots by leaf extract of certain medicinal plants. In : Indigenous medicinal Plant Symp. (Kaushik M.P. ed) Pub. by Today and Tomorrow's Printers and Publishers New Delhi 41-46pp.

Arya A. (2006) Role of fungi in biodeteriorarion of museum objects In; Recent Mycological Researches by Sati S.C.(ed.) Pub by I.K. Int. New Delhi 260-271

Arya C, Arya A (2007) Aeromycoflora of fruit markets of Baroda, India and associated diseases of certain fruits. Aerobiologia, 23: 283-289. doi: 10.1007/s10453-007-9070-2

Baggaley K (2017) The world's art is under attack–by microbes. Science June,. 21, 2017

Bhardwaj N, Bhatnagar IK (2002) Microbial deterioration of paper- painting. In: Biodeterioration of materials- 2. R.B. Srivastava, G.N. Mathur and O.P. Agrawal (eds). Pub. by ICBCP 132-135.

Borrego S, Guiamet P, Gómez de Saravia S, Battistini P, Garcia M, Lavin P, Perdomo I (2010) The quality of air at archives and the biodeterioration of photographs. Int. Biodeter. Biodegr. 64(2): 139–145.

Borregoa S, Guiametb P, Vivara I, Battistonib P (2018) Fungi involved in biodeterioration of documents in paper and effect on substrate. Acta Microscopica. 27(1): 37- 44

Chadeganipour M., Ojaghi R., Rafiei H., Afshar[M., Hashemi S.T (2013) "Biodeterioration of library materials: study of fungi threatening printed materials of libraries in Isfahan University of Medical Sciences in 2011". Jundishapur J. Microbiol. 6(2): 127-131.

Choi S (2007) Foxing on paper: a literature review. J Am Inst Conserv 46: 137–152. https://doi.org/10.1179/019713607806112378

Chivers A H (1915) A monograph of the genera *Chaetomium* and *Ascotricha*. Memoirs of the Torrey Botanical Club, June 10, 1915, 14 (3A): 155-240

Christensen C, Papavizas GC, Benjamin CR (1959) A new halophilic species of *Eurotium*. Mycologia 51: 636–640

Deacon J (2005) Fungal Biology. Blackwell Publishers; Cambridge, Massachusetts: 2005.

Domsch K H, Gams W, Anderson T H (1980) Compendium of soil fungi. 1980, London

Elliot ML, Watkinson SC (1989) The effect of α-amino isobutyric acid on wood decay and wood spoiling fungi. Int. Biodeterioation 25: 355-371

Florian ML (1997) Heritage eaters, insects and fungi in heritage colletions. James and James, London

Florian ML (2000) Aseptic technique: A goal to strive for a collection recovery of moldy archival materials and artifacts. JAIC 39: 107-115

Gallo F, Pasquariello G, Rocchetti F (1998) Biological Investigation on Sizings for Permanent Papers. Restaurator, 19: 61.

Gatenby S, Townley P (2003) Preliminary research into use of the essential oil of Melaleuca alternifolia (tea tree oil) in Museum conservation. AICCM Bulletin. 67-70

Gray W.D. (1959) The relation of Fungi to human affairs. Pub. by Henryholt and Inc. NewYork, 506pp.

Gutarowska B (2016) A modern approach to biodeterioration assessment and the disinfection of historical book collections. Lodz University of Technology

Hassan R.A. and Mansour M.A. (2018) Journal of Polymers and the Environment 26, : 2698–2707

Joshi B.R.(1989) Preservation of palm leaf manuscripts. Con. of Cul. Prop. New Delhi 22: 120-127

Joshi K.B.(1947) Paper making as cottage industry, Maganvadi, Wardha 226pp.

Khan A.A.H., Karuppayil S.M (2012) "Fungal pollution of indoor environments and its management". Saudi J. Biol. Sci. 19(4): 405-426.

Kraková L, Šoltys K, Otlewska A, Pietrzak K, Purkrtová S, Savická D, Puškárová A, Buèková M, Szemes T, Budiš J, Demnerová K, Gutarowska B, Pangallo D (2018) Comparison of methods for identification of microbial communities in book collections: culture–dependent (sequencing and MALDI–TOF-MS) and culture–independent (Illumina MiSeq). Int Biodeter Biodegr 131: 51–59. https://doi.org/10.1016/j.ibiod.2017.02.015

Kulshrestha A, Chauhan S V S(2001) Aeromycoflora of some hospitals of Agra city. Indian J. Aerobiology. 14 (1,2): 33-35.

Lokesha S, Vasantha K, Shetty H S (1986) Effect of plant extracts on the growth and sporulation of *Aspergillus flavus*.Pl. Dis. Reptr. 1(1-2): 7981c f. Rev. Pl. Path. 67(1988): 4864

Melo D, Sequeira S O, Lopes J A, Macedo M F (2019) Stains versus colourants produced by fungi colonising paper cultural heritage: a review. J Cult Herit 35: 161–182. https: //doi.org/10.1016/j. culher.2018.05.013

Mesquita N, Portugal A, Videira S, RodríguezEcheverría S, Bandeira A M L, Santos MJA, Fritas H, (2009) Fungal diversity in ancient documents. A case study on archive of the University of Coimbra. Int. Biodeter. Biodegrad. 63: 626-629.

Micheluz A, Manente S, Tigini V, Prigione V, Pinzari F, Ravagnan G, Varese GC (2015) The extreme environment of a library: xerophilic fungi inhabiting indoor niches. Int Biodeter Biodegr 99: 1–7. https: //doi.org/10.1016/j. ibiod.2014.12.012

Micheluz A, Manente S, Rovea M, Slanzi D, Varese GC, Ravagnan G, Formenton G (2016) Detection of volatile metabolites of moulds isolated from a contaminated library. J Microbiol Methods 128: 34–41. https: //doi. org/10.1016/j.mimet.2016.07.004

Michaelsen A, Pinzari F, Ripka K, Lubitz W, Piñar G (2006) Application of molecular techniques for identification of fungal communities colonizing paper material. Int Biodeter Biodegr 58: 133–141. https: //doi.org/10.1016/j. ibiod.2006.06.019

Michaelsen A, Pinzari F, Barbabietola N, Piñar G (2013) Monitoring the effects of different conservation treatments on paper-infecting fungi. Int Biodeterior Biodegradation. 84: 333–341. doi: 10.1016/j.ibiod.2012.08.005: 10.1016/j. ibiod.2012.08.005

Modica A, Bruno M, Di Bella M, Alberghina MF, Brai M, Fontana D, Tranchina L (2019) Characterization of foxing stains in early twentieth century photographic and paper materials. Nat Prod Res 33: 987–996. https: //doi. org/10.1080/14786419.2016.1180600

Montanari M, Melloni V, Pinzari F, Innocenti G (2012) Fungal biodeterioration of historical library materials stored in compactus movable shelves. Int Biodeter Biodegr 75: 83–88. https: //doi.org/10.1016/j.ibiod.2012.03.011

Nair SM (1972) In : Conservation in Tropics. Proc. Of AST- pacific seminars on con. of Cul. Properties (ed. Agrawal O.P.) Rome: 150-158.

Oetari A, Natalius A, Komalasari D, Susetyo-Salim T, Sjamsuridzal W (2018) Fungal deterioration of old manuscripts of European paper origin. AIP Conference Proceedings 23, Oct. 2018. 020156; https: //doi. org/10.1063/1.5064153

Pinzari F, Pasquariello G, Mico A.D. (2006) Biodeterioration of Paper: A SEM Study of Fungal Spoilage Reproduced Under Controlled Conditions Macromol. Symp. 238: 57–66 DOI: 10.1002/masy.200650609

Pinzari F, Gutarowska B (2021) Extreme Colonizers and Rapid Profiteers: The Challenging World of Microorganisms That Attack Paper and Parchment. In: Microorganisms in the Deterioration and Preservation of Cultural Heritage, E. Joseph (ed.), 79-113. https: //doi.org/10.1007/978-3-030-69411-1_4

Pinzari F, Montanari M (2011) Mould growth on library materials stored in compactus-type shelving units (Chapter 11). In: Abdul-Wahab Al-Sulaiman SA (ed) Sick building syndrome in public buildings and workplaces. Elsevier, Burlingtonpp. 193-206.

Pinzari F, Zotti M, De Mico A, Calvini P (2010) Biodegradation of inorganic components in paper documents: formation of calcium oxalate crystals as a consequence of *Aspergillus terreus* Thom growth. Int Biodeter Biodegr 64: 499–505. https: //doi.org/10.1016/j.ibiod.2010.06.001

Polo A, Cappitelli F, Villa F, Pinzari F (2017) Biological invasion in the indoor environment: the spread of *Eurotium halophilicum* on library materials. Int Biodeter Biodegr 118: 34–44. https: //doi.org/10.1016/j.ibiod.2016.12.010

Prescott S C, Dunn C G (2011) Industrial Microbiology, Agrobios India, Jodhpur

Riecelli A. (1999) Fungal growth on samples of paper: Inhibition by new antifungals, Restaurator20 (2): 97-107

Rakotonirainy M, Hanus J, Bonassies-Termes S, Heraud C, Bertrand Lavedrine B (2003) Detection of fungi and control of disinfection by ATP. biolumlnescence assay. AICCM Bulletin, 28(1): 16-22. http: //dx.doi.org/10.1179/bac.2003.28.1.004

Rakotonirainy MS, Heude E, Lavedrine B (2007) Isolation and attempts of biomolecular characterization of fungal strains associated to foxing on a 19th century book. J Cult Herit 8: 126–133. https: //doi.org/10.1016/j.culher.2007.01.003

Sartory A, Sartory R, Meyer J, Baumli H (1935) Quelques champignons inferieurs destructeurs du papier 38: 529-542.

Sclocchi MC, Kraková L, Pinzari F, Colaizzi P, Bicchieri M, Šaková N (2016) Microbial life and death in a foxing stain: a suggested mechanism of photographic prints defacement. Microb Ecol 73: 1–12. https: //doi.org/10.1007/s00248-016-0913-7

See P (1919) Les maladies du papier piquk. Les champignons que les provoquent, Les modes de preservation O, Doins et Fils, Paris 168pp

Shah NR, Arya A (2000) Studies on some fungal biodeteriogens by Bhartiya Kala Prakashan, New Delhi, pp202.

Sharma N, Srivastava MP (2014) Renovation of cultural heritage with Methanolic plant extracts of *Adenocalymma alliaceum* Miers In: Biodeterioration of cultural property-7. (eds.) S. Dhawan, Abduraheem K. and V. Nath. Pub. by ICBCP, Lucknow.35-46pp.

Sindall RW (1919)The manufacture of paper, Constable and Co. Ltd. 10 and 12, Orango street, Leicester Square. W.C. 276 pp.

Singh A (2014) Biodeterioration of paper manuscripts: A review. Paper presented in National Seminar on Biodeterioration of cultural property and conservation of heritage buildings. Sept. 21-22, 2014.

Sterflinger K (2000) Fungi as geologic agents. Geomicrobiology Journal 17(2): 97–124. 10.1080/01490450050023791

Sterflinger K (2010) Fungi: Their role in deterioration of cultural heritage. Fungal biology review 24: 47-55. Doi: 10.1016/j.fbr. 2010.03.003

Szulc J, Otlewska A, Ruman T, Kubiak K, Karbowska-Berent J, Kozielec T, Gutarowska B (2018) Analysis of paper foxing by newly available omics techniques. Int Biodeter Biodegr 132: 157–165. https://doi.org/10.1016/j.ibiod.2018.03.005

Verma SC, Johri BN (1992) Cellulolysis by different fungi. Proc. Of 79th I.S.C. Baroda. Section- Botany.18pp

Verona O (1938) A propos des causes microbiennes de domage au papier et aux livres, Bull, Soc. Int. microbial.Sez.Ital. 10: 91-92.

Zotti M, Ferroni A, Calvini P (2008) International Biodeterioration and Biodegradation: 569-578.

Chapter 6

Biodeterioration of Natural History Collection with Special Reference to Avian Collection of State Museum Lucknow

Ameeza Zarrin and Abduraheem K.

Department of Museology, Aligarh Muslim University, Aligarh, Uttar Pradesh, India
e-mail: ameeza.zarrin8@gmail.com; ka_rahim_65@hotmail.com

ABSTRACT

State Museum Lucknow harbours one of the oldest, richest, as well as rare collections of Natural History in India with more than six thousand specimens in acquisition. According to the data most of the collection was made in between 1870 and 1880. Due to negligence and inadequate stuff, the natural history collection in museum has come under serious threat of biodeterioration as the collection remained closed for several years which provide sound environment for the sustenance of potential insect pests. Natural history collections are particularly vulnerable to pest attack because much of the collection is composed of edible plant material and animal protein. Damage by insect pests to the avian collection is quite common as many of the insect pests are fond of the protein and keratin found in skin, feathers, nails, and beaks of the birds. Major damage to the avian collection is caused by certain species of dermestid beetles. This paper includes identification of insect pest, their feeding habits and damage extent which is responsible for biodeterioration of the collection. Identifying pest has been the key to successful pest control. Severe damage caused to avian collection due to above mentioned insect pests will be shown, and their control may include non-chemical and chemical measures in case of severe infestation or damage. But different approaches towards the preventive conservation which should be considered while managing to safeguard the significant avian collection are also discussed in detail.

Keywords*: Natural history, Collection, Biodeterioration, Pest identification, Preventive conservation.*

6.1 Introduction

State Museum Lucknow is the fourth oldest Museum of India established in 1863. It comprises of the oldest, richest, as well as rare collections of Natural History in India with more than 6000 specimens in acquisition. Natural History section has specimens belonging to Critically Endangered, Endangered, Vulnerable, Least Threatened category as per IUCN Red Data list. Natural history collections preserved in museums are significant because natural resources reservoir is diminishing continuously due to the impact of human activities such as hunting, poaching, habitat destruction, pollution *etc.* Moreover, in India there is a little scope of further expansion of natural history collection because of different constitutional laws regarding wild life conservation *i.e.* Wild Bird Protection Act (1887), Indian Forest Act (1927), Prevention to Cruelty to Animals Act (1960), Wild life Protection Act (1972), CITES (1976), Forest Conservation Act (1981) *etc.* Due to which there are only few examples of specimens' acquisition in museums and after enactment of above mentioned acts these few examples are results of collaboration of museums with Zoo, Nature reserves, biosphere reserves from where the deceased animals were acquired. Followed by their preservation in museum for the purpose of display, education and research. But since last few decades this practice had also been discontinued which left no scope of further expansion. Such repositories are increasing in value and serving a wide variety and diverse range of disciplines prominently in the field of nature conservation. Natural history collections (NHCs) are important resources for a diverse array of scientific fields (Sara *et al.*, 2020). Therefore, Natural history collection of State Museum Lucknow is very important as it is rich and valuable in terms of research and display. This is only collection of its type in State of Uttar Pradesh. Therefore, care must be taken for its conservation in order to pass on to future generations.

6.2 Natural History Section of State Museum Lucknow

The section has specimens of different genera of Vertebrates and Invertebrates, some of these have crossed Century in terms of age. A wide diversity of birds and animals from Indian Subcontinent are well preserved in Natural History Section through wet and dry preservation techniques. The Collection includes cabinet skin and mounted skin of birds, stuffed animals, eggs, horns, heads, tusk, shells, Ivory objects *etc.* More than three hundreds of horns of various Bovid and Cervids like Wild Water Buffalo, Chiru, Barah Singha, different *species* of Himalayan goats, Nilgiri tahr *etc.* are preserved. Apart from these stuffed heads of Wild boar, Bison, Hippopotamus, Chousingha and a number of dear species and elephant tusk, rhino horn are worth mentioning. A. O. Hume's collection of around 400 eggs belonging to different species ranging from ostrich, emu, and vultures to Sunbird and Munia. Mammals in collection include Asiatic Lion, Royal Bengal Tiger, One Horned Rhinoceros, Duckbilled Platypus (Echidna), Kangaroo, Purple Faced Monkey, Golden Leaf Monkey, Pocket Monkey, Leopard, Jungle Cat, Jackals, Wild Dog, Indian Fox,

Ant Eater, Porcupine, Sloth Bear, Giraffe, Zebra, Mouse Deer, Civets, Wild Goats, Hippopotamus, Bison Head and Various species of Squirrels. Reptiles including Mugger, python, King Cobra, Rattle snake and various other species of snakes, Star Tortoise, Gecko/Indian Monitor Lizard, Mounted Skin of Python. Among wet preserved animals large number of snakes followed by fishes and insects, Foetuses of the mammalian species like human being, sheep, tiger, pig *etc.* are very important. Natural history section also incorporated with some Anthropological as well as Ethnographical collection like ethnic objects of carved ivory and bangles of ivory, objects made of bone, wood, feather, shell, natural fibre, long leaves *etc.* belonging to different tribes of Andaman Nicobar, Bundel Khand, Uttaranchal. Costumes, attire, daily use utensils, agricultural implements, fishery equipment's and other items of ethnic value are also included. The Natural history section of museum Lucknow, also has a mummy of a 13-year-old girl which was purchased in 1952 from a UK National J. J. E. Potter, which belongs to dynasty 22-25. Since last many decades this mummy has been an attraction to many visitors. Continued documentation and examination of biological diversity is essential to successful conservation efforts. Museum collections and researchers will continue to be at the forefront of this research (Winker *et al.*, 1991). The Ph.D. research work was conducted at State Museum of Lucknow during 2011-2014 (Zarrin 2016).

6.3 Avian Collection at Reserve Section

Reserve section of natural history comprises of both kinds *i.e.* dry and wet preserved specimens. As stated, earlier bird collection is a major collection of natural history section therefore out of four rooms of reserve section three are completely occupied by ornithological collection including about four thousand birds and four hundred eighteen eggs where the number of study skins are much greater than that of mounted skins. Bird collection comprises of some critically endangered, endangered, threatened, near threatened, vulnerable birds like Great Indian bustard (*Ardeotis nigriceps*), Baer's pochard (*Aythya baeri*), Siberian crane (*Grus leucogeranus*),White-rumped vulture (*Gyps bengalensis*), Indian vulture (*Gyps indicus*), Bengal florican (*Houbaropsis bengalensis*), Pink-headed duck (*Rhodonessa caryophyllacea*), Red-headed vulture (*Sarcogyps calvus*), Sociable lapwing (*Vanellus gregarius*), Nordmann's green shank (*Tringa guttifer*), Greater Adjutant (*Leptoptilos dubius*), Swamp Francolin (*Francolinus gularis*), Eurasian curlew (*Numenius arquata*), Black and orange Flycatcher (*Ficedula nigrorufa*),Andaman Wood Pigeon (*Columba palumboides*), Nilgiri Wood Pigeon (*Columba elphinstonii*),Grey Headed Bulbul (*Pycnonotus priocephalus*), Indian Spotted Eagle (*Aquila hastate*), Saker Falcon (*Falco cherrug*), Red necked Falcon (*Falco chicquera*),White-naped Tit (*Parus nuchalis*) and so on. Most prized possessed specimen is the Pink headed duck. Salim Ali did not mention about the specimen preserved in State Museum Lucknow, he only mentioned about the specimens of pink headed duck available at Bombay Natural History Society (BNHS) and Zoological Survey of India (Ali 1960). Specimen of pink headed duck preserved in State Museum Lucknow was discovered and reported by the author during her research at the State Museum Lucknow in the year 2014.

6.4 Conservation Status of Avian Collection in the State Museum Lucknow

During the research work dealing with the conservation status of avian collection of State Museum Lucknow (SML), individual condition of the specimens were analysed and recorded. Suggestions were made regarding curative conservation and need of preventive measures. In reserve collection some birds were in good condition and needs only preventive conservation like use of insect's repellent, fumigation, proper cleaning or dusting and more over the proper up keeping. These birds included *Chlidonias hybrida, Hierococcyx various* (Cuckoo/hawk), *Oriolus oriolus* (golden oriole), *Oriolus xanthornus* (black headed oriole), *Acridotheres tristis, Sturnus vulgaris, Garrulus lanceolatus, Sturnus pagodarum, Bubo zeylonensis etc.* On the other hand, some birds were damaged, such damage included falling off feathers, broken/missing/damaged head, tail, wing(s) and leg(s). These specimens required proper curative conservation. There were also some specimens which were badly damaged and beyond the conservation, for instance- *Zosterops palpebrosa* (white eye), *Phylloscopus collybita, Sylvia curruca, Prinia socialis, Chaimarrornis leucocephalus, Molpastes leucogenys, Necterina asiatica, Motacilla alba etc.* Subsequently the Department of Museology, A.M.U. Aligarh undertook a project in 2022 to conserve the avian collection of State Museum at Lucknow.

6.5 Biodeterioration of Avian Collection

Natural history collections are particularly vulnerable to pest attack because much of the collection is composed of edible plant material and animal protein (Pinniger and Harmon 1999). Damage by insect pests to the avian collection is quite common as many of the insect pests are fond of the protein and keratin found in skin, feathers, nails and beaks of the bird. Major damage to the avian collection may be caused by certain species of dermestid beetles. Anon (1885) has reported that 20000 avian specimens of A.O. Hume's collection had been destroyed by dermestid beetles.

Falling off feathers is the most common problem found in the avian collection of State Museum Lucknow. In order to avoid this problem one must understand the structure or composition of feathers. Feathers are composed of about 91 per cent protein, 8 per cent water and 1 per cent lipids. The type of protein found in feathers is known as Keratin, a sulphur containing fibrous protein. The structure of the keratin gives the feather its strength and suppleness.

Due to lack of trained staff, inadequate facilities the collection remain undisturbed or untouched for past several years which provide sound environment for the sustenance of insect pests. No doubt specimens on open display are more vulnerable to pest attack but this may be more serious to the specimens kept undisturbed in the sealed drawers as they allow undiscovered and uninterrupted breeding and feeding activities of pest for long period.

Various dead zones and hidden places are often present in storage areas like corner, bottom of shelves, gaps between the drawers, behind the standing showcases. These places accommodate them with all comfort, usually during day time they become successful in hiding themselves being inactive while become active during night or in dark condition as they are nocturnal species. Thus use of well designed, dust resistant, insect resistant display/storage showcases found very important.

Along with routine surface sweeping, cleaning of such hidden places are very important, so they must be thoroughly cleaned periodically. And pest existing signs like pile of fine dust, powdery substance, exoskeleton of insects, casing, rodent faeces, dead or live insects must be keenly observed near the specimens and removed. *"Pests come into conflict with man because they compete with us for resources. Ever since man first started to store food or make clothing and other artefacts he had faced this battle"* (Pinniger 2015).

6.5.1 Museum Pests

There are various common museum pests like beetles, moths, silverfish, lice, rodents and other mammals. Their feeding habit is the most direct threat to museum collection as they are attracted to proteinaceous materials. In order to flourish pest need a food source, harbourage, warm and damp conditions. All moths and beetle progress through egg, larva, pupa and adult stages.

In many cases rather than the adult, larvae are responsible for major destruction to museum objects mainly of organic nature. However, there are also some exceptions such as silver fish, where adult is responsible. Clothe moths and carpet beetles are among the very few insects, fungi and microorganisms that are capable of digesting keratin, a protein component of feathers, hair, fur, horns, antlers, hooves, nails and beaks. Indirect threats from insects can also be as severe as feeding damage caused by pest like many types of insects like flies and spiders which may not harm the collection but when they die, their dead bodies provide sustenance for Dermestid beetles and other scavengers and decomposing bodies may exude acidic substance that damage precious collection.

6.5.2 Pest Identification

The first step is to recognise and identify the pest seen within the museum area particularly from the area of storage and exhibition. The correct identification of the pests provide us various significant clues about their feeding and breeding, preferable environment, life cycle *etc.* with the help of such information we can control them efficiently. Biological understanding of pest makes us well equipped to handle potential pest within the museum. In addition to feeding destruction, their excretion or secretion can also be harmful to the collection. In this regard we have also inspected showcases and storage cabinets of reserve collection of natural history section of State Museum Lucknow during this research work and successfully collected following museum pests *i.e.* Shiny Spider Beetles (*Gibbium psylloides*), Adult and larvae of Varied Carpet Beetles (*Anthrenus verbasci*), Silver Fish (*Lepisma saccharina*), Larvae of Cloth Moth (*Tinea pellionella*), Furniture Carpet Beetles (*Anthrenus flavipes*) and Black Carpet Beetle (*Attagenus unicolor*).

1. Shiny Spider Beetle (*Gibbium psylloides*): Figure 6.1

It is classified under the Family **Anobiidae.**

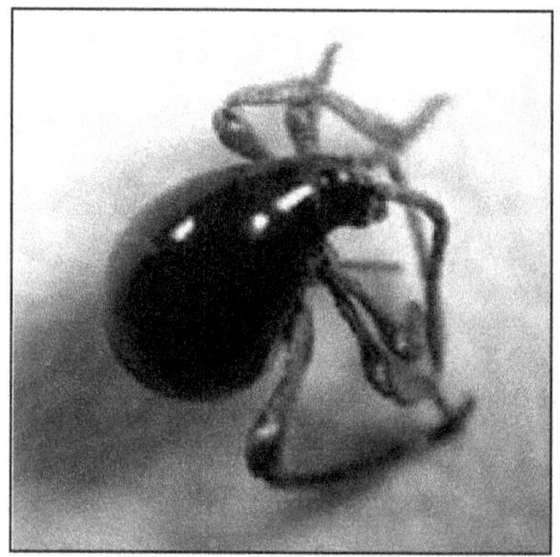

Figure 6.1: Shiny Spider Beetle *Gibbium psylloides.*

☆ It is a small insect which is about 1.5-3.5 mm long. Body colour is completely dark reddish brown. The shiny spider beetle has a round abdomen and very long appendages. The head is relatively smaller and thus sometimes appears headless. Looks superficially like spiders therefore called as spider beetle. The shiny spider beetle is a scavenger. This particular beetle cannot fly but quite is mobile, wander on walls and floor at night. A particular characteristic is their ability to stay alive at low temperature even at freezing point without nourishment.

Feeding Habit and Damage Extent

They have been reported to feed on both animal and plant materials. They can feed on dead insects. The larvae often feed on hair, fur, feathers, wool, paper and a variety of food products. They may feed upon droppings of rodents and even on carcasses of rodent as well. They usually avoid light and are active at night. They can thrive in storage areas and other places where stored items can become damp and humid.

Control Measures

☆ Debris should be removed from nesting birds, rodents and insects particularly nests near the vulnerable collections.

☆ Humidity control is important in order to prevent method of infestations. It is sometimes necessary to use a dehumidifier.

Chemical Control Measures

Residual insecticide like Bendiocarb can be applied for the control.

2. Varied Carpet Beetle (*Anthrenus verbasci*) Figure 6.2

It comes under the Family **Dermestidae**

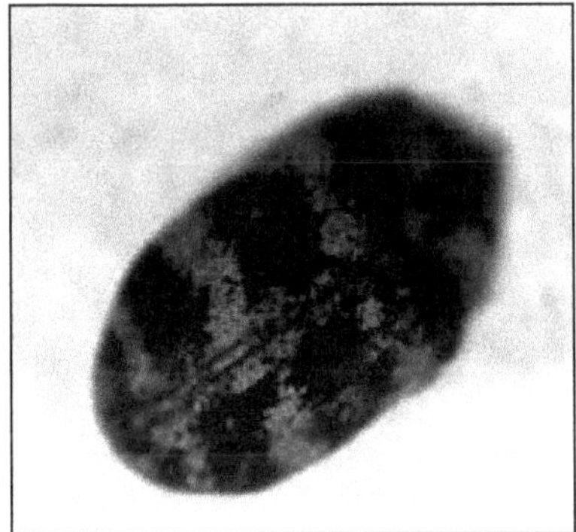

Figure 6.2: Varied Carpet Beetle
Anthrenus verbasci.

> ☆ It measures about 2.5mm long body is covered with scales of different colours like white, brown and dark brown. Scale pattern is irregular. Older adults may appear solid brown or black as the scales wear off with time. Adult may deposit eggs on stuffed animals, leather, animal hides, furs, feathers, animal horns, bones, hair, silk, wool carpets, rugs, dried plant products, as larvae is fond of such materials as its food source. Adults generally appear in spring or early summer; indoors, often found near windows. They may fly fairly high and enter buildings through windows and other opening. Mature larvae are slightly longer than adults and body is covered with dense tufts of hairs. They have alternate stripes of light and dark brown and larval structure being broader in the rear and narrower in front make them distinguishable from other carpet beetle larvae.

Feeding Habit and Damage Extent

Adult feed with chewing mouth parts while larvae feed over a longer period. Some animal products have been reported to be consumed by the larvae of carpet beetles are hair, fur, feather, wool, leather, hides, horn, antlers, stuffed specimens and dead insects.

Control Measures

> ☆ Isolate and sterilise the infested items. Eggs of carpet beetle may be removed by vacuuming and brushing from infested items and nearby surfaces.

> ☆ Accumulated dead insects from windows and other opening should be removed.

> ☆ Remove nest of birds, rodent, wasp and bees.

> ☆ Use nets for sealing or screening entry routes of beetle from outside.

> ☆ Use insect traps or sticky traps for monitoring beetle.

3. Furniture Carpet Beetle (*Anthrenus flavipes*) Figure 6.3

It is included in the Family **Dermestidae**

> ☆ *Anthrenus flavipes* is commonly known as furniture carpet beetle. It looks slightly larger and rounder than varied carpet beetles when

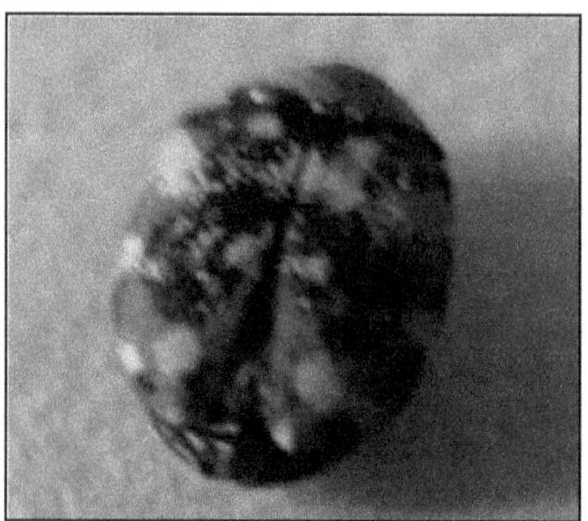

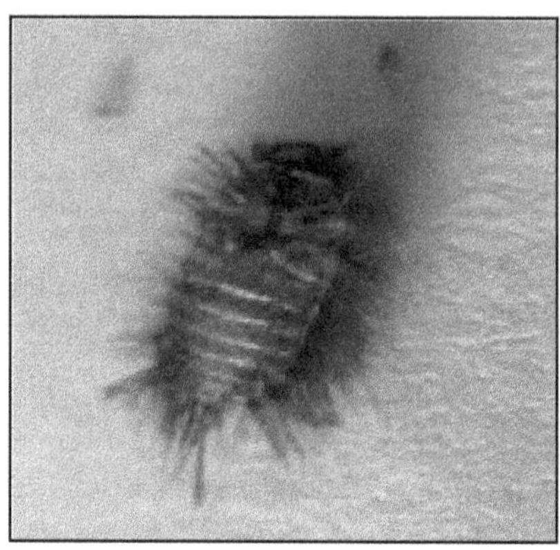

Figure 6.3: Adult Furniture Carpet Beetle *Anthrenus flavipes.*

Figure 6.4: Larvae of Furniture Carpet Beetle *Anthrenus flavipes.*

viewed from above. Colour and markings also vary, the distinguish characteristic of furniture carpet beetle is having mottled appearance where spots of black and white colour intersperse in dark yellow to dark orange scales on their wings cover. The adults may appear solid black when these scales wear off. They have white undersides. Generally larvae appear white at first but as they mature they become dark red or chestnut brown. The larvae are broader in front while narrower at the end as compared to the larvae of other carpet beetles (Figure 6.4).

Feeding Habit and Damage Extent

Larvae of the furniture carpet beetle has been recorded feeding on the same material as the larvae of varied carpet beetle like feather, fur, wool, hair, leather, hides, horn, mounted specimens and dead insects. They may also feed upon the insect collections preserved in museum if not controlled.

Control Measures

☆ Remove furniture carpet beetle and their eggs by vacuuming and brushing from infested items and nearby surfaces.

☆ Remove accumulated dead insects from windows and other opening.

☆ Nest of birds, rodents, wasp and bees should also be removed. Spider web should also be cleaned as insect may got trapped in these webs and later become the food source of these pests.

☆ Sealing or screening entry routes of beetle from outside.

☆ Use sticky traps for monitoring the beetle.

☆ Keep flowering plants and shrubs well away from the building is suggested.

4. Black Carpet Beetle (*Attagenus unicolor*) Figure 6.5

It comes under the Family **Dermestidae**

> ✰ Adults and larvae of the black carpet beetle, *Attagenus unicolor*, are quite different in appearance from other carpet beetles. Black carpet beetles range from 2.5 to 5mm long. They appear dark brown to shiny black in colour and they are bullet-shaped. The length of larvae may reach up to 12mm and colour ranges from golden brown to black. They are shiny, smooth, and hard having short hairs all over the body (Figure 6.5). And body tapers at the end and also having tuft of long hairs at the end.

Figure 6.5: Adult Black Carpet Beetle *Attagenus unicolor*.	Figure 6.6: Larvae of Black Carpet Beetle *Attagenus unicolor*.

Feeding Habit and Damage Extent

The adult of black carpet beetles have been reported to feed mostly on pollen and it may also consume dried meat, plants and dead insects. Larvae of the **black carpet beetle** feed on both plant and animal products. Like other carpet beetle species, they also feed on silk, wool, feathers, leather, dead birds and mammals *etc.*

Control Measures

> ✰ The eggs of black carpet beetles are very fragile and easily removed by brushing.
>
> ✰ A freezing regime of 6 days at –18°C has been reported to kill the adults and larvae.
>
> ✰ Sticky traps are also useful for monitoring larval activity round the year.
>
> ✰ Chemical Measures for Dermestid beetles including Varied, Furniture and Black Carpet beetles reported by Story (1998).
>
> ✰ Sulfuryl fluoride and methyl bromide can be used as conventional fumigants.

☆ Other fumigants like paradichlorobenzene (PDB) and naphthalene have been reported to be effective in enclosed storage containers.

☆ Different residual insecticides like organophosphate, pyrethroids and carbamates. Dust formulation of these is most effective for treating building voids.

☆ Use of atmospheric gases have been successful against carpet beetles: 60 per cent CO_2 for 20 days at 25°C or anoxic conditions *i.e.* argon or nitrogen for two to four weeks may be used.

5. Silverfish (*Lepisma saccharina*) Figure 6.7

It comes under the Family **Lepismatidae**

☆ They are small, wingless insects with silver coloured appearance. Silverfish are also known as paramites and carpet sharks. Silverfish are approximately 10-12 mm in length. Their bodies are carrot shaped, elongated and flattened with a tapered abdomen at the end. Body of adult silverfish have a remarkable metallic shine. They can also be identified by the three sets of appendages called *cerci* which are present on the sides of their body. Two large antennae are present and body is devoid of wings. Generally Silverfish are nocturnal, but they can also be seen during the day if the infestation is large enough. Silverfish can survive in wide range of environment but they prefer highly humid areas.

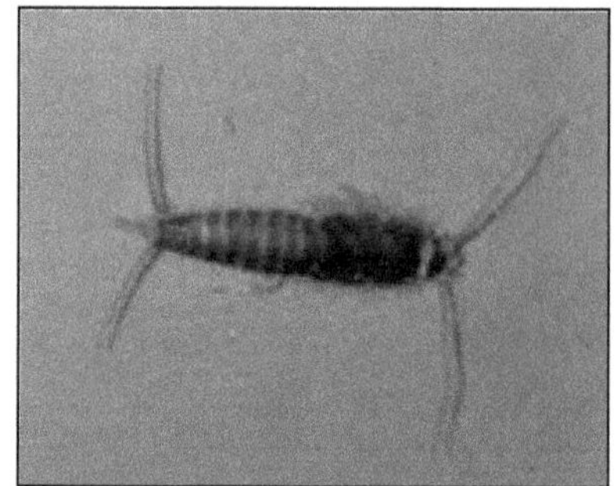

Figure 6.7: Adult Silver Fish
Lepisma saccharina.

Feeding Habit and Damage Extent

Silverfish feed on carbohydrates, particularly starch and sugars. Cellulose, glue in books, linen, silk and dead insects may be food sources.

Silverfish feed on matter containing polysaccharides, especially starches and dextrin.

They may also feed on tapestries, silk, cotton, household dust and synthetic fibres.

They may also feed upon the dead insects and leather during food scarcity.

Control Measures

☆ Humidity control can also be a real solution against silver fish. Freezing easily kills nymphs and adults but their eggs are more resistant.

☆ Sealing crevices or gaps allowing entry from out door.

☆ Checking outdoor supplies especially cardboard boxes from damp condition.

☆ Trapping can also be done by using insect traps.

Chemical Control Measures

The surface, cracks and crevices can be treated with residual insecticides like carbamates, organophosphates and pyrethroids. Boric acid baits may be used for the control. Fumigation with atmospheric gases can be done.

6. Cloth Moths (*Tinea pellionella*) Figure 6.8

It comes under the Family **Tineidae**

☆ The larvae of case-making clothes moths is capsule shaped which is greyish white with a dark head. The most distinguishing characteristic of larvae of this species is to construct a portable case where they may accommodate and used to carry along with them. This portable case is about 6 to 9 mm in length, which is usually constructed of fibres of the same food material they feed upon from that particular area. This peculiar characteristic of making case using the same material they feed upon makes it difficult to identify within that place. Both ends of the protective case are open so that the larvae can move outside and may feed from any of the two ends.

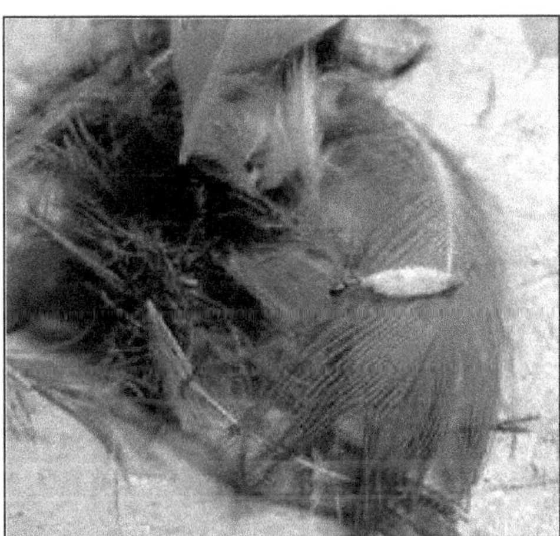

Figure 6.8: Larvae of Cloth Moth *Tinea pellionella* in a Portable Case, Feeding on Feathers.

Feeding Habit and Damage Extent

The case making clothes moth larvae feeds on woolen clothing, carpets and tapestries, causes damage that appears as small holes with webbing. They may also feed on dead bodies of insects or stuffed animals and will readily feed on hair, feather, fur, and lint. They are considered as major pests because of their feeding habits damage fabrics and other items made of natural fibres. Cloth moths larvae have the ability to digest keratin, which is a principal protein in wool, feather, fur, hair, upholstered furniture, animal and fish meals, milk powder, and most animal products, such as bristles, dried hair and leather.

Control Measures

- ☆ Thorough vacuum cleaning is crucially important.
- ☆ Sealing and screening entry path of insects from outside.
- ☆ Sticky traps may also be useful near windows and vulnerable areas.
- ☆ Rodent and bird nests must be removed.
- ☆ Cooling of infested objects to about 9°C to prevent feeding and breeding.

Chemical Control Measures

Use of six different chemicals have been reported by Story (1985) to control Cloth moths, which include use of aqueous formulations of permethrin as an aerosol. Sulfuryl fluoride and methyl bromide have been used as conventional fumigants against cloth moths. Paradichlorobenzene and naphthalene were used as mild fumigants. Pyrethroids, carbamate and chlorinated hydrocarbon methoxychlor as residual insecticide. When specimens or objects are infested with cloth moths then it is advised to keep in atmosphere free of oxygen where oxygen is replaced by nitrogen or argon for pest eradication.

6.6 Conclusion

The museum pests found in the reserve collection are the main causes of biodeterioration of avian collection in museums. Thus there is an urgent need to eradicate them as early as possible in order to ensure the collection preservation. Their control may include non chemical measures discussed in this paper and chemical measures in case of severe infestation or damage. But different approaches towards the preventive conservation should be considered as lack of concern towards its preventive measures will eventually lead collection to the stage where chemical treatment will be a single solution. And chemical applications have not been considered healthy for collection as well as for the persons handling that collection or in direct contact with the places where the chemicals have been used. Thus non chemical measures including debris removal from nesting birds, rodents and insects particularly nests near the vulnerable collection, removal of accumulated dead insects from windows and other openings, use of nets for sealing or screening entry routes and sealing crevices or gaps allowing entry from out door, use of insect traps or sticky traps for monitoring them, thorough vacuum cleaning and humidity control against insect pests infestations should be promoted in museum in order to avoid chemical applications. Windows of the reserve collection should not be wide

Specimens Showing Deteriorated Conditions

Figure 6.9: Specimen with Damaged Head.

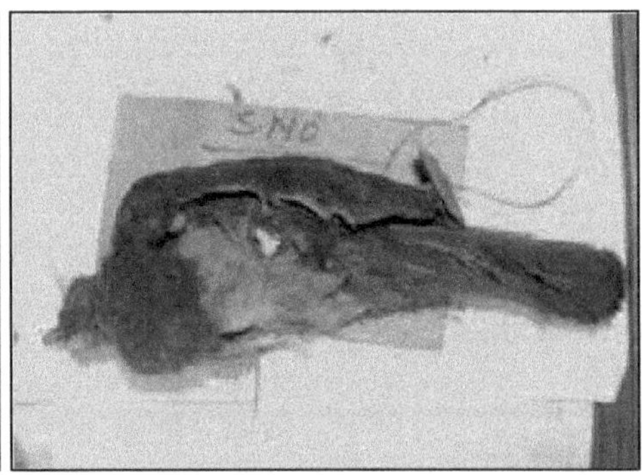

Figure 6.10: Skin along with Feathers have been Eaten.

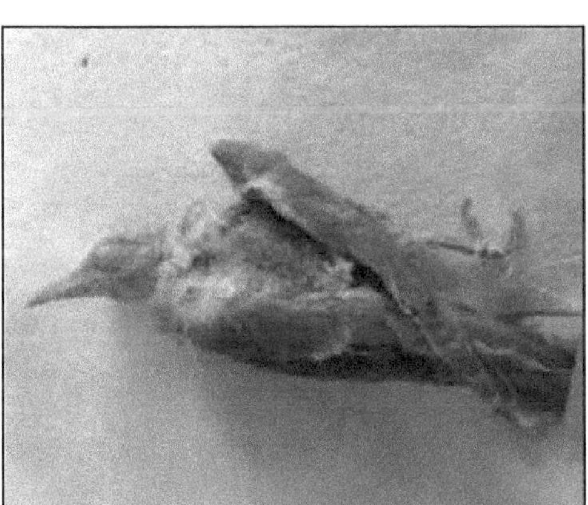

Figure 6.11: Skin Removed and Wing Detached.

Figure 6.12: Skin Splitting, Feathers Falling Badly.

Figure 6.13: Showing Falling of Feathers.

Figure 6.14: Wing Detached, Feathers Eaten, Exposing Stuffed Cotton.

open. It must be covered with micro net to avoid the entry of museum pest and birds through these windows. Windows in the reserve section of the museum were devoid of such net and allowed entry of pests and birds into the section.

☆ Air tight show cases must be used to store study skins in the reserve section, SML has some newly purchased storage cases were having gaps. Display cases should also be air tight in order to preserve the mounted skin displayed in the gallery as well as in the reserve section of SML.

☆ Because of existing inadequacies some museums are now constructing purpose -built stores in which optimal conditions can be maintained. An example is the store at Tring in Hertfordshire which was built to house the ornithological collections in British Museum of Natural history (Stansfield 1984).

REFERENCES

Ali S (1960) www.org.uk/index.php/wildfowl/article/viewFile/120/120 (Pink headed duck by Salim Ali)

Anon (1885) The Hume Collection of Indian Birds. Ibis 3: 456–462

Pinniger D B (2015) Integrated Pest Management for Cultural Heritage. London: Archetype, 2015.

Pinniger D B, Harmon J D (1999). Pest management, prevention and control. In: Carter, D. and Walker, A. (eds). (1999). Chapter 8-Care and Conservation of Natural History Collections Oxford: Butterwoth Heinemannpp. 152-176 URL: http: //www.natsca.org/care and - conservation

Sara E Miller, Barrow LN, Ehlman SM (2020) Building Natural History Collections for the Twenty-First Century and Beyond, *Bio-Science*, 70 (8): 674–687.

Stansfield G (1984), Conservation and Storage: biological collections, In: Manual of Curatorship, Thompson M.A. *et al.* (eds) Butterworths PP 289.

Story K O (1985) Approaches to Pest Management in Museums. Suitland, Md.: Conservation Analytical Laboratory, Smithsonian Institution

Story K O (1998) Approaches to pest management in museums. Retrieved from http: //www.si.edu/mci/downloads/articles/AtPMiM1998-Update.pdf

Winker K, Fall B A, Klicka J T, Parmelee D F, Tordoff H B (1991) The Importance of Avian collections and the need for the continued collecting. The Loon 63: 238-246

Zarrin A (2016) Documentation and Study on the Conservation Status of Avian Collection of State Museum Lucknow. Ph. D. thesis. Aligarh Muslim University, Aligarh.

Chapter 7

Preventive Conservation Strategies for Herbarium Collection

Suboohi Nasrin and Abduraheem K.

Department of Museology, Aligarh Muslim University, Aligarh, Uttar Pradesh, India
e-mail: subuhin22@gmail.com; ka_rahim_65@hotmail.com

ABSTRACT

Since ancient times, India has been storing herbarium as a vast natural heritage in various national and international institutions. This herb is mainly made from dried and pressed plant specimens, which are usually mounted on herbarium sheets. These herbaria are used for educational and research studies, and sometimes scientists and researchers use these collections to classify flowers, in the field of population genetics, to provide historical records of species occurrences prior to habitat fragmentation and verifying the identity of the plant species. It is also helpful for morphological measurements for DNA analysis, habitat fragmentation and comparison studies with fossils. Previous studies have shown that various species of herbaria are stored and preserved in natural history museums, but due to their biological nature, these herbarium collections are at high risk of degradation. Hence there is a need to adopt advanced practices for preventive conservation methods to reduce the risk of deterioration. This paper deals with a detailed report on the best practices of preventive conservation methods of herbarium collection. It includes various environmental controls, biological controls, indigenous controls, physical controls and disaster control to make the herbaria safe and eco-friendly.

Keywords: Herbarium, Natural history museums, Preventive conservation, Deterioration, Heritage material.

7.1 Introduction

India has a vast natural heritage which is stored in the form of herbarium in various national and international institutions. This herbarium is actually a repository of preserved plant specimens which is very useful for scientific and research studies (Thiers 2016). There are more than 3,000 herbs in approximately 170 countries worldwide with millions of specimens collected by the New York Botanical Garden and the Smithsonian Institute. There are some important herbaria located in premier institutes of India like National Botanical Research Institute (NBRI) Lucknow, Forest Research Institute, Dehradun, and Central National Herbarium, Kolkata *etc.* Traditionally used for taxonomic and floristic purposes, scientists and researchers use the collection of herbarium for various purposes (Greev *et al.*, 2016). Herbaria provide convenient sample sweeps in geographically separated collections to identify species that may be useful as bioindicators (Brooks *et al.*, 1977) that have antibacterial properties (Martini *et al.*, 2004). These collections are used to provide knowledge on field-based studies, population genetics and historical records of species occurrence prior to habitat fragmentation (Suarez and Tsutsui 2004; McCallum and Bury 2013). It is an organic material used to confirm the identity of plants, to make morphological measurements, to study flowers, for DNA analysis, and also for comparison studies with fossils. Herbaria also provide reference specimens for identification of plants eaten by animals. These specimens are helpful for microscopic observation, chemical analysis and also provided associated insects collected incidentally with plants. The label present there serves as a means of locating rare and possibly extinct species by recalling the regions listed on the data (Funk 2004). Their preventive protection is of great importance because of their un-imaginable uses in near future.

The herbarium sheets with their affixed plants are organic in nature that's why the risk of deterioration by biotic and abiotic factors in indoor environments have been a great cause of concern all over the world (Pasquarella *et al.*, 2012). Environmental factors like temperature, humidity, light, atmospheric pollution and biological agents like bacteria, fungi, insects, and rodents play a very predominant role in the destruction and deterioration of archival materials (Berliner *et al.*, 2018; Nasrin and Abduraheem 2021). In all these damaging factors biological agents plays a very predominant role, which are responsible for biodeterioration. It is defined as any undesirable change in the properties of materials caused by different activities of organisms (Hueck 1965; 1968). In the beginning it was very common to put herbal leaves with antimicrobial properties in between the archival materials. But after a long passage of time the curators of herbaria adopted chemical control alternatives. These methods were not safe because of their toxicity and potential explosion of hazards to the curators, researchers as well as the general public. These highly toxic synthetic chemicals are very difficult to degrade and remain in the environment, and that's why these are now banned by the government. The residual damage of such materials is unimaginable and it is the duty of museum professional and in-charge of these repositories to adopt best practices including indigenous methods for Preventive Conservation to ensure that the materials are in stable condition and are preserved in an ecofriendly manner.

7.2 Main Risks to Herbarium Preservation

Generally different Herbaria face the same problems to their longevity as books and manuscripts. The main risks which cause herbarium deterioration are listed as follows:

1. Environmental Factors
2. Biological Factors
3. Physical Factors
4. Chemical Factor
5. Disasters

7.2.1 Environmental Factors and their Control

Environmental factors such as temperature, humidity, light and atmospheric pollutions play a very important role in the deterioration or destruction of the herbarium collection. The practice of preventive protection includes the control of fluctuations in temperature, light, relative humidity and pollutants *etc.*

A) Temperature and its Control

One of the main factors in the decline in herbarium collections stored in natural history museums and other institutions related to plant studies is temperature. High temperature deforms objects and increases the speed of chemical reaction and accelerates biological growth of museum deteriorating organisms (Figure 7.1). Therefore, optimum temperature for herbarium collection in the storage areas should be between 15-18°C.

B) Humidity and its Control

Humidity is another main factors of deterioration of herbarium collection. Humidity is defined as the water vapors present in the given volume of air. It is measured as relative humidity (RH), and is expressed in percentage. High relative humidity results in biological activity which causes objects to swell, and in low relative humidity herbarium files lose their water content and shrink periodically. A certain amount of humidity is necessary for the flexibility of the paper, but in high humid conditions (above 70 per cent) for long periods of time, the paper becomes soft and mould growth starts. High RH is the root cause of erosion for organic matter. It weakens the adhesive and book binding of the paper, which weakens the shaping element of the paper resulting in the spread of the ink. Moisture sometimes accelerates the chemical degradation of paper which turns yellow and becomes stained. The ideal relative humidity for the storage of herbarium collection is 50-60 per cent. Fluctuations in temperature and relative humidity can damage delicate plant specimens (Royal Botanic Gardens of Edinburgh, ND). High humidity can also be reduced by the use of de-hydrating agents such as silica gel. Integration of various devices with an alarm system is very helpful in maintaining the temperature and humidity.

C) Effect of Light and its Control

Herbarium sheets degrade when exposed to both natural light and artificial light. Sunlight is the largest emitter of ultraviolet radiation which is mainly responsible for the photochemical degradation of paper which occurs rapidly when exposed to sunlight in the presence of air (oxygen). Dried herbarium specimens are at moderate to high risk of damage from exposure to visible and ultraviolet (UV) light. Exposure to light can cause discoloration and wilting of plant specimens. This damage can be seen on specimens that have been exposed for a long time (Graham 2018). Herbarium sheets become weak and brittle when part of the cellulose is oxidized and the long cellulose chains break. The disappearance of ink in herbarium labels is due to the formation of oxy-cellulose. Fluorescent light also emits a high percentage of ultraviolet radiation that causes herbarium files to deteriorate. The amount of damage caused by light depends on factors such as the intensity of the light, the duration of the exposure, and the distance from the source of the light.

☆ **Intensity of light**: Increase in light intensity also increases the rate of herbarium files spoilage. The optimum light intensity for herbarium collection ranges from 150-200 lux.

☆ **Duration of exposure**: The duration of exposure of paper to light is directly proportional to the deterioration of herbarium files. Damage is seen in herbarium collection when specimens are displayed for a longer time (Graham 2018).

☆ **Distance from the source of light**: When the light source is at a distance, the damage to the herbarium files is less.

To prevent light-related damage, natural specimens or herbarium sheets should not be placed in permanent exhibitions, but may be displayed temporarily or on a rotation basis, if necessary. Therefore, herbarium storage units for these collections should be out of reach of natural day or UV light (Graham 2018). During the display, the windows of such areas should be covered with curtains that would block direct light from falling as well as absorb ultraviolet rays. Window panes may be fitted with green colored glass or lemon yellow glass panes, as these are more effective in blocking ultraviolet rays. Sometimes it's great to fit acrylic plastic sheets to window panes to absorb UV radiation. The ultraviolet rays of fluorescent tube lights must be filtered out by covering the tubes.

D) Effects of Atmospheric Pollution and its Control

Atmospheric pollution plays a very crucial role in the deterioration of herbarium collection. Mainly the dust particles and gases come from outside through the ventilation system or by visitors are major source of this pollution. Botanical specimens need to be protected from dust, as their fragility makes them very challenging to clean and dust also attracts pests (Graham 2018). Dust and dirt particles in the air not only absorb the pollutants but also show an abrasive action on these Natural History specimens. The effects of these pollutants are very serious, which show the close connection between the loss of strength of paper and its acidity resulting from sulphuric acid contamination.

To control these harmful contaminants, the herbarium storage should have good ventilation and air conditioning systems, which can control the amount of dust and entry of other pollutants in the herbarium repositories (Graham 2018). If environmental pollution and dust are already accumulated in or around the surface of the sample, appropriate cleaning methods include using a chemical sponge and a soft brush, though this preventive process typically requires some expertise and training in order to minimize any additional damage (University of Aberdeen, 2017). These natural history specimens must be stored on steel shelves within metal cabinets. Cabinets should ideally include rubber door seals that protect items from exposure to dust and insects (Royal Botanic Gardens of Edinburgh, ND). Cabinets should be sturdy and open and close smoothly to avoid further deterioration. Metal is preferable over wood for storage and display units for cabinet preparations because wood products can increase the risk of insect infestation (British Columbia Ministry of Forestry 1996; Graham 2018).Fragile and breakable items of any kind must be stored in suitable enclosures that are labeled fragile, and have certain restrictions on their access and display, and those are particularly sensitive to light, dust and dirt, hence these should be kept in opaque containers in dark (Graham 2018). During the selection of the building, it should be kept in mind that the building is away from traffic so that dust and dirt can be avoided. The windows of the building should be open during the summer months when the temperature is high. And if the windows are to be kept open then wet curtains should be used. High speed air circulators are used for free air circulation. Floors can be cleaned by wet dusters. There should be a cleanliness program to reduce environmental pollution. Specific instructions should be given to clean the remote corners of cabinets, the backs of cabinets, under desks, chairs and all surfaces that collect dust. A vacuum cleaner may be preferred to remove dust and dirt.

7.2.2 Biological Factors and their Control Method

The presence of excess amount of moisture in the herbarium collection causes the growth of biological agents. These agents such as bacteria, fungi, and insects *etc.* play a very important role in the deterioration of herbarium specimens. As already known that herbarium is organic in nature these are highly susceptible to pests, particularly larvae of cigarette beetles and carpet beetles, moths, booklice and silverfish (Massey1974). These agents are abundant in such places where both the temperature and humidity are uncontrolled. Sometimes human negligence is also responsible for the proliferation of insect growth. Insects are found in natural history museums and herbarium repositories where dust, dirt and food items accumulate. Fungal spores are present everywhere including the e-herbarium repositories (floppy, C.D. *etc.*) until they find suitable conditions for growth. The mold digest the materials on which it has begun to grow, causing staining on the herbarium sheets and damages the specimens. Moulds or fungi play a predominant role in the deterioration and destruction of herbarium collection (Figure 7.2). They are opportunistic microorganisms of varied nature and the number of identified fungal species is over one million (Mueller and Schmit 2007). These degrade the valuable heritage aesthetically, mechanically and chemically. Paper materials show high amounts of organic binders that are particularly susceptible to fungal deterioration.

 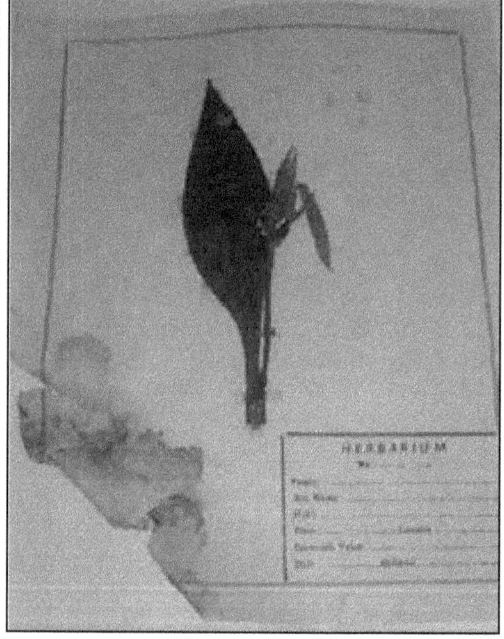

Figure 7.1: Deterioration of Herbarium Sheets Due to Fluctuation of Temperature.

Figure 7.2: Deterioration of Herbarium Sheet by Fungi.

Moulds are well known threatening organisms for various materials including historical materials in libraries and museums (Nitterus 2000a; Allsopp *et al.,* 2004; Capitelli *et al.,* 2009; Mesquita *et al.,* 2009; Pangallo *et al.,* 2009; Sterflinger 2010). Moulds or fungi secrete extracellular enzymes, which digest the archival materials responsible for their weakening (Nitterus 2000). Archival materials such as paper are primarily composed of organic compounds and are cellulosic in nature, so natural environment plays the major source of energy for the microorganisms (Florian 2002). As known that herbarium specimens are very delicate and fragile, and are of scientific importance, they need extra attention to prevent them from getting worse (Massey1974). Firstly there should be the maintenance of proper temperature and humidity, and provision of cross ventilation and exhaust fans to ensure good circulation of air to eliminate moisture but at times with electric fans the air inside the room broadcasting is very important. Freezing is the most common and effective method of pest control. Loose-leaf herbarium specimens must be dried, frozen on a rotating schedule after receiving or returning from credit (Collins 2014). When frozen these specimens should be properly packed in polythene bags and allowed to freeze for one day before being reintroduced into collection. Otherwise, freezing will lead to condensation of moisture, which will weaken the herbarium's dry plants and create a favorable environment for pests and mildew (Figure 7.3). To prevent freezer burn, samples should be kept tightly closed in plastic bags, and further supported with cardboard between the sheets. The bag should be carefully placed in a standard chest freezer at -20°C for 48–72 hours (British Columbia Ministry of Forestry 1996; Collins 2014).

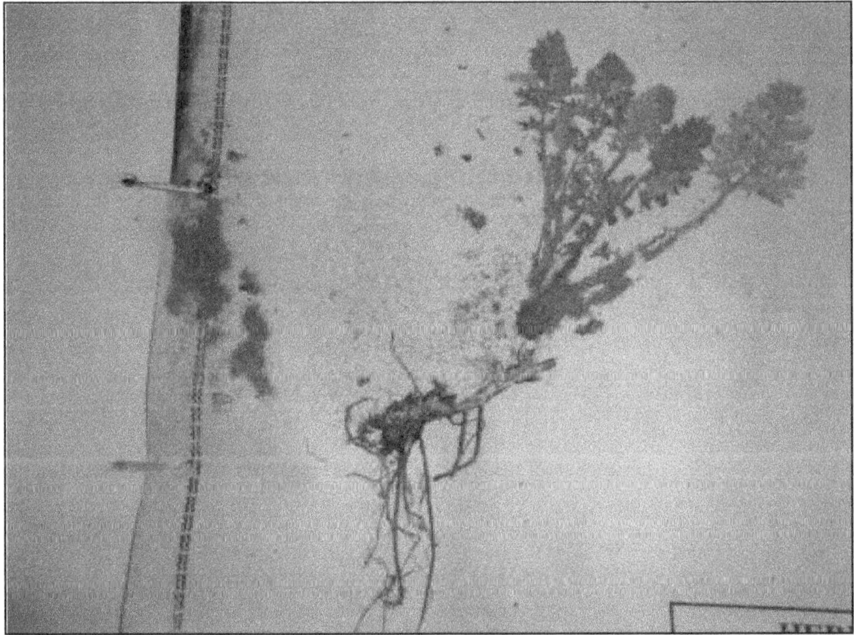

Figure 7.3: Deterioration of Herbarium Sheets by Insects.

7.3 Preventive Conservation with the Use of Natural Products

Curators have long adopted chemical control methods, but these methods are not safe due to their toxicity and potential exposure of hazards to curators, researchers, as well as the general public. Many of these synthetic pesticides are difficult to degrade and has tendency to stabilize in the environment, hence have been banned by the Indian government. Considering all these facts there is a need to develop an alternative to the harmful chemicals at the same time which are safe to the users. The natural products with biocidal activity have shown to be an alternative and useful source for the control of biodeterioration and are more environmentally friendly (Guiamet *et al.*, 2006). These natural products obtained from botanical plants including essential oils hold a good promise (Regnault-Roger 1997; Isman *et al.*, 2001; Isman 2006; Bakkali *et al.*, 2008). For the protection of herbarium collection Natural products like clove (*Syzygium aromaticum*), lemon (*Citrus limon*), lavender (*Lavandula*), pine (*Pinus*), peppermint (*Mentha piperita*), black pepper (*Piper nigrum*), turmeric (*Curcuma longa*), Asafoetida (*Ferula asafoetida*) *etc.* have been used as insect repellents. Researches on these herbals have shown very impressive results.

Turmeric belongs to family Zingiberaceae, which is extensively cultivated for its rhizomes. It is a perennial herb distributed in tropical and sub-tropical regions of the world including India, Pakistan, Bangladesh and Sri Lanka. Its rhizomes are harvested, washed and boiled in mildly alkaline water to soften and dried in the sun or in electric driers. It is used as coloring agent in pharmacy, confectionery, food industry, for dyeing wool, silk, cotton and to obtain combination with other natural dyes to get different shades (Bakshi *et al.*, 1999). Rhizomes are also used as insecticide (Kapoor 1990; Yu *et al.*, 2002). Turmeric has long been used in India and abroad for its antifungal as well as insecticidal properties and for dyeing paper and textiles (Kharbade *et al.*, 2008). When herbarium sheets were dipped in a solution of turmeric, the method was very effective, it preserved the herbarium files for many years. Clove solution acts as an insect repellent, when this solution was used to treat the herbarium sheets, the treated herbarium sheets showed good repellency against insect pests.

Syzygium aromaticum (Syn. *Caryophyllus aromaticum*, *Eugenia aromaticum*, *Eugenia caryophyllata*) commonly known as clove belongs to the family Myrtaceae. It is a precious and valuable spice of the world. It shows insecticidal, antibacterial and repellent properties. Camphor is a white, crystalline substance with a strong odor and pungent taste, derived from the wood of camphor laurel (*Cinnamomum camphora*) and other related trees of laurel family. The camphor tree is native to China, India, Mongolia, Japan and Taiwan and a variety of this fragrant evergreen tree is grown in Southern United States particularly in Florida (Frizzo 2000; Forest *et al.*, 2003). Camphor is obtained through steam distillation, purification and sublimation of the tree's wood, twigs and bark (Zuccerini 2009).Camphor has a significant antifungal activity, showing good results against insect pests (Sattar *et al.,* 1991) when herbarium sheets were treated in camphor solution.

7.4 Physical Factors and their Preventive Methods

Physical factors are the main cause of the deterioration of herbarium specimens. This degradation leads to breakage and loss of plant specimens due to poor handling and storage conditions (Graham 2018). This deterioration is caused by wear, tear and mishandling of the objects, and sometimes the curators are not aware about how to handle, store and use the Collections.

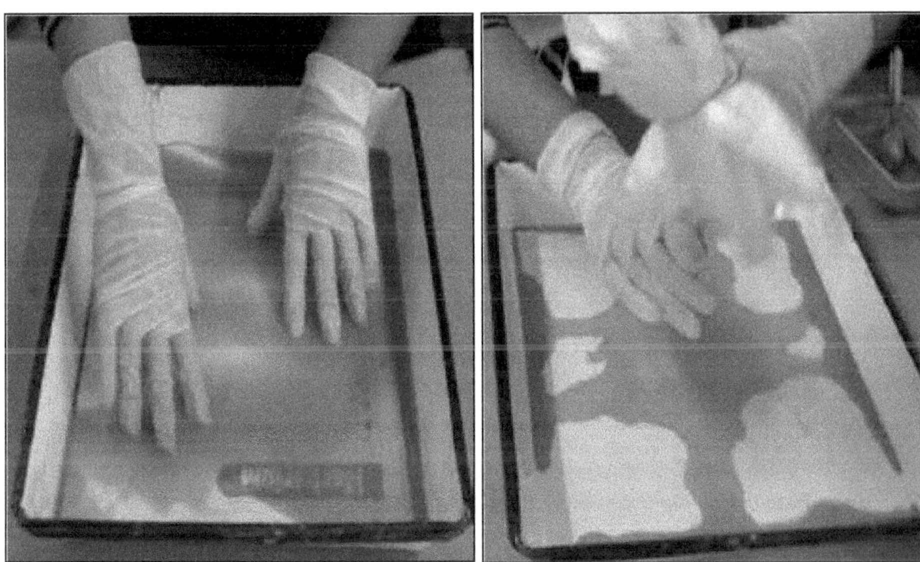

Figure 7.4: Treatment of Herbarium Sheets by Natural Product-Turmeric.

Preventive conservation is essential to avoid physical damage of the collection, and herbarium specimens will last for hundreds of years if properly cared for. The best conditions for storage include low temperature (10-18 °C), low humidity, low light, and infrequent handling. Proper storage of herbarium files in secure and stable shelving will minimize the risk of damage. The dried and preserved plant specimens can be stored best in archival grade boxes or tied in bundles in cardboard folders for a long term storage. The files of herbarium can be placed in archival grade plastic sheets. More over acid free herbarium sheets may be used for long lasting the collection. Care and management should be very important to minimize the physical damage of the collection.

7.5 Chemical Factors and their Preventive Conservation

It is known that herbarium collections are highly vulnerable to insect attack, so it has been previously treated with insecticides (Graham 2018). Several methods have been traditionally used to prevent or get rid of insect pest in herbarium, including poisoning, fumigation and heating (Massey 1974). However, these methods are ineffective for herbarium preservation and raise many potential hazards (British Columbia Ministry of Forests 1996). The specimens which have been previously treated with toxic compounds have some risk to deterioration and adverse effects on humans as well. The herbarium staffs are periodically using insecticidal powder or solution like Lindane at the dark corner walls, beneath the racks and Amirah. Para dichlorobenzene and naphthalene bricks are also used as an insect repellent and insecticides in herbarium.

Figure 7.5: Deterioration by Physical Factors like Mishandling and Improper Storage.

7.6 Disaster and their Preventive Conservation

Disaster is one of the main reasons for the damage of herbarium specimens, which include fire, water, floods, strong winds cyclo, e, earthquakes *etc.* Water damage resulting by flooding can result in staining of plant materials. Sometimes the damage due to water is also caused by leakage from roof and pipes. It is very clear to understand the impact of disasters on these natural history samples. Given the unimaginable uses of the herbarium, it is imperative to prevent the collection from worsening as a disaster. Prevention should have separate alarm systems and a written disaster plan that each staff member should know about emergency equipment and how to operate them.

7.7 Preventive Conservation and their Importance

Preventive conservation is defined as all those measures and actions aimed at avoiding and minimizing future deterioration or loss. They are carried out within the context or on the surroundings of an item, but often a group of items. These preventive conservation methods are indirect actions because they do not interfere with the material and composition of the objects (ICOM-CC). Preventive conservation includes various activities such as risk assessment, development and implementation of guidelines for continued use and care, environmental conditions suitable for storage and exhibition and proper procedures for handling, transportation and use. These preventive responsibilities may be shared by collection managers, conservators, subject specialist and administrators (Society for the Preservation of Natural History collections 1994). This will be of great help in maintaining the herbarium collection and minimizing the risk of further deterioration.

7.8 Conclusion

The Herbarium is the rich source of natural heritage, which is stored in many natural history museums and herbarium institutions. These collection of natural heritage objects are rich sources of knowledge which are spoiled by biotic and abiotic factors. Given their unimaginable uses, it is the duty of custodians of herbarium repositories to adopt best practices on preventive conservation to make it safe. In this study the authors gave a brief report on the preventive protection of herbarium collections. There are many factors that cause herbarium to deteriorate such as temperature, humidity, light, atmospheric pollution, insects, fungi and rodents and disasters. It is recommended to maintain the temperature at 15-18°C and the relative humidity of 50-60 per cent for safeguarding herbarium collection. In terms of conservation, both natural and artificial light are harmful to herbarium collection as they are the largest emitters of ultraviolet radiation responsible for the photochemical degradation of paper. Exposure to light can cause discoloration, and brittleness of plant specimens. To control these problems in herbarium, it is advisable to have green or lemon yellow colored glass panes on the windows. Sometimes acrylic plastic sheets are also used in the panes of windows to filter UV radiation. The herbarium should have the presence of trained staff for the maintenance of the lighting. To control biological agent there should be the provision of cross windows, proper ventilation and exhaust fans. Freezing is the most common and effective method to control pest. To protect the specimen from freezer burn the herbarium specimen should be sealed tightly in plastic bags and these bags were carefully stacked in a standard chest freezer. Sometimes the herbarium specimens were fumigated by various pesticides and these chemical pesticides are toxic and banned by the government, hence the need to develop eco-friendly approach. Some natural products show toxic character on herbarium specimens, such as turmeric, clove and camphor used as killing agents on herbarium sheets have shown good pest control results for many years. Training of herbarium staff to operate electronic gadgets and handle specimens is very important. Periodical monitoring, cleaning and dusting of the herbarium are recommended as best practices.

REFERENCES

Allsopp D, Seal K, Gaylarde C (2004) Introduction to biodeterioration. Cambridge Univ Press. pp. 237.

Bakkali F, Averbeck S, Averbeck D, Idaomar M (2008) Biological effects of essential oils a review. Food, Chem, Toxicol. 46: 446-475.

Bakshi DNG, Sensarma P, Pal D.C. (1999) A lexicon of medicinal plants in India. Naya Prakash, Calcutta.

Berliner A J, Mochizuki T, Stedman K M (2018) Astrovirology: viruses at large in the Universe. Astrobiology, 18: 207-223.

Bridson D, Foreman L (1998) The Herbarium Hand. "Royal Botanic Gardens, Kew, Great Britain". Third Edition.

British Columbia Ministry of Forests (1996) Techniques and procedures for collecting, preserving, processing, and storing botanical specimens (pp. i–vi, 1–39). https: //www.for.gov.bc.ca/hfd/pubs/docs/wp/wp18.pd

Brooks R R, Lee J, Reeves RD, Jaffre T (1977) Detection of nickeliferous rocks by analysis of herbarium specimens of indicator plants. J. Geoch. Explor. 7: 49-57

Cappitelli F, Fermo P, Vechi R, Piazzalunga A, Valli G, Zanardini E, Sorlini C (2009) Chemical physical and microbiological measurements for indoor air quality assessment at the Ca'Granda Historical Archive, Milan (Italy). Water Air Soil Pollut. 201: 109-112.

Collins C (2014) Standards in the care of botanical materials. The Conservation Centre, National History Museum, London. http: //conservation. myspecies.info/node/35#

Florian M L E (2002) Fungal Facts. Solving fungal problems in heritage collections. Archetype Publications Ltd. London.

Forest S, Kim S, Lloyd L (2003) *Cinnamomum camphora*. Rep. Maui, Hawai'i: United States Geological Survey-Biological Resources Division.

Frizzo C D, Santos A C, Paroul N, Serafini L A, Dellacassa E, Lorenzo D, Moyna P (2000) Essential Oils of Camphor Tree (*Cinnamomum camphora* Nees and Eberm) Cultivated in Southern Brazil. Brazilian Archives of Biology and Technology. 43(3).

Funk VA (2004) 100 uses for a herbarium (well at least 72). Ameri. Soc. Plant Taxon. 17: 17-19.

Graham F (2018) Caring for natural history collections– Preventive conservation guidelines for collections. Canadian Conservation Institute.https: //www. canada.ca/en/conservationinstitute/services/preventive-conservation/ guidelinescollections/natural-history.html

Greve M, Lykke A, M, Fagg CW, Gereau R E, Lewis G P, Marchant R, Marshall A R, Ndayishimiye J, Bogaert J, Svenning J C (2016) Realizing the potential of herbarium records for conservation biology. South Afri. J. Bot. 105: 317-323.

Guiamet P S, Gómez de Saravia S G, Arenas P, Pérez M L, de la Paz J, Borrego SF (2006) Natural products isolated from plant used in biodeterioration control. Pharmacology online. 3: 537-544.

Herberling J M, Isaac B L (2017) Herbarium specimens as exaptations: New uses for old collections. Am. J. Bot. 104: 963-965

Hueck H J (1965) The biodeterioration of materials as a part of hylobiology. Mater. Organismen. 1: 5-34.

Hueck V D, Plas H E (1968) The microbiological deterioration of porous building materials. Internat. Biodeterio. Bullet. 4(1): 11-28.

Isman M B (2006) Botanical insecticides, deterrents, and repellents in modern agriculture and an increasingly regulated world. Annu. Rev. Entomol. 51: 45-66.

Isman M B, Wan A J, Passreiter C M (2001) Insecticidal activity of essential oils to the tobacco cutworm, *Spodoptera litura*. Fitoterapia. 72: 65-68.

Kapoor L D (1990) Hand book of ayurvedic medicinal plants. CRC Press, Boca Raton, *FL*.

Kharbade B V, Rajmal SA, Manjunathachari R C (2008) Use of turmeric: Indian traditional materials in preservation of old manuscripts. ICOM-CC Pub., I. Pp. 872-878.

Martini N D, Katerere D R P, Eloff J N (2004) Biological activity of fiveantibacterial flavonoids from *Combretum erythrophyllum* (Combretaceae) J. Ethnopharmacology. 93: 207-212.

Massey J R (1974) The herbarium. In: Vascular Plant Systematics by A. E. Radford, W.C. Dickison, J.R. Massey and C R Bell. https://www.ibiblio.org/uncbiology/herbarium/co

McCallum M L, Bury GW (2013) Google search patterns suggest declining interest in the environment. Biodiv Conserv. 22: 1355-1367.

Mesquita N, Portugal A, Videira S, Rodriguez-Echeverria S, Bandeira A M L, Santos M J A, Freitas H (2009) Fungal diversity in ancient documents. A case study on the Archive of the University of Coimbra. Int. Biodeterior. Biodegrad. 63: 626-629.

Mueller G M, Schmit J P (2007) Fungal biodiversity: what do we know? What can we predict? Biodiversity Conservation. Springer Pub.16: 1-5.

Nasrin S, Abduraheem K (2021) Agents of deterioration of organic museum object and their management: A Review. Internat. J. Curr. Scie. Resea. review. 4(9): 1008-1016

Nitterus M (2000) Ethanol as fungal sanitizer in paper conservation. Restaurator. 21: 101-115.

Nitterus M (2000a) Fungi in archives and libraries. A literary survey. Restaurator. 21: 25-40.

Pangallo D, Chovanova K, Simonovicova A, Ferianc P (2009) Investigation of microbic community isolated from indoor artworks and their environment: identification, biodegradative abilities, and DNA typing. Can. J. Microbiol. 55: 277-287.

Regnault-Roger C (1997) The potential of botanical essential oils for insect pest control. Integr. Pest Manag Rev. 2: 25-34.

Sattar A, Gilani A M, Saed M A (1991) Gas chromatographic examination of the essential oil of *Cinnamomum camphora*. Pak. J. Sci. Ind. Res. 34: 135-136.

Society for the Preservation of Natural History collections (1994) Guidelines for the care of natural history collections. Collection Forum, 10: 32-40.

Sterflinger K (2010) Fungi: Their role in deterioration of cultural heritage. Fungal Biology Reviews. The British Mycological Society.24: 47-55.

Suarez A V, Tsutsui N D (2004) The value of museum collections for research and society. BioScience 54: 66-74.

Thiers B (2016) Index Herbariorum: A global directory of public herbaria and associated staff, New York Botanical Garden's Virtual Herbarium, http: // sweetgum.nybg.org/science/ih/

Yu Z F, Kong L D, Chen Y (2002) Antidepressant activity of aqueous extracts of Curcuma longa in mice. J. Ethno. Pharmacol. 83: 161-165.

Zuccarini P (2009) Camphor: Risks and benefits of a widely used natural product. J. Appl. Sci. Environ. Manage. 13(2): 69-74.

$$Chapter\ 8$$

Experiments with Fungal Deterioration of Museum Collections and Heritage Monuments in Eastern India

Indrani Bhattacharya

Department of Museology, University of Calcutta, 1 Reformatory Street (9th Floor), Kolkata – 700 027, West Bengal India
e-mail: sibmusl@caluniv.ac.in

ABSTRACT

In different museums of Eastern India, whether urban or rural, some common fungal deterioration was observed in palm leaf, paper materials, parchment, bark, basketry, skin, leather and wooden objects. The museum collections consisting of resistant matter or sensitive objects are affected by several environmental parameters that can modify both physical and chemical composition of the matter. They are generally affected by various species of fungi causing acidity and weakness of fibre strength. Sometimes the whole material becomes yellow, brown or black, stiff, brittle and with innumerable tiny holes. In addition, when they are kept into the 'biosphere', 'biotransformation' process by biological mechanisms can cause great damage to the organic objects especially in warm and humid climates where the environmental conditions are extremely favourable to the growth of most microorganisms. Among the various deteriorating agents of cultural property microorganisms play a dominant role in damaging art objects. The microbial biodeterioration is a serious threat to all cultural properties like wooden objects, papers, manuscripts, textiles, decorative art objects, paintings of various types, skin and leather objects. Among the microorganisms fungi play a dominant role in destruction of museum objects. In eastern India where the temperature and humidity (more than 60 per cent) is most suitable for fungal sporulation and spore germination cause severe damage to heritage objects. When the temperature increases, the movement of molecules of both organic and inorganic materials also *increases, the solid particles become softer and chemical reactions speed up. The paper highlights the various fungal species prevalent in different museums of eastern India and certain control measures are suggested.*

Keywords: Experiments, Fungal deterioration, Museum collection, Monuments, Eastern India.

8.1 Introduction

In Eastern India, summer months are extremely hot and have high humidity. The plains of this region have a daily maximum temperature around 35–38 °C. Most of the annual rainfall occurs during the monsoon period. Heavy rainfall above 250 cm is observed in the hilly regions like Darjeeling and Cooch Behar region. In autumn, the southwest monsoon wind returns and clears the sky. So, the states get enough sunshine to be warm in the day and release a lot of heat to be cool at night. At this time, a maximum temperature of 30–33 °C is felt over the plains and 17–19 °C in mountains. Mainly Odisha, Bihar, Jharkhand and West Bengal are the states of Eastern India. The Odisha state has tropical climate, characterized by high temperature, high humidity, medium to high rainfall and short and mild winters in many places. Geographically, capital city of West Bengal – Kolkata is situated at 22°35′ N latitude and 88°2′ E longitude at the elevation of 4.6m high from the sea level. The city lies on an average height of 2.5m from the water level of river Hooghly. There is a dense urban settlement along with industries. Kolkata falls under humid tropic zone. The average temperature ranges between 21°C to 32°C of which the maximum is found in the months of April and May. The annual precipitation ranges between 135 and 150 cm. Some major museums and alike institution of West Bengal are The Indian Museum, Kolkata; The Victoria Memorial Hall, The Asutosh Museum of Indian Art (University of Calcutta, Kolkata); The Jogesh Chandra Purakirti Bhawan, Bishnupur; The Bankim Bhavan Gaveshana Kendra, Naihati; The Gandhi Smarak Sangrahalaya, Barrackpore, Terracotta Temples of Bishnupur, Pancha Ratna Shyam Rai temple, *etc.*

Bhubaneswar is located in the eastern part of the state of Odisha. It is known as "The City of Temples". There are many temples, historic monuments and parks. The State Museum, The State Tribal Museum, Udayagiri and Khandagiri caves, Lingaraja Temple, Radharani Temple, Sri Ram Temple, and so on are present here. The famous sun temple at Konarak is also within the periphery of Odisha. Ranchi the capital of the state of Jharkhand experiences a humid sub-tropical climate with a summer temperature of 20-42 °C, and winter temperatures from 0 to 25 °C. From June to September, the rainfall is about 1,100 mm. On the other hand, climate is warm and humid in Patna the capital city of Bihar. Some important museums of the state include Patna Museum, Patna, Naradah Museum, Nawadah, Bhagalpur Museum, Bhagalpur, Chapra Museum, Chapra, Archaeological Museum, Nalanda and so on.

All these museums house the tangible heritage of humanity, which are also material heritage, and there are movable and immovable properties. These are very rich in the intangible cultural heritage which includes the wealth of knowledge. Inherited from our ancestors and transmitted to our descendants, such as language, oral traditions, performing arts, ways of life, rituals, festive events, knowledge, and practices related to nature and the universe.

All these tangible heritage objects are susceptible to biodeterioration. It is the breakdown of materials by microbial action. A wide range of organisms are covered (including bacteria, fungi, algae, lichens, insects and other invertebrates, birds, mammals and plants). Wood, metal, stone, paper, paintings and leather items, synthetic products such as paint, adhesives and plastics all are attacked by biodeteriogen. Of these fungus is a most serious threat to cultural properties. The high temperature and high relative humidity (RH) in the eastern regions of India favours fungal growth. Biodeterioration due to fungal infestation is acute and it is difficult to ascertain an individual fungal attack on particular museum objects without any proper identification. Cunningham (1873) made a study on air borne microorganisms, mainly fungal spores and pollens in Kolkata. Barat (1969) under the supervision of Dr. P.N. Nandi studied microbial population at different locations in West Bengal. Fungi in libraries, museums and their storage rooms that can seriously threaten the health of the restorers, or the museum personnel and the visitors due to their allergic and mycotoxin producing potential (Crook and Burton 2010).

8.1.1 Fungal Deteriogens

There are three main types of microorganisms participating in this process: bacteria, fungi, and yeasts (Nitterus 2000; Padfield 2002). In India for the first time aerobiological studies inside library were conducted by Tilak and Vishwe (1980). They indicated the presence of *Aspergillus, Torula, Penicillium, Trichoderma* and *Chaetomium* (Figure 8.1a) spores in the atmosphere. The damages caused by fungi on organic objects range from simple chromatic alteration to serious weakening of the material (Nair 1972). Ascomycetes and Deuteromycetes (Fungi Imperfecti) are most devastating fungi for organic objects displayed in museums. The main components of organic objects are cellulose. The cellulolytic fungi produce different organic acids like oxalic, fumaric, or citric *etc.* and utilize cellulose for their nutrition resulting in loss of fiber strength and actual material failure. The complete degradation of cellulose is effected by two groups of cellulases – endoglucanases and exoglucanases that split off mono and disaccharide units from the non-reducing end of a cellulose chain and cause complete degradation of organic objects (Dhawan 1986).

Fungi is a group of eukaryotic, non-phototrophic organisms with rigid cell walls. The cell walls of fungi are unique, they contain large amounts of chitin, a structural component found only in the cell walls of fungi. These fungi are very important to the health of the ecosystem, rapidly breaking down plant

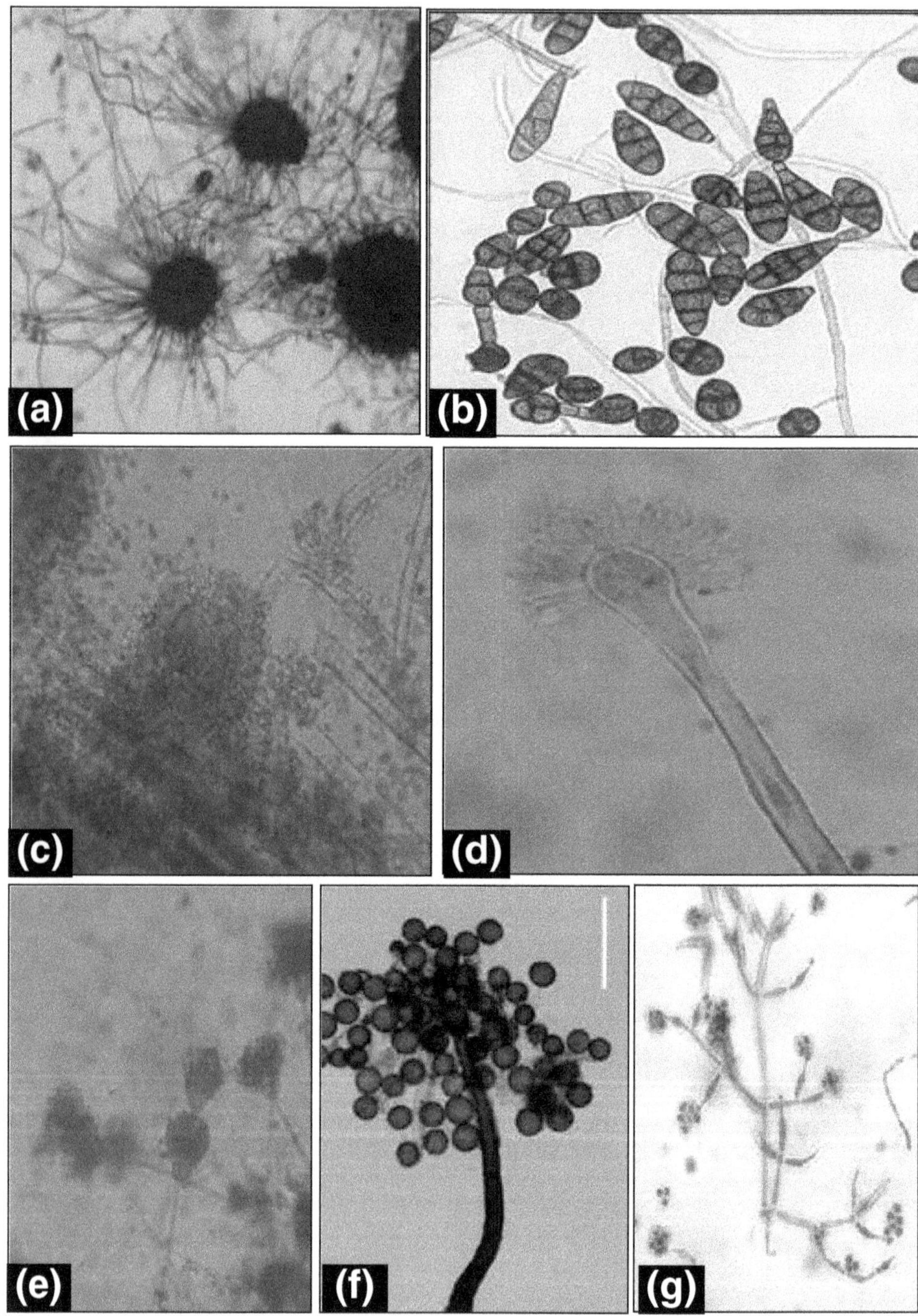

Figure 8.1: Aeromycoflora.

(a) Parithecium and ascospores of *Chaetomium*; (b) *Alternaria*; (c) *Aspergillus flavus*; (d) *Aspergillus*; (e) Conidial heads of *Penicillium*; (f) *Periconia*; (g) Conidia and phialides of *Trichoderma*.

and animal material and returning it to a more usable substances. Fungi may be saprophytic or parasitic. Most fungi reproduce by releasing spores, which are the resistant, reproductive, stage of a fungus. These spores are able to germinate in the high temperatures and pH. Santra and Chanda (1981) and Barui and Chanda (2000) made significant studies on effect of indoor air borne fungal flora of Calcutta. Nair (1982) did the pioneering work in the area of biodeterioration of cultural property (Agrawal and Dhawan 1991). Another study revealed 22 species of fungi belonging to 12 different genera in Baroda Museum (Shah, 1993). The species of *Aspergillus* and *Alterneria* were major air bound decay agents observed in these studies.

A fungus consists of a mass of branched, tubular filaments enclosed by a rigid cell wall. These filaments are called hyphae (singular hypha), branch repeatedly into a complicated, radially expanding network called the mycelium, which makes up the thallus, or undifferentiated body, of the fungus.

Kingdom Fungi are classified into the following based on the formation of spores:

1. Zygomycetes – These are formed by the fusion of two different cells. The hyphae are without the septa.

2. Ascomycetes – They are also called as sac fungi. Example – Saccharomyces

3. Basidiomycetes – Mushrooms are the most common. Another example is *Agaricus*.

4. Deuteromycetes – They are otherwise called imperfect fungi. Example – *Trichoderma*.

Chakraborty (2021) in her thesis entitled " Biodeterioration of Cultural Property: A case study on fungal flora in the museums of West Bengal", identified the following pathogenic fungi predominant in the museums of West Bengal.

☆ *Alternaria* sp.

☆ *Aspergillus* sp.

☆ *Chaetomium globosum*

☆ *Cladosporium* sp.

☆ *Curvularia* sp.

☆ *Emericella* sp.

☆ *Fusarium* sp.

☆ *Ganoderma* sp.

☆ *Helminthosporium* sp.

☆ *Myrothecium* sp.

☆ *Nigrospora* sp.

☆ *Neurospora* sp.

☆ *Penicillium* sp.

☆ *Periconia* sp.

☆ *Pithomyces* sp.

☆ *Pythium* sp.

☆ *Torula* sp.

☆ *Uromyces* sp.

☆ *Ustilago tritici*

Fungal infection of museum materials is mostly airborne. Germination of spores and development of colonies is determined by the chemical composition of the museum material itself and by the environmental factors including temperature, humidity, dust, carbohydrates *etc.* Water availability is the most important factor and the main reason why fungi in museums are predominant as compared to bacteria. Many fungi are able to grow at much lower levels of humidity. Books, manuscripts and scrolls made up of paper, papyrus and parchment often containing considerable amounts of starch paste used as adhesive for mounting. Paper is made up of cellulose, which in natural environment represents the major source of energy for microorganisms and parchment is also made of collagen, rich of nitrogen and therefore, easily degradable by microorganisms, such as proteolytic fungi. Most of the filamentous fungi are associated with paper damage and can dissolve cellulose fibers of the paper with the action of cellulolytic enzymes, or discolor them support by dissolving inks and glues. Types of fungal species vary with climatic and geographical conditions. Type of infection also differs in indoor and outdoor environments. So, fungi have great damaging effect on indoor museum objects as well as outdoor objects like sculptures, monuments, temples *etc.*

In India for the first time aerobiological studies inside library were conducted by Tilak and Vishwe (1980). They indicated the presence of *Aspergillus, Torula, Penicillium, Trichoderma* conidia and *Chaetomium* ascospores.

8.1.2 Damage Caused by Fungal Deteriogens

The damages caused by fungi on organic objects range from simple chromatic alteration to serious weakening of the material (Nair 1972). During the life cycle, fungi produce various types of spores. Which are actively or passively released into the surrounding environment. These are dispersed by air currents to available substrates in a museum. On the other hand, bio aerosol enters into the museum buildings through ventilation systems. These are passively spread through the museum's interior by air circulation. Due to their metabolic activities, numerous fungal species can cause both aesthetic and physical damages to a variety of substrate including paintings, textiles.

Paint, paper, and other types of cultural properties. The application of adequate microscopic techniques, proper identification of organisms and physiological characterization of autochthonous isolates are very important to appropriately assess the potential threats to cultural artworks, especially on those stored in inadequate conditions (Savkovic *et al.*, 2019).

Different types of protein materials are: bone, ivory, silk, organic colors of painting, binding media, adhesive; skin products like wool, leather, parchment *etc.* all are susceptible for mould growth. These materials are considered to be among the most vulnerable to deterioration and the following types of impact may be observed (Brimblecombe 1990; Blades *et al.*, 2000). Fungi damage the supporting materials in which fungal colonies develop as well as make them more susceptible for further infection. Many scientists have reported the problem of deterioration of terracotta temples in West Bengal and stone monuments in Odisha and Bihar caused by fungal and other biological agencies. The presence of these microbes cause flaking of paint, powder formation, appearance of brown black and white spots, patches and cracks *etc.* (Dhawan 1986).

In favorable environmental conditions after landing on host materials, the fungal spores need sufficient nutrients and moisture to germinate. High temperature, presence of dust and dirt particles *i.e.* particulate pollutants, very poor air circulation and dim light are ideal for mold spore germination. In the presence of ideal environmental conditions for growth and development of fungi along with high illumination of light fungal species transform hydroxyl (OH) group to carboxyl (COOH) group of acid. For nutrition, cellulose based organic objects like cotton, linen, paper, wood, adhesives, and others are very prone to direct attack by such microorganisms. Deterioration of materials is an inescapable phenomenon in any given space or time. It is the natural process of recycling the matter.

8.1.3 Biofilms

Biofilms or microbial mats are complex microbial structures in which cells are closely packed and firmly attached to a solid surface of the material. In biofilms, colonies of microbial cells are assembled with extracellular polymeric substances. The importance of biofilm in biodeterioration of cultural heritage has been reported for several decades and is related to: (a) modifications in pH values of ionic concentrations, (b) reduction oxide conditions in the interface of biofilms and substrate, (c) covering the surface and masking its properties, (d) increasing the leaching of additives and monomers outside the polymer matrix by microbial degradation, (e) releasing enzymes that lead to embrittlement and loss of mechanical stability, (f) accumulating water that penetrates the matrix causing swelling and increased conductivity, and (g) excretion of lipophilic pigments among others (Videla 1996).

8.2 Fungal Infestation of different Museum Materials

8.2.1 Wood

Lignocellulose is the major component of wood biomass. It consists of three types of polymers, cellulose (40-55 per cent), hemicellulose (24-40 per cent) and lignin (18-35 per cent) that are chemically bonded by non-covalent bonds and by covalent cross linkages. It is organic in nature. The presence of water content in wooden substrate make it susceptible to microbial attack, especially fungi. High humidity, high temperature, poor ventilation, and poor maintenance accelerate fungal infestation. The fungi that attack wooden material in different museums can be distinguished as white-rot, brown–rot and soft-rot fungi. It degrades all lignin components as well as cellulose and hemicellulose.

8.2.2 Textiles

Apart from synthetic fibers, materials of textile production could be of plant or animal origin. Cotton, hemp, linen, and jute are widely used fabrics of plant origin and hence they are composed of cellulose fibers. Many old fashioned garments, sarees, *kanthas, thankas*, carpets worn by our ancestors were woven from these fabrics and deposited in museum reserves and exhibition rooms, or still worn during traditional festivities. Wool is a textile fibre obtained from various hairy mammals and main constituent of wool is a protein keratin. Silk is obtained from silk worms and main constituent is sericin. Wool is more resistant to fungal attack due to its specific cross- linked structure with disulphide bonds. However, fungi are capable of keratinolysis and can attack wool fibers. Members of genera *Aspergillus, Chaetomium, Fusarium, Microsporium, Penicillium, Rhizopus* and *Trichophyton* are among the most frequent culprits.

8.2.3 Paper

Fungi sometimes attack the sizing materials both protein and starch in nature in paper material. Due to this, they can absorb water more easily and so the paper material becomes deteriorated day by day. The common fungus which are observed on paper object in museums are *Aspergillus niger, A. glaucus, A. flavus, A. restrictus, Alternaria* sp., *Penicillium* sp., *Curvularia* sp., *Trichoderma* sp., *Fusarium* sp., and *Cladosporium* sp. Sometimes due to fungal infections small spots are found on paper and paintings, known as *foxing*. In presence of iron containing inks, the chances of foxing increase. Fungi decompose cellulose in paper either by oxidation which transform cellulose to oxycellulose and turn it to soluble products, which can be absorbed by the mycelium of the fungi, or by hydrolysis in which the glucosides linkages between individual glucose units are hydrolysed by the cellulase enzyme.

8.2.4 Leather

Leather is predominantly a protein material. Cowhide is often used to make traditional leather. It is thick and durable. Pigskin makes comfortable and water-resistant leather. The proteins, fats in the form of glycerides that are present in the leather are ideal source of nutrients having a pH value around 4 that can support fungal growth. Under favourable environmental conditions, fungal spores germinate and form hyphae.

8.2.5 Stone

According to Scheerer *et al.* (2009), when stone of any building is exposed to the environment, fungi may be the most important biodeteriorative organisms because they are extremely erosive. Two main groups of fungi, ecologically and taxonomically separate, have been isolated from stone monuments: (i) Hyphomycetes and Coelomycetes and (ii) black meristematic and black yeasts. The first group is common both in the soil and in the air, they develop in moderate and humid climate.

8.3 Mechanisms of Deterioration

Several research studies have been done on fungal deterioration in stone temples of Odisha and Bihar. Black fungi dwell deep inside marble, granite, calcareous limestone, *etc.* which erode the stone by both chemical and mechanical attacks. They are the main culprits for the phenomenon of bio-pitting. Dematiaceous fungi with yeast like growth patterns can actively penetrate rocks and cause loss of rock material, thus creating bio pitting. Hyphomycetes secrete organic acids and can actively dissolve carbonates, they produce different kinds of pigments. The filamentous structures of fungal hyphae favour their penetration into the substrate, depending on their structure, chemical composition and state of conservation. Increase mechanical pressure give rise to shrinking and swelling. The black yeasts are slow growing and difficult to isolate and identify. Infield observations, culture experiments on marble slices, and further analyses by SEM and EPR demonstrated that black fungi play an important role in the destruction of marble and limestone. It is possible to identify the presence of fungi using only some milligrams of powdered infected marble. The colour changes of antique marbles with time were related to small climatic changes that were caused by solar activity changes and/or microclimate change due to air pollution and atmospheric eutrophication in large and fast growing cities. Microorganisms can establish different relationships with the substrate: epiliths, growing on the stone surface, and endoliths, living in the interior of the stone. Stone biodeterioration is explained by two different mechanisms: biogeophysical or mechanical processes and biogeochemical processes. The members of Hyphomycetes are frequently responsible for the deterioration of temples, caves and restored stone artefacts.

8.4 Deterioration of Heritage Buildings

Observation on the famous Sun Temple of Konarak, Odisha for the growth of lichen revealed that a more than 500 different spots on main temple, small sculptures erected within the temple premises and boundary walls. A total of 15 species belonging 14 genera and 11 families were found growing on the sites surveyed. Ten lichen species tightly adhere to the substrate forming crust (crustose lichen) and produce secondary metabolites were dominated on almost all the sites while only four leafy (foliose lichen) species and one squamulose species were recorded.

"The seasonal variation is an important factor in the life cycle of lichen. During summer the thallus dried and curled up displaced the constituents of surface materials along with its lower surface. When the thallus gets enough moisture it assumes its normal shape" (Nayak *et al.*, 2018). Das (2020) pointed out that the degradation and discolouration of archaeological objects are due to the production and pigmentation of different types of fungi.14 fungal flora have been identified for the samples. 5 per cent -10 per cent solution of ammonium hydroxide have been used for removing moss and algae. Fungicidal treatment for some years is also recommended for complete eradication of fungi. The nature of the stone substrate and the environmental conditions influence the extent of biofilm colonization and the biodeterioration processes. A critical review has been done on microbial biofilms on buildings of historic interest in Bihar, in a paper microbial deterioration of stone monuments – an updated overview (Stefanie *et al.*, 2009). Recent innovations resulting from molecular biology, are presented and microbial activities causing degradation are discussed in the said paper.

The fungal strains were identified by studying the fungal colonial features like texture, colour, and the morphological features like colour, structure of mycelium (septate or aseptate), position of sporangia, shape, colour, spore character *etc.* For isolation and culture of specific fungi some special nutrient medium are required for the proper growth and sporulation of the microbes. For the preservation, microscopic examination and physiological characterization of airborne fungal spores a suitable growth medium is needed. There are various types of culture media, *e.g.* Czapek's Dox Agar, Potato Dextrose Agar (PDA), Malt Agar, Sabouraud Agar, *etc.*, used for fungal culture.

8.5 Conclusion

There are many factors responsible for deterioration of cultural heritage in Eastern India. All the museum objects are prone to decay very fast if proper preventive as well as curative care is not being taken. There is an essential necessity of assessing the air borne micro spores of fungi and its role in biodegradation of our cultural heritage. Fungi play a considerable role in the deterioration of cultural heritage by their huge enzymatic activity as they have the ability to grow in diverse range of temperature and humidity. Deterioration of organic objects like paper, textile, painting, stone, wood, leather is significantly increased by aeromycoflora mainly by epilithic and endolithic fungi. Regular cleaning of museum objects both at galleries and storage room, microbiological monitoring and environmental control are essential for preventing fungal deterioration. But the climatic condition of Eastern India favours fungal growth. Some researches have been done in the field, but a detailed research is very much essential in this respect.

REFERENCES

Agrawal O P, Dhawan S, Garg, K L, Shaheen F, Pathak A (1988) Study of biodeterioration of Ajanta Wall Paintings, Int. Biod. 24: 121-129

Agrawal O P, Dhawan S (1991) Biodeterioration of Cultural Property. Proceedings of the International conference on biodeterioration of cultural property, NRLC, Lucknow, India, pp. 339-352

Bajpai R, Behera P K, Nayak S K, Satapathy K B, Upreti D K (2017) Lichen growth on Sun Temple of Konark in Odisha, India- A curse or blessing, Cryptogam Biodiversity and Assessment 2(02), DOI: 10.21756/cab.v2i02.11119

Barat R (1969) A study on microbial population in air (Ph.D. Thesis), Department of Microbiology, Bose Institute, Kolkata

Barui N, Chanda S (2000) Aeromycoflora in the central milk dairy of Calcutta, India. Aerobiologia 16: 367–372. https://doi.org/10.1023/A: 1026563719385

Bhattacharyya R N (1986) Experiments with Micro-organisms. Emkay Publication, Delhi, India, pp. 34-50

Bhowmik S K (1975) Protection and conservation of museum collection. The Museum and Picture Gallery, Baroda, Gujarat state. The Department of Museum, Gujrat, India. 209pp.

Bisht A S (2003) Conservation Science, Agam Kala Prakashan, Delhi, India, pp.163-168

Blades N, Oreszczyn T, Bordass B, Assar M (2000) Guidelines on Pollution Control in Museums Buildings, Museum Practice, Issue 15, Museum Association, London, 2000.

Brimblecombe P (1990) The composition of museum atmospheres. Atmospheric Environment, 24B: 1–8

Brawne M (1982) The Museum Interior, Thames and Hudson Ltd, London, pp.120-126.

Chakraborty S, Sen S K, Bhattacharya K (2000) Indoor and outdoor aeromycological survey in Bardwan, West Bengal, India. Aerobiologia, 16: 211 219

Chakraborty T (2021) Biodeterioration of Cultural Property: A case study in Fungal Flora in the museums of West Bengal. Ph. D. thesis West Bengal.

Crook B, Burton N C (2010) Indoor moulds, Sick Building Syndrome and building related illness. Fungal Biology Reviews 24: 106-113

Cunningham D D (1873) Microscopic examination of air. Govt. Printer, Calcutta, India

Das G K (2020) Conservation of Temple Architecture of Odisha, Odisha Review, ISSN No.0970-8669, pp.35-39

Dhawan S (1986). Microbial deterioration of paper material - A literature review. Government of India, Department of Culture, National Research Laboratory for Conservation of Cultural Property. Ed. M.M. Khan. Lucknow, India. 1-18.

Garg K L, Garg N, Mukerji K G (1993) Recent Advances in Biodeterioration and Biodegradation, Vol-1. Biodeterioration of cultural heritage. Naya Prokash, Kolkata, India, 496 pp.

Garg K L, Garg N, Mukerji K G (1994) Recent Advances in Biodeterioration and Biodegradation, Vol-2. Naya Prokash, Kolkata, India,

Ramesh B, Jeyaraj V (2005) Preventive conservation of Information materials with special reference to manuscripts, The director of museums, Government museum, Chennai, pp. 23-45

Kamalakar G, Rao V P (1995) Conservation, Preservation and Restoration: Tradition, Trends and Techniques. Birla Archaeological and Cultural Research Institute, Hyderabad, India, 316 pp.

Kundu A (2012) Heritage of Gandhi Smarak Sangrahalaya, Gandhi News, bulletin of Gandhi Museum, January-March, pp.25-32

Mills S J, White R (1987) The Organic Chemistry of Museum Objects. Butterworth and Co Ltd. pp.1-19

Mitra J N, Chaudhuri S K (2002) Studies in Botany. Moulik Library, Calcutta, pp.522-539

Nair S M (1972) Conservation in the tropics. Proceedings of the Asia Pacific seminar on conservation of cultural property, New Delhi, Feb. 7-16, p. 150-158

Nair S M (1982) Biodeterioration of Natural History specimens- Causes and Control. Jour. Ind. Mus.38: 27

Nair S M (2011) Biodeterioration of Museum Materials. Agam kala Prakashan, Delhi, India, pp.2-15, 38-40

Nayak S, Behera P, Bajpai R, Satapathy K, Upreti D (2018) A Need for Lichen Bio-deterioration Study on Ratnagiri and Udayagiri Excavation Site of Jajpur, Odisha. 20-23.

Nitterus M (2000) Fungi in archives and libraries. 21(1): 25-40

Padfield T (2002) The window in context: the interplay between building components.pp. 1-42

Plenderleith J H, Werner A E A (1971) The Conservation of antiquities and works of art. Oxford University Press, London, pp.125-142

Santra S.C. and Chanda S.: 1981, Indoor airborne fungal spores of Calcutta, West Bengal. Proc Nat Conf Env Biol, 45–48.

Savkovic Z, Stupar M, Unkovic N, Ivanovic Z, Blagojevic J, Vukojevic J (2019) In vitro biodegradation potential of airborne Aspergilli and Penicillia. The Science of nature 106(3-4): 8. DOI: 10.1007/s00114-019-1603-3

Scheerer S, Morales O, Gaylarde C (2009). Chapter 5 Microbial Deterioration of Stone Monuments-An Updated Overview. Advances in applied microbiology. 66: 97-139. 10.1016/S0065-2164(08)00805-8.

Shah N R, Arya A (2000) Studies on some fungal biodeteriogens. Bharatiya Kala Prakashan, Delhi, India, 202pp.

Sengupta S R, Nigam S S, Tandon R N (1950) A New Wool-Destroying Fungus– Ctenomyces Species. Textile research Journal, 20: 671-673.

Shaver C L, Cass G R, Drujik, J R (1983) Ozone and the deterioration of works of art. Environmental Science and Technology. 17(12): 748-752

Stefanie S, Ortega M O, Christine G (2009) Microbial deterioration of stone monuments-An updated overview advances in applied Microbiology 66: 97-139

Tilak S T (1991) Biodeterioration of Paintings at Ajanta. In: Biodeterioration of Cultural Property. Agrawal O P, Dhawan S, (eds.) Macmillan: New Delhi, India 1991: 204–212.

Tilak S T, Vishwe D B (1980) Aeroallergenic pollen at Aurangabad. Adv Pollen Spore Res 7: 145, 1980

Videla HA (1996) Manual of Bio-corrosion. Boca Raton, PL, USA: Lewis Publishers/CRC   Press; 1996. pp. 288.

Chapter 9

Aerobiological Researches for the Prevention and Conservation of Cultural Heritage

Hina Upadhyay and Debjani Choudhary

School of Agriculture, Lovely Professional University, Phagwara – 144 001, Punjab, India
e-mail: hina_boston@yahoo.com

ABSTRACT

Our cultural legacy is essential for maintaining our identity. It connects us to the past – to specific cultural ideas, beliefs, customs, and traditions – helping us to identify with others and improve our sense of belonging, and helps us in bringing national pride. Buildings, monuments, landscapes, works of art, and artefacts are examples of tangible culture; folklore, traditions, language, and knowledge are few examples of intangible culture; and National parks, wildlife, natural heritage is an example of natural heritage. Cultural heritage sites include hundreds of old structures and ancient town sites, noteworthy archaeological sites, and monumental sculptures or painting works. Cultural assets are destroyed due to physical, chemical, physicochemical, and biological reasons. Many studies look at biological harm to cultural material, but just a few look at aerobiological assessments as a way to prevent it. Any country's Archaeological Survey department, which is part of the Ministry of Culture, is responsible for conservation of heritage monuments. To restore monuments to their aesthetic, cultural, and historic worth, two basic approaches are used. Structural preservation and chemical preservation are two of these options. The present paper aims to highlight the scope of aerobiological studies and preventive measures to control the deterioration of cultural heritage by certain biodeteriogens.

Keywords: Aero-bio components, Cultural heritage, Sustainable management, Eco-friendly methods, Samplers.

9.1 Introduction

Aerobiology has been a popular applied field in India for the last fifty years, as visualized by a large amount of literature generated in the field. Although the quantity of fungi appeared to rise in wet weather when the amount of inorganic dust reduced, there was no direct link of the deterioration caused. Aero-mycological researches in India restarted over half a century after Cunningham's pioneering work, but this time by plant pathologists. While Chatterjee (1931) and Mehta (1933) explored the dissemination and distribution of wheat rust spores, Padmanabhan and Ganguly (1954) investigated the prevalence of *Helminthosporium oryzae* conidia in the air over paddy fields in Cuttack, which was responsible for the legendary Bengal famine. *Alternaria, Curvularia, Fusarium*, and *Helminthosporium* were discovered to be the most prevalent in army supplies in Kanpur; *Mucor, Aspergillus, Helminthosporium* and smut spores *Alternaris, Fusarium*, hyphal fragments were found abundantly in the air of Barshi, Mahrashtra (Patil *et al.,* 2016).

Sreeramulu and Ramalingam (1966) work done in Visakhapatnam with the Hirst Volumetric Spore Trap can be considered to be the beginning of systematic and intensive investigations of Aerobiology in India. Hirst's (1952) invention of an "Automatic Volumetric Spore Trap" signalled a new era in aerobiological research around the world, especially in India. In this area, the AIA (Italian Aerobiology Association) working group "Aerobiology and Cultural Heritage," established in 1992, has produced scientific papers, manuals, and reports at national and international congresses (*e.g.* ICA), as well as supported the development of technical standards (UNI and CEN) for both the preservation of artefacts and the health of operators.

The bioaerosol contains different types of bio components among them fungal spores are the dominant aeromycoflora present in such type of environment, Fungi with tall conidiophores that penetrate into or through the laminar boundary layer of specific liberation mechanisms that forcibly project spores through this layer are frequently well adapted to airborne dispersal of their spores. Fungi have probably used the wind for dispersal more thoroughly than any other species of organisms throughout evolution, and as a result, they dominate the aerospora (80-90 percent). Many cultural works are affected by environmental factors such as temperature, humidity, and light and UV light exposure. Taking adequate precautions to protect goods in a controlled setting with such variables. Preventive conservation is defined as conservation within a range of damage-limiting levels. The most common cause of information resource degradation was discovered to be dust and particulate particles. The deterioration of cultural materials is also influenced by relative humidity, excessive acidity levels, and high temperatures.

9.2 Aeromycological Studies in India

Over the previous 50 years, three distinct trends have emerged: (i) Outdoor aero mycology – monitoring airborne fungal spores in the atmosphere of metropolitan cities and towns, including simultaneous comparative research on urban and rural locations; (ii) Indoor aero mycology –mold enumeration in indoor environments; and (iii) crop field air mycoflora. Although fungal spores are almost constantly present in the air, their number and kind vary depending on the time of day, weather, season, vegetation, and geographic location. They were dedicated to monitoring the airborne spores in the environment of numerous cities and towns using various methodologies, and they played an essential part in the etiology of respiratory allergy illnesses.

9.3 Sampling of Airborne Bioparticles

Many researchers used gravity slides and settle plates, while others used vertical cylinders, rotarod samplers, and aeroscopes, and still, others used volumetric samplers such as Hirst and Burkard traps, Andersen samplers, and Tilak samplers for trapping of airborne bio-components. The sampling period varied from place to place, ranging from a few months to a couple of years. The data on airborne fungal spores was connected with meteorological conditions in long-term surveys.

Figure 9.1: Burkard Sampler.

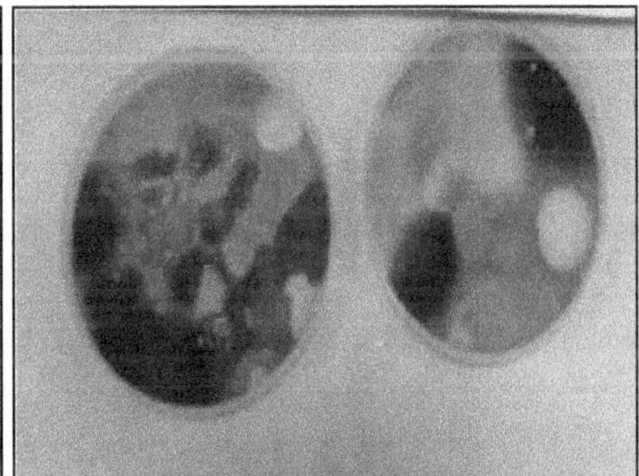

Figure 9.2: Airborne Culture of Mould on SDA Medium.

(Trapping of airborne fungal spores)

9.3.1 The Common Equipment for Trapping Viable Particles

Generally Surface Air System Sampler (flow rate of 90–180 l/min) and Andersen Microbial Air Sampler (flow rate of 28.3 l/min). The data will be commonly recorded in CFU/m3 (colony-forming units per cubic meter). Non-viable bioaerosol (*i.e.,* fungal spores and pollen) is collected using the Hirst Sampler and Burkard traps (Flow rate 10 l/min), (Figure 9.1), which should then be recognized morphologically using microscopy.

For example, sampling viable bioaerosol necessitates extra care in order to keep organisms alive for further cultivation and analysis. Before being put into a culture medium, microorganisms are collected in a solid or liquid culture medium. Passive sampling is used to determine the rate at which microorganisms settle on surfaces by exposing Petri dishes with an appropriate culture medium for a given period of time (usually 10 min.). The results are generally reported in colony-forming units (CFU/dm^2) per dm^2 of surface area (Tilak 1989).

9.4 Biodeterioration of Cultural Heritage

Both microbes and insects have been found responsible to deteriorate cultural heritage which is organic in nature. The effect of biodegrading organisms/components on ethnological artefacts differ depending on the type of organic material present. It may be wood, paper, cloth, leather, palm leaf or parchment. Fungal organisms disfigure, decolorize and make the objects fragile.

Scholars also noted a number of insect species that can harm cultural or heritage objects such as cellulosic materials, leather, placed in art galleries and libraries. Insects can be damaging at various stages of their life cycle. Certain insects eat during the larval stage, which is when they cause undetected damage. Though many insects, such as silverfish, cockroaches, wood borers, carpet beetles, and others, are known to have a global distribution. Not only had they feed on organic objects the insects may act as carrier in spread of microbes from one object to other one.

9.4.1 Microorganisms and Insect Pests

Bacteria, fungi, other microbes and insect pests, are engaged in the destruction of cultural property; fungi, in particular, play a crucial role, but the degree of infestation is dependent on the specific conditions and nutrients available (Walston, 1980). Monthly, weekly, and daily, the makeup of these bacteria in the air changes. Various organic materials, such as library materials, leather, wood, museum, and historical material, provide suitable conditions for the growth of microbes. Fungi can withstand high temperatures, and Walston (1973) studied the deterioration of ethnographic artefacts in Australian museums, reporting that mould has always been a problem in such a setting.

9.5 Aeromycoflora and its Impact on the Deterioration of Cultural Heritage

Many studies look at biological harm to cultural material, but just a few look at aerobiological assessments as a way to prevent it. The development of instrumentation and technologies for biological monitoring of air and surfaces, the definition of biological risk thresholds for different types of artworks, and the creation of a "biodeterioration risk card" could open up new perspectives for aerobiology in the field of cultural heritage. The greenery in the immediate vicinity. Dead and decomposed stuff has also been discovered to be a major source of multiplication and spread of biodeteriogens.

The prevalence of deuteromycetous fungus in the aerospora has been confirmed by all scientists, regardless of sample location, length, or technique. In some regions, basidiospores were the second most common spore type, whereas in others, aero spores were the second most common spore type. Deuteromycetous conidia have been found throughout the year, with the highest concentrations in November-December and the lowest concentrations in April-May, both of which are dry and hot months.

Aero-spores were found to be most prevalent from July to September, which corresponds to the rainy season, and they were most numerous at night. When the temperature is high in April and May, the concentration of ascospores is significantly reduced. There appears to be a strong link between rainfall and ascospore discharge. Along with *Ganoderma* (Thakur 2018) and *Coprinus*, rust fungi's basidiospores, urediniospores, and teleutospores, as well as smuts chlamydospores were frequently seen. Comparative simultaneous research on the aeromycoflora of urban and rural locations revealed that the dispersion of airborne spores varied in both variety and magnitude from place to place and year to year.

Spores can arise from a variety of places within a building. They could be caused by fungi growing in condensation on walls, food, or mouldy stored spores that build in house dust. Human activity within the building could spread them. When opposed to the outdoors, there are limited studies on the mycoflora of indoor air in India. Tilak and colleagues found high concentrations of *Cladosporium, Penicillium, Paecilomyces*, and *Aspergillus* in libraries. These moulds have been discovered to be cellulolytic and biodeteriogenic. After agitation of books, increased amounts of *Aspergillus niger, Penicillium* spp., and *Cladosporium* were observed in the atmosphere (Singh *et al.*, 1990). Nadimuthu and Vittal (1995) found high amounts of cultivable mould in the indoor air of libraries. When compared to traditionally ventilated libraries, airborne spore concentrations were low in air-conditioned libraries. The stack room had a higher number of species, which was attributed to dust and its dispersion during the handling of books. By exposing culture plates to the air of the majority of the indoor habitats investigated, the preponderance of *Aspergillus, Penicillium*, and *Cladosporium*, as well as several moniliaceous fungi, has been confirmed. Singh *et al.* (1990) found sixteen species of *Aspergillus* in the bakery environment, with *A. flavus* being the most common in the storage area and *A. niger* in the packing section.

9.6 Organic Materials Available as a Food Source

These tiny creatures have distinct dietary preferences; some prefer carbs and proteins, while others prefer both proteins and carbohydrates. Carbohydrate diets include cellulose, starch, wood, gums, cotton, and other cellulosic materials, which degrade the substance of organic things associated with cultural heritage.

9.7 Conservation of Monuments

✰ Adopting monuments for their long life is a good idea.

✰ Strict government action and a plans are needed to safeguard sites, including the establishment of Archaeological departments, and bodies like Archaeological survey of India (ASI), Indian Conservation Institute (INTACH).

✰ A public awareness campaigns should be organized time to time to educate the public about the cultural wealth present in historical monuments, temples and different museums of India.

✰ Collection and generation of funds and donations for the maintenance of cultural heritage and their renovation if desired.

✰ Chemical treatments and repair of historical Monuments.

9.8 Eco-friendly Methods for the Conservation of Cultural Heritage

Essential oils (EOs) have recently yielded successful natural biocides for use as an alternative to chemicals in the restoration of cultural assets. In the first experiment, EOs from *Lavandula angustifolia* Mill. and *Thymus vulgaris* L. were evaluated against cyanobacterial biofilms at a concentration of 5 per cent. (Ranaldi *et al.,* 2022). Bio cleaning with live bacterial cells or hydrolytic enzymes is now recognized as a valuable resource for restoring cultural property while minimizing dangers to artworks and human health. In the removal of black crusts from stone surfaces or organic materials such as glue and/or adhesives, from paintings and other substrates, new approaches based on sulphate-reducing bacteria or bioactive compounds with hydrolytic activity have been used as selective and safer cleaning methods (Balloi and Palla 2017).

Neem leaves: Neem leaves are now commonly used to preserve paper and textiles from fungi and insects. Azadirachtin and Quercetin, the secondary metabolites, possesses antibacterial and insceticidal effects.

Pepper: *Piper nigram* has been used as an insecticide for cellulosic material stored in these types of indoor locations.

According to the literature, there is a need to propose preventive, control, and novel eradication techniques for the management of deteriorating Heritage items, such as buildings, caused by the proliferation of bacteria, and hazardous fungi.

9.9 Recommendations and Scientific Approach of Preservation

Protection and conservation of cultural heritage:

1. Inform as many people as possible about the thefts; Raise public awareness of the country's and other countries' cultural heritage;

with the help of cultural institutions, provide training courses for law enforcement, customs, and judicial agencies.

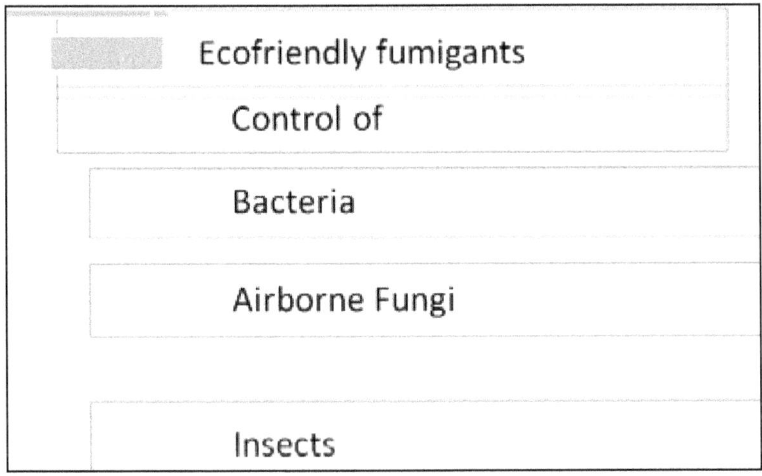

Figure 9.3: Eco-friendly Methods for the Preservation of Cultural Heritage.

2. Integrating education about the costs and consequences of global antiquities trade into school curricula and university syllabi, as well as local outreach in community organizations, libraries, churches, and other public places, is one of the most important ways to address the problem of the ongoing loss of global heritage.

3. According to UNESCO, there are two ways to preserve cultural heritage: one is to document it in concrete form and save it in an archive, and the other is to keep it alive by assuring its transfer to future generations. SLT was established in reaction to the second approach.

4. The first technique to preventing biodeterioration entails inspecting and maintaining heritage surfaces, reducing biological particles in their environs, and actively altering the bioreceptivity of heritage surfaces.

5. Biofouling begins when microorganisms migrate from free-floating planktonic to stationary sessile lifestyles, resulting in the creation of a biofilm. They use an adhesive termed extracellular polymeric material to stick to one other and to a hard surface (EPS). A metal (copper alloy) anti-fouling sensor guard that fits on the end of the sonde and shields the water quality probes can be used for effective biofouling deterrence. This guard is used instead of the standard plastic guard that comes with the instrument.

6. Storage cabinets with a unique design,

7. Use of dehumidifiers if required

8. Active regulation for the control of microenvironment

Researches on Aerobiology are being utilized to prevent and conserve valuable cultural heritage.

9.10 Conclusion

The present paper focuses on the application of Aerobiological research for cultural heritage preservation and prevention. On the basis of the present study, we can conclude how aerobiology is applicable to the preservation and conservation of cultural assets as a preventive measure. The development of instrumentation and technologies for biological monitoring of air and surfaces, the definition of biological risk thresholds for different types of artworks, and the creation of a "biodeterioration risk card" could open up new perspectives for aerobiology in the field of cultural heritage. The greenery in the immediate vicinity, dead and decomposed stuff has also been discovered to be a major source of biodeteriogens.

REFERENCES

Balloi A, Palla F (2017) Bio cleaning. In Biotechnology and Conservation of Cultural Heritage (pp. 67-84). Springer.

Chatterjee G (1931) A note on an apparatus for catching spores from the upper air. Indian J. Agr. Sci. 1: 306-308.

Hirst J (1952) An automatic volumetric spore trap. Annals of Applied Biology 39 (2): 257-265.

Fidanza M R, Caneva G (2019) Natural biocides for the conservation of stone cultural heritage: A review. Journal of Cultural Heritage, 38: 271-286.

Mehta K C (1933) Rusts of wheat and barley in India. A study of their annual recurrence, life histories, and physiologic forms. Indian J. Agr. Sci. 3: 939-962.

Nadimuthu N, Vittal BPR (1995) A volumetric survey of culturable molds in library environment. Ind J Aerobiol 8: 12-16.

Padmanabhan S Y, Ganguly D (1954) Relation between the age of rice plant and its susceptibility to *Helminthosporium* and blast diseases. Proc. Indian Acad. Sci. Sect. B. 1954: 39: 44–50. doi: 10.1007/BF03050372.

Palla F (2020) Biotechnology and cultural heritage conservation. In : Heritage. London, UK: Intech Open.

Patil MT, Kanade MB, Mali NS (2016) Present Status of fungal diseases of jowar fields at Barshi area, Maharashtra. Proceedings of International Conference Plant Research and Resource Management organized by Tuljaram Chaturchand College, Baramati, Dist. Pune (M.S.),India, 166-168

Ranaldi R, Rugnini L, Gabriele F, Spreti N, Casieri C, Di Marco G, Bruno L (2022) Plant essential oils suspended into hydrogel: Development of an easy-to-use protocol for the restoration of stone cultural heritage. International Biodeterioration and Biodegradation, 172, 105436.

Singh A, Singh AB, Bhatnagar AK, Gangal SV (1990) Prevalence of Aspergilli in bakery environment. Ind J Aerobiol 3-15.

Sreeramulu T, Ramalingam A. (1966) A two year study of the airspora of paddy field near Visakhapatnam. Indian J Agaric. Sci, 36: 111-132

Thakur V A (2018) Aerobiological studies with special reference to airborne basidiospores of *Ganoderma* Karst. at Pune, Maharashtra, India. Anals of Plant Sciences7(5): 2209-2212

Tilak, S. T. (1989) Atlas of airborne pollen and fungal spore. Vaijayanti Prakashan, Aurangabad, 1: 316.

Tilak S T (1989) Atlas of airborne pollen and fungal spore. Vaijayanti Prakashan, Aurangabad, 1 (1989): pp 316.

Walston S (1973) Preservation of ethnographic collections in the Australian Museum, Sydney. Proceedings of the National Seminar on Conservation of Cultural Material, Perth ed. C. Pearson: 81-87.

Walston S (1980) The conservation of Cultural Materials in Humid Climate Regional Seminar on Conservation of Cultural Materials in Humid Climate.

Chapter 10

Biodeterioration of Wall Paintings of Archaeological Monuments in India

D.P.Tewari[1] and Shashi Dhawan[2]

[1]*Professor (Rtd), Department of Ancient Indian History and Archaeology,
University of Lucknow, Lucknow – 226 007, Uttar Pradesh, India
e-mail: tewaridplu@gmail.com*
[2]*Late, Senior Scientist, NRLC, Aliganj, Lucknow, Uttar Pradesh, India*

ABSTRACT

Wall Paintings form an important parts of nation's cultural heritage. It is well known that the paintings because of their complex structure are easily affected by biological agencies like microorganisms, bats and birds etc. The microbes growing on wall paintings include actinomycetes, algae, bacteria and fungi. Fungi are exceptionally abundant on the earth and have variety of fermentative abilities and high ecological adaptabilities which give them a highest place among other bio deteriorating agencies.

Bats and birds due to excreta damages paintings by encouraging the fungal activity and also cause bad smell, which in turn affects the number of visitors to that particular site. Fungal activity causes damage through discoloration of paint, flaking of paint and by imparting different types of colored spots which diminishes the esthetic beauty of the paintings. The present paper deals with the fungi present on some wall Paintings as well as their control

Keywords: Biodeterioration, Wall paintings, Monuments, Microbial, Fungi, Preservation.

This paper was presented by first author in ACHAM, Annual Conference at Jishou, China on 21 October, 2014.

10.1 Introduction

Biodeterioration is a serious problem throughout the world. Similarly, like other objects of art, wall paintings are also prone to deteriorate with biological agencies. The study of biodeterioration is a very young field of science in the field of conservation. It has drawn the attention of conservators as they have always been interested in chemical deterioration. At the initial stage of attack biological agents are relatively easy to control, but once such an attack is wide spread it is extremely difficult to deal with. The microorganisms develop on the paintings only if the relative humidity of the surrounding exceeds more than 75 per cent and temperature ranges 20-35°C. Since early times wall paintings are supposed to be one of the highest forms of art. The art of wall painting in India reached its zenith during the period starting from 2nd Century BC to 6th century AD., wall painting which are the integral part of the building structure are

Considered as sources of invaluable information of the contemporary society and occupy a place of pride in the repository of any cultural heritage of archaeological monuments/buildings. Wall paintings are generally classified as Fresco or Stucco and Tempera depending upon on the technique of execution (Agrawal 1969-70). In India we generally come across tempera paintings. At National Research Laboratory for Conservation of cultural property Shashi Dhawan and her team studied number of wall paintings of India and experimented with number of aspects related to the biodeterioration (Dhawan *et al.*, 1992; Dhawan and Sharma, 1994; 2005). In the present paper two archeological monumental sites were selected in two different states.

1. **Nagaur Fort, Rajasthan (Figures 10.1–10.2):** Nagaur, one of the oldest towns in Rajasthanis situated near to the north-east of Jodhpur. According to legend, Nagaur is said to have been founded by Nagvansis Rajputs of ophidian race and have been the capital city of Mahabharat period. According to another local tradition it was founded at the command of Prithaviraj Chauhan by his sardar, named Rai Vishal of the Pandrali Estate. According to some other historians in 1560, the Mughal emperor Akbar occupied Nagaur and after 16 years *i.e.*, 1576 gave it to Rai Singh of Bikaner as gift.

   Although the fort is old, there are hardly any structures of early period inside it, because every new ruler changed various parts according to his own likings. Most of the present structure is of Maharaja Bakhat Singh's time (1724-1750 AD). Inside it most of the walls are having paintings of that period (Maharaja Bhakat Singh, Agrawal 1977). In Sheesh Mahal on the ceiling flying winged celestial figures and other scenes are depicted (Figure 10.3)

2. **Kali Ji ka Mandir (Temple) of Ram Nagar Fort, Varanasi, Uttar Pradesh.**

Figure 10.1: General View of Nagaur Fort, Rajasthan.

Figure 10.2: General View of Sheesh Mahal, Nagaur Fort, Rajasthan.

10.2 Materials and Methods

10.2.1 General Observation of the Wall Painting Site

It was visually observed for action of birds, bats excreta, water seepage, building structure dust and dirt *etc.*, These studies are to be encouraging to understand the problem

Figure 10.3: Sheesh Mahal Paintings Covered with Dirt and Dust.

10.2.2 Visual Observations

Visual observations of the selected site were made with the help of magnifying glass. Fungal growth was confirmed by lightly pressing the cello tape over the suspected affected are, and then examine under the microscope.

10.2.3 Mycological Study

As the paintings were fragile and delicate a nondestructive method was adopted to isolate fungi (Dhawan and Agrawal 1986). The methods applied were:

i. Cotton Swab Method

Sterilized cotton swabs were gently rubbed over the painted surface and then pressed on Petri plate of Czapek Dox Agar media (CDA) and Malt Extract Agâr media (MEA). The plates were sealed with tape and wrapped to allow further microbiological growth.

ii. Paperstick Method

A 1cm square piece of filter paper slightly moistened with water and sterilized, was lightly pressed over the affected surface, placed in a flask with 5 ml of sterile water and shaken well. Then 1ml of the suspension was pipette in to sterile Petridish CDA media pour plates were prepared. After four days at room temperature isolations were made and pure isolations were carried out in the respective media slants.

Fungal species were identified with the help of standard works (Raper and Fennell 1965; Pitt 1980; Onions *et al.*, 1981).

10.3 Results and Discussion

There was lot of dust, bird's excreta *etc.* in the area of different Mahals of Nagaur Fort painting site. While observing wall paintings it was found complete blackening of faces and at some places hands also (Figures 10.4–10.6). Quite a good area was having yellowish brown color patches which may be due to the water seepage (Figures 10.7–10.8).

Figure 10.4: Sheesh Mahal Paintings, Blackening on Faces due to Fungal Infection.

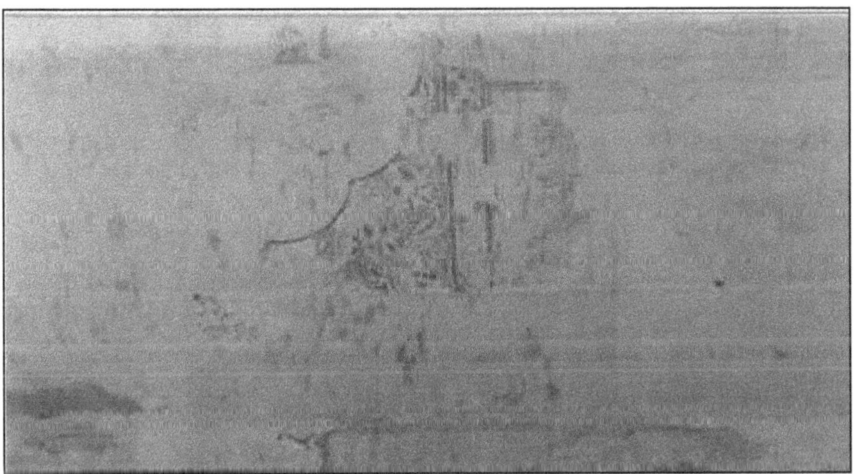

Figure 10.5: Sheesh Mahal, Badly Deteriorated Paintings.

During the present investigation sites, the fungi isolated from Nagaur Fort wall paintings, Rajasthan, were *Alternaria alternata, Aspergillus flavus. A. fumigatus., A. niger, Cladosporium cladosporioides, Chaetomium* sp. *Curvularia lunata, Emericella nidulans* and *Humicola* sp.

Figure 10.6: Sheesh Mahal, Cracks in Paintings.

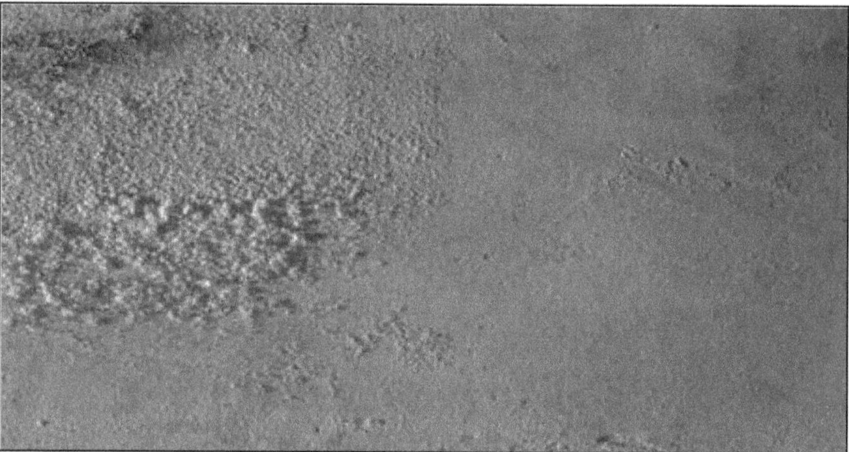

Figure 10.7: Sheesh Mahal, Biodeterioration and Salt Efflorescence.

The Kali ji ka Mandir (Temple) of Ram Nagar Fort, Varanasi (Figures 10.9–10.10), and building structure was badly affected by ground water capillary action (Figures 10.11–10.12). It was told that there are a lot of visitors during festive seasons otherwise ambiance was quite neat and clean. There were patches of green, cream brown on paint layer at some places paintings were completely disappear, however at some places white paint showed blisters. Fungal forms (Figure 10.14) isolated were, *Alternaria alternata; Chaetomium* sp. *Curvularia pallescens, Emericella ruber, Fusarium culmorum, Penicillium* sp. *P. granulatum* and *Humicola* sp.

It may be concluded that there is a wide range of fungi growing on wall paintings but *Alternaria alternata* and *Chaetomium* sp. were commonly occurring forms. It is well known that some fungi having hard fruiting bodies represented by sclerotia, pycnidia, perithecia, claustrophobia and stroma could

Figure 10.8: Fading of Paint Layers due to Fungus.

organize themselves inside the painted layer. The fungi having such structures were identified from our experimental sites *viz.* like *Emericella nidulans* and *Chaetomium* sp. These structures with little increase in humidity may cause swelling in the paint film, finally fungi grow profusely beneath and over the surface and powdering, flaking *etc.* could be noticed. Besides the physical damage of the paintings the fungi have also been responsible for chemical decay through their metabolites. Generally, both the processes occur simultaneously playing important role in deterioration of wall paintings. This was confirmed by (Dhawan and Sharma 2005)

10.4 Recommendations for Preservation

On the basis of present studies conducted following points were recommended:

**Figure 10.9: Ramnagar Fort, Varanasi Built by
Raja Balvant Singh in 1742-50 A.D.**

Figure 10.10: Kali Mandir, Ramnagar Fort, Inner Varanda and Courtyard.

i. Preventive Methods

☆ Periodic inspection and good housekeeping.

☆ Elimination of birds by putting wire mesh in windows and passages without doors should have doors.

☆ Removal of plants and grasses from the roofs.

Prohibit visitors to eat on site.

☆ If at all it is not possible to save paintings from the above building defects then it may be transferred to some other place by special transfer methods.

Figure 10.11: Kali Mandir, Fungal Infection, Breaking of Plaster, Fading of Color.

Figure 10.12: Deterioration in Paintings Caused by Seepage and Ground Moisture.

Figure 10.13: Kali Mandir, Fungal Impact on Paintings and are being Weathered.

ii. Curative Methods

☆ One of the biggest problems is not only to prevent but also to kill them without damaging the paintings. On the basis of in vitro studies carried out (Dhawan *et al.*, 1992) on the efficacy of p-chloro-m cresol

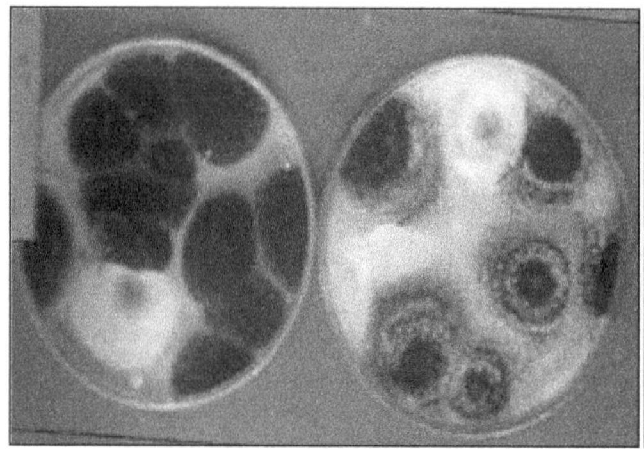

Figure 10.14: Fungi Growing on Culture Media, 1.CDA, 2. MEA.

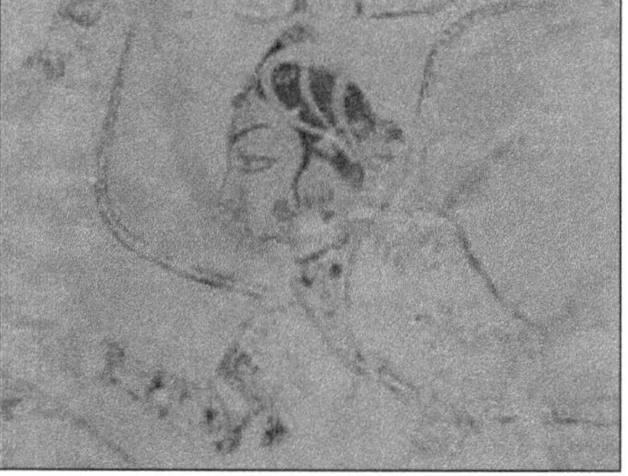

Figure 10.15: Nagaur Fort, Paintings in the Hadi Rani Mahal, after Conservation.

against different fungal forms isolated from wall paintings, 0.02 per cent solution in rectified spirit can be sprayed on the wall.

☆ 0.1 per cent solution of phenyl mercuric acetate in rectified spirit can also be used. This biocide has been used in several countries. (Strzelczyk and Ochrona 1978; Roznerska 1973 and Jaffries 1986). However, we have used p-chloro-m cresol and the results were satisfactory (Figure 10.15).

It is to be noted strictly that before using any biocide it should be tested on the unimportant side of the painting before applying any biocide on wall painting.

REFERENCES

Agrawal O P (1969-70) A Study on technique of wall paintings, Journal of Indian Museums, 25-26: 99-118.

Agrawal O P (1989) Examination and Conservation of wall Paintings of Sheesh Mahal, Nagaur. pp.1-58. (A Programme under National Project on wall Paintings).

Agrawal O P, Dhawan S, Garg KL, Shaheen F, Pathak N, Mishra A (1988) Study of Biodeterioration of Ajanta wall Paintings, Int. Biodet. 24: 121-129.

Agrawala R A (1977) Marwar Murals, Agam Kala Prakashan, Delhi, PP. 16-17.

Dhawan S, Agrawal O P (1986) Fungal flora of miniature paper paintings and Lithographs, International Biodeterioration 22 (2): 95-99.

Dhawan S, Sharma N (2005) Influence of temperature and pH on the cellulolytic activity of Ajanta wall paintings fungi: In Biodeterioration of cultural Property-4 (Eds. O.P. Agrawal and S. Dhawan) pp. 251-259, Publisher, International Council of Biodeterioration of Cultural Property, Lucknow, India.

Dhawan S, Garg K L, Pathak N (1992) Microbial analysis of fungal deterioration of Ajanta wall paintings and their control in situ. In proceedings of 2ndInternational Conference on Biodeterioration of Cultural Property, 5-8 October, Yokahama, Japan, pp. 59-60.

Dhawan S, Garg K L, Shaheen F (1994) Biological decay of wall Painting- A case study of Ajanta wall paintings, In Conservation of wall Paintings in India- Achievements and Problems (Ed. O.P. Agrawal), pp. 135-140.

Gaylarde C, Gaylarde P M (2000) Biodeterioration of Painted walls and its Control, First International RILEM workshop on Microbial Impact on Building Materials (Eds.) Ribas Silva, RILEM Publications, pp. 28-33.

Gorbushina A A, Mertens J, Petersen K (1999) Distribution of microorganism on ancient mural paintings as related to associated fungal elements, International Biodeterioration and Biodegradation, 44: 159-160.

Jaffries P (1986) Growth of *Beauvaria* on mural painting on Canterbury Cathedral Bio-lab University of Kent. Canterbury, Kent, U.K.

Jeffries P (1991) Biodeterioration of wall Paintings in Canterbury Cathedral; Agrawal, O.P. Dhawan. S. (Eds.) Biodeterioration of Cultural Property, Macmillan, New Delhi, India, pp. 287-293.

Onions A H S, Allsopp D, Eggins H O W (1981) Smiths's introduction to industrial Mycology. London E. Arnold Ltd. 398 p.

Pitt J l (1980) The genus *Penicillium* and its telomorphic states *Eupenicillium* and *Talaromyces*, London, Academic Press.

Raper K B, Fennell D I (1965) The Genus *Aspergillus*. Williams and Wilkins, Philadelphia, 686 p.

Roznerska M (1973) Acta Universitatis, N. Copernici, 5 (52): 183-201.

Strzelezyk A, Ochrona Z (1978) English Summary, 31 (2): 128-131.

Strzelezyk A (1981) Paintings and Sculptures, In Rose, A. H. (Eds.) Microbial Biodctcrioration, Academic Press, London, 20-34.

Zammit G, Psaila P, Albertano P (2008) An Investigation in to Biodeterioration Caused by microbial Communities Colonizing Art works in three Maltese Palaeo- Christian Catacombs, 9th international Conference on NDT of Art, Jerusalem, Israel, 25-30 May.

Part – II

Biodeterioration of Heritage Monuments

अप्रतिम स्थापत्य कला

कला को प्रणाम
प्रणाम भारतीय स्थापत्य कला के उत्कृष्ट
नमूनों को
बुलाती हैं मदुरई मीनाक्षी मंदिर की
मनोहारी मनभावन मूर्तियां
बोलते हैं पत्थर पद्मनाभ मंदिर के
विराजे हैं भगवान विष्णु यहां तो
तिरुपति में बालाजी
और स्थापित है शिवलिंग बनारस में
हमारी हिंदू संस्कृति के दर्शन
अयोध्या और मथुरा वृंदावन में
अमृतसर के स्वर्ण मंदिर की ख्याति देश-विदेश में
हमारी प्राचीन सभ्यता के प्रतीक रहे हैं
सांची और सारनाथ के बौद्ध स्तूप
खजुराहो की मूर्तियां और
अजंता एलोरा की गुफाएं
माउंट आबू के जैन दिलवाड़ा मंदिर की
अद्भुत है वास्तुकला
निराली है शान दिल्ली और आगरा की
जामा मस्जिद की
ख्वाजा अजमेर शरीफ का अपना ही महत्व है
वास्तुकला के निशां फैले हैं
राजस्थान के महलों में
आगरा का ताज है अप्रतिम प्रेम की निशानी
विश्व की अनमोल धरोहर है यह।

Chapter 11

Disaster Preparedness Strategies for Taj Mahal

Mohd. Daniyal and Abduraheem K.

Department of Museology Aligarh Muslim University, Aligarh – 202 002, Uttar Pradesh, India
e-mail: mohddaniyal723@gmail.com; ka_rahim_65@hotmail.com

ABSTRACT

An immensely beautiful mausoleum of white marble was built in Agra between 1631 and 1648 AD by the order of the Mughal emperor Shah Jahan in the memory of his favourite wife Mumtaz Mahal. The Taj Mahal is a jewel of Muslim art in India and one of the universally admired masterpieces of the world's heritage. The Taj Mahal was included in the UNESCO World Heritage Site in 1983, and millions of tourists visit the Taj Mahal every year. The management of Taj Mahal complex is carried out by the Archaeological Survey of India and the legal protection of the monument and control over the regulated area around the monument is through the various legislative and regulatory frameworks that have been established, including the Ancient Monument and Archaeological Sites and Remains Act 1958 and Rules 1959 Ancient Monument and Archaeological Sites and Remains (Amendment and validation); which is adequate to the overall administration of the property and buffer areas.

There are many probabilities of disaster at Taj Mahal, as it comes under 2nd and 3rd grades of earthquake zone, moreover the chances of flood are high because it is situated at the right bank of Yamuna River. The recent thunderstorm damaged the parts of marble railing and minaret of East gate. There are frequent threats of demolition and vandalism from anti-social elements.

Keywords: Disaster preparedness, Strategies, Earthquake, Floods, Terrorist attack, Taj Mahal

11.1 Introduction

One of the Seven Wonders of the World, Taj Mahal is facing several problems, causing concern to the entire world. It is increasingly exposed to the disasters caused by natural and human-induced acts, such as fire, flood, earthquakes, thunderstorms, vandalism, terrorist attack *etc.* Therefore, Taj Mahal needs a disaster preparedness plan with the latest strategies to protect it from various factors of disasters. The heritage is our legacy; it is our duty to protect our heritage in all possible way and preserve it for future generations. These disasters not only affect the immovable-built heritage component such as monuments, archaeological sites and historic urban areas but also cause damage to the museum buildings and heritage objects that are in active use, such as in education, cultural and religious purposes, and other artefacts of significance to the local community. There are many probabilities of disasters on Taj Mahal at Agra as it comes under second and third grade of an earthquake zone moreover it is situated at the right bank of Yamuna river. There are possibilities of the flood, earthquake, on Taj Mahal and a recent thunderstorm damaged the parts of the marble railing and minaret of the eastern gate. Some other possible threats may arise from antisocial elements to demolition and vandalism.

A survey on the disaster preparedness plan for Taj Mahal was conducted as part of our project work and the study was undertaken per the guidelines prepared by the National Disaster Management Institute (NDMI), New Delhi. Certain details of the survey work and findings have been discussed in the paper. The safety and preservation of Taj Mahal from all kinds of disasters have been a great concern to not only for India but the whole world as it is a world heritage site recognized by UNESCO.

Disaster preparedness provides a platform for designing effective realistic, and coordinated planning and increase the overall effectiveness of national societies, households and community members in disaster preparedness and response efforts. The disaster may be natural or man-made and complete prevention of the natural disaster may be impossible, but it is certainly possible to take precautionary measures to avert losses on a large scale.

11.2 Disaster

A serious disruption, occurring over a relatively short time, of the functioning of a community or a society involving widespread human, material, economic or environmental losses and impacts, which exceed the ability of the affected community or society to cope using its own resources are comes under the disaster. Though often caused by nature, disasters can have human origins. Major natural disasters are, Flood, Fire, Earthquake, Thunderstorm, Lightning, Cyclone, Land slide, Tsunami *etc.*

11.3 Disaster Preparedness

Disaster preparedness refers to measures taken to prepare and reduce the effects of disasters. That is, to predict and, wherever possible, prevent disasters, mitigate their impact on vulnerable populations, and respond to and effectively cope with their consequences. Disaster preparedness has the potential to save the maximum number of lives and property during a disaster, and it aims to return the affected populations to normalcy as quickly as possible *i.e.* disaster preparedness includes administrative decision and operational activities that involve, Prevention, Mitigation, Preparedness, Response, Recovery, Rehabilitation *etc.*

11.3.1 Significance of Disaster Preparedness

☆ Disaster preparedness provides a platform to design effective, realistic and coordinated planning, reduces duplication of efforts and increase the overall effectiveness of national societies, household and community members disaster preparedness and response efforts.

☆ Disaster preparedness activities are embedded with risk reduction measures, and can prevent disaster situations and result in saving maximum lives and livelihoods during any disaster situation, enabling the affected population to get back to normalcy within a short span.

☆ Disaster preparedness is a continuous and integrated process resulting from a wide range of risk reduction activities and resources rather than from a distinct sectoral activity by itself. It requires the contributions of many different areas ranging from training and logistics to healthcare, recovery, livelihood to institutional development *etc.*

11.4 Recent Disasters Occurred in and around Taj Mahal

11.4.1. Damage due to Thunderstorm

Thunderstorm occurred on first May, 2022 and damaged some parts of the Taj Mahal and nearby monuments. The North West side sunshades of Taj Mahal was broken and fell down due to heavy storms. Some trees have been uprooted at Ram Bagh, Sikandra, Mahtab Bhagh, and few heavy branches of the trees had fallen on Itemad-ud-Daulah monument and caused damage to some parts of the monument.

11.4.2. Minaret Damage by Thunderstorm

A minaret at the entry gate of the Taj Mahal was crashed as heavy rains and high winds lashed in Agra on 11ᵗʰ April, 2018. The 12-feet metal pillar at the entry gate, referred to as Darwaza-e-Rauza collapsed as wind speed during the thunderstorm crossed 100 km per hour. Sources said that the pillar fell down just past midnight on Thursday and the minaret and dome attached to it was broken into several pieces.

Figure 11.1: Broken Minaret due to Thunderstorm, which Occurred on April 11, 2018.

11.4.3. Taj Mahal Area Waterlogged due to Heavy Rains

On July 27, 2018 heavy rains have caused water logging inside the Taj Mahal complex. The following visual clearly shows that the garden area of the main mausoleum was completely inundated on 27th July due to continuous downpour, causing visitors distress and throwing the authorities out of gear.

Figure 11.2: Water Logging Condition at Taj Mahal on July 28, 2018.

11.4.4. Damage to Marble Railing Due to Thunderstorm

A thunderstorm that swept through Agra on Friday night of 31st May 2020, killing at least three persons and it also damaged the marble railings of the city's most famous monument – the Taj Mahal. There has been damage to the premises of Taj Mahal due to rains and thunderstorm. Part of the marble railing of the main mausoleum towards the river Yamuna fell down.

**Figure 11.3: Marble Railing Damaged by Thunderstorm
(Picture on June 01, 2020).**

11.4.4. Bomb Threat Call

The iconic Taj Mahal was evacuated at 9 AM on March 4th, 2021, following a bomb threat call received by police helpline. Security officials conducted searches in the premises of the monument and later found the call to be a hoax, reported news agency PTI. An unidentified person called the 112-emergency response number of the Uttar Pradesh police at around 9 AM. and claimed that a bomb was kept inside the monument. It is conserved by the Archaeological Survey of India (ASI) and protected by armed personnel of the Central Industrial Security Force (CISF). The Uttar Pradesh police immediately informed the CISF personnel, who asked the visitors to vacate the premises of the iconic 17th -century monument and launched anti-sabotage checks around 9:15 am.

In the view of above few incidents in recent past, it is vivid that there are many chances of various types of disasters, which may occur. Therefore, a survey on disaster preparedness for Taj Mahal was conducted and the following are the major findings observed.

11.4.5 Survey Checklist of Risk Factors on Taj Mahal

Sl.No.	Risk Factors	Yes	No
1.	Heritage site/precinct falls in active earthquake zone (Zone IV/V)		✔
2.	Heritage site/precinct is located near an identified geological fault or in an area with known history of landslides		✔
3.	Heritage site/precinct is near a water body such as a lake, river, ocean (closer than 500 meters)	✔	
4.	The site/precinct is near an industry/industrial area, particularly heavy industries	✔	
5.	The site/precinct has high number of visitors or huge mass gathering on certain days/months making precincts at specific times	✔	
6.	The site/precinct has insufficient entrances and exits and poor circulation networks		✔
7.	The site/precinct has been impacted due to previous disasters and has a history of recorded hazards	✔	
8.	The site/precinct is inaccessible/poorly accessed/partially accessible, especially with respect to differently able, elderly and children	✔	
9.	Regular monitoring and maintenance of the site/precinct is absent/poorly managed/insufficient		✔
10.	Presence of inflammable liquids/materials, multiple ignition points or harmful chemicals used near the heritage site/precinct		✔
11.	Large trees, vegetation and other potential obstructions are present	✔	
12.	Security arrangement is inadequate/poor	✔	
13.	Individual buildings/structures have gone through structural changes in the past, changing the structural loading/behavior		✔
14.	There are structural cracks in the heritage structure		✔
15.	There are leakages in the heritage building		✔
16.	Ventilation system of the building is inadequate/absent	✔	
17.	Doors, windows, fences, gates are in poor condition		✔
18.	There is biological growth/vegetation or mould growth within the structure	✔	

Sl.No.	Risk Factors	Yes	No
19.	Storm water system in the vicinity is inadequate/absent		✔
20.	Sewage and wastewater management system of the site is inadequate	✔	
21.	Regular inspection and testing of electrical installations, faulty equipment replaced and no overloaded circuits	✔	
22.	Building security staff were not well-trained for all kinds of possible emergency, especially evacuation, emergency response *etc.*	✔	
23.	Several structures are near each other along with other potential obstructions	✔	
24.	Chances of fire A. Lightening earth or net B. Electric short circuit	✔	
25.	Level of vandalism	✔	
26.	Threat to demolition from anti-social element	✔	
27.	Chances of flood	✔	
28.	Pollution/automobile pollution/dust/dirt	✔	
29.	Presence of bird, honeybee *etc.*	✔	
30.	Heritage site/precinct falls in active earthquake zone (Zone 2nd and 3rd)	✔	

11.5 Conclusion

An immensely beautiful mausoleum of white marble was built in Agra between 1631 and 1648 AD by the order of the Mughal emperor Shah Jahan in the memory of his favourite wife Mumtaz Mahal. The Taj Mahal is a jewel of Muslim art in India and one of the universally admired masterpieces of the world's heritage. The Taj Mahal was included in the UNESCO World Heritage Site in 1983, and millions of tourists visit the Taj Mahal every year. The management of Taj Mahal complex is carried out by the Archaeological Survey of India and the legal protection of the monument and control over the regulated area around the monument is through the various legislative and regulatory frameworks that have been established, including the Ancient Monument and Archaeological Sites and Remains Act 1958 and Rules 1959 Ancient Monument and Archaeological Sites and Remains (Amendment and validation); which is adequate to the overall administration of the property and buffer areas.

There are many probabilities of disaster at Taj Mahal, as it comes under 2nd and 3rd grades of earthquake zone, moreover the chances of flood are high because it is situated at the right bank of Yamuna River. The recent thunderstorm damaged the parts of marble railing and minaret of East gate. There are frequent threats of demolition and vandalism from anti-social elements.

A survey was conducted on disaster preparedness for Taj Mahal to prepare a checklist of risk assessment and analyse it for further necessary actions and suggestions have been given to concerned authorities. Based on the findings of the survey of Taj Mahal, it is concluded that the Taj Mahal is under various threat factors of disasters as mentioned in the checklist and it should be protected by disaster preparedness strategies.

It was also suggested that on the river side of the Taj Mahal, the boulders of granite stone should be spread along with the walls of Taj Mahal to control the erosion of soil in the foundation of Taj Mahal. The seismic and earthquake sensor should be installed to set advance signals if any disaster occurs due to earthquake. The Archaeological Survey of India should implement all possible measure for the security at the Taj Mahal and should take necessary and appropriate action according to the National Disaster Management Guidelines. It is significant to mention that, the heritage is our legacy and it is our duty to protect our heritage by all possible ways and preserve it for the future generations.

REFERENCES

Agarwal O P (1982) Precautions for the use of insecticides and notes on chemical used as termiticides" 1982 Butterworth and Co (Publishers) Ltd.

Daniyal Md, Abdurahim K (2020) Disaster Preparedness Plan for Cultural Heritage with special reference to Taj Mahal. Dissertation submitted in the department of Museology, Aligarh Muslim University

https://thecsrjournal.in/csr-conservation-of-taj-mahal/

https://www.indiatoday.in/india/story/taj-mahal-agra-bomb-scare-call-turns-out-to-be-hoax-1775451-2021-03-04

https://www.unisdr.org/files/7817_UNISDRTerminologyEnglish.pdf

https://english.jagran.com/india/taj-mahal-agra-bomb-threat-bomb-threat-tourists-evacuated-search-operations-live-news-updates-10024065

ICOMOS (1964) international charter for the conservation and restoration of monuments and sites 2nd International Congress of Architect and Technician of Historic Monuments, Venice, 1964.

Khan T, Abdurahim K (2019) Disaster Preparedness in Indian Museum with special reference to Raza Library Museum. Dissertation submitted in the department of Museology, Aligarh Muslim University.

NDMA (2017) NATIONAL DISASTER MANAGEMENT GUIDELINES September 2017 (Cultural Heritage Sites and Precincts) A publication of National Disaster Management Authority, Government of India NDMA Bhawan A-1 Safdarjung Enclave New Delhi – 110029

UNISDR M (2009) UNISDR Terminology for disaster risks reduction. United nation international strategy for disaster risk Reduction (UNISDR) Geneva, Switzerland.

Chapter 12

Biodeterioration of some Archaeological Monuments Due to Algae, Lichens and other Plants at Gwalior (Madhya Pradesh), India

Ashok K. Jain

School of Studies in Botany, Jiwaji University, Gwalior – 474 011, India
e-mail: asokjain2003@yahoo.co.in

ABSTRACT

World's valuable cultural heritage and historic buildings deteriorate due to natural and anthropogenic contributors in which role of flora, fauna and microorganisms is well documented. Pollutants also play a significant role. Deteriorating processes have been studied. And it has been found that surfaces of the cultural heritage can destruct the underlying materials through their influences on physical, chemical and biochemical reactions. The stone surfaces can initiate a number of subsequent physical and chemical changes of the substratum materials, including the discoloration in appearance, alternation of porosity and accumulation of organic substances, which is further used by mosses and plants to grow. The great fort of Gwalior situated in Central India (M.P.) contains various objects of historical and antiquarian interest. The deterioration of monuments of the gigantic Jain sculptures carved along the rock face of the fort. And Sas – Bahu temple situated on a picturesque position on the eastern rampart as well as Telika Mandir are described in details. The paper presents a list of plants, lichens, bryophytes and microbes present on these stone structures. Members of Cyanobacteria and Chlorophyta groups and crustose and foliose lichens, members of Fungi imperfecti and Ascomyctes and a large number of flowering plants were observed. The remedial methods are suggested to protect the cultural heritage of Gwalior.

Keywords: *Biodeterioration, Archaeological monuments, Algae, Lichens, Plants, Gwalior.*

12.1 Introduction

Cultural properties existing in the present period are our human heritage produced from individual or complexes of various materials such as plants, animals and minerals of different ages (Arai 1995). Cultural properties are spread all over the world in the form of monuments, paintings, sculptures, arts and craft, books, archives, archaeological objects, wooden objects, ethnological objects and various other forms. All over the world, the decomposition and deterioration of rocks and stones have become the matter of great concern. The parts which are vulnerable to destruction are the outer surface of stone elements, which of course determine their artistic expression. Algae, fungi, lichens and higher plants happen to be the chief biodeteriorating agents of these rock cut monuments (Hueck 1958). Biological deterioration can be found in the full range of these materials because they have been made by various materials like; wood, paper, cloth, leather, soil, stone and metal. It is well known understandable that tropical countries suffer through biodeterioration more than the countries of temperate climates, because of the hot and humid environment prevailing in the tropics.

India is a country having a vast reservoir of natural, cultural, historical, religious, ethnic and archaeological resources. A large number of monuments, museums, stone sculptures *etc.* of great importance are found at several places in the country. In India the damage to cultural property due to biodeterioration is enormous. Surfaces of several archaeological monuments and other important ones turn green, brown or black due to growth of algae, lichens and other higher plants.Central India occupies an important position having great cultural wealth. Historical monuments at Khajuraho, Gwalior, Chanderi, Mandav, Sanchi, Vidisha, Ujjain and other places are world famous and popular tourist spots.

12.1.1 Gwalior

The present observations were made at Gwalior, a cultural and historical city of Central India. The city is situated between 260 – 14' N latitude and 780- 15' E longitude at an elevation of 212 meters above mean sea level. It lies between two great geological formations, Plateau of Deccan on the south and plains of Ganges and Yamuna on the north. It is situated at a point where Vindhyan hills gradually merge with Ganges alluvium. The peculiar feature with regard to the geology of greater Gwalior and its vicinity is that the area is made up of four distinct type of geological systems *viz.* (i) Archeans (ii) Cuddapah (iii) Vindhyans and (iv) Recent alluvium.

The city of Gwalior falls under the semi-arid category of climate and is generally dry. A wide gap lies between the maximum (approx. 46°C) during the summer season and the minimum (approx. 20°C) in the winter. The mean Relative Humidity is as low as 15 per cent in summer while maximum Relative Humidity is 82 per cent during rainy season in August.

12.2 Historical Monuments

The city of Gwalior comprises a large number of monuments and palaces that have defied the ravages of time and still stand out in all their strength and charm. The fort of Gwalior, aptly called the 'Pearl in the necklace of the castles of India' stands on a long and narrow rocky hill of sandstone which rises abruptly 90 M.S.L. The great fort contains various objects of historical and antiquarian interest. The notable monuments are the gigantic Jain sculptures carved along the rock face of the fort. A large number of standing and seated statues of Jain *Tirthakaras* (God) have been carved on rocks inside and outside a large number of caves. The height of these images varies from 6 cm to 80 feet. According to historical evidences these beautiful sculptures were built between 8th to 15[th]century A.D (Jain 1966) by the then rulers. These sculptures which are in five groups, represent the nude figures of 24 *Tirthankars* in standing and seated postures (Figure 12.1).

**Figure 12.12: The Sculptures Representing the Nude Figures of
24 *Tirthankars* in Standing and Seated Postures.**

Of the many temples and shrines which still stand in the fort and deserve special mention. The twin structures which are known as *Sas – Bahu* temple occupies a picturesque position on the eastern rampart. The third temple of interest is now called the Telika Mandir. It is the loftiest of all the existing buildings on the fort, over 30 m high.

To the east of the town stands the mausoleum of Mohammad Ghaus, a very fine example of early Mughal architecture. It is built in the form of a square with hexagonal towers at its corners surrounded by small domes. Close by the tomb of Mohammad Ghaus stands the more modest but perhaps better known tomb of the great musician, Tansen.

Another noteworthy edifice is the 'Jai Vilas Palace', the residence Sindhias, former ruler of Gwalior state. It was designed on the plan of an Italian palazzo. Other places of interest are Moti Mahal, Kampoo Kothi, Imambara and Gorkhi. Not only Gwalior but the entire northern region of Central India consists of a large number of monuments which date back the 9th to 19th century A.D.

12.3 Biodeterioration

These unparallel man made creations are now getting deteriorated and ruined due to several natural and man made factors. The surface of almost all images have been replaced by algae and lichens (Figure 12.2). Several crevices in the rocks have allowed the plants to penetrate their roots; plenty of honey combs, bird's and bat's excreta inside the caves provide most suitable surface for the growth of microorganisms. Continuous seepage of water from the rocks keeps several statues and caves in moist condition round the year (Figure 12.3). The beautiful carvings at many sites have been completely vanished and only remnants can be seen. In one of the caves a 42 feet high statue of *Tirthankar* (Lord) Parshvanath is supposed to be the highest seated rock cot statue in India. On the other side of the fort slope, a 84 feet standing statue of Lord Adinath continuously gets water from the top of the fort. A large number of bats have also made their shelter points inside the caves. Honey combs and their bees on several statues are also causing deterioration to this cultural property (Figure 12.4).

Figure 12.2: Continuous Seepage of Water from the Rocks and the Growth of Algae and Lichens can be seen on all the Images.

Figure 12.3: Growth of Plants in moist Condition round the Year.

12.4 Methodology Adopted

1. Samples of algae and lichens growing on the surface of statues and other sculptures were taken very carefully with the help of forceps

and small sized brushes in small polythene bags. The samples were taken to the laboratory and identified with authentic literature. Only qualitative observations were made.

Figure 12.4: Honey Combs and their Bees on Several Statues causing Deterioration to this Cultural Property.

2. The Angiospermic plant species growing around the monuments, penetrating roots in the rock crevices were collected and identified with the help authentic literature.

3. For fungal studies petri plates with Potato Dextrose Agar media were exposed at different sites inside the caves. Petri plates were randomly exposed during all the seasons in a year. After incubation, fungal colonies appeared on the media. Different fungal types were identified with the help of reference slides and authentic literature.

12.5 Results

12.5.1 Occurrence of Algae, Bryophytes and Lichens

The common algae found on the surface of statues and monuments belong to the species of *Anabaena, Nostoc, Scytonema, Lyngbya, Gloeocapsa, Oscillatoria* of blue green group or Cyanobacteria and *Chlorella, Ulothrix and Spirogyra* of Chlorophyta or green algae.

The predominant lichens belong to the species of *Lecanora, Caloplaca, Parmelia, Dirinaria, Diploschistis* and *Peltula*. Pure and extensive growth of these species can be seen on monuments and nearby rocks. Besides above species the bryophytes like *Riccia* and certain mosses also exhibit luxuriant growth during rainy season.

12.5.2 Angiosperms

The following Angiospermic plant species have been observed either growing or penetrating their roots in the crevices of rocks and even hanging down inside the caves:

- ☆ *Ailanthus excelsa*
- ☆ *Anogeissus latifolia*
- ☆ *Capparis decidua*
- ☆ *Clerodendrum phlomidis*
- ☆ *Dichrostachys cineria*
- ☆ *Ficus religiosa*
- ☆ *F. benghalensis*
- ☆ *Mimosa rubicaulis*
- ☆ *Pongamia pinnata* (syn. *Derris indica*)
- ☆ *Tamarindus indica*
- ☆ *Holoptelia integrifolia*

On the extensively withered surface where litter and soil has been deposited, several species of grasses and small herbs are growing profusely. Some common examples are:

- ☆ *Abutilon indicum*
- ☆ *Boerhaavia diffusa*
- ☆ *Cassia tora*
- ☆ *Chenopodium album*
- ☆ *Convolvulus pluricaulis*
- ☆ *Eclipta alba*
- ☆ *Aristida funiculata*
- ☆ *Cyperus rotundus*
- ☆ *Setaria glauca*
- ☆ *Heteropogon contortus*

The following species of climbers are also found to be growing with the support of statues and other sculptures :

- ☆ *Argyreia nervosa*
- ☆ *Clitoria ternatea*
- ☆ *Cryptostegia grandiflora*
- ☆ *Ipomoea pestigridis*

☆ *Lindenbergia indica*

☆ *Pergularia daemia*

12.5.3 Fungi

Fungal incidence inside the caves indicated the occurrence of 19 different types of fungi during the study period. Maximum count was recorded in March, June, August and November while minimum in May. The common fungal types belong to the species of *Alternaria, Aspergillus, Cladosporium, Fusarium, Penicillium, Syncephalastrum etc.* However species of *Curvularia, Drechslera, Rhizopus and Trichothecium* were also recorded in certain months. Species of *Fusariella* exhibited its presence in February only. Considering the total percentage of distribution *Cladosporium herbarum* showed maximum dispersal (40.63 per cent) followed by *Aspergillus niger* (17.50 per cent) and *Aspergillus flavus* (9.92 per cent). Maximum growth of fungi was noticed at places where the fecal matter of birds and bats was maximum. At honeycomb sites also the growth of fungi was recorded.

12.6 Discussion

Several beautiful historical sites all over the world are getting deteriorated due to various biotic, man made or natural factors. Decomposition and deterioration of cultural properties is an universal phenomenon. It is usually the outer surface of stone elements which is vulnerable to destruction. The microflora leaving on rocks and stones constitute a community of organisms capable of carrying on processes of rock weathering and stone decomposition (Sarbhoy 1994).The caves at Gwalior fort continuously get seepage of water. The recent study reveals the presence of a good number of fungal spores in the air of caves. Normally fungi does not grow on rocks and stones. However, different species of fungi have been isolated from weathered stones also. It is mainly because of the ability of fungi to sustain itself on the organic residues left by other decomposed microorganisms on the surface of stones (Jaln *et al.*, 1993). These fungi mostly belong to the species of *Aspergillus, Alternaria, Cladosporium, Penicillium,* and *Curvularia etc.* Honey drops and bird's excreta on monuments also provide suitable medium for the growth of several fungi. Chromatographic studies have indicated that fungi produce many organic acids responsible for dissolution of rocks and stones (Sarbhoy 1994).

The luxuriant growth of algae is noted under conditions of nearly full saturation of rocks with water. Algae have been found to produce and secrete a variety of metabolic products, among which predominate organic acids include, lactic, oxalic, succinic and pyruvic acids. These acids either directly dissolve rock surface and stone components or increase their solubility.

Lichens cause physical damage as a result of deep hyphal penetration (Gehrmann *et al.*, 1989). The deterioration of stones by lichens is predominantly biochemical in nature, although mechanical disruption is also possible (Awasthi 1989). In epilithic lichens the upper surface (*i.e.* upper cortex) of thallus expands and contracts with changes in atmospheric humidity (Jain *et al.*, 1993).Similar observations have also been made in Italy (Seward *et al.*, 1989). At present site also several species of lichens are growing on monuments (Figure 12.5).

Figure 12.5: Growth of Crustose Lichens seen on Monuments.

Figure 12.6: Growth of Higher Plants, Causing both Biophysical and Biochemical Decay.

In case of root penetration and growth of higher plants, both biophysical and biochemical decay play an important role (Figure 12.6). Growth and radial thickening of roots of plants are mainly responsible for biophysical decay. Their mean pressure can be as much as 15 atmosphere (Winkler 1975). In case of decay on account of root exudates the acidic substances have a tendency to form chelateswith stone minerals. Agarwal *et al.* (1994) have suggested some methods for the removal of plants from historic buildings.

The present monuments of Gwalior require urgent steps of conservation because of fast rate of deterioration. In absence of conservational steps such cultural heritages of glorious past would be converted in to ruins. Now a day's several scientific methods have been evolved for conservation.

REFERENCES

Agarwal O P, Mishra A K, Jain K K (1994) Removal of plants and trees from historic buildings. INTACH Pub. Lucknow.

Arai H (1995) Biological researches in conservation of cultural property.In Aranyanak C and Singhasiri C (eds) Biodeterioration of cultural property-3. Proc. of the 3rd Int. Conf. Pub. Conservation Science Div, Bangkok.

Awasthi D D (1989) Lichens and monuments. In Proc. Int. Conf. on Biodeter. of Cult. Prop. Lucknow, 1989, pp. 207-211.

Gehrmann C K, Petersen K, Krumbein W E (1989) Silicole calcicole lichens on dewish tomb stones- Interactions with environment and biocorrosion. In VIth International Congress on Detereoraments Conservation Della Pietra Venejia. 3: 289-299.

Hueck H J (1968) Thebiodeterioration of materials – An appraisal. In Biodeterioration of materials. Pp. 6-12. Elsevier Pub. Co. Ltd. London

Jain A K, Mishra R (1988)Airborne pollen grains and other biocomponents at Gwalior. Ind. J. Aerobiol. 1(1): 30-35.

Jain K K, Mishra A K Singh T (1993) Biodeterioration of stones: A review of mechanism involved. In Recent Advances in Biodeterioration and Biodegradation. (eds) K.L. Garg, Neelima Garg and K.G. Mukherjee. 1: 323-354. Naya Prakash Pub. Calcutta.

Jain R (1966) Gwalior Gaurav Gopachala. GopachalNyas Pub. pp.108.

Sarbhoy A K (1994) Biodeterioration of ancient stone work. In Current Trends in Life Sciences- Advances in Mycology and Aerobiology. Ed U.K. Talde, 20: 161-169.

Seward M R D, Giacobini C, Giuliani M R, Roccardi A (1989) The role of lichens in the biodeterioration of ancient monuments with particular reference to Central Italy. Int. Biodet. 25: 49-55.

Winkler E M (1975) Stone properties durability in man's environment. pp. 154-161. Springer Verlog, Wien.

Chapter 13

Beauty of Certain Historical Monuments in Agra and Problems of Deterioration of Mughal Architecture in Agra Fort

Arun Arya

Department of Environmental Studies, Faculty of Science, The Maharaja Sayajirao University of Baroda, Vadodara – 390 002, Gujarat
e-mail: sarojarun10arya@rediffmail.com

ABSTRACT

Agra was an important place in Northern India during Mughal dynasty. Today it is famous worldwide for the beauty of Taj Mahal and various other monuments. The historical buildings are part of our biosphere and their biodeterioration is well recognized phenomenon in tropical regions. Changing environment like prevalence of high heat during summer associated with high humidity during rains and subsequent months, offer suitable conditions for the growth of bacteria, cyanobacteria and dematiaceous fungal flora. These biodeteriogens along with air pollutants cause severe damage to world monumental heritage. The Supreme Court of India took drastic action in 1994 on a petition filed by advocate M.C. Mehta and suggestions by NEERI, Nagpur by banning various coal based small scale iron foundries and smelting industries in the Agra city and restricting plying of petrol or diesel driven vehicles around the Taj area. Few studies have indicated role of polluted river Yamuna and growing bacteria and cyanobacteria as another damaging agents to Taj. The chapter gives a detailed account of deterioration caused to the UNESCO heritage site of Red Fort of Agra. The chapter also presents causes of deterioration and modern technologies available for restoration of such historical stone monuments.

Keywords: *Beauty, Historical monument, Agra Fort, Biodeterioration, Cyanobacteria, Fungi, Taj Mahal, Jama Masjid, Radhaswami Temple, Pollution, Environmental factor.*

13.1 Introduction

Agra a well-known city of Uttar Pradesh was known during Mahabharata age as Agravana, later on came to be recognized as the seat of Mughal monuments. The mention of Agra itself takes us to the epic Taj Mahal, the delicious Mughlai dishes, and entire kingship period of the Mughal. But, some of the less noticed things in Agra are the temples of the place. Mughal ruler Akbar built the famous red fort of Agra and 15 km away another city known as Fatehpur Sikri. This was built in memory of his Muslim priest Sheikh Salim Chisti, who blessed him with a son, and Buland darwaza was built in 1601 to celebrate victory over Gujarat. Shah Jahan built the beautiful Taj Mahal in the memory of his wife, Mumtaz Mahal. Unlike his grandfather, Shah Jahan tended to have buildings made from white marble. He destroyed some of the earlier buildings inside the fort to make his own.At the end of his life, Shah Jahan was deposed and restrained by his son, Aurangzeb, in the Agra fort. It is rumoured that Shah Jahan died in Muasamman Burj, a tower with a marble balcony with a view of the Taj Mahal. The Fort was invaded and captured by the Maratha Empire in the early 18th century. Thereafter, it changed hands between the Marathas and their foes many times. After their catastrophic defeat at Third Battle of Panipat by Ahmad Shah Abdali in 1761, Marathas remained out of the region for the next decade. Finally Mahadji Shinde took the fort in 1785. It was lost by the Marathas to the British during the Second Anglo-Maratha War, in 1803.The fort was the site of a battle during the Indian rebellion of 1857, which caused the end of the British East India Company's rule in India, and led to a Century of direct rule of India by Britain. Chapter describes in detail about the deterioration of Agra Fort besides Architectural beauty of certain other monuments like Jama Masjid, Radhaswami Temple, Akbar's Tomb at Sikandra and beautiful World famous Taj.

13.1.1 Jama Masjid

The Jama Masjid is one of Agra's oldest mosques; Emperor Shah Jahan built it in 1648 in honour of his daughter Jahan Ara Begum. Its name Friday mosque, is in reference to the large congregation held each Friday. The architecturally impressive Masjid stands proudly on a plinth in the heart of an energetic and traditional bazaar (Figure 1). On the eastern side of the mosque is the main entrance, adorned with carved arches and pillars. The gateways lead to a vast courtyard, which has a capacity for about 10,000 people. Marvel at how the atmosphere changes from the chaos of the surrounding bazaar to a feeling of peace and tranquillity. On western side of the courtyard is prayer hall crowned by three domes. Check out the domes' inverted lotus decorations and zigzagging white-marble designs. Peek inside to see an attractive pulpit and the mihrab, an alcove that worshippers face during prayer time. Stroll through the arcades that frame the courtyard, where pilgrims find shade from the afternoon sun. Lining the walls of the arcades are Mughal-era paintings, inscriptions and colourfully decorated tiles. The mosque is situated amid Agra's lively Kinnari Bazar. Jama Masjid, one of the incredible places to visit in Uttar Pradesh, was constructed using red sandstone along with marble decorations to accompany it.

Figure 13.1: Showing Agra's Oldest Mosque Jama Masjid.

13.1.2 Radhaswami Temple of Dayalbagh

One of the very famous and oldest temples of Agra is Radhaswami Temple of Dayalbagh (Figures 13.2 and 13.3). This temple is under construction for last 116 years. Dayal Bagh is the headquarters of the Radha Swami sect. It is a self-sufficient colony. The tomb of their master is a highly impressive monument. Devout followers believe in the service to humanity that purifies the soul. This huge and magnificent temple complex would be on a tourist's list of top places in Agra. The name Dayalbagh is derived from the name of the garden of the merciful lord. The temple houses 'samadhi' of Swamiji Maharaj, the founder of Radha Swami Faith. This is a very famous temple of Sant Satguru, the 8th Satguru who used to stay here for all his 'satsangs'. The chief working committee is known as Radhaswami satsang sabha. Followed by a large number of people across India, Radhaswami Faith is not a religion but a sect led by various satgurus. The shrine is likely to be completed soon.

13.1.3 Akbar's Tomb: Sikandra

Although planned by Akbar, his son Salim completed the construction during 1605–1613. It is located at Sikandra, in the suburbs of Agra, on the Mathura road (NH2), 8 km west-northwest of the city. During the reign of Aurangzeb, Mughal prestige suffered a blow when Jats ransacked Akbar's tomb, plundering and looting the gold, jewels, silver and carpets. The grave was opened and the late king's bones were burned. As Viceroy of India, George Curzon directed extensive repairs and restoration of Akbar's mausoleum, which

Figures 13.2 and 13.3: Radhaswami Temple of Dayal Baug, Agra.

were completed in 1905. Curzon discussed restoration of the mausoleum and other historical buildings in Agra in connection with the passage of the Ancient Monuments Preservation Act in 1904, when he described the project as "an offering of reverence to the past and a gift of recovered beauty to the future". This preservation project may have discouraged veneration of the mausoleum by pilgrims and people living nearby. Tourists can see the deer and monkeys here. The south gate is the largest, with four white marble chhatri-topped minarets which are similar to (and pre-date) those of the Taj Mahal, and is the normal point of entry to the tomb. The tomb itself is surrounded by a walled enclosure 105 m square. The tomb building is a four-tiered pyramid, surmounted by a marble pavilion containing the false tomb. The true tomb, as in other mausoleums, is in the basement. The buildings are constructed mainly from a deep red sandstone, enriched with features in white marble (Figures 13.4 and 13.5). Decorated inlaid panels of these materials and a black slate adorn the tomb and the main gatehouse.

Figure 13.4 and 13.5: Akbar's Tomb at Sikandra, Note the deterioration in one of the walls.

Figure 13.6: Entrance of Agra Fort through Amar Singh Gate.

13.1.4 Marvellous Taj Mahal

The beautiful Taj Mahal is a perfect symmetrical planned building (Figures 13.8 and 13.9), with an emphasis of bilateral symmetry along a central axis on which the main features are placed. The building material used is brick-in-lime

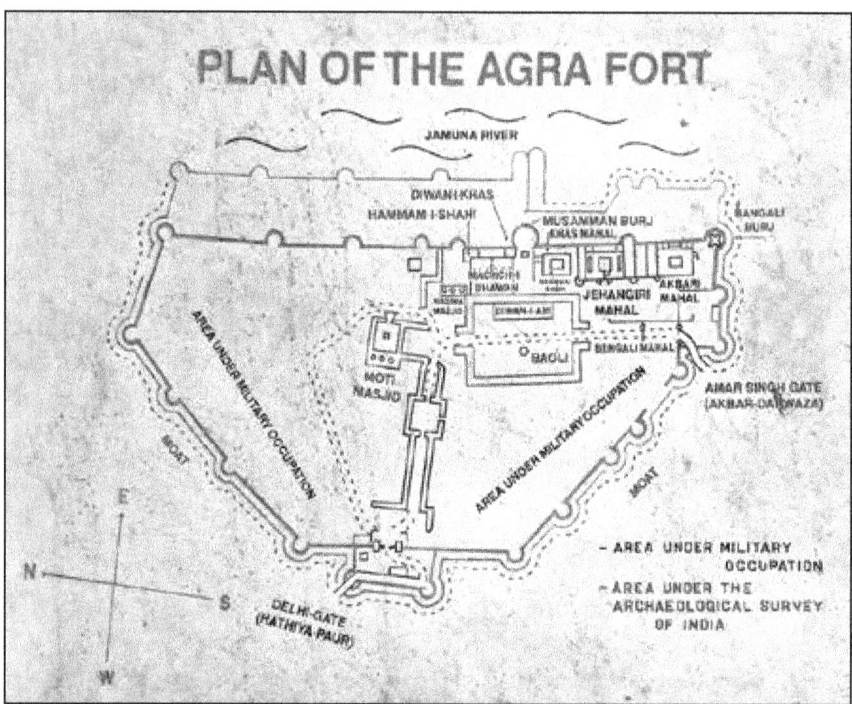

Figure 13.7: Site Plan of Agra Fort Showing Position of different Buildings.

Figure 13.8: Showing River Yamuna and Beautiful View of Taj Mahal at Agra.

mortar veneered with red sandstone and marble and inlay work of precious stones. The mosque and the guest house in the Taj Mahal complex are built of red sandstone in contrast to the marble tomb in the centre. Both the buildings

Figure 13.9: Tourist's Paradise, Beautiful Taj Mahal at Agra in 1970 (Front view).

have a large platform over the terrace at their front. Both the mosque and the guest house are the identical structures. They have an oblong massive prayer hall consist of three vaulted bays arranged in a row with central dominant portal. The spandrels are filled with flowery arabesques of stone intarsia and the arches bordered with rope molding. Taj Mahal represents the finest architectural and artistic achievement through perfect harmony and excellent craftsmanship in a whole range of Indo-Islamic sepulchral architecture. The physical fabric is in good condition and structural stability, nature of foundation, verticality of the minarets and other constructional aspects of Taj Mahal have been studied and continue to be monitored. To control the impact of deterioration due for atmospheric pollutants, an air control monitoring station is installed to constantly monitor air quality and control decay factors as they arise.In 1982, the government of India declared the formation of the Taj Trapezium Zone (TTZ) – a trapezoid-shaped area of 10,400 sq km around the monument where setting up or expansion of polluting industries was prohibited.Attemptshave been made by ASI to clean the monument, earlier with an ammonia-based chemical solution and distilled water, and chemical coatings, and later with repeated mud packs (using Bentonite clay or Fuller's earth).

Most accounts of the monument talk about its aesthetics, architecture, beauty or symbolism. Very few have mentioned about its colour. Lahauri, a historian at Shah Jahan's court, gave a detailed account of the Taj complex on the occasion of its formal completion in 1643, when the tomb and gardens were completed, on the 12th anniversary of Mumtaz Mahal's death. He uses the expression "white marble" several times to describe the façade, platform and the dome. The Viceroy, Lord Curzon, who first saw the Taj in 1887, found it to be a "snow-white emanation starting from a bed of Cypresses, and backed by a turquoise sky..."

According to historical sources, the marble (*sang-e-marmar*) of the Taj, was sourced from Makrana in Rajasthan. This marble is considered to be white (some say milky white) with grey or black streaks. Art historian Ebba Koch, who has authored *The Complete Taj Mahal*, says that due to its capacity to transmit and refract light, the white marble reacts differently to different atmospheric conditions at different times of the day. The Taj thus appears pinkish in the morning, dazzling white at noon, golden orange in the evenings and another kind of white at night. In his work, *Marble in India*, scholar Coggin Brown emphasises the point that white marble's translucency makes it react very interestingly to atmospheric changes. Perhaps the Mughal poet Kalim also noticed it:

....Nay, not marble: because of its translucent colour (ab-u-rang)

The eye can mistake it for a cloud....

The whole idea of 'white' is relative in the case of the Taj. The colour of the marble varies according to the hour, day and season, and this phenomenon adds to the mystical aura of the monument. In any case, not only has the marble undergone a natural wear and tear process over four centuries; the climatic conditions too have changed.

The debate on the colour of the marble has once again sidelined the many real threats being faced by the Taj – what is the state of the sandstone which constitutes a large part of the monument complex? Pollution of Yamuna is affecting the monument? *Goeldichironomus holoprasinus*, a particular species of insect has been breeding in the dirty waters of river and causing green spots on the Taj? The blotches on the walls are in fact the excreta or cyanobacteria deposited revealed ASI experts. These stains if not removed can ultimately ruin the intricate designs and floral motifs embossed on the mosaics of the mausoleum. Besides the restriction of vehicular pollution there is a need to restrict the stay time of visitors there.

13.2. Red Fort of Agra

Agra Fort or Red Fort is located on the right bank of the river Yamuna in the city of Agra(270 10' 47"N and 780 1' 22 E). It was originally a fortress made of bricks, owned by the Chauhan Rajputs. When first battle of Panipat was fought in 1526 Ibrahim Lodhi moved to Agra and was living in Fort. He was killed by Babur, the first Mughal ruler, who then captured the Fort and built a Baori or step well in it. Akbar rebuilt the Fort with red sand stone and bricks in inner core. Though construction was done by Akbar, many of its buildings which we see today were made by his grandson Shahjahan. The fort complex was designated as UNESCO world heritage site in 1983. Agra Fort features in 'The Sign of the

Four", a Sherlock Holmes mystery novel written by Sir Arthur Canon Doyle. In 2004 the Agra Fort had won the Aga Khan award for architecture and to commemorate the win, a special stamp was issued by India Post (Figure 13.10).

Figure 13.10: Release of Two Postage Stamps on the Occasion of Aga Khan Award for Architecture.

13.2.1 Agra Fort: Splendid Architecture

Made from red sand stone and marble in Mughal style of architecture the fort is semi circle in shape. Massive red stone doublemented walls are 21 m high and 2km in perimeter. The fort has four gates namely Delhi gate, Amar Singh gate, Hathi pol and Ghazni gate. Delhi gate is the largest but visitors enter through Akbar Darwazza renamed by Shah Jahan as Amar Sigh gate (Figure 13.6). The fort comprises several palaces halls and monuments. Important being Jahangir mahal (Figures 13.7 and 13.17), Bengali mahal, Khash mahal, Sheesh mahal and Shah Jahan mahal. The main structure include Diwan-i-Am and Diwan-i-Khash. Mughal writer Abul Fazal recorded about 500 buildings in fort but hardly 30 Mughal buildings have survived on South-eastern side facing the river.

13.2.2 Agra Fort: The Muthamman Burj (Shah-Burj) and Jharokha

This beautiful palace surmounts the largest bastion of Agra Fort on the riverside, facing the East. It was originally built of red stone by Akbar and owing to its octagonal plan, it was called 'Muthamman Burj'. It has also been mentioned as 'Shah-Burj'(the imperial or king's tower) by Persian historians and foreign travellers. Its name jasmine tower or 'Samman-Burj' as recorded by the contemporary historian Lahauri is a misnomer. It was rebuilt with white marble by Shah Jehan around 1632-1640 A.D. It is an octagonal building, five

Figure 13.11: Towards Right Hand Side in Next Gate Situated to Amar Singh Gate in Agra Fort (Note the blackening of stone and damage or loss of some stones towards lower side in picture).

Figure 13.12: Blackening of Red Stone Caused due to Microbes in a Gate Towards Diwan e am in Fort.

external sides of which make a dalan overlooking the river. Each side has pillar and bracket openings, the easternmost side projects forward and accommodates a jharokha majestically. On the western side of this palace is a spacious dalan with Shah-Nasin (alcoves). A shallow water-basin is sunk in its pavement. It is profusely inlaid. This dalan opens on a court which has a chabutara projected by a jali screen, on its northern side, a series of rooms leading to Shish Mahal on its

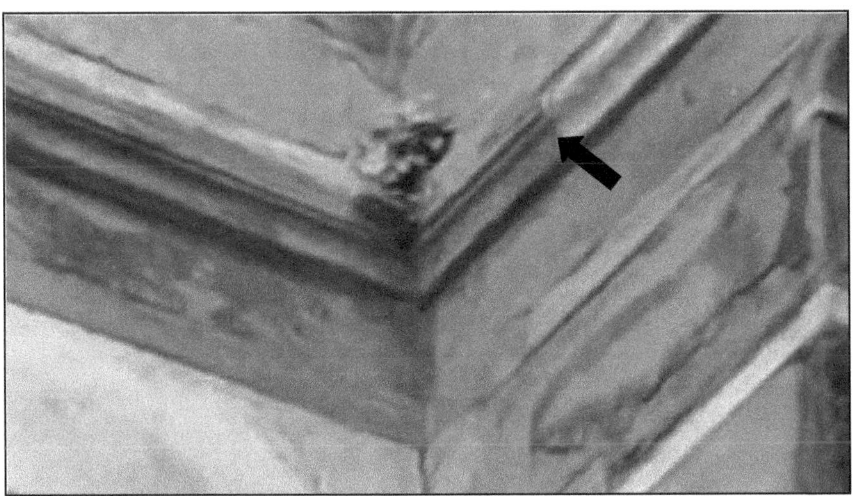

Figure 13.13: Note the Growth of a Plant in Corner of a Wall after Amar Singh Gate in Agra Fort.

Figure 13.14: Blackening of Wall on Right Hand Top Due to Moisture during Rains in Agra Fort.

western side; and a colonnade (dalan) with a room attached to it on the southern side (Figure 13.15). It is, thus, a large complex entirely built of white marble. Dados have repetitive stylized creepers inlaid on borders and carved plants on the centre pillars, brackets and lintels also bear exquisitely inlaid designs and it is one of the most ornamented buildings. This palace is directly connected to the other palaces and it was from here that the Mughal emperor governed the whole country. This burj offers full and majestic view of Taj Mahal and Shah Jahan spent eight years (1658-1666 A.D.) of his imprisonment in this complex.

Figure 13.15: Picture Showing Open Space at Jharokha burj. A Chabutra is Located Here.

13.3 Deterioration of Stone Monuments

Stone objects deteriorate continuously as a result of physical, chemical and biological processes. The durability of stones is mainly dependent on its internal structure and petro graphic composition and also to the environment to which they are exposed. The various types of deterioration are:

- ☆ Decay due to quarrying: - the decay in the stone objects can be due to the method of quarrying or dressing. The micro cracks developed will further deteriorate the stone objects.

- ☆ Decay due to pollutants: - the decay in the stone objects can be due to the method of quarrying or dressing. The micro cracks developed will further deteriorate the stone objects

- ☆ Decay due to environmental factors: - high temperature and dampness are two major agents of the decay. The moisture absorbed will help the stone to take in the salts, which would result in surface damage of the stone objects. Rapid changes of heat due to sun and the sudden rain cause strain between the outer and inner portion of the rocks or stones, which result in breaking of the specimens

☆ Decay due to soluble salts: - the salt absorption by the stone objects creates crumbling of the objects due to the crystallization internally.

☆ Biodeterioration due to activities of bacteria, cyanobacteria, microalgae and fungi *etc.*

☆ The decay due to moss and lichen *etc.*- is caused only in the case of stone objects exposed to rains. Acids generated by moss and lichens not only damage carbonaceous stones but will also attack silica and cause damage on the surface. The growth of Angiosprmic vegetation may occur later on where light is available in different stone monuments. The vegetation growth withdraws water and retains the moisture inside the structure there by damaging the monuments.

☆ Damage due to human vandalism

☆ Damage due to poor maintenance.

13.3.1 Decay Due to Atmospheric Pollutants

Atmosphere consists of pollutants such as carbondioxide (CO_2), nitrogenoxides (NOx), hydrogen chloride (HCl), hydrogen fluoride (HF), hydrogen sulphide (H_2S), aerosol, suspended particulate matter *etc.*, which get dissolved in the moisture and are absorbed by stone objects once again resulting in the crystallization thereby in the surface crumbling. Air pollution problem caused by industrial development is the reality, which the modern world is facing. A huge amount of sulphur-dioxide and other pollutants like nitrogen oxides, carbon monoxide, *etc.* are discharged into the atmosphere near red fort of Agra (Sharma and Sharma 1982; Lal 1978). Consequently, the atmospheric acidification, with high SO_2 concentrations in the air and acid rains, becomes more and more serious, especiallyin developing countries (Maeda *et al.*, 2001). Damage to architectural and cultural constructions caused byacidifying atmospheric pollution imposes huge socio-economic costs (Soon-Bok *et al.*, 1996).

Therefore, considerable attention has been paid to the effect of air pollutants on architectural material, and especially on the cultural assets made of carbonate stones. Many reports concerning deterioration of carbonate stones were published (Alonso and Martinez 2003; Boke and Gauri 2003). It was found that SO_2 caused corrosion of carbonate stones by sulfation, *i.e.* the reaction of SO_2 with carbonate calcium and carbonate calcium was consequently changing to more soluble gypsum ($CaSO_4$ $2H_2O$). The deterioration due to sulfation was found more severe in case of marble (Lan *et al.*2005).

Figure 13.16a–d

(a-c) The loss of aesthetic beauty due to removal of precious stones from the marble building of Muthamman Burj. Efforts are taken by ASI to restrict the public interaction by putting an aluminium grill; (d) The beautiful marble building of Muthamman Burj (Shah-Burj).

13.3.2 Decay Due to Environmental Factors

Cooling and heating due to changing temperatures in summer may cause cracks in buildings. The effect of moisture in rains may cause havoc. Leaching of alkali metals and iron may occur, which often forms black crusts on the stone surface, causing discoloration. Török *et al.* (2011) investigated the effect of the environment on the formation of these crusts in urban and rural areas and Farkas *et al.* (2018) studied the crusts in different countries. Gibeaux *et al.* (2018) determined the pollution rates from the speed of colour change measured by fixed-point observations. Efficient automotive combustion is an important contributor of nitrogen oxides, which are readily converted to corrosive nitric acid in the presence of oxidants. Although it is also a strong acid, nitric acid is less damaging to carbonate rock than sulfuric acid and sulphates due to the greater reactivity of sulphates with stone. All these atmospheric pollutants, after undergoing emission, transmission, and deposition, can cause devastating salt weathering hazards. The environmental factors not only cause weathering of stones but high humidity promotes growth of a number of microorganisms. Blackening of stones and plaster of buildings can be seen in (Figures 13.17 and 13.18a,b) at Agra Fort. There is a need to isolate the microbes if any and restoration is required to improve the aesthetic ambiance.

**Figure 13.17: Deterioration (Blackening) due to
Rain Rater Flow in Jahangir Palace of Agra Fort.**

Figure 13.18a-b: Note the Blackening due to Dampness and Growth of Microbial Flora in Plastered Wall of Agra Fort.

13.3.3 Decay due to Soluble Salts

In general, the presence of salts in stone is due to: natural sources (salts dissolved in groundwater that has permeated the stone before quarrying), pollution or miscellaneous factors (residues from stone cleaners, salt decay products from mortars or other materials in the structure. Water also carries dissolved salts into the stone. Some soluble salts are present in the stone when it is quarried. Other sources of salts include groundwater, sea spray, deicing salts, and cements used in mortars (Winkler, Er 1994). It has been demonstrated that damage to sandstone from salt occurs when dissolved salts are deposited in the pores near the surface of the stone (Lewin 1982). The salts expand when they crystallize and cause a thin layer of the stone to blister or turn to powder. Decay will continue under this initial layer. One published study explains how this type of decay, ranging from dislodging of individual grains to scaling up to 2 cm thick, can occur (Snethlage and Wendler 1997). Salt concentrates and crystallizes in the part of the stone where water is retained in the pores the longest, not necessarily near the surface. The depth of the zone of maximum water content varies according to the pore size and distribution of the particular stone, as well as conditions on the building. Surfaces exposed to sun and wind will dry quickly, meaning the area that remains wet and attracts salts lies beneath the surface. Damage to the stone will be manifested by the formation of scales. Surfaces that are sheltered and retain moisture will exhibit granular detachment, or "sanding off". The Figures 13.19a-d depicted here show decay due to salt formation in Agra Fort.

11.3.4 Microbial Deterioration

Due to technological innovations, a plethora of techniques and their

Figures 13.19a-d

(a) Broad sloppy pathway leading to Jahangir palace. Deterioration due to salt formation is visible on right side wall; (b) Deterioration of arch shaped gate of building in Jahangir palace; (c) Courtyard of Darbar-I-aam; (d) Brick tiles on floor showing damage due to weathering.

combination to identify and remove unwanted microorganisms have opened up new opportunities to both microbiologists and conservators (Teixeira *et al.*, 2018; Rusu *et al.*, 2020; Cappitelli *et al.*, 2020). The drawbacks to control the biodeterioration of heritage materials, with a special focus on outdoor stone heritage. Several authors have reported high numbers of microorganisms in decaying stones, and they concluded that microorganisms cause the deterioration (Doehne and Price 2010). However, an alternative and likely explanation could be that the decayed stones supplied a preferred habitat for microbiological growth (Doehne and Price 2010). Indeed, not even endolithic growth can be definitely associated with biodeterioration. Alteration of marble surfaces was caused by Biofilm Formation (Villa *et al.*, 2020)

A) The Activity of Cyanobacteria

Figures 13.14, 13.17 and 13.18 show the blackening or deterioration caused by pollution and microbes. The Cyanobacteria like *Oscillatoria, Lyngbya,* and *Gloeocapsa* may be responsible for such damage to old heritage buildings. The black crust coating the surfaces of building materials located in urban (polluted) environment may contain all kinds of organic compounds present in aerosols and particulate matter. Wet and dry deposition processes combined with plaster component result in dirty, grey-to-black crust formation, in which aerosols, spores, pollen, dust and every class of particulate matter are entrapped in the mineral matrix. Analysis of the organic compounds extracted from black crust has demonstrate mainly the characteristics of petroleum derivatives. The composition of each crust is governed by the composition of the particular airborne pollutants in the area. The chief factor determining microbial growth on constructional materials is moisture. Thus it is important for conservators to consider critical points for the better conservation and preservation. Damp surfaces are readily colonized by microbial cells settling from the air.Villa *et al.* (2020) and Hauer *et al.* (2015)reported diversity of cyanobacteria on rock surfaces. These can damage the stone structure or form biofilms which cause deterioration. Adhikary (2001) reported occurrence of *Tolypothrix, Plectonema, Calothrix, Phormidium* and *Nostoc* on Jagannath, Sun temple Udayagiri caves and Lingraj temples of Orissa. Some of these cyanobacteria were resistant to dry period of one year, few formed endospores for survival in adverse heat. He suggested control by heat, H_2O_2, and urea. The illumination in caves can promote growth of cyanobacteria, algae, mosses, ferns and sometimes flowering plants referred as lamp flora. The 15 per cent concentration of H_2O_2 was used for the destruction of lamp flora (Faimon *et al.,* 2003).

B) The Activity of Fungi

The importance of microbial impacts in the alteration and deterioration of cultural artefacts made of mineral, metallic or organic materials has been widely acknowledged in the course of many recent investigations. While in the past biodeterioration problems on cultural artefacts were often handled without a detailed analysis and, as a consequence, simply controlled by biocidal treatments, a much deeper interdisciplinary understanding of the environmental factors and material properties regulating the biogenic damage factor would allow more specific and adequate interventions. Fungi like *Alternaria* and *Curvularia* produce black pigments and discolour the objects. Figures 13.10, 13.16 and 13.17 show blackening of stone and plaster on walls. A detailed study is required to analyse the associated fungal flora with these buildings. The fungal organisms may show the fast growth at increase temperature. They show increased production of enzymes at higher temperature. Usually the temperature in summer and rainy months is too high. The fungi which grow at higher temperature are termed thermophilic fungi. The fungi most commonly recovered from air are *Aspergillus niger,* and *Penicillium citrinum,* the percentage occurrence of these two fungi was more in 2018. Research on the biodeterioration of stonework has increased considerably over the second half of the last century (Warscheid 2000; Scheerer *et al.*, 2018). From a biological point of view, stone represents an extreme habitat with large variations in environmental factors such as temperature, water availability, UV radiations and nutrients. Moreover, vertical or sub vertical surfaces are even more difficult to be colonized due to higher desiccation conditions (Caneva *et al.*, 2008).

Microorganisms can establish different relationships with the substrate: epiliths, growing on the stone surface, and endoliths, living in the interior of the stone. Endoliths are subdivided according to, their ecological niche in chasmoendoliths colonizing preexisting fissures and cavities, cryptoendoliths, colonizing structural cavities within porous rocks, and euendoliths, actively dissolving and penetrating into the stone. Some species are known to be strictly endolithic (endolithic lichens), while other microorganisms can live on and/or within the rock. The chemical and structural characteristics of lithotypes, surface roughness, porosity and their state of conservation, as well as environmental conditions and climate are the key factors influencing the establishment and growth of microorganisms (Caneva, *et al.*, 2008).

Two main groups of fungi, were isolated from stone monuments: (i) Hyphomycetes and Coelomycetes and (ii) black meristematic or MCF, and black yeasts (Sterflinger 2010). Hyphomycetes include species with colourless or brightly coloured colonies (hyaline Hyphomycetes) and species producing dark brown, green-black, or black colonies (dematiaceous Hyphomycetes). Hyphomycetes excrete organic acids and can actively dissolve carbonates, they produce different kinds of pigments, but only a few species are melanin producers (Sterflinger, 2006). They generally require high levels of humidity and organic substances to develop. In spite of their presence in biofilms on outdoors stone monuments they are frequently responsible for the deterioration of wall paintings, caves and restored stone artefacts The MCF belongs to Ascomycetes, principally to the orders *Chaetothyriales*, *Dothideales*, and *Capnodiales*, and were isolated mostly, but not exclusively in arid and semi-arid environments.

The deterioration of stone and minerals induced by fungi is generally ascribed to biogeophysical and biogeochemical processes that are not mutually exclusive but are assumed to act together, and the latter are believed to be more important. The filamentous structures of fungal hyphae favour their penetration into the substrate, depending on its structure, chemical composition and state of conservation. Fungi can also perforate intact minerals (Santo *et al.*, 2021). Penetration can be also favoured by turgor pressure inside hyphae and melanin. Moreover, EPS produced by fungi facilitate fungal biofilm formation and the attachment to the rock, and increase mechanical pressure giving rise to shrinking and swelling (De Leo *et al.*, 2019).

Biogeochemical processes involve the production of metabolites which react with stone to form secondary minerals. Fungi excrete a large number of organic acids (oxalic, citric, acetic, formic, gluconic, glyoxylic, fumaric, malic, succinic, and pyruvic), which can act as chelators (Marvasi *et al.*, 2012). Moreover fungi produce siderophores, low molecular weight structures generally classified into two structural groups hydoxamates and catecholate compounds, which have a high specificity for chelating or binding iron (De Leo *et al.*, 2019). Their role in the *etc.*hing of microfractures on olivine and other silicate has been shown in laboratory experiments (Sterflinger *et al.*, 1997).

Microcolonial fungi together with cyanobacteria, algae and lichens, are poikilohydric microorganisms and can have active metabolic or dormancy periods according to water availability. As they are resistant to multiple and variable stress factors, and have a wide range of tolerance, they are considered poikilo-tolerant. Usually, in oligotrophic conditions they form small black clump-like colonies on stone, consisting of isodiametrically dividing cells, from which hyphae can branch.In presence of carbohydrate and suitable environmental conditions, meristematic fungi can grow profusely on and into the stone, causing a large dark colouration. It is likely that this kind of patina is more common on stone monuments than suggested in the literature. It is important to note that, where analyses have not been performed, grey-black patinas are often misinterpreted and ascribed toMCF growth is not limited to the stone surface, on marbles black fungi penetrate into the intercrystalline spaces contributing to the loosening and detachment of crystals and leading to the formation of pitting (or biopitting) (Sterflinger and Krumbein 1997).The pigmentation of fungal cells decreases with greater depth and the hyphae appear colourless, underling the light protective function of black pigments (Hoppert *et al.*, 2004). Sterflinger and Krumbein (1997) considered dematiaceous fungi as being one of the principle agents responsible for biopitting on Mediterranean marble and limestone. They described four different types of pitting and isolated fungal strains from all of them.

C) The Damage to Buildings by Lichens, Moss and Angiosperms

It is well known that lichen colonization on outdoors monuments depends on many different factors: the environment, the orientation, the type of substrate, and the availability of nitrogen compound. A correct identification of lichen species is a fundamental preliminary step of whatever study, moreover the ecological study of lichen communities growing on monuments permits to investigate the most important factors favouring their growth and spreading, therefore it is very useful to evaluate if preventive measures can direct methods to eliminate them. A conceptual model of bio protection limestone by epilithic lichens was proposed by Carter and Viles (2005). Lichens act as an umbrella protecting the substrate from rainfall and atmospheric deposition. Lichens can absorb/reflect solar radiation protecting the stone from temperature variations and thermal stress, depending on the colour of the thalli (dark thalli, of crustose lichen *e.g. Verrucaria nigrescens*, absorb heat causing an increase of temperature fluctuations and thermal gradient in the rock surface) Carter and Viles (2004), also as a barrier protecting stone from water penetration, and reducing or neutralizing the negative action of wind, rain, pollutants and marine aerosols. Alterations in environmental factors bring marked changes in morphological and physiological features of lichens. Even changes in chlorophyll can occur due to pollution (Kumari and Nayaka 2018). An index of Lichen Potential Biodeteriogenic Activity (LPBA) was proposed to quantify lichen impact on stonework taking in account of cover, reproductive potency, depth of hyphal penetration, physical and chemical action, hyphal spread and bioprotection (Gazzano *et al.*, 2009). The application of the index to some case studies highlights its effectiveness for a more thorough evaluation of biodeterioration induced by lichens on different stoneworks with respect to cover only

Italy is one of the richest countries in historical and artistic heritage. The Central Institute for Catalogue and documentation of Italy lists 781 archaeological areas or parks and 622 main monumental complexes. Natural flora in monumental areas is seen with some concern because plants can damage monuments with their roots, can give the appearance of neglect, obstruct site access for visitors or conceal the monuments (Kanellou *et al.*, 2017). In some case, trees have to be prudentially kept in the site as they provide stability to the structures and their removal could be dangerous. In the past (Romantic era) or in some peculiar situations (archaeological area of Angkor, in Cambodia, or Tikal, in Guatemala).The interest on the flora of archaeological sites in Italy dates back to the XVII Century. Panaroli (1643) reports 337 different plantsgrowing on the Flavian Amphitheatre in Rome. The scope of floristic inventories can more focus on biological features of the plants growing into archaeological areas (Ceschini *et al.*, 2006) or the list can be accompanied by the indication of their dangerousness for the monument with the indications for their management (Signorini 1995).

13.3.5 Damage to Buildings by Human Vandalism

The monuments can be spoiled by environmental conditions like temperature and rains, but it has been found in certain cases that deterioration may be caused by visitors. A glance to the wall paintings present in Jahangir palace show that paintings are disfigured due to cumulative effect of dampness and human interference (Figures 13.19a and b). The repairing done by ASI has also caused aesthetic damage to wall paintings. These paintings are present on lower side of wall and are within reach of people. There are no security persons seen at this place to restrict any such damage to heritage structure. Figures 13.16a and b, shows the loss of aesthetic beauty due to removal of precious stones from the building of Muthamman Burj. Efforts are taken by ASI to restrict the public interaction by putting an aluminium grill at this place. The large number of visitors also increase humidity thus changes in microclimate occurs. As such there is no restriction of numbers present at any spot in Fort. The time slot 2hours is given for Taj Mahal, it helps to restrict the number of visitors.

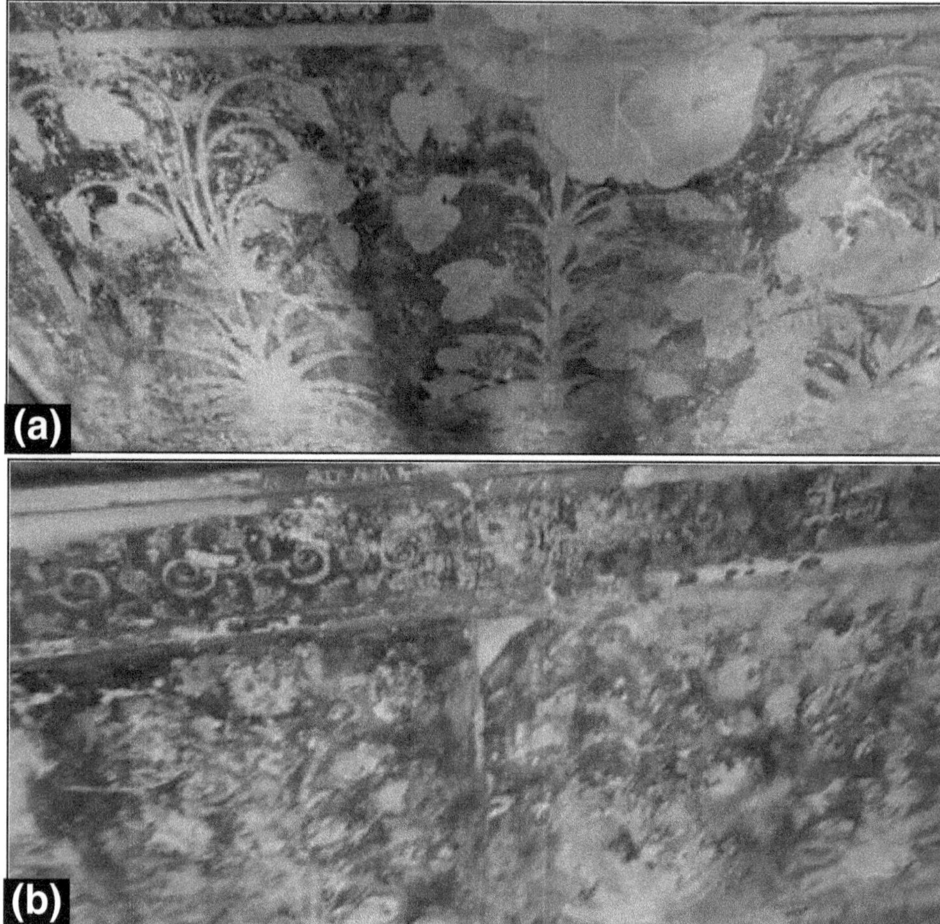

Figure 13.20a-b: Deterioration of Wall Painting in Jahangiri Palace due to Humid Environment and Human Vandalism.

13.3.6 Damage to Buildings by Poor Maintenance

Maintenance of old buildings is yet another major issue for ASI in different monuments of India. Figure 31.19a shows how the damage has been repaired by ordinary cement. This type of repairing further damages the buildings. Figures 13.21b and c shows how the marble stones have fallen from wall and repair with cement is done. There is need of immediate repair at certain places and it should be aesthetically good.

13.4 Biological Weathering

Biological weathering includes the actions of a wide range of organisms. In arid regions, 'higher' plant assemblages tend, for the most part, to be sparsely distributed or absent altogether from areas that experience particularly extreme

Figure 13.21a-d

(a) Ceiling of a palace showing damage in centre and borders; (b) Part of a wall and (c) A stone wall on the side of door needs repair as broken parts can be seen; (d) A stone in front top is missing in centre and right side wall is broken, plaster damaged.

environmental conditions. Consequently, in the context of biological weathering in arid regions, the emphasis is on the more 'primitive' but nonetheless ancient organisms that comprise the lithobionts – a group that includes cyanobacteria, algae, fungi, and lichen that inhabit rock surfaces (Warke 2013). The type of organisms that colonize rock surfaces reflects the combined effects of the mineralogical and structural characteristics of the rock – its 'bioreceptivity' (Guillitte and Dreesen 1995; Hutchens 2009) and the limitations imposed by the prevailing environmental conditions. Bioreceptivity is mainly dependent on structural properties such as rock surface roughness and porosity and the dominant mineralogical characteristics. Large pore spaces facilitate surface colonization by algae, fungi, and lichen (Guillette and Dreesen, 1995).

The chemistry of the mineralogical constituents of rock has significant implications for the type of species able to colonize, with, for example, species adapted to alkaline conditions being most successful on calcareous rock such as limestone, while those with a preference for more acidic conditions preferring rock types rich in silica such as sandstone (Lisci *et al.*, 2003). Because of the severity of environmental conditions in arid regions, many organisms are endolithic in habit.

Viles (1995) identified the optimum environmental conditions for biological growth and weathering as occurring in regions with abundant moisture; thus, in arid regions, biological activity will necessarily be more restricted because of the stress imposed by aridity. However, with regard to colonization of rock, organisms can occupy either epilithic or endolithic niches (Viles and Goudie 2004). The former describes organisms that live on rock surfaces, while the latter includes organisms that have evolved to cope with extreme environmental conditions by living below the rock surface in pre-existing microfractures and cracks or within pore spaces and boreholes. Endolithic forms become increasingly dominant as moisture availability declines (Viles 1995). Arai and Wada (2014) found association of fungus *Cladosporium* in blackening of plastered wall in Himeji castle in Japan. Biofilm Formation: Effects of Chemical Cleaning Coatings (Villa *et al.*, 2020).

13.4.1 Biochemical Effects

The iron-chelating ability of lichens and fungi is well documented (Seaward 2004; Allsopp *et al.*, 2004) but it is a mechanism that is more commonly associated with environmental conditions that are more humid than those typically associated with contemporary arid regions. It is important to remember that the complexity of micro-environmental conditions across desert surfaces creates micro scale ecological niches in which organisms can comfortably live and where biochemical mechanisms can actively weather rock (Friedmann and Galun 1974). Biochemical weathering is also typically associated with the secretion of organic acids during normal metabolic functions. For example, a by-product of cellular respiration by algae, fungi, and lichens is carbon dioxide, which combines with organic moisture to form carbonic acid that in turn can contribute to localized small-scale dissolution of susceptible minerals (Chen *et al.*, 2000). Calcareous rock types are particularly susceptible because of the high solubility of elements such as calcium (Smith *et al.*, 2000), although the significance in terms of rock weathering is only evident at the sub-millimeter scale but comprising a hierarchy of micro-dissolution forms associated with isolated microbial organisms as well as those created by larger colonies (Figures 13.21a–c). Another commonly secreted organic acid of lichen is oxalic acid. Which is a chelating agent, and reacts with various minerals to form a range of different oxalates.

13.4.2 Biophysical Effects

Biophysical weathering effects consist of those that are directly associated with rock breakdown and those that play an indirect role. Direct physical weathering effects are most efficiently demonstrated by lichen where stresses arising from the periodic wetting and drying of lichen thalli are sufficient to 'pluck' fragments and individual mineral grains from rock surfaces (Cooks and Otto 1990; Moses and Smith 1993; Seaward 2004). The penetration of fungal hyphae from beneath lichen into the fabric of rock may also facilitate physical breakdown by providing a mechanism for moisture and salt ingress and by the gradual separation of mineral grains or parts of grains as the number of hyphae and the extent of their penetration increase. At a smaller scale, algae also contribute to the direct physical disruption of rock through the expansion and contraction of the algal mucilage associated with repeated wetting and drying of individual algal bodies and colonies located beneath rock surfaces in pore spaces and/or micro-fractures (Hall 1990).

Indirect biophysical weathering effects include changes to rock surface temperatures caused by the presence of lichens that can alter albedo characteristics of the rock surface. Dark-coloured lichens have been shown to significantly increase heating of underlying rock and it is suggested that these lichens make an indirect contribution to rock breakdown through enhancement of thermal stress between surface and subsurface rock layers, leading eventually to fatigue-related failure (Schwartzman *et al.*, 1997; Carter and Viles 2004; Hall *et al.*, 2005). The higher temperatures established may also enhance the effectiveness of any biochemical weathering mechanisms associated with acids secreted during lichen metabolism. Any physical deterioration of rock surfaces or subsurface material arising from biophysical actions will leave points of weakness that may subsequently be exploited by other agents of weathering such as salt and moisture.

The connection between biochemical and biophysical mechanisms is highlighted by Paradise (1997), who reported the operation of both beneath lichen on sandstone in Arizona, where biochemical weathering through the effects of acid secretion and mineral dissolution weakened the microstructure of the sandstone, which in turn aided subsequent hyphal penetration and related biophysical weathering as the lichen grew and the thallus expanded outward to continue the process on 'fresh' rock. In the last few decades, technological advances in microscopy have enabled better understanding of the micro-scale changes caused by a range of organisms from lichens to algae and has demonstrated their presence in even the harshest of environmental settings. Field evidence indicates that these organisms are well adapted to thrive under extreme environmental conditions, entering a state of minimal metabolic activity during periods of desiccation and becoming more active in response to episodes of moisture availability. Their role in the weathering of rock in arid regions may appear to be somewhat restrictive and dependent on the availability of moisture but nevertheless biological weathering is another component of the desert weathering system and, as such, warrants serious consideration.

13.5 Control of Biodeterioration

Biodeterioration involves a combination of physical and chemical damages together with aesthetic alteration to cultural heritage. Due to technological innovations, a plethora of techniques and their combination to remove unwanted microorganisms have opened up new opportunities to both microbiologists and conservators (Teixeira *et al.*, 2018; Rusu *et al.*, 2020). Before removing microorganisms, those causing biodeterioration should be clearly identified. Indeed, when the efficacy of treatment is uncertain, *e.g.*, not targeting the actual biodeteriogens as these are not known, any treatment should be carefully considered or avoided (Nugari and Salvadori 2003). Studying the microbial community of the Preah Vihear temple in Cambodia using DNA and RNA sequencing. Meng *et al.* (2020) showed that the active community revealed by RNA sequencing was different from the whole community revealed by DNA sequencing. Thus, active members in the community need to be studied based on RNA rather than DNA to provide significant information for the conservation.

13.5.1. Traditional Chemicals

Biocides have been used for any kind of cultural heritage material both indoors and outdoors. The active ingredients should be known, according to Bajpayee and Kumari (2002) 'many biocide products available in the Asia Pacific market do not disclose their ingredients.' They suggested use of Acticide-14 for wet and Acticide EP/EP in dry or paste form for the control of microorganisms. The Biocides Directive 98/8/EC on placing biocidal products on the market gives the framework for biocide policy in the EU. In this directive, active substances are evaluated. Nugari and Salvadori (2003) stated that the only effective gas against insects and fungi for fumigation was ethylene oxide, which has been banned in several countries due to carcinogenic and mutagenic features. However, more recent research (Wawrzyk *et al.*, 2020) proved that vaporized hydrogen peroxide disinfection is useful. If a biocide approach is planned to control microbial growth on a heritage surface, in situ pilot tests to calibrate biocidal treatments on the particular study case (species, site) often need to be run. Studying five biocides (namely, BiotinR, BiotinT, DesNovo, Lichenicida 464, and Preventol RI80) against lichens, Favero-Longo *et al.* (2017) showed that different biocidal products and application methods have different efficacies against each species tested. However, biocidal treatments have a brief duration and must often be repeated frequently, creating a repeated threat to the heritage material and the environment (Li *et al.*, 2019; Becerra *et al.*, 2020). In addition, repeated biocidal treatments can cause resistance in target biological agents, and they can modify biofilm structures favouring the growth of more harmful biodeteriogens (Salvadori and Charola 2011). As an alternative to traditional biocides, dimethyl sulfoxide (DMSO) in a solvent gel has been used as the

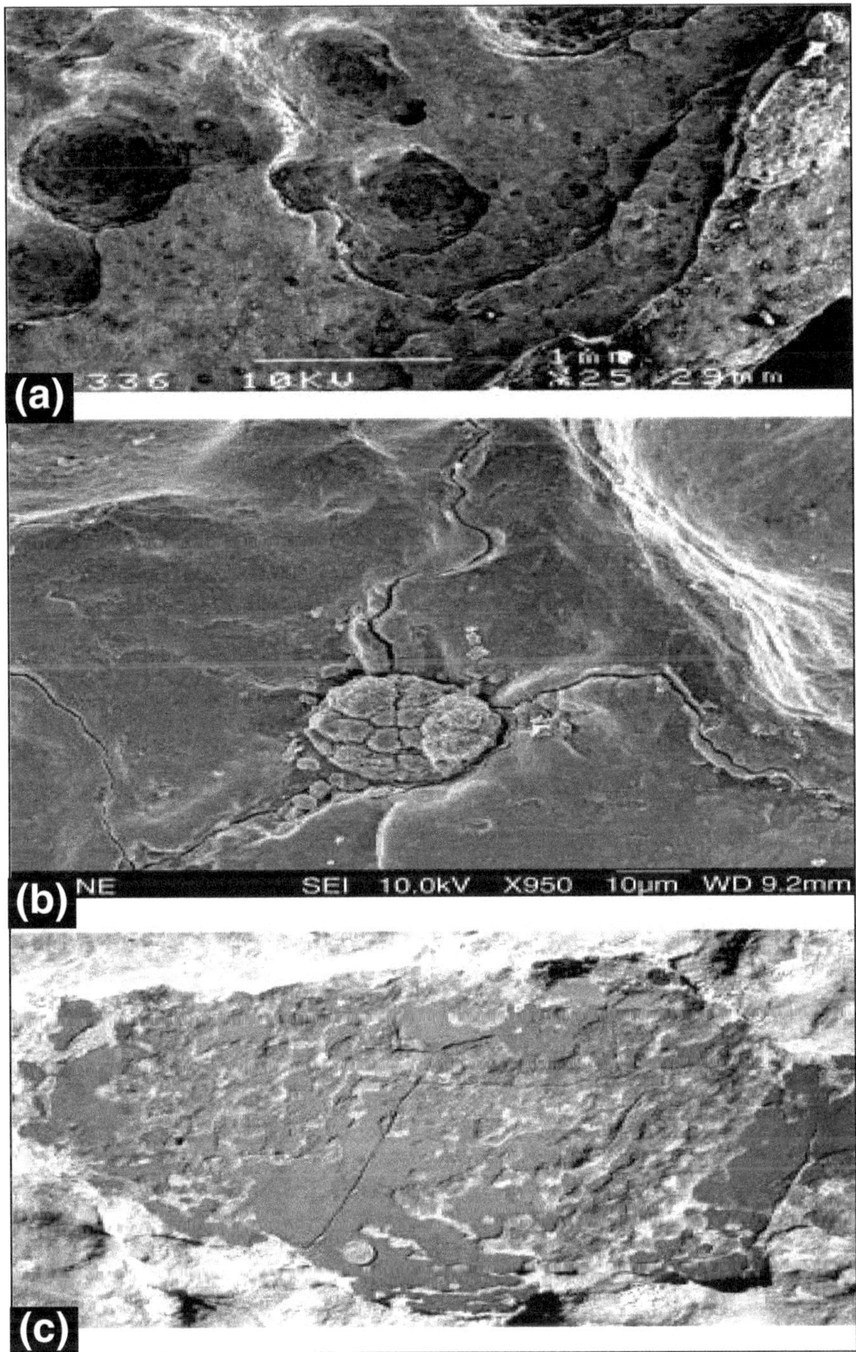

Figure 13.22a-c

(a) Scanning micrograph showing biochemical dissolution of limestone by organic acid of microbes; (b) Scanning image of cluster of epilithic organisms in a Manganese rich rock; (c) Remnant of rock varnish covered sand stone.

active substance to remove biological colonization on marble and compared to biocides currently used in conservation (Toreno *et al.,* 2018). In 1963, regarding the 'maladie verte de Lascaux', the scientist Dobat believed that fighting algal invasion with chemicals was, in the long run, doomed to failure because it was impossible to kill all algae and spores that were present in the cave, new microorganisms would always be introduced, and chemical treatments would be frequently needed with high risk of harming the painting (Lefèvre1974).

Agrawal (2007) suggested use of 10 per cent aqueous ammonium hydroxide solution in which small amount of neutral detergent and hydrogen peroxide were added to clean the blackened microbial growth in sculptures of Mathura. Application of CMC paste and peeling off the dried layer cleaned such sculptures completely. Dust and dirt can be removed by mechanical method using dilute solution of a neutral detergent. In all the cases washing with clean water suggested.

13.5.2. Natural Biocides

The use of natural alternatives for commercial biocides is gaining importance in the field of conservation. Natural biocides are considered safer for human beings and greener for the environment and have been used both for organic and inorganic materials (Tayel *et al.,* 2016). Many of these products are derived from plants and could be employed in their pure form as well as crude extracts or as essential oils. Tayel *et al.* (2016) claimed that plant smouldering fumes could be used as an effective alternative to chemical fungicides, completely sterilizing the fumigated archival repository and leaving no modifications to colour or surface structure of paper specimens. To date, the use of natural substances has largely focused on organic materials, such as wood and paper, in indoor environments (Jeong *et al.*2018). However, inorganic materials have also been treated with these natural compounds (Fidanza and Caneva 2019). Essential oils from *Thymus vulgaris, Origanum vulgare,* and *Calaminthanepeta* and their major components (thymol, carvacrol, pulegone) have been prepared within a hydrogel matrix (made of Gelrite, poly-vinylacetate, $CaCl_2$, and Acemoll CC) to remove biofilm from a travertine wall of the Sapienza University in Rome (Genova *et al.,* 2020). Volatile extracts from *Illicium verum,* the flower buds of *Syzygium armoaticum, Quercus infectoria, Coptis chinensis,* and *Phellodendron amurense* were profitably tested with fungi isolated from biofilm on the World Heritage Yeongneung stone monument (Jeong *et al.,* 2018). The best antifungal effects were shown for Eugenol isolated from clove extract (Mak *et al.,* 2019). The volatile organic compound was used with the stable emulsifiers Tween and Span. *Thymbra capitata* essential oil was proposed for use against cyanobacteria and green algae on historical monuments (Candela *et al.,* 2019). The major components of the essential oil were the phenolic monoterpene carvacrol (73.2 per cent) and its biogenetic precursor's γ-terpinene (6.9 per cent) and p-cymene (4.3 per cent).

13.5.3. Nanoparticles

Nanoparticles (NPs) display unique physicochemical properties including an ultra-small size, large surface-to-mass ratio, and a peculiar reactivity with organisms, and they are appealing to use both for organic and inorganic materials. Some NPs composed of silver (Ag), copper (Cu), titanium dioxide (TiO_2), or zinc oxide (ZnO) display interesting biocidal features including disruption of the cell wall and the plasmatic membrane, the inhibition of protein synthesis and DNA replication, (Becerra *et al.*, 2020).

13.5.4 Physical Methods: Mechanical Removal

Traditional mechanical methods remove biodeteriogens on organic and inorganic objects with tools such as brushes, scalpels, spatulas, scrapers, air abrasives, high-pressure blasting, low-pressure washing, and vacuum cleaners. Phototrophic biofilms should be completely dry before any mechanical cleaning, because dry crusts often readily detach from the materials and can be easily removed by using brushes, sand blasting, or low-pressure washing (Sterflinger and Sert 2006). Generally, mechanical methods have the benefit of not using additional compounds in the heritage substratum. For instance, Lee *et al.* (2016) treated a standing Buddha engraved on a shale wall (South Korea) by dry cleaning followed by wet cleaning using distilled water. The researchers observed the removal of persistent lichens by soaking them in distilled water and using soft brushes and wooden knives to completely remove them. Efficacy of the dry brushing method, among other chemical methods, in removing an algal biofilm formed on a granite-built historical monument in Galicia, Spain was tested (Sanmartín *et al.*, 2020).

A commercial product called Hydrogommage–a sand blasting, low-pressure (0.5–1.5 bar) air/water mixture with SiO_2 particles (0.5–0.1 mm)–was applied on granite specimens to remove thick phototrophic biofilms composed of the green algae *Trebouxia* sp. and the cyanobacteria *Gloeocapsa* sp. and *Chroococcus* sp. under laboratory conditions (Pozo et al 2013). Although the treatment efficiently removed the biofilms, it caused textural changes to granite including an increase in roughness and microfissures (Pozo *et al.*, 2013). The marble statues in the gardens of the National Palace of Queluz, Portugal were subjected to a grit-blasting treatment to remove biological colonization (Delgado *et al.*, 2011). Mechanical treatment increased the surface roughness as compared to the uneven and pitted surface left behind by the chemical cleaning methods. The surface roughness resulting from the cleaning intervention affected the subsequent microbial recolonization of the marble (Delgado *et al.*, 2011).

13.5.5. UV-C Irradiation

UV-C irradiation has been applied to stone surfaces. Regarding the 'maladie verte de Lascaux', in the 1960s, in addition to the modification of the physicochemical parameters of the Cave (CO_2, temperature, humidity), the scientist Dobat envisaged the use of appropriate filters to monitor the quality of light as well as ultraviolet rays to kill the phototrophic microorganisms (Lefèvre 1974). Species able to live in aquatic and terrestrial habitats, *e.g.,* cyanobacteria, have adopted survival strategies when exposed to UV light. The terrestrial species of *Tolypothrix, T. byssoitka,* isolated from the stone surface of the Sun Temple, Konark, India, could survive UV-C irradiation up to 1 day, whereas the aquatic *Tolypothrix* UU 2434 was killed after exposure to UV-C for half an hour (Adhikary and Sahu 1998).

13.5.6 Laser Cleaning

In the last decade, this technique has been amply studied on outdoor stone surfaces due to its advantages over traditional cleaning methods: it is controllable, selective, contactless, and environmentally friendly (Pozo-Antonio2019; Barreiro *et al.,* 2020). The main drawback is that every study case needs to be optimized. The Nd:YAG laser (0.5 ms pulse, 5 J:pulse, 1 Hz) yielded a satisfactory surface lichen cleaning outcome on outdoor sculptures at the Seattle Art Museum but at a comparably slow rate (Leavengood *et al.,* 2000).

13.5.7 Heat Shocking, Microwaves, and Dry Ice Treatment

Many microorganisms living on outdoor stone are thermotolerant (up to 65–70 °C) when dry but are thermosensitive when wet, and heat shock treatments on wet, metabolically active lithobionts cause a loss of membrane permeability and denaturation of proteins (Tretiach *et al.,* 2012). The method is simple, easy, and fully substratum-, operator-, and environment-friendly, and it employs plastic foils, thermal blankets, and infrared lamps. The feasibility of the in situ thermal treatment was verified with several lichen species under laboratory conditions (Tretiach *et al.,* 2012). Six-hour treatment at 55 °C was able to kill the fully hydrated lichens. At 40 °C, thermal treatment damaged lichens and was combined with a concentration 10-fold lower than that normally used.

13.5.8 Other Biological Methods

The following biological methods are environmentally friendly, low cost, and present low to no risk to human beings. These methods have been applied to different materials. Valentini *et al.* (2010) applied a new method based on the glucose oxidase enzyme to remove biofilms from travertine and peperino substrata of the Villa Torlonia in Rome,Italy. A comparative study was also performed to validate the enzyme-based cleaning procedure by using a saturated solution of $(NH_4)_2CO_3$ and ethylenediamine-tetraacetic acid (EDTA) and the enzyme lipase. Among all, the cleaning procedure using glucose oxidase showed the best results. Glucose oxidase produces H_2O_2 when glucose is present. Since the release is highly controllable, this method is rather safe for the substratum. Chitinases were also used to reduce the growth of filamentous fungi developed on word walls and canoes at the Cross Lake Bridge ruins in Xiaoshan (Wu and Lou2016).

13.5.9 Combination of Control Methods

There are many papers that compare the efficacy of different control methods on the same object or site. The comparison is extremely useful if for example considerations including environmental safety are taken into account (Mascalchi *et al.*, 2020). The following literature proves that the combination of different control methods is often a successful strategy. Prior irradiation of the granite specimens with UV-B and UV-C light enhanced the cleaning efficacy of benzalkonium chloride (Pozo-Antonio and Sanmartín 2018). Gamma irradiation has been used at sub-lethal doses followed by treatment with a biocide. This was the case of mural paintings in the Tell Basta and Tanis tombs exposed to various doses of gamma radiations (5, 10, 15, 20, and 25 kGy) and then treated with tricyclazole (Sakr *et al.*, 2019). This combined treatment completely inhibited melanin production in the investigated biodeteriogens *i.e. Streptomyces* spp.

In this context, it is generally accepted that sessile microorganisms are more resistant to antimicrobial agents than their planktonic counterparts. This also is due to the development of resistant phenotypes upon adhesion to substrata and within the biofilm (Meyer2003). In investigating the effects of TiO_2 nano-powder and thin film, a decrease of one order of magnitude of *Pseudomonas aeruginosa* planktonic cells after 2 h and an almost complete removal of *P. aeruginosa* planktonic cells was observed. Contrarily, photocatalytic treatment with TiO_2 film or TiO_2 nano-powder did not affect *P. aeruginosa* sessile growth (Polo *et al.*, 2011). Green algae isolated in three caves grown in a mineral medium and inoculated on limestone specimens were subjected to UV-C irradiation (Borderie *et al.*, 2011). This research reported that UV-C exposure was more deleterious when algae grew in liquid medium (planktonic growth) than on solid medium, simulating an in situ situation. Consequently, lab experiments are better performed on biofilms rather than planktonic cells, and therefore, the development of further biofilm lab models is needed (Villa *et al.*, 2015). Removing water from various stone related monuments was suggested by (Arai and Wada 2014), they also recommended use of "Water ceramic" as long term water proofing agent to control the biodeterioration.

13.6 Conclusion

The beautiful city of Agra is on tourist map of India due to its heritage monuments and world famous Taj. Because of the historical significance Red fort Agra complex was designated as UNESCO world heritage site in 1983. The chapter discusses various architectural details of certain heritage structures and damages caused to heritage site of Red fort in long span of five centuries. The consciousness about conservation has increased but much is needed from agency like ASI and visitors to protect its beauty. Various problems of deterioration and alternative suggestions for restoration are discussed for conserving this monument.

India's archaeological horizons expanded considerably in the twentieth century as research revealed the temporal depth and richness of its past. India needs a comprehensive, feasible national policy for the care of its cultural heritage. Its rapidly growing population and the attendant sprawl of cities and towns, industrial development, and even changes in agricultural technology are destroying archaeological sites in every part of India (Lewis 2014). It is unquestionably true that the prosperity and well-being of today's India is of far greater concern than the care of its ancient sites, but, while it is easy to concede this point, it is equally clear that the destruction of the nation's cultural heritage must be controlled as much as possible for the public good and for the understanding of India's past by future generations. In 2010, a major amendment to The Ancient Monuments and Archaeological Sites and Remains (Amendment and Validation) Act of 1958 created the National Monuments Authority, which adds teeth to existing legislation that protects the immediate vicinity of national monuments and sites from unauthorized encroachments. Researches are needed to suggest effective control measures for biodeterioration of heritage monuments.

REFERENCES

Adhikary S P (2001) Survival strategies of lithophytic cyanobacteria on the temples and monuments. In: Studies in Biodeterioration of Materials-1. O.P. Agrawal, S. Dhawan and R. Pathak (eds) Pub. by INTACH and ICBCP, 29-44pp.

Adhikary S P, Sahu J K (1998) UV protecting pigment of the terrestrial cyanobacterium *Tolypothrix byssoidea*. J. Plant Physiol. 153: 770–773.

Agrawal O P (2007) Study and conservation in spotted red stone of Mathura. In : Essentials of Conservation and Museology. Pub. By Sandeep Prakashan, Delhi. 50-56 pp.

Allsopp D, Seal K, Gaylarde C (2004) Introduction to Biodeterioration. Cambridge University Press, Cambridge.

Alonso E. and Martìnez L (2003) The role of environmental sulfur on degradation of ignimbrites of the Cathedral in Morelia, Mexico.Building and Environment 38(6): 861-867

Arai H, Wada K (2014) Blackening on monuments and its control: Long term prevention of biodeterioration. In : Biodeterioration of Cultural Property-7 (eds) Dhawan S. Abduraheem K. and Nath V. Pub. By Sukriti Nikunj, Lucknow 1-10pp 89-94

Bajpayee M, Kumari P K (2002) Eco-friendly control of biodeterioration of coatings and perfect biocides. In: Biodeterioration of Materials (eds)R.B. Srivastava, G.N. Mathur and O.P. Agrawal. Pub. by ICBCP. 180-193.

Barreiro P, Andreotti A, Colombini M P, González P, Pozo-Antonio J S (2020) Influence of the laser wavelength on harmful effects on granite due to biofilm removal. Coatings. 10: 196.

Becerra J, Ortiz P, Zaderenko A P, Karapanagiotis I (2020) Assessment of nanoparticles/nanocomposites to inhibit micro-algal fouling on limestone façades. Build. Res. Informat. 48: 180–190.

Bo¨ke H, Gauri KL (2003) Reducing marble–SO_2 reaction rate by the application of certain surfactants. Water, Air, and Soil Pollution 142 (1–4): 59–70.

Borderie F, Laurence A-S, Naoufal R, Bousta F, Orial G, Rieffel D, Badr A-S (2011) UV-C irradiation as a tool to eradicate algae in caves. Int. Biodeter. Biodegr. 65: 579–584.

Candela RG, Maggi F, Lazzara G, Rosselli S, Bruno M (2019) The essential oil of Thymbra capitata and its application as a biocide on stone and derived surfaces. Plants 8: 300.

Caneva G, Gasperini R, Salvadori O (2008). Endolithic colonization of stone in six Jewish cemeteries in urban environment. Proceedings of the 11th Int. Congress on Deterioration and Conservation of Stone, September 15-20, Torun, Poland. 49-56.

Cappitelli F, Cattò C, Vill F (2020) Review The Control of Cultural Heritage Microbial Deterioration. Microorganisms 8: 1542; doi: 10.3390/microorganisms8101542

Carter NE, Viles HA (2004) Lichen hotspots: raised rock temperatures beneath *Verrucaria nigrescens* on limestone. Geomorphol. J. 62: 1-16. [http://dx.doi.org/10.1016/j.geomorph.2004.02.001]

Carter NE, Viles HA (2005) Bioprotection explored: the story of a little known earth surface process. Geomorphol. J., 67: 273-281. [http://dx.doi.org/10.1016/j.geomorph.2004.10.004]

Chen J, Blume HP, Beyer L (2000) Weathering of rocks induced by lichen colonisation – a review. Catena 39. 121–146.

Ceschin S, Caneva G, Kumbaric A (2006) Biodiversità ed emergenze floristiche nelle aree archeologiche romane. – Webbia 61(1): 133-144.

De Leo F, Antonelli F, Pietrini A M, Ricci S, Urzì C (2019) Study of the euendolithic activity of black meristematic fungi isolated from a marble statue in the Quirinale Palace's Gardens in Rome, Italy. Facies 2019, 65, 18.

Delgado R J, Vale Anjos M, Charola A E (2011)Recolonization of Marble Sculptures in a Garden Environment. In Biocolonization of Stone: Control and Preventive Methods: Proceedings from the MCI Workshop Series; Charola, A.E., McNamara, C.J., Koestler, R.J., Eds.; Smithsonian Institution Scholarly Press: Washington, DC, USA, 2011; pp. 71–85.

Doehne E, Price C A, (2010) Stone Conservation–An Overview of Current Research, 2nd ed.; The Getty Conservation Institute: Los Angeles, CA, USA, 2010.

Faimon J, Štelcl J, Kubešová S, Zimák J (2003) Environmentally acceptable effect of hydrogen peroxide on cave "lamp-flora", calcite speleothems and limestones. Environ. Pollut. 2003, 122, 417–422.

Farkas O, Siegesmund S, Licha T, Török Á (2018) Geochemical and mineralogical composition of black weathering crusts on limestones from seven different European countries. Environ Earth Sci 77: 211. https: //doi.org/10.1007/s12665-018-7384-8

Favero-Longo S E, Benesperi R, Bertuzzi S, Bianchi E, Buffa G, Giordani P, Loppie S Malaspina P, Matteucci E, Paoli L. (2017) Species- and site-specific efficacy of commercial biocides and application solvents against lichens. Int. Biodeter. Biodegr. 2017, 123, 127–137.

Fidanza MR, Caneva G (2019) Natural biocides for the conservation of stone cultural heritage: A review. J. Cult. Herit. 2019, 38, 271–286.

Friedmann E I, Galun M (1974). Desert algae, lichens and fungi. In: Brown, G.W. (Ed.), Desert Biology. Academic Press, London, vol. 2, pp. 165–212.

Gazzano C, Favero-Longo S E, Matteucci E, Roccardi A, Piervittori R (2009) Index of Lichen Potential Biodeteriogenic Activity (LPBA): A tentative tool to evaluate the lichen impact in stonework. Int. Biodeterior. Biodegradation. 63: 836-43.

Genova C, Grottoli A, Zoppis E, Cencetti C, Matricardi P, Favero G (2020) An integrated approach to the recovery of travertine biodegradation by combining phyto-cleaning with genomic characterization. Microchem. J. 2020, 156, 104918.

Gibeaux S, Vázquez P, De Kock T, Cnudde V, Thomachot-Schneider C (2018) Weathering assessment under x-ray tomography of building stones exposed to acid atmospheres at current pollution rate. Constr Build Mater 168: 187–198. https: //doi.org/10.1016/j.conbuildmat.2018.02.120

Guillitte O, Dreesen R, (1995) Laboratory chamber studies and petrographical analysis as bioreceptivity assessment tools of building materials. The Science of the Total Environment 167, 365–374.

Hall K (1990) A note on biological weathering of nunataks of the Juneau Icefield, Alaska. Permafrost and Periglacial Processes 1, 189–196.

Hall K, Arocena J M, Boelhouwers J, Liping Z, (2005) The influence of aspect on the biological weathering of granites: observations from the Kunlun Mountains, China. Geomorphology 67, 171–188.

Hauer T, Mühlsteinová R, Bohunická M, Kaštovský J, Mareš J (2015) Diversity of cyanobacteria on rock surfaces. Biodivers. Conserv. 2015, 24, 759–779.

Hoppert M, Flies C, Pohl W GA B, Schneider J (2004) Colonization strategies of lithobiontic microorganisms on carbonate rocks. Environ Geol., 2004, 46, 421-428. [http://dx.doi.org/10.1007/s00254-004-1043-y]

Hutchens E, (2009) Microbial selectivity on mineral surfaces: possible implications for weathering studies. Fungal Biology Reviews 23(4), 115–121.

Jeong SH, Lee H J, Kim DW, Chung Y J (2018) New biocide for eco-friendly biofilm removal on outdoor stone monuments. Int. Biodeter. Biodegr. 131: 19–28.

Kanellou E, Economou G, Papafotiou M, Ntoulas N, Lyra D. Kartsone E, Knezevic S (2017) Flame weeding at archaeological sites of the mediterranean region. Weed Technology 31(3) 396-403. DOI 10.1017/wet.2016.31

Kumari K, Nayaka S (2018)Lichen use as a biomonitoring toolfor the assessment of air quality. 6th Int. Conf. on plants and environmental pollution,NBRI, Lucknow27-30 Nov. 2018

Lal B B (1978) 'Weathering and preservation of stone monuments under tropical conditions', International Symposium on Deterioration and Protection of Stone Monuments, Paris, pp. 1-9

Lan T T N, Nishimura R, Tsujino Y, Satoh Y, Thoa N T P, Yokoi M, Maeda Y(2005) The effects of air pollution and climatic factors on atmospheric corrosion of marble under field exposure. Corrosion ScienceV 47 (4) : 1023-1038https://doi.org/10.1016/j.corsci.2004.06.013

Leavengood P, Twilley J, Asmus JF (2000) Lichen removal from Chinese Spirit path figures of marble. J. Cult. Herit. 2000, 1, s71–s74.

Lee C H, Lee M S, Kim Y T, Kim J (2006) Deterioration assessment and conservation of a heavily degraded Korean Stone Buddha from the ninth century. Stud. Conserv. 2006, 51: 305–316.

Lefèvre M (1974) La 'Maladie Verte' de Lascaux. Stud. Conserv. 19: 126–156.

Lewin S Z (1982) The Mechanism of Masonry Decay Through Crystallization, in Conservation of Historic Stone Buildings (Washington: Committee on Conservation of Historic Stone Buildings and Monuments), 120.

Lewis B (2014) India: Historical archaeology. In: Encyclopedia of Global Archaeology, (ed.) C. Smith, pp. 3751-3760.

Li T, Hu Y, Zhang B (2019) Evaluation of efficiency of six biocides against microorganisms commonly found on Feilaifeng Limestone, China. J. Cult. Herit. 43: 45–50.

Lisci M., Michela M., Pacini E. (2003). Lichens and higher plants on stone: a review. International Biodeterioration and Biodegradation 51, 1–17.

Maeda Y., Morioka J., Tsujino Y., Satoh Y., Zhang X., Mizoguchi T., Hatakeyama S. (2001) Material damage caused byacidic air pollution in east Asia. Water, Air, and Soil Pollution 130, 141–150.

Mak K.-K., Kamal M.B., Ayuba S.B., Sakirolla R., Kang,Y.-B., Mohandas K. Balijepalli M.K., Ahmad S.H. Pichika M.R. (2019) A comprehensive review on Eugenol's antimicrobial properties and industry applications: A transformation from ethnomedicine to industry. Pharmacogn. Rev. 13: 1–9.

Marvasi M., Donnarumma F., Frandi A., Mastromei G., Sterflinger K., Tiano P., Perito B. (2012) Black microcolonial fungi as deteriogens of two famous marble statues in Florence, Italy. Int. Biodeterior. Biodegrad. 2012, 68, 36–44.

Mascalchi M., Orsini C., Pinna D., Salvadori B., Siano S., Riminesi C. (2020) Assessment of different methods for the removal of biofilms and lichens on gravestones of the English Cemetery in Florence. Int. Biodeter. Biodegr. 154: 105041

Meng H., Zhang X., Katayama Y., Ge Q., Gu J.-D. (2020) Microbial diversity and composition of the Preah Vihear temple in Cambodia by high-throughput sequencing based on genomic DNA and RNA. Int. Biodeter. Biodegr. 149: 104936.

Meyer B. (2003) Approaches to prevention, removal and killing of biofilms. Int. Biodeter. Biodegr. 51: 249–253.

Nugari MP, Salvadori O (2003) Biodeterioration control of cultural heritage: Methods and products. In Molecular Biology and Cultural Heritage; Saiz-Jimenez, C.(ed) Swets and Zeitlinger: Lisse, The Netherlands. pp. 233–242.

Panaroli D (1643) Jatrologismi sive Medicae Observationes quibus additus est in fine Plantarum Amphitheatralium Catalogus. – Romae.

Polo A, Diamanti M V, Bjarnsholt T, Høiby N, Villa F, Pedeferri M P, Cappitelli F (2011) Effects of photoactivated titanium dioxide nano-powders and coating on planktonic and biofilm growth of Pseudomonas aeruginosa. Photochem. Photobiol. 2011, 87: 1387–1394.

Pozo-Antonio J.S. Sanmartín P. (2018) Exposure to artificial daylight or UV irradiation (A, B or C) prior to chemical cleaning: An effective combination for removing phototrophs from granite. Biofouling 2018, 34, 851–869.

Pozo-Antonio J.S. Barreiro P., González, P. Paz-Bermúdez G., Nd: YAG Er: YAG laser cleaning to remove Circinaria hoffmanniana (Lichenes, Ascomycota) from schist located in the Côa Valley Archaeological Park. Int. Biodeter. Biodegr. 2019, 144: 104748.

Pozo S., Montojo C., Rivas T., López-Díaz A.J. Fiorucci M.P. López De Silanes M.E. (2013) Comparison between Methods of Biological Crust Removal on Granite. Key Eng. Mater. 548: 317–325.

Rusu D.E., Stratulat L., Ioanid G.E., Vlad A.M. (2020) Cold high-frequency plasma versus afterglow plasma in the preservation of mobile cultural heritage on paper substrate. IEEE Trans. Plasma Sci. 48: 410–413.

Salvadori O, Charola A E (2011) Methods to Prevent Biocolonization and Recolonization: An Overview of Current Research for Architectural and Archaeological Heritage. In Biocolonization of Stone: Control and Preventive Methods: Proceedings from the MCI Workshop Series; Charola, A.E., McNamara, C., Koestler, R.J., Eds.; Smithsonian Institution Scholarly Press: Washington, DC, USA, 2011; pp. 37–50.

Sanmartín P, Rodríguez A, Aguiar U (2020) Medium-term field evaluation of several widely used cleaning-restoration techniques applied to algal biofilm formed on a granite-built historical monument. Int. Biodeter. Biodegr. (2020) 147, 104870

Sakrt A, Ghaly M F, Edwards H G M, Elbashar Y H (2019)Gamma-radiation combined with tricycloazole to protect tempera paintings in ancient Egyptian tombs (Nile Delta, Lower Egypt). J. Radioanal. Nucl. Chem. 321: 263–276.

Santo A P, Cuzman OA, Petrocchi D, Pinna D, Salvatici T, Perito B (2021) Black on white: Microbial growth darkens the external marble of Florence cathedral. Appl. Sci, 11: 61-63.

Scheerer S, Schütze E, Curbach M (2018). Strengthening and Repair with Carbon Concrete Composites – the First General Building Approval in Germany. In: Mechtcherine.

Seaward M R D (2004) Lichens as subversive agents of biodeterioration. In: St. Clair, L., Seaward, M. (Eds.). Biodeterioration of Stone Surfaces. Kluwer Academic Publishers, The Netherlands, pp. 9–18.

Sharma J S, Sharma D N (1982) Atmospheric contamination of archaeological monuments in the Agra Region (India). Sci Total Environ 23: 31–40. https://doi.org/10.1016/S0166-1116(08)70988-9

Signorini M A (1995) Lo studio e il controllo della vegetazione infestante nei siti archeologici. Una proposta metodologica. Pp. 41-46 in: Marino, L. and Nenci, C. (eds), L'area archeologica di Fiesole. Rilievi e ricerche per la conservazione. – Firenze

Snethlage R E, Wendler (1996) "Moisture Cycles and Sandstone Degradation." In Report of the Dahlem Workshop on Saving Our Architectural Heritage: The Conservation of Historic Stone Structures Held in Berlin 3-8 March 1996, edited by N.S. Baer and R. Snethlage, 7-24. Chichester, New York,

Weinheim, Brisbane, Singapore, and Toronto: John Wiley and Sons, 1997.17-18pp.

Soon-Bok L, Cowell D, Apsimon H, (1996) Estimating the cost of damage to buildings by acidifying atmospheric pollution in Europe. Atmospheric Environment 30 (17), 2959–2968

Sterflinger K. (2010) Fungi: Their role in deterioration of cultural heritage. Fungal Biol. Rev. 2010, 24, 47–55.

Sterflinger K, de BaereR, de Hoog GS, de Wachter R, Krumbein W E, Haase G.(1997) *Coniosporium perforans* and *C. apollinis*, two new rock-inhabiting fungi isolated from marble in the Sanctuary of Delos (Cyclades, Greece). Antonie Van Leeuwenhoek J. Microb. (1997) 72, 349–363.

Sterflinger K, Krumbein W E(1997) Dematiaceous fungi as a major agent for biopitting on Mediterranean marbles and limestones. Geomicrobiol. J., 1997, 14, 219-230. http://dx.doi.org/10.1080/01490459709378045

Sterflinger K., Sert H. (2006) Biodeterioration of buildings and works of art–Practical implications on restoration practice. In heritage, weathering and conservation; Fort, R., Alvarez de Buergo, M., Gomez-Heras, M., Vazquez-Calvo, C., Eds.; Taylor and Francis Group: London, UK, 2006 (1): 299–304.

Tayel A.A., Ebeid M.M., ElSawy E., Khalifa S.A. (2016) Fungicidal effects of plant smoldering fumes on archival paper-based documents. Restaurator 37: 15–28

Teixeira F.S., dos Reis T.A., Sgubin L., Thome L.E., Bei I.W., Clemencio R.E., Correa B., Salvadori M.C. (2018) Disinfection of ancient paper contaminated with fungi using supercritical carbon dioxide. J. Cult. Herit. 30: 110–116.

Toreno G, Isola D, Meloni P, Carcangiu G, Selbmann L,Onofri S, Caneva G, Zucconi L (2018) Biological colonization on stone monuments: A new low impact cleaning method. J. Cult. Herit. 30: 100–109.

Török A, Licha T, Simon K, Siegesmund S (2011) Urban and rural limestone weathering; the contribution of dust to black crust formation. Environ Earth Sci 63: 675–693. https://doi.org/10.1007/s12665-010-0737-6

Tretiach M, Bertuzzi S, Candotto C. F. (2012) Heat shock treatments: A new safe approach against lichen growth on outdoor stone surfaces. Environ. Sci. Technol. 46: 6851–6859.

Valentini F, Diamanti A, Palleschi G (2010) New bio-cleaning strategies on porous building materials affected by biodeterioration event. Appl. Surf. Sci. 256: 6550–6563.

Viles H A (1995) Ecological perspectives on rock surface weathering: towards a conceptual model. Geomorphology 13: 21–35.

Viles H A, Goudie A S (2004) Biofilms and case hardening on sandstones from Al-Quwayra, Jordan. Earth Surface Processes and Landforms 29: 1473–1485.

Villa F, Pitts B, Lauchnor E, Cappitelli F, Stewart P S (2015) Development of a laboratory model of a phototroph-heterotroph mixed-species biofilm at the stone/air interface. Front. Microbiol. 6: 1251.

Villa F, Gulotta D, Toniolo L, Borruso L, Cattò C, Cappitelli F(2020) Aesthetic alteration of marble surfaces caused by biofilm formation: Effects of chemical cleaning coatings 10: 122; doi: 10.3390/coatings10020122

Warke P A (2013) Weathering in Arid Regions, In: (ed): John F. Shroder,Treatise on Geomorphology, Academic Press,2013,pp 197-227, ISBN 9780080885223,

Warscheid T H (1997) 'Biodeterioration mechanisms of historical monuments', in 'Deterioration of Concrete and Natural Stone of Historical Monuments', Proceedings of an International Seminar, Brasília, May 1997 (Departamento de Eng. Civil da Universidade de Brasília, 12–17.

Warscheid T H (2000) Integrated concepts for protection of cultural artefacts against biodeterioration, in: 'Of Microbes and Art, the Role of Microbial Communities in the Degradation and Protection of Cultural Heritage, Pub. Kluwer/Plenum, New York, 303–420.

Wawrzyk A, Rybitwa D, Rahnama M, Wilczynski S (2020) Microorganisms colonising historical cardboard objects from the Auschwitz-Birkenau State Museum in O´swi ²ecim, Poland and their disinfection with vaporised hydrogen peroxide (VHP). Int. Biodeter. Biodegr. 152: 104997

Winkler E M (1994) Stone in Architecture: Properties, Durability, 3d ed. Berlin, Heidelberg, New York: Springer-Verlag. 157-59

Wu J, Lou W (2016) Study of biological enzymes' inhibition of filamentous fungi on the Crosslake Bridge ruins. Sci. Archaeol. Conserv. 3: 25–29.

Chapter 14

Postage Stamps Depicting Heritage Buildings and Caves in India and Issues Related to their Microbial Decay

Arun Arya

Department of Environmental Studies, Faculty of Science, The Maharaja Sayajirao University of Baroda, Vadodara – 390002, Gujarat
e-mail: sarojarun10arya@rediffmail.com

ABSTRACT

Microbes can cause biological weathering. Early investigations on biodeterioration were mainly focused on the isolation and identification of the cultural microbes. The application of polymerase chain reaction (PCR) technique has helped to cultural heritage microbiology research, and has revealed the complex microbial communities, even non-culturable ones as well. This new approach also has some major weakness in its inability to identify the active microbes from the genomic DNA-based community analysis and, in addition, the active deteriorating ones are not identified from the community. Because of these, the advances made on knowing the community better have not translated to useful results on the deteriorating processes for a more effective management and restoration activities.

Indian postal department has issued a large number of stamps on world heritage sites to generate awareness about these unique places. They are also depicted on currency notes as well. The cultural heritage of a country is a spectacular memorandum of its glorious past. It is that precious inheritance which should be preserved to be passed on from generations to generations.

The paper brings out a review of studies done on microbial association with some of these important monuments and Ajanta and Ellora like world heritage sites. Certain preventive and control strategies are also discussed.

Keywords: *Postage stamps, Heritage buildings, Taj Mahal, Ajanta caves, Microbes, Deterioration.*

14.1 Introduction

We are aware that historical monuments, buildings and archaeological sites, all of which are cultural analogues of the biosphere, are at risk from human activities and in need of conservation. According to Hammond (1995) microbes form largest biomass and maximum number of species on this planet. Architectural beauty is lost, due to deterioration caused by bio-colonization of microbes and plant communities (Veneranda *et al.*, 2018). This phenomenon takes place on structures exposed to an environment characterized by specific local conditions, such as high moisture, high salinity, and abundance of organic nutrients (Gaylarde 2020; Mascaro *et al.*, 2022). The damage to historic buildings is a result of both natural and anthropogenic contributors involving flora, fauna, and microorganisms as well as pollutants. According to the colonizing organisms involved, the characteristics of the inhabited area (surfaces, cracks, or pores), the severity of the damage changes, thus varying from exterior damage to irreversible disintegration of the inner substrates (Charola *et al.*, 2011). Biofilms can be found in archaeological sites and hypogean spaces, but also on statues and buildings (Casaletto *et al.*, 2017). Considerable aesthetic and structural damage to culturally significant materials and monuments – such as the 12th-century Hindu Temple at Angkor Wat (Cambodia) – can be caused by the growth of biofilms and the production of harmful metabolites by microorganisms (Venkataraman 2008). Biodeterioration is caused by the formation of active biofilms, due to interacting microbiota growth constituting a complex ecosystem. Contamination depends on different factors, such as environmental conditions–relative humidity (RH per cent), light, and temperature–and the physicochemical properties of the substrate, *i.e.*, roughness, absorbency, hydrophobicity, porosity, and chemical composition (Young 1997). Biofilms growing on stone monuments generally include photolithoautotrophs, such as algae, cyanobacteria, mosses, and higher plants. As the biomass expands, it releases organic nutrients fostering the growth of other microorganisms, such as chemolithoautotrophic bacteria, chemoorganotrophic bacteria, and fungi (Crispim and Gaylarde 2004). Biogenetic pigments lead to aesthetical damage, causing the formation of stains on the stone artifacts. Furthermore, biofilms produce extracellular polymeric substances (EPS) that induce mechanical degradation due to both the penetration of hyphae and filaments, and the shrinking-swelling cycles of biological colloidal particles in the material's pores. An alteration in the size and distribution of the pores, moisture circulation, pH, and water permeability can cause biocolonization. It has been demonstrated that the presence of biogenic patinas on exposed surfaces accelerates the accumulation of atmospheric pollutants (Macchia *et al.*, 2022).

Indian postal department has issued various stamps on world heritage sites to generate awareness about these unique places. They are also depicted on currency notes. The cultural heritage of a country is a spectacular memorandum of its glorious past. It is that precious inheritance which should be preserved to be passed on from generations to generations. There might be a number of architectural prodigies in a country but only a few of them hold universal importance. These are the cultural heritages which are so exceptional that the protection of these sites becomes the concern of the entire mankind. Such manifestations of history have been categorized as the World Heritage Sites by UNESCO and their preservation is of utmost importance to these international bodies. At present, 137 nations are a part of this resolution and India is one of them. There are 40 such properties in the country which have been touted as the World Heritage Sites by the UNESCO. Out of them 25 are important cultural heritages. Mawmluh Cave, also known as Krem Mawmluh, explored by Lt. Yule (British) in 1844, in Meghalaya has been listed as UNESCO'S one of the 'First 100 IUGS (International Union of Geological Sciences) Geological Sites'. The cave is known for its stalagmite formations. These prominent caves in Meghalaya have now been selected by the International Union of Geological Sciences (UNESCO). The collection and display of stamps on a particular topic is termed thematic philately. Braun Lucien of France, founding father of thematic, differentiated for the first time between traditional and thematic philately in 1949. According to Mackay (2007) in thematic collection use of other postal materials, mint stamps, special covers, picture postcards and even pictorial cancellations with any messages are permitted.

Our knowledge of the mycobiome, which forms the environment, is still incomplete, and most surveys available in the literature have focused on airborne molds. These molds grow on indoor damp materials and can produce abundant conidia that get easily aerosolized. The black yeasts or fungi (BF), characterized by the conspicuous dark pigmentation due to melanin may be found. A total of 83 species of extremophilic fungi were analyzed in relation to their known ecophysiology and updated phylogeny. When colonizing rocks, BF can induce a chemical deterioration by secreting siderophore like compounds, however the most relevant damages are believed to be due to a mechanical action. Hyphal morphology and the strong mechanical turgor pressure (up to 12.39 bar in penetrating silicate and carbonate rocks) allow them to dig cavities on the heritage stone surfaces at depths ranging from a few hundred microns to several millimeters (Tonon *et al.*, 2021). Furthermore melanin, the most important factor in BF stress resistance, is also responsible of the aesthetic alterations imparting a dark, blackish-brown appearance to the lithic surface (Toreno *et al.*, 2018).

A little work has been done on occurrence of algae in air. Generally, algae dominate upto a height of 2 me-ters. The most common algae found in air are the species of *Chlorella, Chlorococcum, Chlamydomonas, Aulosira, Nostoc* and *Phormidium*. Ramalingam (1971) reported some of the algal types (*e.g.* diatoms, *Protococcus, Spirogyra, Oscillatoria, etc.*) from Mysore. Lichen component of aerospora with special reference to al-lergenic ones, from hill districts of Uttar Pradesh. The most common lichens were the species of *Cladonia, Heteroderma, Parmelia, Usnea, etc.* (Dubey and Maheshwari 2000). In rainy season beauty of the world fame Taj Mahal fades due to the growth of algae on it. However, it is properly cleaned regularly. A National Laboratory for Conservation of Cultural Properties has been established at Lucknow that takes care of developing methods for conservation of monuments, archives, *etc.* Although fungi and algae may colonize surfaces faster than lichens but the damage is lesser. Brock (1987) defined microbial ecology as the study of microorganisms in their natural environment, and emphasized that the researcher should examine the microorganisms *in situ*. Genera *Chroococcus* and *Nostoc* were forming brilliant green patina, red powdery granules and black biofilm on wall paintings of Holy Saviour's cave in Vallerano, Italy. *Gloeocapsa* caused discoloration of architectural acrylic paints in Sao Paulo and Ubatuba, Brazil, as well as formation of black biofilm. Published data on the distribution of different taxa of photosynthetic microorganisms do not indicate a clear relationship between the organisms present and stone composition and the major influence is considered to be climate, rather than substrate (Gaylarde and Gaylarde 2005).

14.2 Problems of Deterioration of Heritage Monuments

Early investigations on biodeterioration were mainly focused on the isolation and identification of the cultural microorganisms for a description of them without specific function to the biodeterioration (May *et al.*, 2000). Such practice had been persisted before the application of polymerase chain reaction (PCR) to cultural heritage microbiology research to reveal the complex microbial community without culturing and isolation (Rölleke *et al.*, 1998). The latest high-throughput sequencing and metagenomics provide a much deeper description of the microbial community and composition without culturing or isolation (Zhang *et al.*, 2019). This new approach also has some major weakness in its inability to identify the active microbes from the genomic DNA-based community analysis and, in addition, the active deteriorating ones are not identified from the community. Because of these, the advances made on knowing the community better have not translated to useful results on the deteriorating processes for a more effective management and prevention (Liu *et al.*, 2020). Amman *et al.*, (1995) found that many fungi inhabiting rock substrate cannot be isolated or cultured. These microbes were viable but non-culturable. This study presented by Gutàrowska and Zakowska (2010) showed an estimation of mould contamination of 31 samples, as well as a comparison of two methods of mould estimation: dilution plate method and UV-determination of ergosterol.

Many fungi which fail to be cultured can be identified by this method. Cecchi *et al.*, (1996) proposed fluorescence lidar technique for the monitoring of biodeteriogens by remote sensing method.

Cyanobacteria are known to participate in deterioration of limestone walls of Jerusalem, marble monuments in Rome, Great Jaguar Pyramid in Guatemala, various historical buildings in Brazil, and many other monuments. They are known to degrade synthetic polymer materials and wood. They cause fouling of painted and unpainted concrete structures. Genera *Chroococcus* and *Nostoc* were forming brilliant green patina, red powdery granules and black biofilm on wall paintings of Holy Saviour's cave in Vallerano, Italy (Singh *et al.*, 2010). *Gloeocapsa* caused discoloration of architectural acrylic paints in Sao Paulo and Ubatuba, Brazil, as well as formation of black biofilm (Boopalan and Sasikumar 2001). Biodeterioration of mortar was prevented by addition of anatase to the formulation that was used on two external walls of Palacio Nacional da Pena in Sintra, Portugal (Ferreira 2010).

Scientific observations of a wide variety of natural habitats have established that the majority of microbes predominantly live in complex sessile communities known as biofilms (Bridier *et al.*, 2017). Biofilms are highly structured assemblages of microbial cells attached to a surface and entwined in a matrix of self-produced extracellular polymeric substances (EPS) (López *et al.*, 2010). Surface biofilms are microbial cells embedded in extracellular polymeric substances (EPS). The simple presence of a biofilm has aesthetic, chemical, and physical effects on the stone. EPS, produced by the cells to allow their adhesion to a given surface, facilitate entrapment of airborne particles, aerosols, minerals, and organic compounds, increasing the dirty appearance of the substrate (Kemmling *et al.*, 2004). Biofilms are areas of high metabolic activity, where digestive enzymes excreted by microorganisms are concentrated. The mechanisms involved in biodeterioration of natural sandstone and man-made materials include both abiotic and biological ones (Mitchell and Gu 2000).The structured biofilm ecosystem enables microbial cells to resist stress and increase their tolerance to stressors, both at the level of individual microorganisms and collectively as a community (Flemming *et al.*, 2016). The microorganisms within a biofilm can survive and thrive in harsh environments characterized by desiccation, low-nutrient concentrations, large temperature variations, and high exposure to wind, UV radiation, and physical damage (Jacob *et al.*, 2018). Biofilms are present in natural, industrial, medical, household environments and, from the human point of view, they can be either beneficial or detrimental. The characteristics of microbial cells forming biofilms are distinct from those of their planktonic counterparts; their higher resistance to antimicrobial agents and ultraviolet radiation (UV), the development of physical and social interactions, an enhanced rate of gene exchange, and the selection for phenotypic variants are traits related to the structural characteristics of the community (Hentzer *et al.*, 2003; Mittelmann 2018). Cell-to-cell biochemical signals influence each step in the process of biofilm formation and enhance the persistence of both individual species and the biofilm (Katharios-Lanwermeyer *et al.*, 2014). Multispecies biofilm is a result of cell–cell and cell–environment interactions such as cooperation, competition or exploitation (Liu *et al.*, 2016). The spatial organization of biofilms is driven by the specific interactions between species. The first step in the development of biofilms is the formation of a conditioning layer. The layer is formed by organic substances like polysaccharides and proteins which improve the attachment of initial colonizers. In the second step, microbial cells attach to these substances due to their size and net negative charge by nonspecific interactions, such as electrostatic, hydrophobic, and van der Waals forces (Mittelmann 2018). Hydrophobicity and electric charge of the conditioning substances are of importance in bacterial adhesion (Diao *et al.*, 2014). EPS play significant roles in the attachment of microorganisms to the mineral surfaces and in the protection of the microbial community from toxic compounds (Diao *et al.*, 2014).

Degradation of monuments may be due to microclimate *i.e.* temperature, humidity, darkness or biological agents like microorganisms, plants, algae, birds, bats, and insects. India being rich in cultural and historical heritage has many historical buildings and heritage monuments, mostly administrated and conserved by the Archaeological Survey of India (ASI).

14.3 Preventive and Control Strategies for Biodeterioration

Regular survey of monument is required to assess the damage caused by microorganisms. During rainy season growth of algae, lichens or bryophytes can be seen. An early treatment is suggested. Nugari and Salvadori (2003) stated that the only effective gas against insects and fungi was ethylene oxide, which has been banned in several countries due to carcinogenic and mutagenic features. However, more recent research (Wawrzyk *et al.*, 2020) proved that vaporized hydrogen peroxide disinfection for new and historical cardboard at the Auschwitz–Birkenau State Museum in O´swiecim, Poland, has comparable effects to that of ethylene oxide. Use of such gases for buildings has practical difficulties. Recolonization on outdoor heritage is generally to be expected and depends on several factors, such as the antimicrobial agent employed and the way it was applied, the nature of the material and its state of conservation, the type and degree of colonization before the use of the biocide, and the micro- and macro-environment (Salvadori and Charola 2011). In 1963, regarding the 'maladie verte de Lascaux', the scientist Dobat believed that fighting algal invasion with chemicals was, in the long run, doomed to failure because it was impossible to kill all algal spores that were present in the cave, new microorganisms would always be introduced, and chemical treatments would be frequently needed with high risk of harming the paintings (Lefèvre 1974). The problems of using chemicals and recolonization have not been solved yet. The site of Feilaifeng, which includes Buddhist statues, was inscribed in the

UNESCO list in 2011. The statues affected by biodeterioration were successfully treated with the biocides AW-600 and octhilinone, but recolonization was observed shortly thereafter (Li *et al.*, 2019). Advantages and disadvantages of few treatment methods are described in Table 14.1.

Table 14.1: Principal Control Methods: Advantages and Disadvantages

Control Strategy	*Advantages*	*Disadvantages*
Chemical Methods		
Traditional Biocides	A wide variety of compounds are available, which are economical and easy to apply. Effective against a broad range of microbes. Application in remote areas possible.	These are toxic for the operators and also to the environment. Have no long-term protection of monuments. Often not selective against specific biodeteriogens. Possible modification of biofilm structures which may cause growth of more harmful biodeteriogens. Repeated use may be harmful to structure.
Nano particles	These are effective at very low concentration. A wide variety of compounds are available in the market. These are easy to apply.	Harmful to the operators and the environment. Not selective against specific biodeteriogens. Help to evolve biocide-resistant communities. Lack of experiments discussing the interference with the heritage materials. These are very costly.
Physical Methods		
Mechanical removal	Efficient method on surfaces with good state of conservation. We get instant results. No requirement of the toxic chemicals. Do not generate toxic products.	These provide no long-term effectiveness. Repeated use is harmful to the heritage monuments. Biological contaminants can be pushed deep into the heritage material and spread in the environment.
UV-C irradiation	Do not cause any harm to humans, and to the environment. Do not generate any toxic residual element in the environment. Easy to use.	Repeated use may damage organic heritage material such as wood, leather, parchment, and textiles *etc*. Low penetration in substratoo. Not oclcotive against specific biodeteriogens. Can't be used in remote areas.
Gamma radiation	This application also not cause any harmful effect to humans, to environment, or to the heritage material. High penetration in substratoo and in very thick biofilms.	Repeated use may damage organic heritage material such as wood, leather, parchment, and textiles. Require specialized staff to operate. Applications are limited to artworks of smaller size. No longer possible to carry out radioluminooocnoc dating after irradiation. Limited application in heritage buildings. Is expensive.

Control Strategy	Advantages	Disadvantages
Laser cleaning	Controllable, selective, contactless, and environmentally friendly. Do not introduce any harmful chemicals to humans, to environment, or to the heritage material. Instant results can be seen. Do not generate any toxic residual elements.	Repeated use may damage the heritage material. Not selective against specific biodeteriogens. Limited application in remote areas. Required specialized staff to use. Treatment is costly.
Heat shock, microwaves, and dry ice treatment	Instant results with highly localized effect. Do not require the use of toxic compounds. Do not generate toxic products.	Microwaves and dry ice treatment equipment complicated to transport and apply, require continued access to energy supply, and are costly. Hazardous to handle. Not selective against biodeteriogens. Repeated use may damage some fragile surfaces. Can't be used in buildings. Costly
Biological Methods		
Biocidal treatments with compounds of natural origin	Generally safer for human beings and greener for the environment than traditional biocides. Usually easy to use. Effective against a broad range of microorganisms. Application on monuments possible.	The extract composition depends on the harvesting season, geographical location, and other agronomic factors. Only few market products are available. Not selective against specific biodeteriogens. Not much data available on experiments discussing the interference of the natural compounds with the heritage materials.
Other biological methods	Harmless for humans and the environmental health. Relatively easy to set up and improve. Effective against a broad range of microorganisms. Selective for the target microorganism. Application in remote areas possible.	Lack of experiments discussing the interference with the heritage materials. Lack of experiments assessing the persistence over time of the treatment.

14.4 Microbial Deterioration in Sanchi

About 48 km from Bhopal, the Buddhist monuments at Sanchi are located in Vidisha, on a serene and picturesque forested plateau, also considered to be the sacrosanct Cetiyagiri in the Sri Lankan Buddhist chronicles, where Mahindra, the son of Emperor Aœoka, stopped prior to undertaking his journey as a missionary to Sri Lanka. The stupas, temples, viharas, and stambha at Sanchi in central India are among the oldest and most mature examples of aniconic arts and free-standing architecture that comprehensively document the history of Buddhism from the 3rd century BCE to the 12th century CE.

Postal department has issued stamps on Buddha, Sanchi, and 2500 years of Figure 3 Picture of Buddha on folder and postage stamp of 2 Anna depicting leaf

Figure 14.1

(a) Postage stamp of 2 Anna depicting leaf of Bodhi tree (Peepal) released by postal dept. on 24/5/1956 to commemorate 2500th year of Buddha's birth; (b) A 3 Annas brown orange stamp of East gate of Sanchi; (c) Picture of Buddha on Rs.5 stamp; (d) and (e) Stupa of Sanchi; (f) A set of silk stamps of Buddha from Bhutan.

of Bodhi tree (Peepal) released by postal dept. on 24/5/1956 to commemorate 2500th year of Buddha's birth. A 3 Annas brown orange stamp of East gate of Sanchi was released on 15th August, 1949. Nasik Press of post and telegraph department on 4/4/1994 printed a Rs. 5 stamp in Bright Siena and dark blue green colour of UNESCO heritage Buddhist site Sanchi. Geography is not a boundary in "Buddhist Philately". There are more than 4,200 stamps issued by nearly 151 countries and territories around the world (including former colonies). Bhutan Post, which released only 2 new issues last year, issued a spectacular set of 2 miniature sheets (20 different stamps, 10 in each sheet) and an additional single stamp miniature sheet on 24 March as its first issue of 2021. The sheets are printed on "Tara-rayon silk" and have a very local flavor to them and are extremely attractive.

Preliminary examination of stone pillar situated near main stupa of Sanchi showed the blackening caused by Cyanobacteria like *Nostoc, Scytonema* and more importantly by the activities of certain Dematiaceous fungi like *Cladosporium* and *Drechslera*. In certain places the pillars formed dark coloured biofilm due to growth of microbes. Microorganisms colonizing surfaces of cultural heritage can destruct the underlying materials through their influences on the physical, chemical, and bio-receptibility of the substratum materials, especially metabolic activities and biochemical reaction. The pioneering microorganisms form biofilms on rock and stone surfaces and cause the discoloration in appearance, alteration of porosity and moisture diffusivity in and out of the stone. Adhikary and Satapathy (1996) reported *Tolypothrix byssoidea* from temple rock surfaces of coastal Orissa. In only two of the six sites sampled in Angkor Wat were filamentous cyanobacteria, *Microcoleus, Leptolyngbya,* and *Scytonema*, found; the first two detected by sequencing of 16S rRNA gene library clones from samples of a moist green biofilm on internal walls in Preah Khan, where *Lyngbya* (possibly synonymous with *Microcoleus*) was seen by direct microscopy as major colonizer. *Scytonema* was detected also by microscopy on an internal wall in the Bayon. This suggests that filamentous cyanobacteria are more prevalent in internal (high moisture) areas (Gaylarde *et al.*, 2012).

14.5 Biodeterioration of Ajanta and Ellora Caves

Many of these ancient heritage monuments are located far from human habitations inside forests, pristine areas including in secluded caves. Ajanta, Ellora, Elephanta, Guntupalle, Bhaja, Karla, Bedse, Kanheri, Saptaparni, Udayagiri, and Khandagiri are some of the famous cave monuments of India. The Ajanta caves, a world-famous UNESCO heritage site, is located at latitude 20%33012" N to 75%42001" E longitude, at 33.5 m AMSL in Aurangabad district of Maharashtra, India. The caves are carved out of flood basalt rock of a cliff, part of the Deccan Traps formed by successive volcanic eruptions at the end of the Cretaceous period. The inhomogeneity in the rock has also led to cracks and collapses in the centuries that followed, as with the lost portico to cave 1.

Excavation began by cutting a narrow tunnel at roof level, which was expanded downwards and outwards; as evidenced by some of the incomplete caves such as the partially-built *vihara* caves 21 through 24 and the abandoned incomplete cave 28 (Uno *et al.*, 2012; Gontareva *et al.*, 2015).

HYDERABAD STATE in 1935, issued a stamp depicting entrance gate to Ajanta Caves. Official Stamp, of 8 Annas. Orange colored (Figure 14.2a). Indian postage stamp showing an elephant carved in Ajanta caves, release date: 15/8/1949 (Figure 14.2b), To celebrate 25th anniversary of foundation of UNESCO a commemorative stamp was released on 4th Nov. 1971, depicting a painting in Ajanta caves, red brown stamp is of 20 P (Figure 14.2d).

Figure 14.2: Four Stamps on Ajanta Caves and Paintings.

A stamp on Ellora caves was released by Nevis in Belarus (Eastern Europe) on 6/7/ 2017. My stamp on Ellora can be seen in Figure 14.3. Some cancellations available at post offices in Maharashtra are given below.

Figure 14.3: A Photograph of Ellora Cave and Stamp Released on Caves.

Buddhist caves showcase the excellence of artistic and technical achievement (Singh and Arbad 2012). Buddhist monks constructed Ajanta caves long back in the Vakatakas period. These caves are renowned for the wall paintings of Hinayana and Mahayana beliefs of Buddhism (Brimblecombe and Lankester 2013). Before 1819, the caves were lost in anonymity and it was only in the year 1819 when the caves were rediscovered by the Officer of British Battalion Mr. John Smith (Bharti 2013). In 1995, the Government of India declared Ajanta caves the monument of national and heritage value (Brimblecombe and Lankester 2013). In 1953, the Archaeological Survey of India undertook protection and

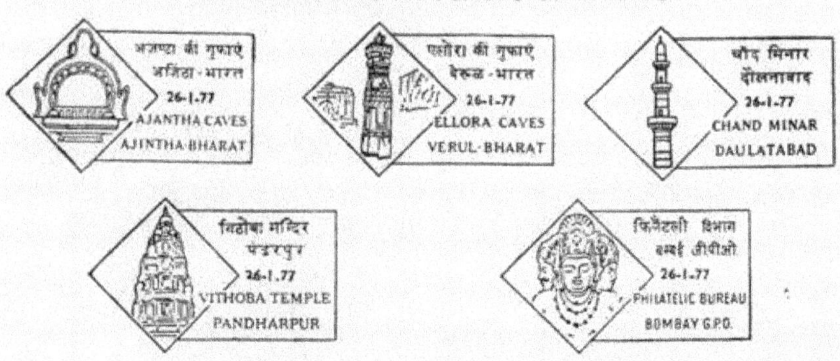

The Maharashtra Postal Circle has issued a memento at places of interest viz. Ajantha Caves, Ellora Caves, Daulatabad Fort, Pandharpur and Bombay by providing pictorial cancellations from 26th January 1977 onwards on a special designed cover depicting Apsara 'of Ajantha, Victory Pillar' of Ellora, 'Chand Minar' of Daulatabad fort and 'Vithoba Temple' of Pandharpur. A separate cover depicting the building of Bombay G.P.O. with a cancellation of Elephanta Cave has also been issued from the same day. The designs of both the covers and special cancellations with relics of caves or places are very attractive. Blank covers will cost 25 P each and postage stamp of 25 P will be affixed on each cover. These covers will also be — available for the visitors/Collectors with pictorial cancellations on all days except on Sundays and Holidays at these Post Offices.

preservation of Ajanta caves, rock carvings as well as paintings (Umadi *et al.*, 2019).

Dampness in the cave atmosphere is provided by rainwater and water from Waghura River, which increases growth of algae, fungi, and variety of insects and microbes (Barton 2006). One quarter of Ajanta paintings are reported to be lost due to damage caused by algae, fungi, insects, and pests. Many anthropogenic activities have attracted biological agents (Deutsch *et al.*, 2018). Penetration of woody roots of vegetation on cave roofs have resulted in crack formations (Bankar and Bhosle 2017). A mixture of hemp, clay, and lime plaster was considered efficient for preserving paintings and carvings in nearby Ellora caves. Lime plaster and hemp is reported to regulate humidity and control insects in Ellora whereas, hemp has been not used in Ajanta caves and has likely resulted in deterioration of painting and cave walls to algae, fungi and insect presence (Deutsch *et al.*, 2018). Algae, fungi, and microbes provide enough food to insects for their survival inside damp cave monuments. It is noticed that in some caves at Ajanta, there is a superficial encrustation of white and sometimes brown or black layers over the paintings (Lal 1966).

Insects are reported to be frequent visitors in the Ajanta caves due to the location of the monuments in the densely forested area, with high humidity, low temperature in the caves in comparison to outside dry weather and almost negligible sunlight penetrating the caves (Bharti 2013). Warmer climate and humidity increase the number of insect pests close to historical properties (Barton 2006). Future and ongoing changes in climate variability and rainfall patterns are observed to increase presence and degradation of monuments due to the rise in insect and pest populations (Barton 2006). Research on protecting heritage monuments against biological agents especially insect pests and factors that accelerate the insect pest population has been underexplored and mostly understudied. Considering the background of the heritage importance of the Ajanta caves in India and ongoing damage to the structure due to a variety of factors that have affected the heritage structure study provides an overview of factors that have substantially affected the structure and paintings with a special focus on insect pests. Silverfish were the predominant insects. Although on the first visit (April 1986) silverfish were very few in number, they were abundant in July and October 1986. Presumably conditions were then favorable for multiplication. On the other hand, beetles and Lepidopteran larvae which were collected in April were not found during July and October; the reason for this is not apparent (Agrawal *et al.*, 1988).

Fungi in the genera *Aspergillus, Alternaria, Acrophialophora, Acremonium, Curvularia, Cladosporium, Chaetomium, Cunninghamella, Drechslera, Epicoccum, Fusarium, Gliomastbc, Mucor, Memnoniella, Macrophomina, Nigrospora, Paecilomyces, Papularia, Penicillium, Rhizopus, Rhizoctonia* and *Sordaria* were isolated and identified. In addition to this, it was observed that the number of fungal species was higher in the month of July than in April and October. This was probably due to high atmospheric humidity and moisture in the substratum during monsoon season creating favorable conditions for the development of microbes. These fungi spoiled the appearance of the paintings (Agrawal *et al.*, 1988). The illumination in caves can promote growth of cyanobacteria, algae, mosses, ferns and sometimes flowering plants referred as lamp flora. The 15 per cent concentration of H_2O_2 was used for the destruction of lamp flora (Faimon *et al.*, 2003). According to Adhikary (2001) heat treatment with 60°C, exposure to H_2O_2, and urea can restrict the growth of cyanobacteria.

According to Hueck-Van der Plas (1968), fungal development on the inorganic substratum (plaster) would be due to the fact that fungi can live on very small quantities of organic substances contaminating the surface of the substratum. Savulescu and Ionita (1971) demonstrated that in many cases biological degradation of frescoes accompanies physical or chemical deterioration. This was explained by physico-chemical degradation favoring the establishment of fungi, as the colour layer loses its integrity and the organic ingredients of the fresco become more easily attacked by fungi. Organic fragments in the fissured portions also facilitate their establishment.

14.6 Airborne Microbial Communities in an Outdoor Environment of Taj

The Taj Mahal is one of the seven wonders in the world, but at present, it is endangered due to heavy loads of aerosol pollution over the past few decades (Rohra *et al.*, 2018). Uncontrolled large scale construction and proliferation of illegal factories are deteriorating beauty of the Taj due to the increased presence of microbial components. Agra attracts tourist from all over the world.

Isolation, enumeration, and identification of airborne bacteria and fungi in the outdoor environment were experimentally investigated by Raghav *et al.*, (2020). Ambient aerosol samples were collected from March 2013 to February 2014. A total number of 20 bacterial strains and 15 fungal strains were obtained. The concentration of bacteria is higher as compared to slow-growing fungi. The maximum concentrations of bacteria and fungi were found in PM_{10} as compared to $PM_{2.5}$ samples. Some of the identified bacteria are pathogenic and causes various types of respiratory and infectious diseases. A total number of 15 fungi were isolated. *Aspergillus* and *Penicillium* have a maximum percentage occurrence of 98 per cent and 95 per cent, respectively. *Botrytis* recorded a minimum percentage of 22 per cent, followed by *Neurospora* 20 per cent. The species of *Alternaria, Penicillium, Aspergillus*, and *Cladosporium* are known for their causing allergic respiratory disease. The urban populations are at risk all over the world. Bioaerosols may be associated with a wide range of health effects *viz*, respiratory and lung function impairment and allergy to skin. The survey conducted during January- December, 2000 showed the presence of major spore types of *Aspergillus, Cladosporium, Penicillium, Alternaria, Curvularia* and *Fusarium*. Variation in different months was also observed and attributed to changes in the climatic conditions (Kulshrestha and Chauhan 2001). Airborne fungal spores have been widely recognized as major allergens capable of causing asthma, allergic rhinitis and other allergic diseases (Baruah 1961; Sharma and Mathur 2020).

The Taj Mahal, is pictured on six stamps issued July 2016 by the United Nations Postal Administration. The stamps are part of the UNPA World Heritage series that began in 1997. The Taj Mahal was inscribed on the UNESCO World Heritage list in 1983.

The UNESCO website provides a description of the structure: "An immense mausoleum of white marble, built in Agra, India between 1631 and 1648 by the Mughal emperor Shah Jahan in memory of his favourite wife, the Taj Mahal is the jewel of Muslim art in India and one of the universally admired masterpieces of the world's heritage."The stamps, two for each UNPA post office, reproduce photographs of the Taj Mahal. Animals are included in the photographs shown on the two stamps for use from the post office at the International Vienna Center in Vienna, Austria. The €0.90 stamp pictures a man leading a camel in Yamuna River beside the Taj Mahal. An elephant saddled for a ride is grazing near the

**Figure 14.4: Certain Stamps Released on
Taj Mahal by India, UNO and other Nations.**

Taj Mahal on the 01.70 stamp. In the almost two decades it took to complete the mausoleum, approximately 1,000 elephants carried material for the 22,000 laborers.

UNESCO said: " … For its construction, masons, stone-cutters, inlayers, carvers, painters, calligraphers, dome builders and other artisans were requisitioned from the whole of the empire and also from Central Asia and Iran. Ustad-Ahmad Lahori was the main architect of the Taj Mahal." The two stamps for use from the post office at U.N. headquarters in New York City are denominated 49¢ and $1.15.A southern view of the white marble tomb and the four minarets surrounding it is featured on the 49¢ stamp. The large white dome in the center of the tomb is approximately 115 feet tall. The minarets are 130 feet tall. The Taj Mahal's main chamber houses the false tombs of Mumtaz Mahal, the beloved wife whose death inspired the building of the mausoleum, and Shah Jahan. The actual graves are at a lower level. The photograph on the $1.15 stamp shows the Taj Mahal and its mosque reflected in the river, all bathed in a rosy glow.

The website of *National Geographic* said: "The mausoleum's clean white marble shifts in color and tone to match the mood of the world outside – a transformation so enchanting that it's worth lingering to gaze at the building in different conditions, such as the rosy glow of dawn or the magical light of a full moon."The sky is filled with orange hues on the 1.40- Franc stamp for use from the post office at the Palais des Nations in Geneva, Switzerland. The 1.90fr stamp shows the tomb and its reflecting pool. Rorie Katz designed the stamps, using photographs from Age fotostock and Getty Images. Each stamp measures 50 millimeters by 35mm and is perforated gauge 13. Lowe-Martin of Canada printed the stamps by offset with gold foil in sheets of 20.

14.7 Microbial Deterioration of Agra Fort

14.7.1 The Activity of Cyanobacteria

The Cyanobacteria like *Oscillatoria*, *Lyngbya*, and *Gloeocapsa* were found responsible for damage to old heritage buildings of Agra fort. The black crust coating the surfaces of building materials located in urban (polluted) environment may contain all kinds of organic compounds present in aerosols and particulate matter. Wet and dry deposition processes combined with plaster component result in dirty, grey-to-black crust formation, in which aerosols, spores, pollen, dust and every class of particulate matter are entrapped in the mineral matrix. Damp surfaces are readily colonized by microbial cells settling from the air. Villa *et al.*, (2020) and Hauer *et al.*, (2015) reported diversity of cyanobacteria on rock surfaces. These can damage the stone structure or form biofilms which cause deterioration. Adhikary (2001) reported occurrence of *Tolypothrix, Plectonema, Calothrix, Phormidium* and *Nostoc* on Jagannath, Sun temple Udayagiri caves and Lingraj temples of Orissa. Some of these cyanobacteria were resistant to dry period of one year, few formed endospores for survival in adverse heat. He suggested control by heat, H_2O_2, and urea.

A commercial product called Hydrogommage–a sand blasting, low-pressure (0.5–1.5 bar) air/water mixture with SiO_2 particles (0.5–0.1 mm)–was applied on granite specimens to remove thick phototrophic biofilms composed of the green algae *Trebouxia* sp. and the cyanobacteria *Gloeocapsa* sp. and *Chroococcus* sp. under laboratory conditions (Pozo *et al.*, 2013). Although the treatment efficiently removed the biofilms, it caused textural changes to granite including an increase in roughness and microfissures (Pozo *et al.*, 2013). The marble statues in the gardens of the National Palace of Queluz, Portugal were subjected to a grit-blasting treatment to remove biological colonization (Delgado *et al.*, 2011). Mechanical treatment increased the surface roughness as compared to the uneven and pitted surface left behind by the chemical cleaning methods. The surface roughness resulting from the cleaning intervention affected the subsequent microbial recolonization of the marble (Delgado *et al.*, 2011).

14.7.2 Deterioration of Agra Fort Caused by Fungi

A detailed study of blackening of stone and plaster showed the association of fungal flora. The fungal organisms may show the fast growth at increase temperature. They show increased production of enzymes at higher temperature. Usually the temperature in summer and rainy months is too high. The fungi which grow at higher temperature are termed thermophilic fungi. The fungi most commonly recovered from air are *Aspergillus niger*, and *Penicillium citrinum*, the percentage occurrence of these two fungi was more in 2018. Research on the biodeterioration of stonework has increased considerably over the second half of the last century (Scheerer *et al.*, 2018). From a biological point of view, stone represents an extreme habitat with large variations in environmental factors such as temperature, water availability, UV radiations and nutrients. Moreover, vertical or sub vertical surfaces are even more difficult to be colonized due to higher desiccation conditions (Caneva *et al*, 2008) Caneva, G.; Nugari, M.P.; Salvadori, O. Two main groups of fungi, were isolated from stone monuments: (i) Hyphomycetes and Coelomycetes and (ii) black meristematic or MCF, and black yeasts. Hyphomycetes excrete organic acids and can actively dissolve carbonates, they produce different kinds of pigments, but only a few species are melanin producers (Sterflinger and Sert 2006). They generally require high levels of humidity and organic substances to develop. In spite of their presence in biofilms on outdoors stone monuments they are frequently responsible for the deterioration of wall paintings, caves and restored stone artefacts The MCF belongs to Ascomycetes, principally to the orders *Chaetothyriales*, *Dothideales*, and *Capnodiales*, and were isolated mostly, but not exclusively in arid and semi-arid environments.

The filamentous structures of fungal hyphae favour their penetration into the substrate, depending on its structure, chemical composition and state of conservation. Fungi can also perforate intact minerals (Santo *et al.*, 2021). Penetration can be also favoured by turgor pressure inside hyphae and melanin. Moreover, EPS produced by fungi facilitate fungal biofilm formation and the attachment to the rock, and increase mechanical pressure giving rise to shrinking and swelling (De Leo *et al.*, 2019).

14.8 Biodeterioration in Baroda Museum and Picture Gallery

Baroda Museum and Picture gallery was established in 1887 by the Maharaja Sayajirao Gaekwad III. The building of the museum was designed by Major Mant in association with R.F. Chisholm who refined some of Mant's finest works to make genuine Indo-Saracenic architecture. To maintain proper humidity a trench is present surrounding this building. A large number of microbes have been reported affecting objects placed indoor (Arya *et al.*, 2001). The Building is old at various places blackening can be seen due to presence of microbes like *Nostoc, Anabaena,* and fungi like *Alternaria, Curvularia, Periconia etc.* These microbe form black crest, which needs to be removed physically through brushing. A silicon coating of building is suggested.

The museum is famous for a variety of objects like Tibatan wooden work with precious stones, skeleton of blue whale and Egyptian mummy. The collection of European Paintings is another attraction. The museum building was completed in 1894, and opened to the public. Construction of the art gallery commenced in 1908, and was completed in 1914, but did not open until 1921 as the First World War delayed transfer of pieces from Europe intended for the gallery.

A special cover was released by Baroda Philatelic society on 16/1/1989 (Figure 14.5a). Exclusive photo of the cancellation of exquisite shows Jain bronze statue of Chaurie Bearer (Chamar Dharini) which was found in Akota area of Vadodara and belong to 8th century A.D. It is now exhibited in the Baroda Museum.

14.9 Occurrence of Microbes on Sun Temple

A 30 m high chariot with immense wheels and horses, all carved from stone can be seen in the Hindu Sun God Surya temple. The structures and elements that have survived are famed for their intricate artwork, iconography, and themes, including erotic kama and mithuna scenes. Also called the *Surya Devalaya*, it is a classic illustration of the Odisha style of architecture or Kalinga architecture. The cause of the destruction of the Konark temple is unclear and still remains a source of controversy. The temple that exists today was partially restored by the conservation efforts of British India-era archaeological teams. Declared a UNESCO world heritage site in 1984, it remains a major pilgrimage site for Hindus. Growth of algae, cyanobacteria and at certain places occurrence of Dematiaceous fungi like *Alternaria* and *Drecshlera* can be seen on the walls of this monument.

Figure 14.5

(a) A special cover which was released by Baroda Philatelic society on 16/1/1989 showing the cancellation of exquisite Jain bronze statue of Chaurie Bearer; (b) A photograph showing Baroda Museum and picture gallery; (c) and (d) Showing stamps of Konark Sun temple.

A stamp of 5 paisa of a powerful horse of Sun Temple was issued in 1949- as First Independent India Archaeological series. A Se-tenant block of 4 stamp of sun temple Konark was issued on 1/12/2001 (Figures 14.5c and d).

REFERENCES

Adhikary S P (2001) survival studies of lithophytic cyanobacteria on the temples and monuments. In: Studies in Biodeterioration of Materials-1 (eds.) O.P. Agrawal, S. Dhawan and R. Pathak. INTACH, Lucknow 29-44.

Adhikary S P and Satpathy D P (1996) Tolypothrix byssoidea(Cyanophyccae/Cyanobacteria) from temple rock surfaces of coastal Orissa, India. Nova Hedwigia 46: 419-42

Agrawal O P, Dhawan S, Garg K L, Shaheen F, Pathak N, Misra A (1988) Study of Biodeterioration of the Ajanta Wall Paintings. International Biodeterioration 24: 121 – 129

Amman R I, Ludwig W, Schleifer K H (1995) Phylogenetic identification and in situ detection of inidividual microbial cells without cultivation. Microbiological Reviews 59: 143-169.

Arya A, Shah A R, Sadasivan S (2001) Indoor aeromycoflora of Baroda Museum and deterioration of Egyptian mummy. Curr. Sci. 81(7): 793-799.

Bankar M V, Bhosle N P (2017) Ethnobotanical Survey of Medicinal Plants in Ajanta Region (MS) India. J. Pharm. Biol. Sci. 12: 59–64.

Baruah H K (1961): The air spora of cowshed. J. Gen. Microbiol., 25: 483-491.

Barton H A (2006) Introduction to cave microbiology: A review for the non-specialist. J. Cave Karst Stud. 68: 43–54.

Bharti G (2013) Ajanta caves: Deterioration and Conservation Problems (A Case Study). Int. J. Sci. Res. Publ. 3, 1–3. Available online: www.ijsrp.org.

Boopalan M, Sasikumar A (2001) Silicon, 3: 207-14, 2001.

Bridier A, Piard J-C, Pandin C, Labarthe S, Dubois-Brissonnet F, Briandet R (2017) Spatial organization plasticity as an adaptive driver of surface microbial communities. Front Microbiol 8: 1364

Brimblecombe P, Lankester P (2013) Long-term changes in climate and insect damage in historic houses. Stud. Conserv. 58: 13–22.

Brock T D (1987) The study of the microorganisms in situ: progress and problems. In: Ecology of Microbial Communities (M. Fletcher, T.R.G. Gray and J. G. Jones ed): 1-17. Society for General Microbiology, Cambridge.

Caneva G, Gasperini R, Salvadori O (2008). Endolithic colonization of stone in six Jewish cemeteries in urban environment. Proceedings of the 11th Int. Congress on Deterioration and Conservation of Stone, September 15-20, Torun, Poland. 49-56

Charola A E, McNamara C, Koestler R J (2011) Biocolonization of Stone: Control and Preventive Methods Proceedings from the MCI Workshop Series A; Smithsonian Institution Scholarly Press: Washington, DC, USA. 2: 1–116.

Casaletto M P, Privitera A, Testa M L, La Parola V, Mazzaglia A, Zagami E R (2017) Hybrid â-cyclodextrin/silica systems for the preservation of biodeterioration of stone materials. In Proceedings of the 9th European Conference on Ciclodextrines, Lisbon, Portugal, 3 October 2017.

Cecchi G, Pantani L, Raimondi V, Tirelli D, Luisa Tomaselli L, Lamenti G, Bosco M, Tiano P (1996) Fluorescence lidar technique for the monitoring of biodeteriogens in cultural heritage studies. Proc. SPIE 2960, Remote Sensing for Geography, Geology, Land Planning, and Cultural Heritage; doi: 10.1117/12.262463

Crispim C, Gaylarde C (2004) Cyanobacteria and biodeterioration of cultural heritage: A review. Microb. Ecol. 49: 1–9.

De Leo F, Antonelli F, Pietrini A M, Ricci S, Urzì C (2019) Study of the euendolithic activity of black meristematic fungi isolated from a marble statue in the Quirinale Palace's Gardens in Rome, Italy. Facies 65: 18.

Delgado R J, Vale Anjos M, Charola A E (2011) Recolonization of Marble Sculptures in a Garden Environment. In Biocolonization of Stone: Control and Preventive Methods: Proceedings from the MCI Workshop Series; Charola, A.E., McNamara, C.J., Koestler, R.J., Eds.; Smithsonian Institution Scholarly Press: Washington, DC, USA pp. 71–85.

Deutsch C A, Tewksbury J J, Tigchelaar M (2018) Increase in crop losses to insect pests in a warming climate. Science. 361: 916–919.

Diao M, Taran E, Mahler S, Nguyen A V (2014) A concise review of nanoscopic aspects of bioleaching bacteria–mineral interactions. Adv Colloid Interf Sci 212: 45–63

Dubey R C, Maheshwari D K (2000) a Text Book of Microbiology. A B books, 265pp

Faimon J, Štelcl J, Kubešová S, Zimák J (2003) Environmentally acceptable effect of hydrogen peroxide on cave "lamp-flora", calcite speleothems and limestones. Environ. Pollut. 122: 417–422.

Ferreira C, Rosmaninho R, Simoes M, Pereira M C, Bastos S M, Nunes O C, Coelho M, Melo L F (2010) Biofouling control using microparticles carrying a biocide. Biofouling, 26(2): 205-212.

Flemming H C, Wingender J, Szewzyk U, Steinberg P, Rice SA, Kjelleberg S (2016) Biofilms: an emergent form of bacterial life. Nat Rev Microbiol 14: 563–575

Gaylarde CC (2020) Influence of Environment on Microbial Colonization of Historic Stone Buildings with Emphasis on Cyanobacteria. Heritage. 3: 1469–1482.

Gaylarde C C, Gaylarde P M (2005) A comparative study of the major microbial biomass of biofilms on exteriors of buildings in Europe and Latin America. Int. Biodeterior. Biodegrad. 55: 131–139.

Gaylarde C, Hernández-Rodriguez C, Navarro-Noya Y, Morales O (2012) Microbial Biofilms on the Sandstone Monuments of the Angkor Wat Complex, Cambodia. Current microbiology. 64: 85-92. 10.1007/s00284-011-0034-y.

Gontareva E F, Ansari M K, Ruban D A, Ahmad M, Singh T N (2015) Geological dimension of the cultural heritage: A case example of the Ajanta caves (Maharashtra, India). Cuad. Lab. Xeol. Laxe. 38: 67–78.

Gutarowska B, akowska Z (2010) Estimation of fungal contamination of various plant materials with UV-determination of fungal ergosterol. Ann Microbiol 60: 415–422. https://doi.org/10.1007/s13213-010-0057-9

Hammond P M (1995) Described and estimated numbers: an objective assessment of current knowledge. In : Allsopp D. Colwell R.R. and

Hawksworth D.L. (eds.) Microbial Diversity and Ecosystem Function. 29-71. CAB International, Wallingford.

Hauer T, Mühlsteinová R, Bohunická M, Kaštovský J, Mareš J (2015) Diversity of cyanobacteria on rock surfaces. Biodivers. Conserv. 24: 759–779

Hentzer M, Eberl L, Nielsen J, Givskov M (2003) Quorum sensing: a novel target for the treatment of biofilm infections. BioDrugs 17: 241–250

Hueck-Van der Plas E H (1968) The microbiological deterioration of porous building materials. International Biodeterioration Bulletin 4: 11-28.

Jacob J M, Schmull M, Villa F (2018) Biofilms and lichens on eroded marble monuments. APT Bull 49: 55–60

Katharios-Lanwermeyer S, Xi C, Jakubovics N S, Rickard A H (2014) Mini-review: microbial coaggregation: ubiquity and implications for biofilm development. Biofouling 30: 1235–1251

Kemmling A, Kamper M, Flies C, Schieweck O, Hoppert M (2004) Biofilms and extracellular matrices on geomaterials. Environ. Geol. 46: 429–435.

Kulshrestha A, Chauhan S V S (2001) Aeromycoflora of some hospitals of Agra City. Indian J. Aerobiol. 14: 33-35

Lal B B (1966) The Murals- their composition and technique. In The Murals: The Preservation of the Ajanta Murals; Ghosh, A., Ed.; Director-General of Archaeological Survey of India: New Delhi, India.

Lefèvre M (1974) La 'Maladie Verte' de Lascaux. Stud. Conserv. 19: 126–156.

Li T, Hu, Y, Zhang B (2019) Evaluation of efficiency of six biocides against microorganisms commonly found on Feilaifeng Limestone, China. J. Cult. Herit. 43: 45–50.

Liu W, Order H L, Madsen J S, Bjarnsholt T, Sorensen S J, Burmolle M (2016) Inter-specific bacterial interactions are reflected in multispecies biofilm spatial organization. Front Microbiol 7: 1366

Liu X, Koestlker R J, Warscheid T, Katayama Y, Gu J (2020) Microbial deterioration and sustainable conservation of stone monuments and buildings. Environ. Sci. Eng. 3: 991–1004.

López D, Vlamakis H, Kolte R (2010) Biofilms. Cold Spring Harb Perspect Biol 2: a00398

Mackay J (2007) The illustrated encyclopedia of stamps and coins. Pub. by Hermes House, London. pp512.

Mascaro M E, Pellegrino G, Palermo A M (2022) Analysis of Biodeteriogens on Architectural Heritage. An Approach of Applied Botany on a Gothic Building in Southern Italy. Sustainability. 14: 34.

May E, Papida S, Hesham A, Tayler S, Dewedar A (2000) Comparative studies of microbial communities on stone monuments in temperate and semi-arid climates. In "Of microbes and art. The role of microbial communities in the degradation and protection of cultural heritage" (O. Ciferri, P. Tiano and G. Mastromei, Eds.), pp. 49–62. Kluwer Academic/ Plenum Publisher, New York.

Mitchell R, Gu J D (2000) Changes in biofilm microflora of limestone caused by atmospheric pollutants. Int. Biodeterior. Biodegrad. 46: 299–303.

Macchia A, Strangis R, De Angelis S, Cersosimo M, Docci A, Ricca M, Gabriele B, Mancuso R, La Russa M F (2022) Deep Eutectic Solvents (DESs): Preliminary Results for Their Use Such as Biocides in the Building Cultural Heritage. Materials 15: 4005. https: //doi.org/ 10.3390/ma15114005

Mittelmann M W (2018) The importance of microbial biofilms in the deterioration of heritage materials. In: Mitchell R, Clifford J (eds) Biodeterioration and preservation in art, archaeology and architecture. Archetype Publications, London, pp 3–15

Nugari M P, Salvadori O (2003) Biodeterioration control of cultural heritage: Methods and products. In Molecular Biology and Cultural Heritage; Saiz-Jimenez, C., Ed.; Swets and Zeitlinger: Lisse, The Netherlands, pp. 233–242.

Pozo S, Montojo C, Rivas T, López-Díaz A J, Fiorucci M P, López De Silanes M E (2013) Comparison between Methods of Biological Crust Removal on Granite. Key Eng. Mater. 548: 317–325.

Raghav N, Mamta, Shrivastava J N, Satsangi G P, Kumar R (2020) Enumeration and characterization of airborne microbial communities in an outdoor environment of the city of Taj, India. Urban Climate. 32: 100596, https: // doi.org/10.1016/j.uclim.2020.100596.

Ramalingam A (1971) Air spora of Mysore. Proceedings: Plant Sciences. 74(5): 227-240

Roelleke S, Witte A, Wanner G, Lubitz W (1998). Medieval wall painting–A habitat for Archaea: Identification of Archaea by denaturing gradient gel electrophoresis (DGGE) of PCR amplified gene fragments coding 16S rRNA in a medieval wall painting. Int. Biodeterior. Biodegrad. 41: 85–92.

Rohra H, Tiwari R, Khare P, Taneja A (2018) Indoor-outdoor association of particulate matter and bounded elemental composition within coarse, quasi-accumulation and quasi-ultrafine ranges in residential areas of northern India. Sci. Total. Env. 631-632

Salvadori O, Charola A E (2011) Methods to Prevent Biocolonization and Recolonization: An Overview of Current Research for Architectural and Archaeological Heritage. In Biocolonization of Stone: Control and Preventive Methods: Proceedings from the MCI Workshop Series; Charola,

A.E., McNamara, C., Koestler, R.J., Eds.; Smithsonian Institution Scholarly Press: Washington, DC, USA, pp. 37–50.

Santo A P, Cuzman O A, Petrocchi D, Pinna D, Salvatici T, Perito B (2021) Black on white: Microbial growth darkens the external marble of Florence cathedral. Appl. Sci, 11: 61-63.

Savulescu A, lonita I (1971) Contributions to the study of the biodeterioration of the works of art and historic monuments. I. Species of fungi isolated from frescoes. Revue Romaine Biologie. Serie de Botanique 16: 201-6.

Scheerer S, Schütze E, Curbach M (2018) Strengthening and Repair with Carbon Concrete Composites – the First General Building Approval in Germany. In: Mechtcherine, V., Slowik, V., Kabele, P. (eds) Strain-Hardening Cement-Based Composites. SHCC 2017. RILEM Book series, vol 15. Springer, Dordrecht. https://doi.org/10.1007/978-94-024-1194-2_85

Sharma N, Mathur P K (2020) Aeromycoflora Analysis in Outdoor Environment in Various Localities at Agra. Asian Journal of Agriculture and Life Sciences 5(2): 1-5

Singh M, Arbad B R (2012) Conservation and Restoration Research on 2nd BCE Murals of Ajanta. Int. J. Sci. Eng. Res. 3: 1–8.

Singh M, Arbad B R (2015) Characterization of 4th–5th century A.D. earthen plaster support layers of Ajanta mural paintings. Constr. Build. Mater. 82: 142–154.

Singh N K, Purkayastha B D, Roy J K, Banik R M, Yashpal M, Singh G, Malik S, Maiti P (2010) Nanoparticle-Induced Controlled Biodegradation and Its Mechanism in Poly(å-caprolactone). Appl. Mater. Interfaces, 2(1) 69-81.

Sterflinger K, Sert H (2006) Biodeterioration of buildings and works of art– Practical implications on restoration practice. In heritage, weathering and conservation; Fort, R., Alvarez de Buergo, M., Gomez-Heras, M., Vazquez-Calvo, C., Eds.; Taylor and Francis Group: London, UK (1): 299–304

Tonon C, Breitenbach R, Voigt O, Turci F, Gorbushina A A, Favero-Longo S E (2021) Hyphal morphology and substrate porosity -rather than melanization-drive penetration of black fungi into carbonate substrates. J. Cult. Herit. 48: 244–253.

Toreno G, Isola D, Meloni P, Carcangiu G, Selbmann L, Onofri S, Caneva G, Zucconi L(2018) Biological colonization on stone monuments: A new low impact cleaning method. J. Cult. Herit. 30: 100–109.

Umadi R, Dookie S, Rydell J (2019) The Monumental Mistake of Evicting Bats from Archaeological Sites–A Reflection from New Delhi. Heritage 2: 553–567.

Uno T, Shimazdu Y (2012) Thermal environment in Ajanta caves. In Proceedings of the Archi-Cultural Translations through the Silk Road, 2nd International Conference, Nishinomiya, Japan, pp. 191–196.

Venkataraman B (2008) Microbes eating away at pieces of history. The New York Times. http://www.nytimes.com/2008/06/24/science/24micr.html?_r=0.

Veneranda M, Blanco-Zubiaguirre L, Roselli G, Di Girolami G, Castro K, Madariaga, J.M (2018) Evaluating the exploitability of several essential oils constituents as a novel biological treatment against cultural heritage biocolonization. Microchem. J. 138: 1–6.

Villa F, Gulotta D, Toniolo L, Borruso L, Cattò C, Cappitelli F (2020) Aesthetic alteration of marble surfaces caused by biofilm formation: Effects of chemical cleaning. Coatings 10: 122.

Wawrzyk A, Rybitwa D, Rahnama M, Wilczy´nski S (2020) Microorganisms colonising historical cardboard objects from the Auschwitz-Birkenau State Museum in O´swi ²ecim, Poland and their disinfection with vaporised hydrogen peroxide (VHP). Int. Biodeter. Biodegr. 152: 104997

Young M E (1997) Biological growths and their relationship to the physical and chemical characteristics of sandstones before and after cleaning. Ph.D. Thesis. The Robert Gordon University Aberdeen, UK

Zhang K Y, Gao Y Z, Du M Z, Liu S, Dong C, Guo F B (2019) Vgas: A Viral Genome Annotation System. Front. Microbiol. 10: 184.

Chapter 15

Conservation of Wooden Monuments of Maldives: An Endeavour of Padmasri O.P. Agrawal Gains Reputation to the Nation

Sanjay Prasad Gupta and Iliyas Ahmed

National Research Laboratory for Conservation of Cultural Property, Lucknow – 226 024, Uttar Prdesh, India
e-mail: guptasanjayprasad@gmail.com

ABSTRACT

The buildings built of organic materials like bamboo and wood has an average life span much less than one built of stone, brick, cement or other relatively stable inorganic substance. Old farm-houses which used large timbers lasts longer, and Shinto shrines and some Buddhist temples longer still. Nevertheless, with normal maintenance, a wooden building rarely lasts longer than 250 years; Biological agents such as bacteria, moulds and insects live off wood. Fire is a constant hazard. Wood also reacts to the presence or absence of moisture-swelling, shrinking and cracking and other climatic changes. In countries where wooden structures are important to their cultural heritage the necessities for their preservations and restoration have imposed criteria to take these factors into account.

Out of numerous achievements of visionary Padmasri Dr. O.P. Agrawal, the father of conservation in India was setting up of the National Research Laboratory for Conservation of Cultural Property (NRLC), in 1976 under the Ministry of Culture. NRLC is the premier organization for the researches in conservation of the cultural property including monuments and sites, as well as museums, library and archive collections nationally and internationally. Conservation of buildings like Friday mosque and Dharumavantha mosque, which were made from coral stone and timber was done in Maldives during 1984-2001. Details are provided in the chapter.

Keywords: Conservation, Wooden monuments, Friday mosque, Dharumavantha mosque, Maldives

15.1 Introduction

Maldives is a group of islands extending nearly 750 km from North to South in the Indian Ocean in the S-W direction from India. Today there are twenty-eight mosques in Male, the capital of Maldives, and Dharumavantha Rasgefaanu Mosque of Male is considered to be the oldest existing mosque in the Republic of Maldives (Gupta and Ahmed 2022). The Maldives boasts a cultural fusion with a history that extends to 300 BCE, and an interesting interaction between different religions and importantly between Buddhism and Islam. The local people practiced Buddhism until the conversion Islam in 1153 CE. Construction in ancient Maldives was mainly dependent on the local availability of materials. Coral stone and timber were the only long lasting materials available and coral stone became the primary building material for monumental buildings.

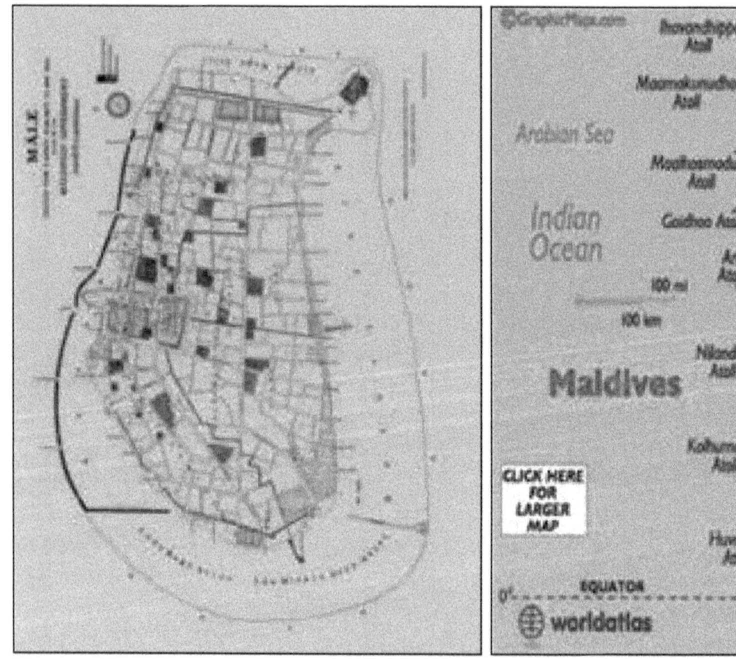

Figure 15.1: Location of Dharumavantha Rasgefaanu Mosque at Male, Maldives.

Conservation of monuments in the Maldives was a diplomatic as well as pragmatic contribution of the Indian Government through Ministry of Foreign affairs and Ministry of Culture to preserve the physical heritage of Maldives. It may gives us great satisfaction that a laboratory started in National Museum with two staff members only, emerged and evolved into three beautiful campuses. A Research Laboratory at Aliganj Lucknow and training institute and Guest House at Jankipuram, Lucknow and Regional Conservation Laboratory at Mysore in 1984, 2008 and 1986 respectively (Agrawal 1972; Agrawal and Dhawan 1991).

The conservation of monuments done by NRLC in Maldives, a strategic neighbor of India in the last few decades especially Dharumavantha Rasegfaanu Mosque at male' in which one of the author was the key member of the execution team. First project was that of conservation of Friday mosque in 1984 at male the capitol of Maldives and conservation of Friday mosque in 2001 at Fenfushi island and finally conservation of Dharumavantha Rasegfaanu mosque the oldest mosque of Maldives at Male. This initiative of conservation of Monuments of Maldives was provided with the friendly relation of India and Maldives.

Figure 15.2: Friday Mosque (a) Before Restoration (b) After Restoration.

15.1.1 Conservation Issue

The mosque was examined by two experts from NRLC, Lucknow in February 2003, and, accordingly, a plan for its conservation was prepared. The conservation material, tools, *etc.* required for the project were procured in India and shipped to Male, and execution of the conservation of the mosque was started in May 2004 after the material sent from India was delivered at Male. Part of the material was procured at Male or provided by NCLHR. The conservation work was completed in seven months and the mosque was opened for prayer on 30th November 2004 by High commissioner of India and chairman of NCHLR, Maldives.

The Dharumavantha Rasgefaanu Mosque is a one-room structure measuring approx. 13x6.6m with a 1.8m wide covered veranda in the front, and an open corridor of 1.5m wide all around the main area. The main building is on the podium of about 3.5 feet height from the ground level. There was a well inside the campus for drawing water for ablution, *etc.* Main building of mosque is constructed out of coral stone blocks up the height of six feet. The walls have been white washed with several layers of plastic paint in the past.

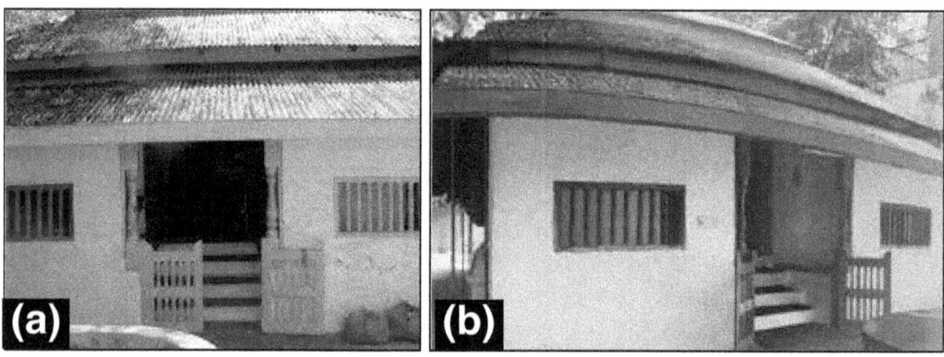

**Figure 15.3: Dharaumavantha Mosque
(a) Before Restoration (b) After Restoration.**

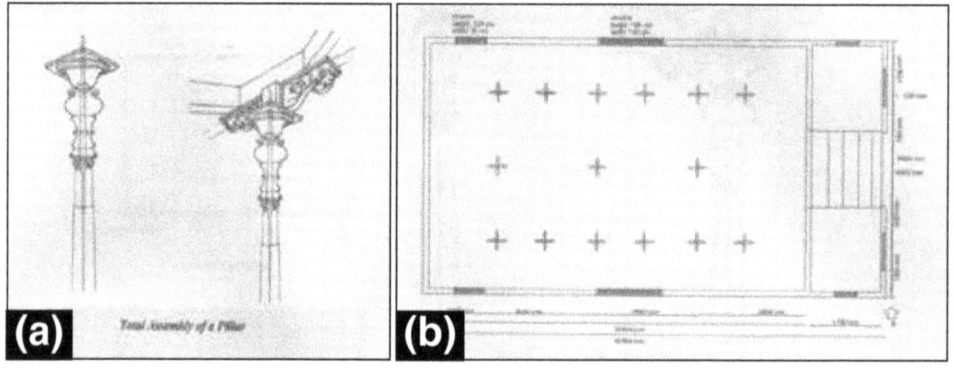

**Figure 14.4: (a) Pillar Assembly; (b) Layout of Mosque Showing
Pillars and Window.**

The interventions to conserve Dharaumavantha mosque, one of the oldest and the historic made up of timber structure was initiated in the 80 per cent of the structural components. Interventions were based on proper assessments. The most important phase of the conservation of the mosque was the assessment of deterioration near to exact assessment and the extent of deterioration lays the foundation for proper execution of the project. The mosque was beautifully planned and was architecture marvel of the Maldivian style. Layout of the mosque clearly indicates that whole mosque erected on the 17 wooden pillars assembly of pillar bearing the load of ceiling. The components of the mosque *i.e.* load bearing beams, frames and lacquered panels were affected badly by subterranean termite (Shrivastava 1997). Most of the wooden components were severely damaged by termites from the inner side and only a thin outer layer was remaining there. Load bearing beams, which had become hollow, had got de-shaped and big cracks were seen on them. It was eaten in such a way by termite that only paint layer was remained there at many places (Eaton and Hale 1993).

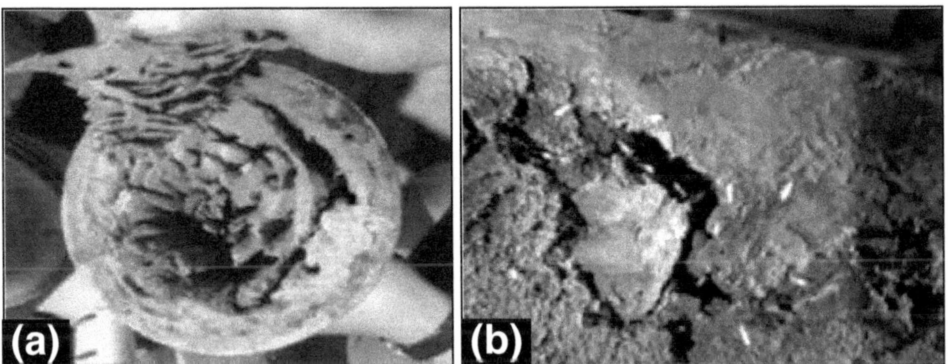

Figure 15.5: Deterioration by Termites
(a) Load Bearing Component; (b) Painted Panel.

Some other components though appearing in sound condition from outside but when undergo acoustic method were sound like hollow or near to hollow were recorded (Findlay 1985). By using the acoustic method most of the components were tested for the strength and it was noticed that most of the load bearing beams, vertical components painted panels, frames of the painted panels and many rafters were sound like hollow or nearly hollow panels were found and the extent of damage was really horrible and beyond the expectation and imagination, even in some of the cases *i.e.* load bearing beams were completely converted into debris (Charles and Charles 1986; Unger *et al.*, 2001) except the painted surface along with some of the wooden layer varying with the thickness from 2 mm to 15 mm. Insects had generally not eaten the painted surfaces however; paintings on some of the beams and panels were also damaged by physical means.

The carved lacquered components along with the full assembly which placed between the load bearing beams and roof rafters were also got severely damaged. Some of the components were missing or might not be replaced after damage *i.e.* engraved wood on the wall which acts as base for the frames of painted panels written with Quranic verses.

Repeated application of oil had made the paintings obscure and the bare wood blackish. Practice of burning oil lamps for illumination inside the mosque might have also contributed to the darkening of the wood surface. Paintings on the beams nearer the window on the southern side had developed chalkiness on them and had suffered extensive deterioration of the paint layer.

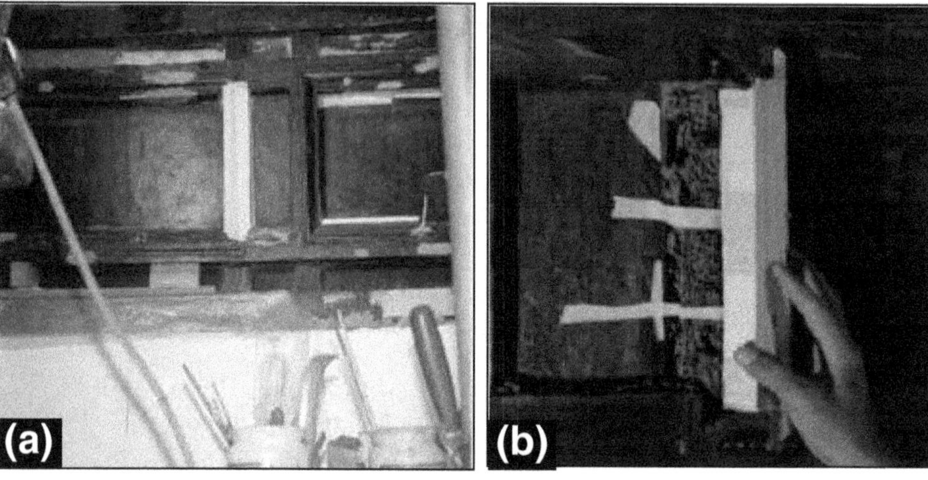

Figure 15.6: Deteriorated by Termites
(a) Vertical Component (Lateral view); (b) Painted Panel.

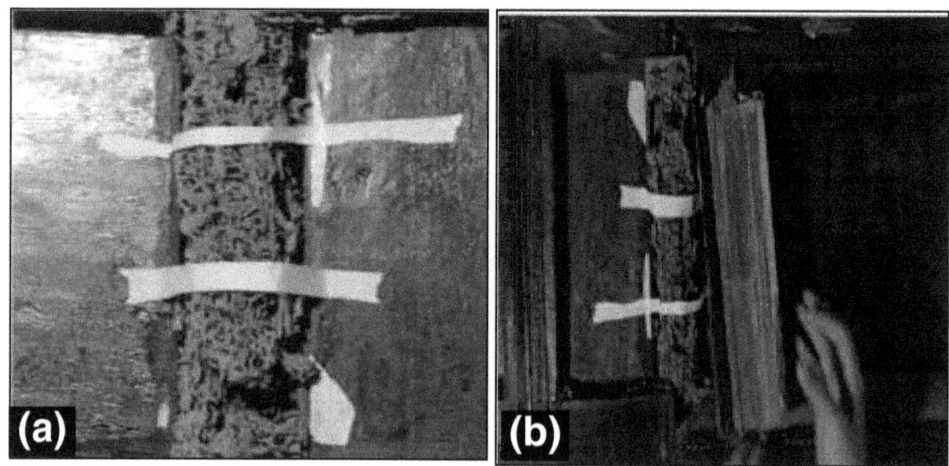

Figure 15.7: Deteriorated by Termites
(a) Vertical Component (Inner view); (b) Painted Panel.

The frame of the painted panels might have originally placed on the inner wall at the far end of the hall as it is evident from the remains at the floors, was now supported on iron poles painted green with modern paint. Due to replacement of wall by iron poles, downward force per unit area had become very high which might lead to stress failure. At the same time, the material and its green colour were not in unison with the old structure of historical importance. Wooden window frames and grills and two pillars at the outer entrance were also painted similarly. Electric lines running on the wooden beams *etc.* were a potential fire hazard, and tube lights were fixed directly on the beams.

The exterior shade on windows and the pillar at outer gate were painted with several layer of green paint. The painted exterior shades at windows and pillars were not unison with the mosque. Load bearing beams were cracked and bulged outside horizontally due to the pressure applied vertically. The lower and upper surfaces of the beams have shrinked inside. The bulging and shrinkage was due to the force applied per unit area was more at the vertical pillars.

15.2 Materials and Methods

15.2.1 Conservation Techniques

The main objective of conservation work or treatments was to strengthen the structure of mosque and to reveal the aesthetic beauty by cleaning the painted areas and bare wood surfaces without compromising with architectural entity of the mosque. Subsequently, the mosque was to be made presentable and fit for performing prayers. As mentioned above most of the wooden members were in bad state of preservation. Therefore, to carry out the treatments smoothly it was necessary to ensure the stability of the structure and security of the working professionals. To achieve this, ceiling and all the components were firmly supported on several jacks (145 jacks), which were placed under the beams and some of the sturdy rafters. The structure was evenly supported by putting wooden planks between the jack and beams to avoid any damage to the painted surfaces. A sheet of felt was also placed between the plank and beam/rafter so as to avoid any chance of jacks skipping from their positions or abrasion of painted surface. The electric wiring and fittings laid on wooden structure were removed.

Though no live insect was found, there had been intense insect activity in the wooden structure of the mosque in the past. Therefore, the wooden structure was thoroughly sprayed with a commercial termicide, which is a 20 per cent aqueous solution of chlorophyriphos. There was a lot of dust accumulated on the planks placed on top of the ceiling. The dust was vacuum cleaned and the planks were taken down, and kept aside. The structure was strengthened part by part and reassembled. The painted and other bare wooden members were cleaned and a protective coating was given. The floor was treated with insecticide and made even, and finally a carpet was laid. The treatment is detailed below:

15.2.2 Selection of Adhesives, Consolidants and Gap Fillers

Adhesives are non-metallic materials which join solids (adherents) to each other by adhesion and cohesion, without causing any significant changes in the surfaces to be joined. The cohesion drives from internal, inter-molecular forces and the molecular size of the adhesives. The adhesion refers to a molecular forces between the solid surfaces of the adherent and second phase, the adhesive, which may consists of individual particles such as droplets or powders, or a continuous liquids or solid film. The adhesion of adhesives is affected primarily by Van-der-wall's forces, but also by electrostatic forces.

There are many parameters which an adhesive should meet:

1. Reversibility of the adhesive and the adhesive joints
2. Absence of darkening or colour changes of the joined pieces
3. Compatibility of the adhesive with wood preservatives and consolidants
4. Adequate strength of the adhesive joints
5. Permanence and flexibility of the joints
6. Resistance of the adhesive to pests
7. Evidence of adhesive in objects after conservation treatments.

Reversibility of the adhesive joints is especially important where old, open joints are to be re-adhered, or where a new and different adhesive is to be used for an old joint, so that the adherent will not be damaged further.

The selection of adhesive in the case of Dharumavantha mosque was to be more selective as the wooden parts were damaged partially and completely by insects. Because the density and strength of the damaged and deteriorated woods was reduced, which increased the permeability and capacity for absorbing adhesive, and ultimately increased permeability of the wood to absorb excessive adhesives or starved joints when the adhesives tends to disappear into the interior of the wood. This can make corrections or newer joints with or without removing the old adhesive more difficult.

In principle it was decided not to select the piece of wood for joining or replacing the damaged wood and the selected new wood should not differ too much in their anatomic and structural characteristics. The joining parts of different components of different degrees of decomposition or destruction increases the danger of pre mature joints failure because permeability, density and strength can differ significantly from each other. The grain direction can also have a significant effect on the life of joints. Cross-sectional surfaces of wood generally absorb adhesives much faster and in greater quantities, which can lead to poor joints, while side grain joints are also important factors. The stronger the wood, the stronger the joints must be. Wood of higher density will take up less adhesive than low density woods, and the latter may even take up too much.

There were many woods available in the Maldives and several pieces were selected for replacements of damaged components. There were three category of wood which were low, medium, and high density and it was studied that wood of higher density are more difficult to adhere than less dense ones. It was learnt during the experimentation of selection of wood for the different use in the mosque that wood moisture content is also decisive factor in determining the stability and longevity of the joint. It was also learnt with the experimentation that there must be no more difference in the moisture content of the woods needed to be joint otherwise failure percentage will be high because wood with moisture content that is too high will delay the adhesive setting and leads to weaker joints or even separation of surfaces.

There are many ways to classify the adhesives such as

1. Chemical composition and origin,
2. Type of setting,
3. Manner of use, and end use.

Most commonly it is done on the basis of origin (in-organic and organic) or of chemical nature (natural, semi synthetic or synthetic) as well as according to whether it sets physical or by chemical reaction. Physically setting adhesives which are usually single-component can be free of solvents (hot-melt adhesives) or contain solvents (protein glues, and solvents based, dispersion, pressure sensitive and contact adhesive). Adhesion is achieved by changing from liquid to solid phase, or by evaporation of solvents either before or during the process of adhesion.

Adhesion process can be classified according to the process i.e pressure sensitive or contact adhesive; the state of adhesive at the time of gluing *e.g.* wet or melted adhesives; the temperature *e.g.* cold, warm or hot; the applied pressure *e.g.* with or without; and the wood surface of the adherents; flat, edge or end-grain surfaces.

Several adhesives were tried and tested for the proper joining of the joints are listed below.

1. Lac derived from shellac
2. Poly vinyl acetate emulsion
3. Carbohydrates derived from wheat floor
4. Epoxy resins
5. Epoxy Putty
6. Plastic wood paste.

15.2.3 Gap Fillers

Gap fillers are knead able to highly viscous masses which can be divided into gap-filling adhesives and fillers, gap filling adhesives are adhesives can be deformed plastically at normal temperatures, contains only volatile or no solvents and fill material, and are used to bridge thick glue lines. Gap fillers used in the conservation of wooden objects to fill cracks, splits, holes such as insects exit holes and for loss compensation should meet the following requirements.

1. Significant shrinkage during solidification is most undesirable in patching compounds used to fill the gaps and holes. Good Adhesive strength, but this should only be high enough so that any failure during swelling and shrinking of the wood will be in the filler and not in the wood.

2. Cohesive strength commensurate with that of the wood, tp prevent excessive resistance during swelling and shrinkage which might lead to damage to both wood and filler

3. Sufficient permeability to water vapour

4. Long term flexibility if it is required

5. Good light fastness if the surface need to be exposed

6. Good ageing characteristics and reversibility

7. Receptivity to in-painting or colouring

8. Adequate pot life

9. Adequate shelf life and lack of separation of components

10. No swelling of wood by solvents in the filler, and no stress cracks in the filler upon evaporation of solvents

11. Can be worked with spatula and sanded after curing without creating toxicity.

Following combinations in the different ratio of adhesives and fillers were used to get the proper strength and desired shape of the collapsed components

1. Glue and sawdust

2. Ready to use Wood filler

3. Plastic wood resin

4. Actual wood piece

5. Kaolin

6. Marble dust

7. Epoxy wood putty

15.2.4 Consolidation

It was a foremost challenge before the team to strengthen the weaker and deteriorated parts of the mosque. Some of the team members were of the opinion that the deteriorated components may be consolidated in situ as they did in the case of Fenfushi and Friday mosque. But the team lead and some other members were opted for few more alternatives and finally, the dismantling of the all possible deteriorated components was decided for the purpose of consolidation.

After deciding the methodology for consolidation the material and method was the next to get upon to a conclusion. It was decided to first start the process of consolidation from the painted panels and their assembly after getting an idea of degree of deterioration.

The totally damaged frame of the painted panels were restored by inserting a full real wood by engraving the desired design to take on the left over skin and then pasting of skin of original wood left by the termite to retain the original design. Patching material was a mixture of PVA:Saw dust:Kaolin so that any shrinkage may be avoided and to further avoid the deformation of finally pasted layer of left over skin of original wood.

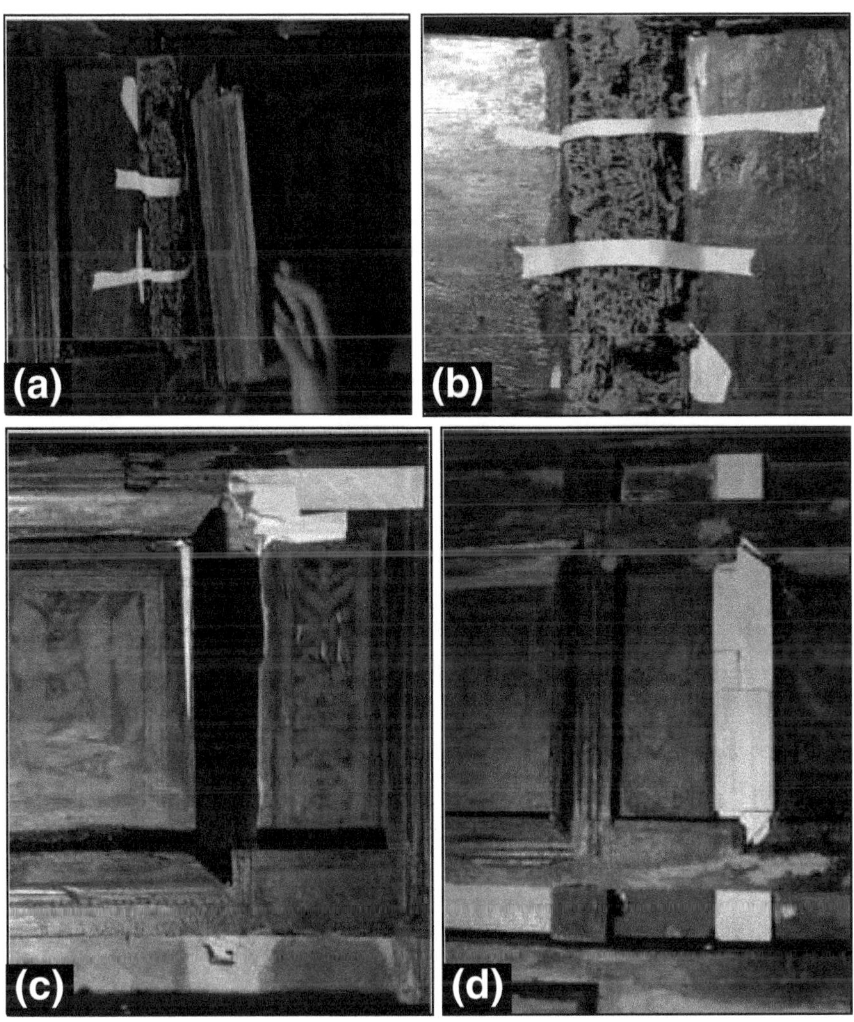

Figure 15.8: Deteriorated by Termites
(a) and (b) Vertical Component (Inner view); (c) Hollowed Space;
(d) New Wooden Vertical Component.

15.2.5 Strengthening of Painted Panels

The panels painted with lacquer work and written upon Quaranic Verses were cleaned off the debris from the damaged portion. Pre-wetting with 50 per cent solution of chloropyriphos in water as insecticidal was applied prior to applying of filling and patching material. A diluted solution of ply vinyl acetate emulsion was infused to the holes and insects tunnels to get the proper consolidation and strength.

15.2.6 Pre Wetting of Filling Holes and Gaps

Before penetrating the adhesive into the holes which were deep-rooted and flowing to the interior core of the wooden components, a solution of 60:40 termicide: water was poured or sprayed or injected into and let it absorbed by wood then a mixture of saw dust and PVA emulsion of sufficient viscosity was poured and forced to reach to the core in sufficient quantity. Excess mixture was wiped out and after 4 hours the process was repeated and every time the viscosity was increased by increasing the saw dust. The outer layer was mixture of PVA, saw dust and inorganic filler (kaolin powder) in the sufficient ratio to hold it on the surface applied on. The final layer was of 60:30:10 ratio of Kaolin:PVA:Saw dust to achieve any re-integration.

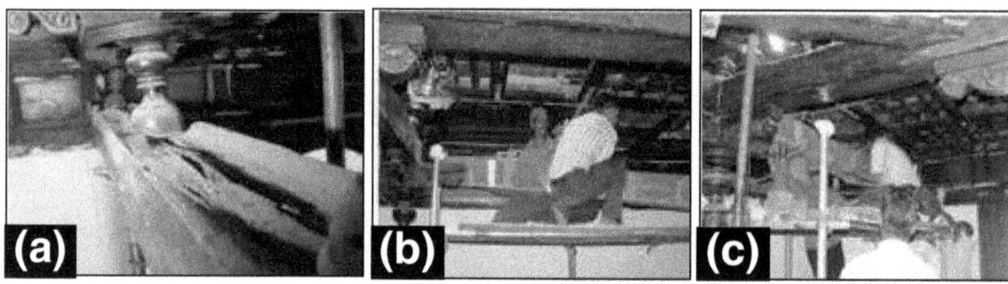

Figure 15.9: Deteriorated by Termites
(a) Compression of Beam; (b) and (c) Removal of Beam.

15.2.7 Reinforcement of Beams

All the beams were badly deteriorated by subterranean termite and the extent of damage was such that only wood left was in mm or just the paint layer. This was the great challenge towards the strengthening of the mosque, because the load of the mosque was entirely on these beams. The beams were collapsed from upper and below and bulged from sides due to hollowing from inside and the whole structure was retained in the shape due to the strength in vertical pillars only.

1. Thick rafters on both sides of the beam were further supported on additional jacks and lifted about an inch, and the assemblies of supporting legs placed on the beams were removed cautiously part by part and one by one. After removing the supporting legs, the jacks supporting the rafters were brought back to their original position.

2. The next step was to remove the pillar supporting the beam. The pillar was found to be embedded in the cemented floor, which was laid at a later date. The cemented floor around the pillar was dig to make the pillar free. The beam was raised by about 5 cm, and the lower end of

the pillar was moved away from its position, and the pillar was taken out. The components at the top end of the pillar were then dismantled one by one.

3. Then the beams and other load bearing members were taken down for strengthening. The painted beam located on the right hand side at the far end of the hall was taken up first for strengthening, as the extent of damage was most serious in this beam. Insects had eaten the core of the beam, and, with the result, the beam was not having any load-bearing capacity. It was decided to reinforce the beam by providing a new core of new seasoned wood, and then place back in its original position. The work was carried out in the following steps:

Figure 15.10: Process of the Removal of Debris from Beam.

15.2.8 Protection of Painted Surface

Application of Paste on the Beam

Before putting down the deteriorated beams down to floor for treatment it was mandatory to fix and consolidated the painted surface on the pillar. To achieve this target it was decided to prepare paste which will not only give strength but also easy to remove after treatment. A mixture of 60:40 ratio of wheat starch paste: Carboxymethyl Cellulose (sodium salt) was found suitable to achieve the desired strength and swelling for easy removal (Kryg *et al.*, 2020). Application of multi layer cotton gauge: multi layer cotton gauge was applied over paste in such a way that each layer may be embedded with the paste to get desired strength. Three layers embedding of paste and cotton gauze was found to be worked perfectly to wear all jerk and risk of scratch *etc.*

Supporting of Beam by Coupling with Plank from Lower Side

A plank of sufficient thickness (1.5″) and width of beam and 12″ short of length of original beam. Supporting plank was fixed with the original beam with the help of nylon fastening belts with all precautions not to damage the beams edges and painting as well.

Loosening of End Joints

In the due course of time beam was applied with several layers of protective oils and varnishes which almost act as binding media for edges to fix with surrounding wood. All the corner of beams was cleaned off the varnishes and debris with spatula and solvents.

Removal of Beam

Hydraulic jacks along with chain pulley were used to lift the beam which helped in fulfilling the objectives. First the corner towards the central dome was lifted up to an inch and after that another end towards the entry gate was lifted. Simultaneously the whole beam was lifted. Once the beam freed from joints it was brought down manually and placed on floor.

15.2.9 Cleaning of Debris

Cutting of Upper Side which was not Painted as well as Visible

After brought down the beam was inspected thoroughly and as expected the beam was fully converted into debris from inside without any wooden core. It was not possible to clean off the debris from 18 feet beam. To clean it off from inside the upper portion of the beam was incise from the edges with the help of electric zig saw fitted with blade of T218A Bosch to minimise the impact of cut. The incision was made in such a way that there should remain an edge of at least one inch of original wood at the corners of beam. After opening of upper face of beam, debris was cleaned manually and a hollow channel of uneven surface was achieved.

15.2.10 Insertion of New Wood into the Hollowed Beam

Elimination of Uneven Surface from Inside of Beam

After removing debris it was noticed that the beam was uneven from inside and contours of 1 to 3 inch was seen inside. It was great challenge to make it even from inside to insert a wooden beam of new wood.

Figure 15.11: (a) and (b) Process for Removal of Debris from Beam and (c) and (d) Insertion of New Wood and Pasting of Remains of Beam.

Application of Filler to Achieve even Surface Inside of Beam

The contours inside of the beam were filled with mixture of PVA:Saw dust:Kaolin in several layer one over other after drying of each layer for better and shrinkage free surface. A new beam of locally available hard wood was made. The thickness of the new beam was such that it was just filling the gap in the old beam and its length was bigger than the old beam. Slurry of PVA emulsion and sawdust was coated inside of the old beam, and PVA emulsion was applied on the new beam,. While the PVA was still wet, the new beam was placed inside the old beam keeping the extra length of the new beam projecting equally on both the sides.

Preparation of New Wood

New wood which was to be inserted into the hollow channel of the original beam was made to fit into the original groove. The ends of the reinforced beam projecting outside the old beam were shaped as tenon to fit in the grooves of the receiving beams to fit and the surface of new wood was abraded with zig zag cut marks so that a groove can be created to hold the fillers and channel of old beam skin. Prior to this new wood was applied with anti termite chemical and wrapped with polythene to let it impregnate.

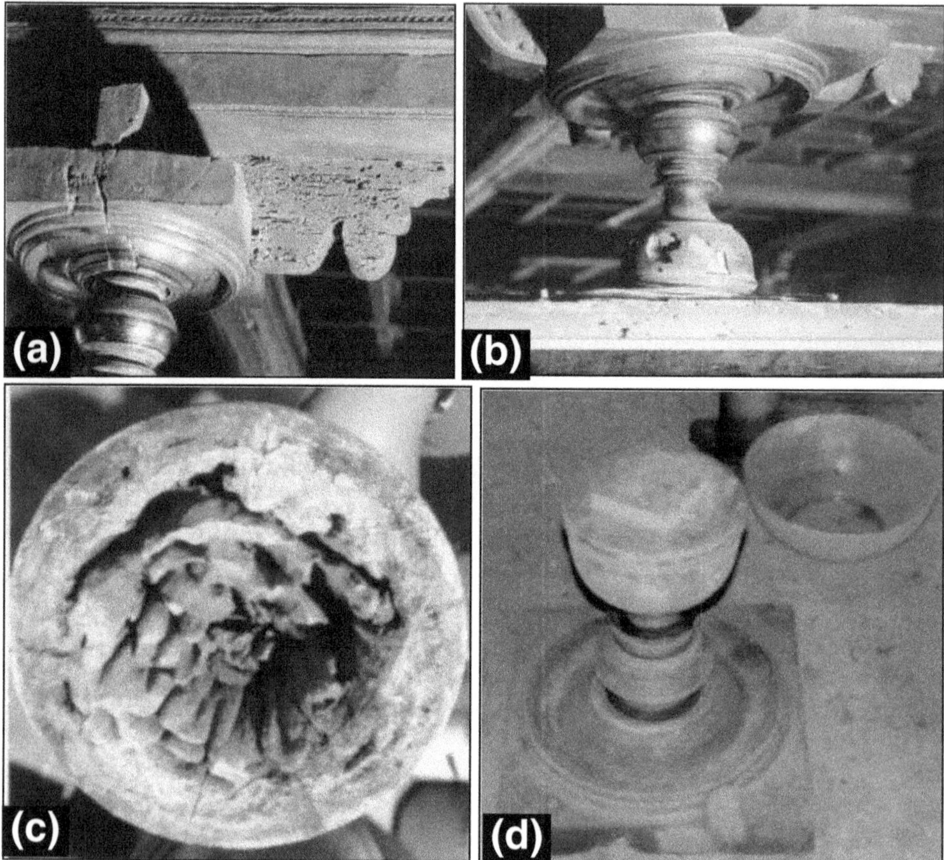

Figure 15.12: (a) (b) and (c) State of Deterioration of Turned Lacquered Component; (d) Restoration of Turned Lacquered Component.

Clamping of Skin to Achieve the Perfect Contact and Adhesion

The beam was clamped hard from all the sides after putting polythene sheet, felt and planks on it, and left to dry. The clamps were used at every six inch to get an even surface.

Removal of Cotton Gauge and Placing of Upper Surface

After the PVA emulsion was semidry, the clamps were removed, and the excess PVA emulsion was cleaned from the surface after removing the gauze. The top layer of the old beam was then replaced using sawdust and PVA emulsion as adhesive. Care was taken that the top layer was exactly at the old level; wherever required extra sawdust and PVA emulsion putty was used. After it was completely dry, grooves present in the upper layer to receive the legs on top of the beam were further deepened into the new beam. Any lacunae still present in the surface of the beam were filled with PVA emulsion and whiting.

Reinforcement of Small Turned Pillars

Before removal of beam it was mandatorily feel necessary to remove the carved small pillars placed above the beam for supporting the rafters in ceiling. The pillars were mostly cracked and some parts had been eaten away by insects. The components of the supporting legs to be placed on the reinforced beam were consolidated with sawdust and PVA putty as well as new wood as per their requirements and where openings were restricted PVA emulsion was injected with hypodermic needle.

In most cases, the base of the legs had been eaten away by insectcompletely. New base to such legs, which exactly fit in the groove in the beam, was provided by inserting a conical shape wood piece while maintaining the length of the legs.

Reinforcement of Circular and Turned Components

The circular component placed at the top of the small turned pillars in most cases had been eaten away by insects from inside to the extent that the gaps and cracks in the components were filled with sawdust and PVA putty even some times new wood was inserted to achieve required strength. The carved wood placed over the circular components to support the ceiling through rafters was also badly eaten away by termites and wood borer. In some cases where the wood was thought not to provide sufficient strength, the components were replicate to replace with new wood.

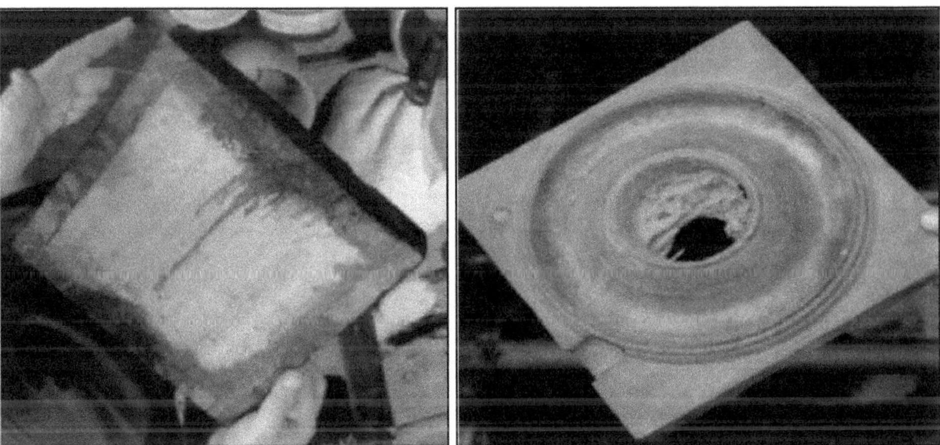

Figure 15.13: Restoration of Horizontal Turned Lacquered Component.

All the supporting legs and the pillars were consolidated as above. The thicker rafters present above the beam were raised with jacks, and upper ends of the supporting legs were put into the place in the rafters. The pillar was placed in its position and the reinforced beam was lifted and placed in its original position. The rafters along with the supporting legs were then lowered so that all the components were in place as they were originally.

The beams and majority of pillars and supporting legs were treated one by one in the above manner as per their requirements. Nine segments of the main beams were thus reinforced, and one missing beam at the left far end was fabricated with seasoned wood.

15.2.11 Replacement of Iron Pillars with Wooden Pillars

The wall supporting the painted panels at the far end of the hall might have been decayed beyond repair, and three iron poles had been provided instead in the past to support the painted panels. The iron poles painted with modern green paint were not in unison with the monument of historical importance and the architecture of the interior of the mosque apart from the fact that downward force per unit area was also very high. Therefore, three pillars along with all their components similar to other pillars of the mosque were fabricated with seasoned wood and placed instead of iron poles.

Figure 15.14: (a) Iron Pillars; (b) Replacement of Iron Pillars with Wooden Pillars.

The rafters were taken up next for conservation. Most of these were eaten by insects or were rotten, and many were missing. Deteriorated ones were cleaned and consolidated with sawdust and PVA putty and sometimes with epoxy putty and the missing ones were fabricated new. Epoxy putty filling was smoothened with electric tools.

15.3 Cleaning and Protective Coating

It is evident that mosque was painted with several layers of plastic paints were applied on the exterior wooden components *i.e.* window and grill made from wooden strips with wooden shade. Two pillar at small gate along with two doors. It was noticed that the entire component inside the mosque, painted and unpainted were coated with multi layers of oils. The application of multilayers of oil in the due course of time darkens with the cumulative effect of light, dust and dirt.

Cleaning was divided broadly into two categories:

15.3.1 Lacquered Wooden Components

a. Turned component with thick layer of lacquer
b. Quranic Versace and floral design with thin layer of lacquer

15.3.2 Bare Wood and Plastic Paints Coated Components

It was very important to test and devise a safe and effective method and solution for cleaning of the surfaces. To achieve this several solvents/solvent mixtures were tasted for their efficacy for cleaning. Different solvents were tried:

a. Acetone,
b. Tri Chloro Ethylene
c. Carbon Tetra Chloride
d. 2-Ethoxy Ethanol or Celosolve
e. Methanol
f. Turpentine oil

Individually and in the combination with solvents, different formulations were tried. It was found that some solvent were working good on plastic paint and bare wooden surfaces and some were effective on lacquer painted surfaces.

15.3.3 Cleaning of Turned Component

Turned component were applied with thick multi layer of lacquer. It was found that the painted areas were cleaned with 1:3 alcohol and turpentine mixture, and the bare wood was cleaned with 1:1 mixture of the same reagents. The contact time was observed for each type of component to clean. This was achieved after leaving application of cleaning chemical and wrapped up for 2 to 5 minutes. This time interval was enough to swell the multilayer of oil applied on the component and slowly the deposits were removed off. The white patches of putty used for consolidation were reintegrated, and finally a protective coating of paint.

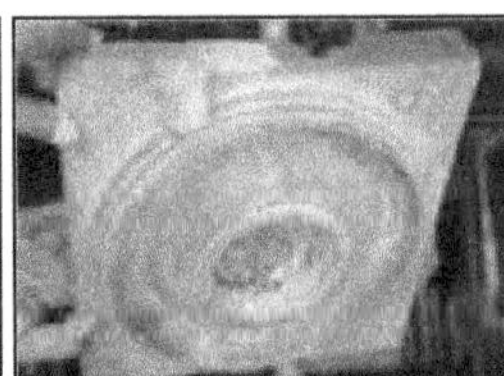

Figure 15.15: Cleaning of Lacquered Components.

15.3.4 Cleaning of Beam and Panel with Quranic Verses

The painting on beams and panels was thinner than the turned component and need careful application of cleaning formulations and it was calculated that the application of 1:3 alcohol and turpentine for 2 minutes was enough to swell the coating.

Figure 15.16: Cleaning of Beam and Panel with Quranic Verses.

Wooden window grills and two pillars in the front corridor were painted with several layers of modern paint. Bulk of the paint was removed with a commercial paint stripper, and final layer was removed with 1:1 alcohol and turpentine mixture.

Figure 15.17: Cleaning of Front Gate (a) before and (b) after.

The cleaning of green plastic paint from the windows and small entry gate was done by Methyl Ethyl Ketone based paint strippers. Seven layers of plastic paint were applied one over other to protect the underneath wood. During cleaning and removing of paint a beautiful lacquer work was found on the pillars at gate.

15.4 Results and Discussion : Maintenance of the Mosque

As is the practice with the people of Maldives to keep the places of worship very neat and clean, the Dharumavantha Rasgefaanu Mosque should also be maintained in the same manner. Regular and proper use and general cleaning of the mosque and its campus will go a long way in the preservation of the mosque. Taking into consideration the ethics of conservation of monuments and their preventive conservation needs, further suggestions are offered below for the attention of the concerned authorities:

1. Original character of the mosque should be maintained to the maximum extent possible.

2. Modern paint with bright colours does not go well with the old monument. Refrain from using them on the wooden components, as has been the practice, even if the wood appears dirty.

3. The 220 volts electric lines should not be laid on the wooden parts of the mosque.

4. Interiors of the mosque should be evenly illuminated with UV-free light. Normal fluorescent lamps presently used in the mosque emit high amounts of UV radiations.

5. Some oils may provide protection to bare wood against vagaries of the nature. However, they should not be applied on the painted wood, as they become dark on ageing and attract dirt and change appearance of the paintings. Unpainted wood in the mosque has been provided with a protective coating as a part of the conservation treatment and application of oils there also is unwanted.

6. Speed breaker are to be provided on the road passing by the mosque so that particulate matter (dust) in the surrounding air, is reduced.

7. A work force of motivated and dedicated individuals should be created to look after the old mosques and other cultural property of Maldives, and they should be trained well in the scientific conservation of cultural property.

Acknowledgements

Authors are grateful to Dr Tej Singh, Former Director, NRLC, Shri Atul Kumar Yadav, Former Scientist, NRLC for providing technical support and scientific evaluation and Shri Anil Risal Singh, Former Photography officer, NRLC for the complete documentation process.

REFERENCES

Agrawal O P (1972) Problems of Preservation of Indian Murals. Conservation in the Tropics. 87-90pp.

Agrawal O P, Dhawan S (1991) Biodeterioration of the Cultural property. Macmillan, New Delhi, India 300 pp.

Charles F W B, Charles M (1986) Conservation of timber buildings. London, England: Hutchinson. 256pp.

Eaton R A, Hale M D C (1993) Wood: Decay, pests and protection. Chapman and Hall, London 546 pp

Findlay W P K (1985) Preservation of the timbers in tropics. Nijhoff/Junk, Dordrecht, Boston, Lancaster. 273pp.

Garg K L, Garg N, Mukerji K G (1993). Recent advances in biodeterioration and biodegradation: volume 1: biodeterioration of cultural heritage. Recent advances in biodeterioration and biodegradation: 496pp.

Gupta S P, Ahmed I (2022) Selection of adhesives and fillers in conservation of wooden monument deteriorated by termite with reference to Dharumavantha Rasegafanu mosque at Male Maldives. International Journal of Current Research.14 (2): 20596-20600, DOI: https: //doi. org/10.24941/ijcr.43063.02.2022

Kryg P, Mazela B, Broda M (2020) Dimensional Stability and Moisture Properties of Gap-Fillers Based on Wood Powder and Glass Microballoons. Stud. Conserv. 65: 142–151.

Larsen K E, Marstein N (2000) Conservation of historic timber structures: Pub by Butterworth-Heinemann, USA 140pp.

Marianne Webb M (2000) Lacquer: Technology and conservation: a comprehensive guide to the technology and conservation of Asian and European lacquer. Oxford: Butterworth-Heinemann. 182 pp.

Phillips M W, Selwyn J E (1978) Epoxies for wood repairs in Historic Buildings. Office of Archaeology and Historic Preservation. Report. 88pp.

Shrivastava M B (1997) Wood Technology. Vikas Publishing house, New Delhi 180pp.

Unger A, Schniewind A P, Unger W (2001) Conservation of wooden artefacts. Springer Science and Business Media, 578pp.

Chapter 16

Need for Holistic Approach in Lichen Mediated Biodeterioration Studies on Indian Monuments and their Conservation

Sanjeeva Nayaka and Dalip Kumar Upreti*

Lichenology Laboratory, CSIR – National Botanical Research Institute, Rana Pratap Marg, Lucknow – 226 001, Uttar Pradesh, India
e-mail: nayaka.sanjeeva@gmail.com

ABSTRACT

India is a culturally rich country and over the period of civilization has accumulated numerous heritages in the form of art and architecture. Therefore, it has become necessary to conserve these monuments by potential damage causing organisms such as lichens. In India biodeterioration studies with respect to lichens have been initiated during 1980 but still it is limited to few monuments of select states. The studies are focused mostly on inventory of lichens growing on the monuments and indirectly attributed their weathering potential to the type of growth form or secondary metabolites produced by them. Several reviews are already available on lichen biodeterioration studies in India and hence in this chapter we try to emphasize on the aspects that are unexplored earlier. The endolithic lichens are least studied groups which are considered as potential deteriorant than epilithic ones. The lichens become more virulent when they grow in association of other microbes or lithobionts, however it remains as an untouched aspect in Indian scenario. Among the techniques in assessing the biodeteriortion advanced microscopy and Raman spectroscopy need to be extensively used while Index of Lichen Potential Biodeteriogenic Activity and Pixel-based image supervised classification may provided specific information needed for conservation of monuments. Owing to the rich cultural heritages in India *and limited studies carried out so far there is a need for holistic approach in lichen mediated biodeterioration incorporating modern tools and techniques. Also, unstudied organism such as endolithic lichens, and lichens associated microbes should be studied.*

Keywords: Endolithic fungi, Lithobiont, LPBA, Raman spectroscopy, Weathering.

16.1 Introduction

The lichens are the symbiotic association of a fungus and an alga or cyanobacteria (sometimes both). In lichens fungus and photosynthetic partners undergo dramatic morphological and physiological changes and behave as single organism. The lichens enjoy a worldwide distribution and occur in every habitat which is well supported with moderate temperature, humidity and shade. The common and natural lichen substrata include rocks, live barks, dead wood, evergreen leaves, both mineral and organic enriched soils. They also colonize on man-made substrates such as rubber, glass, stone works, concrete plaster, ceramic artifacts, clothes, canvas and various metals (Brightman and Seaward 1998).

The 'biodeterioration' is a natural process occurring in the nature ever since life originated. It can be defined as "any undesirable change in the properties of a material caused by the vital activities of organism" (Hueck 1986). In other words, it can also be referred as 'biological weathering' which literally means weakening and subsequent disintegration of rock by plants, animals and microbes (Finaly *et al.*, 2019).Among varieties of organisms causing damage to valuable materials lichens have emerged as serious threat to buildings, monuments and other exposed artifacts. The lichen requires favourable environmental conditions for their colonisation. In addition, lichens also depend on surface texture, moisture content, pH and chemical composition of the substratum. The lichens are hardy organism having ability to grown on hostile substratum such as barren rocks. Through the process of biodeterioration or weathering the lichens contribute to formation of soil and the process is called 'pedogenesis'.The lichens convert the rock into a suitable habitat for growth of other organisms. Therefore, lichens are called as pioneer of xerosere, a type of ecological succession involving rocks. However, it is a slow process and hard to visualize but an important process going-on in nature.

The process of weathering of rock or rocky monuments by lichens is well explained by Seaward (2015). The weathering takes place mostly through physical or chemical methods. In physical weathering the rhizines and hyphae present on the lower side of the lichen thallus penetrate in to the small, invisible crevices in the rock for anchorage. During the growth they exerts pressure and mechanically disintegrate the rock surface into particles. The monuments built out of limestone (calcium carbonate) or sandstones are highly porous compared to the other rocks and hence facilitates deep penetration of rhizines. Secondly, lichen thallus expands wherever moisture is available and contracts during harsh conditions and lose all the water. Such expansion and contraction take place at least twice daily (morning and evening) which create pressure and pulling tension on rock and thus detaches rock particles from surface. The crustose lichens, and tightly attached squamulose and foliose lichens (*Dirinaria, Peltula, Pyxine*) are the most effective in physical weathering. In case of chemical weathering the carbon dioxide produced by lichen during respiration dissolves in the moisture present in the thallus and produce H^+ ions, which in turn participate in the chemical reaction with the minerals of the rock and displaces it. The lichens also produce wide range of secondary metabolites which interacts with the rock minerals. For example, oxalic acid produced by lichens reacts with the rock minerals such as calcium to form white powdery mass called as chelates or calcium oxalates. This powdery mass can be seen settling on the outer surface of lichen thallus. This process is technically called as 'chelation'. The magnesium, copper, ferric or manganese oxalates can also be formed based on the minerals present in the stone. The lichens can also deteriorate minerals such as botite, quartz, feldspars (Adamo and Violante 2000). The lichen producing such secondary metabolites are more harmful to sculptures and monuments made of stone or rock.

The studies have indicated that recent environmental changes including air pollution are conducive for dense growth of lichens, especially toxitolerant species that have severe detrimental effect on monuments (Seaward *et al.*, 1989,Seaward 1997). These lichens cause much damage as the monuments provide excellent niche for growth of such lichens. Some lichens (*Dirina massiliensis* f. *sorediata* (Müll. Arg.) Tehler) are capable of biodeteriorating stone in relatively short time scale that thought earlier (Edwards *et al.*, 1997).

16.2 India as Cultural Capital

India is one of the world's oldest civilizations and a confluence of religions, traditions and customs. Over the period of centuries India has accumulated rich heritage in the form of art, architecture, classical dance, music, flora and fauna. According to Archeological Survey of India our country has more than 3650 ancient monuments, archaeological sites and remains of national importance. In addition, there are 32 World Heritage Properties notified of which 25 are cultural properties (https://asi.nic.in). These monuments belong to different periods, ranging from the prehistoric period to the colonial period and are located in different geographical settings. They include temples, mosques, tombs, churches, cemeteries, forts, palaces, step wells, rock-cut caves, and secular architecture as well as ancient mounds and sites which represent the remains of ancient habitation. According to CEOWORLD magazine 2021 report India is positioned at fourth spot in terms of cultural heritage influence while Italy, Greece and Spain occupy first to third spots respectively (Anonymous 2021).

16.3 Lichen Biodeterioration Study in India

In India lichens in relation to biodeterioration of monuments are being studied since 1980s. Gayathri (1980) was first to describe the effects of lichens on granite statues in India. Thereafter, very few prominent monuments of Assam, Karnataka, Madhya Pradesh, Maharashtra, Odisha, Uttar Pradesh and Uttarakhand were studied. A collection of more than 1000 specimens and systematic account of 112 species growing on these monuments are available (Bajpai and Upreti 2014; Choudhry *et al.*, 2016; Uppadhyay *et al.*, 2016; Deshmukh *et al.*, 2017; Behera *et al.*, 2020).This is just a drop in the ocean when compared to 3650 monuments available in the country. A great portion of studies being carried out in India are limited to inventory of lichens occurring on the monuments or old buildings. In many cases the biodeterioration potential of lichens are indirectly attributed to their growth form and type of secondary metabolites produced by them. For example, Upreti *et al.* (2004) studied the lichens growing over rock shelters of Bhimbetka World Heritage site in Madhya Pradesh and listed 14 specie of lichens. They also documented lichen thallus characters along with chemicals produced by the lichens as a prediction to biodeterioration by lichens. Bajpai *et al.* (2008) provided the account of lichens occurring over major monuments of Madhya Pradesh and studied the metal complexing activity of some common lichens of monuments. The crustose lichens and the species producing secondary metabolites are considered as potential biodeteriorants. However, there is no empirical evidence for such a conclusion. The studies addressing the method for estimating the damage caused by lichens is almost lacking in India and here is the urgent need for quantitative assessment of biodeterioration. In Indian studies the conservation of monuments is advised by removal of the lichens either mechanically or by using biocides (chemical or herbal) (Chandra *et al.*, 2019), however, there are few better methods available. Upreti *et al.* (2009), Bajpai *et al.* (2012), and Bajpai and Upreti (2014) already reviewed the lichen related biodeterioration studies in India and emphasized the need for more such studies. In this chapter instead of repeating the same narration we suggest Indian researchers possible unexplored aspects in lichen mediated biodeteriation studies.

16.4 Some Modern Methods for Estimation of Biodeterioration

Advanced Microscopy

One of the common tools used for assessing the biodeterioration are microscopy techniques. The microscopy is not onlyuseful for observing the morphology of lichen taxa, but also to study the relationship between associated organism and the substratum by applying nondestructive methods. Some of the most common microscopy techniques used in studying lichens include stereo-zoom microscopy, optical microscopy or light microscopy, scanning electron microscopy (SEM) with back-scattered electron imaging (BSE), transmission electron microscopy (TEM), confocal laser scanning microscopy (CLSM), and environmental SEM (ESEM). The stereo-zoom and light microscopes will help in species identification of lichens, whereas electronic microscopes allow surface characterization at a higher resolution and depth of fieldwith 3D effect to the images. The light microscopy alone or in combination with scanning electron microscopy can used for the study (de losRìos and Ascaso 2002).CLSM is extremely useful for revealing the chemical and biological relationship between a lichen and its substratum. CLSM is nondestructive and "real-time" technique which can be used *in situ* to detect the growth, metabolism, penetration of hyphae and its microenvironment surroundings (Di Carlo *et al.*, 2017).Further, the observation of samples under reflected light, polarized microscopy, powderX-ray diffraction (XRD), F-T Raman spectroscopy and ESEM for analysing the lichen-substratum interface are considered very useful techniques. The SEM coupled with EDX has been extensively used in biodeterioration studies. These techniques helpful in determining the elemental chemistry of thebase material in the attacked areas and abiotic deposits, and also the presence, morphology, and distribution of microorganisms (Gholipour-Shahraki and Mohammadi 2017).

Raman Spectroscopy

Edwards *et al.* (1991) are the pioneer to utilize Raman spectroscopy technique to study biodeterioration of monuments by lichens. The Raman microscopic analysis has proved effective in the interpretation and characterization of both the physical and chemical nature of the lichen and substratum system. The detailed chemical analysis of very small samples-substratum encrustations formed lichens can be investigated. The material analysed provides excellent Raman spectra from light scattered by different molecular species, particularly in respect of oxalates(s). By means of this technique it has been possible to investigate microscopically the chemical nature of the gradient through the lichen thallus, the interface and its substratum and incorporation substratum material into the encrustation (Edwards and Seaward 1993). Currently, there are about 75 research articles available on lichen mediated biodeteriation study and the modern Raman spectrometers use the principle of diffraction from gratings.

Index of Lichen Potential Biodeteriogenic Activity (LPBA)

Guzzano *et al.* (2009) developed LPBAwhich quantifies the overall lichen impact on stonework on the basis of the volume of influence of each species, quantified both on the surface of and within the substratum, and of other parameters related to reproduction, physico-chemical action, and bioprotection. LPBA is calculated using formula,

$$\text{LPBA} = \log \sum_{i,j=1}^{n} \left\{ a_{i,j} b_i \left[c_{i,j} (d_{i,j} + e_i) f_{i,j} \right]^{g_{i,j}} \right\}$$

where, for the *i*th species on the *j*th lithotype, a is the cover, b is the reproductive potency, c is the depth of hyphal penetration, d and e are the physical and chemical action, respectively, f volume of rock where penetrating hyphae spread, and g is the bioprotective effect. The parameter a is defined on a continuous scale between 0 and 100, while parameters b–g are defined with reference to ordinal scales fixed on the basis of experimental/literature data. A logarithmic reduction is applied to obtain LPBA values which could be interpreted with user-friendly scales to satisfy management requests. The cover (a) refers to the total or species percentage cover, *i.e.* the overall surface colonized by thalli or the surface colonized by each species, respectively. The reproductive potency (b) is the reproductive strategy of each species or development of reproductive structures of individuals of the different species in the examined site. The species producing vegetative propagules are considered to have a higher reproductive potency than species displaying sexual reproductive structures and value ranges from 1 – 10. The depth of hyphal penetration (c) is measured by observing the polished or thin cross-sections of lithic substrate stained using periodic acid-Schaffer method under light microscopy. The value can range from 1 – 10 corresponding to <50µm and >200 µm of penetration. The physical action (d) is the mechanical actions of lichens on lithic substrate scaled from 1 – 3based on the quantity of disaggregation, from no disaggregation to detachment of rock fragments. The chemical action (e) is scaled 0 – 5 ranging from species not secreting secondary metabolites (0) to species secreting various secondary metabolites and having pitting effect caused by unknown chemical process (3) and secreting oxalis acid (5). The hyphal spread (f) is related to hyphal penetration, however here the volume of rock occupied by the hyphae is measured form 0.1 per cent to >50 per cent. The bioprotection (g) is the protection of lithic surface by lichens acting as protective barrier or shielding. It is based only on the available literature and examination of species collected from study site. Here the value 1 means no reports about bioprotective effect and 0 mean report available.

Pixel-Based Image Supervised Classification

Marques *et al.* (2016) developed the protocol of pixel-based supervised image classification. This method helps to quantify the extent of hyphal penetration, as well as the size of other important weathering-related features (e.g. weathering rind). Here the series of images of Periodic Acid-Schiff stained cross-sections of colonized schist were acquired and submitted to the protocol of pixel-based supervised classification using color and texture features. It includes following steps - Image pre-processing in program *ImageJ*, including the resize of original images to 40 per cent of the initial size using bilinear resampling, contrast and sharpening enhancement; texture feature extraction based on run-length, co-occurrence, image histogram and gradient matrices in program *MaZda*; color features extraction based on several color spaces namely RGB, HIS, YUV, YIQ and XYZ using *'adimpro'* package; generation of training input data in *ImageJ*, by manually assigning XY coordinates of pixels in the original images to the corresponding structures of interest, namely 'lichen thallus', 'hyphae', 'weathering rind' and 'rock core'; combination of the extracted features and training data, to calibrate a Random Forest classifier in *R* with 'ntree'(200), 'mtry' (6), nodesize' (5) and the remaining parameters kept as default;evaluation of classifier's performance through Monte-Carlo cross-validation with 70 per cent for training and 30 per cent for testing and a total of 100 replicates; prediction of the labels of the structures of interest for the whole image, using the RF classifier with highest overall test accuracy. Many studies are available which determined the maximum or average depth of hyphal penetration for a wide range of lithotypes, but usually ignore hyphal spread. Image analysis of colonized cross- sections after Periodic Acid-Schiff staining can retrieve accurate values for the hyphal spread of each species inside the rock. These values could then be used to determine if surface orientation had any effect on species ability to spread into the rock interior and induce mineral breakdown.

Multi-Temporal Mapping of Biodeterioration Pattern

Here the study includes systematic comparison of occurrence of biodeteriorating organism over a period of time. The data on the presence of species and its cover values is organized in a matrix for a time series and then multivariate and cluster analyses are carried out. In addition available ecological information incorporated. The ordination models are developed with ecological information about the species and environmental factors affecting the biodeterioration is deduced. The multi-temporal mapping helps in identifying biodeterioration pattern which can be used as bioindicators for mapping humidity, moisture level and ecological parameters in heritage building or site for better management or conservation. By monitoring development of biodeteriorating organism over the time scale can be useful for long term maintenance of cultural heritage and provide an evidence-based model for preventive interventions (Caneva *et al.*, 2019).

16.5 Prospects in Biodeterioration Study

The lichens are pioneer organism to occupy to a barren substratum, especially rocks. Although lichens represent the most prominent component of lithobiontic communities, biodeterioration processes in a given site not caused only by them. Associated with lichens there are several microorganism such as cyanobacteria, green algae, free-living fungi and heterotrophic bacteria contribute significantly to biodeterioration. In fact there is a succession in the microbial community on exposed rock surface. Initially autotrophs colonize recently exposed rock surfaces and as they die, they are decomposed by saprophytes. Simultaneously, the lichens can colonize the rock surface and start altering the surface. The other micro and macro organisms gradually start colonizing the same substratum after substantially alteration. The areas of attachment of lichens to rocks is important. Here significant geochemical transformations occur and free living cyanobacteria, bacteria, algae and fungi start growing and involve in altering a rock substrate (de los Rios *et al.*, 2002; Gadd 2017).

Most of the studies in India and abroad are mostly restricted to epilithic lichens while endolithic lichens are neglected. The endolithic lichens are usually inconspicuous, appear through only fruiting bodies and hence overlooked by researchers. Euendolithic lichens actively dissolve and penetrate rock and thus integrate the substratum as part of their thalli. Although endolithic lichens grow in a variety of rock types including sand-stones and granites, it is generally suggested that euendolithic growth and dissolution of the substratum is restricted to calcites (Bungartz *et al.*, 2004). The endolithic growth is often interpreted as an adaptation to extreme habitats such as the cold deserts of Antarctica (Kappen *et al.*, 1981). However, endolithic lichens are not restricted to extreme environments as they also occur in temperate regions (Tretiach and Geletti 1997).These endolithic lichens can be potential threat to the monuments (Salvadori and Municchia 2016).Besides the lichens, several other lithobiont microorganisms have been detected inside the stones. The occurrence of these endolithic microorganisms is highly influenced by the presence of the epilithic and endolithic lichen thallus. The lichen screate conducive microenvironment for colonization of microorganism. Similar to the lichens the lithobiontmocrobes interact both physically and chemically with the stone substrate causing biodeterioration to the monuments they are growing (Gholipour-Shahraki and Mohammadi 2017).The microbial activity also often results in the acidification of the surrounding habitat as in case of accumulation and excretion of tricarboxylic acid pathway metabolites is widespread in fungi. These byproducts, particularly oxalic and citric acids, are linked to the dissolution of inorganic substrates necessary for fungal growth. Apart from interaction of substrate minerals and the oxalate minerals that accumulate in lichen thalli, the dissolution of respiratory CO_2 in water held by lichen thalli can lower the pH at the substrate-thallus interface, accelerating the chemical weathering of the rock.

As discussed earlier microscopy techniques such as SEM-EDS or a full range of multi-microscopic techniques (CLSM/SEM) can be used to detect microbial communities on samples from monuments or cultural heritage to confirm the presence of certain organisms, especially to detect dominant organism is causing deterioration or to detect the chemical accumulated (Di Carlo *et al.*, 2017). For example, at Bandelier in United States, the role of the microbial community as a source of oxalic acid is demonstrated by dense accumulations of 1–3 µmoxalate crystals in lichen thalli. These crystals are identified as weddellite ($CaC_2O_4 \cdot 2H_2O$) based on microprobe and SEM-EDS analyses(Porter *et al.*, 2017).

The biodeterioration of rock surface also influenced by the mineralogy of rock surfaces. The minerals of rock can also supports richness of particular species of lichens. In turn the mechanism of biodeterioration is often determined by the type of substrate on which the lichens are growing, the availability of nutrients, mineral composition and water permeability (Cwalina and Dzier¿ewicz 2007).

In case of historic buildings autotrophs such as algae and cyanobacteria are first to colonize as they require small nutrients and their growth is dependent on carbon dioxide, light, and appropriate moisture levels (Crispin and Gaylarde 2005, Herrera *et al.*, 2004). However, algal growth results in the formation of bright green, gray-green or blackish patches and streaks on the buildings giving a shabby look. Further, they degrade construction materials (mostly brick, stone, concrete, mortar, *etc.*) and make the place suitable for colonization of lichens, which are resistant to dryness and extreme temperatures. They further stain the stone substrate and alter the appearance of historic buildings. The lichens in turn intensively deteriorate concrete and stone. The presence of lichens on the external walls of monuments and buildings also depended on the exposure of the wallsto light and moisture.

The other aspects Indian researchers have to attempt are - detecting geophysical and geochemical changes at the lichen-rock interface associated with the growth of 'individual species' including the occurrence of organic and mineral by-products of lichen activity. Further, methods should be developed to quantify the weathering rates induced by individual species or by a limited set of the most representative ones on the surface of interest (Marques *et al.*, 2016). Addressing the influence of human activities on such changes is again an important aspect needing researcher's attention.

16.6 Bioprotective Role of Lichens a Neglected Aspect

Several studies have shown that lichens alone or with associated lithobionts (bryophytes, cyanobacteria, green algae, black yeasts, and/or meristematic fungi) provide bioprotection to the stone or monument surface (Carter and Viles 2005). The bioprotection can take place on both natural and man-made mineral substrates under aggressive environmental conditions (Favero-Longo *et al.*, 2009; De La Rosa *et al.*, 2013) such as high atmospheric pollution or in coastal areas with risk of salt-induced weathering (Carballal *et al.*, 2001).

The type of bioprotection includes erosion protection to relatively porous and unconsolidated rock strata of monuments by lichens along with another microflora. The lichen cover can function more like a biological soil crust, forming a shield to rock surfaces and protecting it from water flow, wind abrasion and temperature variation, resulting in reduced weathering rates. Sometimes the protective effects appear to outweigh the biodeteriorative effects in terms of surface durability.

The generalized bioprotective actions of lichens are still a controversial subject needing additional studies (De La Rosa *et al.*, 2013). Further, evaluations of biodeterioration vs. bioprotection is very essential as patterns and effects of the interaction between lithobionts and lithic substrates may be different for different organisms. Some same species of lithobionts may behave differently depending on the type of substratum and environmental condition (Salvadori and Municchia 2016). Along with biodeterioration studies Indian researchers should also study bioprotection role of lichens and associated microbes.

16.7 Improvised Methods for Conservation of Monuments

The first step to protect cultural heritage must be the detection and identification of deteriorating agents and consequently, the best decision should be selected for the conservation of this marvelous monument. The lichens apart from causing deterioration they also give untidy look to the monuments and old buildings. Therefore, it is advisable to remove these lichens for the maintenance of the monuments. Mechanically the lichens can be carefully scrapped in regular intervals using sharp edged tools. Also, the lichens can be removed using wire brush and washed. The agricultural fungicides and herbicides (Ex. Butachlor 5 per cent, Calixin 2 per cent, Propachlor, Simazine 1 per cent and Zebtane 5 per cent) can be used to void lichens for longer period (Garg *et al.*, 1988; Singh *et al.*, 1999). The stone paints can be applied to create a protective covering over the monuments to prevent fungus and algal growth. Sometimes the herbal formulations made of plant extracts such as neem (*Azadirachta indica* A. Juss.) and chukri (*Rumex hastatus* D. Don.) may also be attempted (Chandra *et al.*, 2019). However, researchers have shown that removal of lichens with either chemical or physical methods is a short term solution and recolonization will rapidly occur again. According to Aptroot and James (2002) the nitrophytic lichens are notorious and colonize the many monuments and elimination of them, without identifying the source of nitrogen compounds will be useless. Another study by Caneva *et al.* (2019) carried out in Italy suggested that biodeterioration processes should be analyzed with an integrated ecological approach, including the multi-temporal monitoring of biodeterioration phenomena, for a long-term maintenance of cultural heritage. Various ecological models can be developed for preventive interventions based on lichen species as bioindicators of environmental conditions, and of their variations in space and time within a given site.

Following methodology can be considered for conservation of monuments (i) the autoecology of deteriogenic/dominant species, to understand factors determining their local distribution, (ii) the specific interactions with the rock-substrate and the associated microbial consortia, to evaluate the overall impact on the stone durability, (iii) the optimization of biocidal and other chemical and/or physical removal approaches, to chose the most appropriate treatments. Such knowledge is still widely missing for Indian condition.

16.8 Conclusion

Indian researchers have tremendous scope for carrying out biodeterioration studies using lichens. India being a climatically and culturally diverse country one can find varieties of lichen species growing on monuments in different parts of the country. The assessment of lichen diversity inhabiting these monuments should be accompanied with quantification of biodeterioration using modern techniques. Identification of lichen species potentially causing damage to the monuments is necessary for the conservation. Also, a look into the deterioration caused by lichen associated organisms on the substratum is important to quantify the contribution of lichens alone or in combination. The endolithic lichens being more virulent than epilithic counter parts a serious study is required in this area. The mineralogy of the substratum and bioprotective nature of lichens require equal importance. The knowledge regarding the lichens and their potential role as deteriorate or protectant is unknown to the layman while the archeologist needs to further be educated. In one of the studies carried out by the authors in Odisha neither the administration nor Archaeological department have taken any measures to control the growth of lichens on the temples of the state. In fact, temple admiration is ignorant regarding the problem caused by lichens (Behera *et al.,* 2020). Therefore, it is the role of scientists to create an awareness regarding the seriousness of lichen mediated bioderioration to both lay man as well as archaeologists.

Acknowledgements

We are thankful to Director, CSIR-NBRI, Lucknow, for providing infrastructure facilities for the research under the project OLP 101 and Dr. Rajesh Bajpai for providing useful literature.

REFERENCES

Adamo P, Violante P (2000) Weathering of rocks and neogenesis of minerals associated with lichen activity. Appl. Clay Sci. 16: 229–256.

Anonymous (2021) Revealed: World's best countries for cultural heritage influence, 2021 https: //ceoworld.biz/2021/01/31/best-countries-for-cultural-heritage-influence-2021 (Accessed on 1 Aug. 2022)

Aptroot A, JamesP W (2002) Monitoring lichens on monuments. In: Monitoring with Lichens - Monitoring Lichens (eds) Nimis P L, Scheidegger C and Wolseley PA. Pub. By Kluwer Academic Publishers, Netherlands. 239–253 pp.

Bajpai R, Upreti DK, Dwivedi SK (2008) Distribution of lichens on some major monuments of Madhya Pradesh. Geophytology 37: 23–29.

Bajpai R, Upreti D K (2014) Lichens on Indian Monuments: Biodeterioration and Biomonitoring. Bishen Singh Mahendra Pal Singh, Dehra Dun. 222 pp.

Bajpai R, Upreti D K, Nayaka S, Dwivedi S K (2012) Lichen biodeterioration studies in India: An overview. In: Bioremediation of Pollutants (eds) Dubey R C and Maheshwari D K. Pub. By I.K. International Publishing House, New Delhi. 63–73 pp.

Behera PK, Nayaka S, Upreti DK (2020) Lichens on monuments of Odisha – Are they causing biodeterioration? NeBIO 11(2): 71–78.

Brightman FH, Seaward MRD (1998) Contribution to the lichen flora of southeast Ireland – II.Proc. R. Ir. Acad.B 77: 119–134.

Bungartz F, Garvie LAJ, Nash III, T H (2004) Anatomy of the endolithic sonoram desert lichen *Verrucaria rubrocincta* Breuss: Implications for biodeterioration and biomineralization. The Lichenologist, 36(1): 55-73.

Caneva G, Bartoli F,FontaniM,Mazzeschi D, Visca P (2019) Changes in biodeterioration patterns of mural paintings: Multi-temporal mapping for a preventive conservation strategy in the Crypt of the Original Sin (Matera, Italy). J. Cult.Herit. 40: 59–68. https://doi.org/10.1016/j.culher.2019.05.011

CarballalR, Paz-Bermúdez G, Sánchez-Biezma MJ, Prieto B (2001) Lichen colonization of coastal churches in Galicia: biodeterioration implications. Int.Biodeter.Biodeg. 47(3): 157–163.

Carter NEA,Viles HA (2005)Bioprotection explored: the story of a little known earth surface process. Geomorphology 67: 273–281.

Chandra K, Joshi Y, Upadhyay S, Bisht K (2019) Can *Rumex hastatus* D. Don. be used as a biological agent for removing lichens colonizing monuments? A case study from Kumaun Himalaya. Natl. Acad. Sci. Lett. https://doi.org/10.1007/s40009-018-0757-4

Choudhry MP, Sarma M, Nayaka S, Upreti DK (2016) Distribution of lichens on few ancient monuments of Sonitpur district, Assam, North East India. Int. J.Biod. Con. 8(11): 291–296.

Crispim CA, Gaylarde CC (2005) Cyanobacteria and biodeterioration of cultural hertigate a review. Microbial Ecol. 49: 1–9.

Cwalina B, Dzier¿ewicz Z (2007) Factors contributing to biological corrosion of reinforced concrete structures. Przegl Bud 7: 52–59.

De La Rosa, J P M,Warke P A, Smith B J(2013) Lichen-induced biomodification of calcareous surfaces: Bioprotection versus biodeterioration. Prog. Phys. Geogr. 37(3): 325–351. http://dx.doi.org/10.1177/0309133312467660

de los Ríos A, Ascaso C (2002) Preparative techniques for transmission electron microscopy and confocal laser scanning microscopy.In: Protocols in Lichenology. Culturing, Biochemistry, Ecophysiology and Use in Biomonitoring (eds.) Kranner I, Beckett R P, Varma, AK. Pub. By Springer-Verlag, Berlin, Heidelberg. 87–117 pp.

Deshmukh VP, Bajpai R, Upreti DK, Wagh VV, Rajurkar AV,Bondarkar SG (2017) Lichen diversity of Gawilgarh fort, Amravati district, Maharashtra, India. Cryptogam Biodiversity and Assessment 2(2): https://doi.org/10.21756/cab.v2i02.10818.

Di Carlo E, Barresi G,Palla F (2017) Biodeterioration. In: Biotechnology and Conservation of Cultural Heritage (eds) Palla F and Barresi G. Pub. By Springer International Publishing. 1–30 pp. https://doi.org/10.1007/978-3-319-46168-7_1

Edwards H G M, Farwell D W, Seaward MR D, Giacobini C (1991) Preliminary Raman microscopic analyses of a lichen encrustation involved in the biodeterioration of renaissance frescoes in Central Italy. International Biodeterioration27(1): 1–9.https://doi.org/10.1016/0265-3036(91)90019-n

Edwards HGM, Farwell DW, Seaward MRD (1997) Raman spectroscopy of *Dirina massiliensis* f. *sorediata* encrustations growing on diverse substrata. *Lichenologist* 29(1): 83–90.

Edwards HGM, Seaward MRD (1993) Raman microscopy of lichen-substratum interfaces. J.Hatt. Bot. Lab. 74: 303–316.

Favero-Longo S E, Castelli D, Fubini B,Piervittori R(2009) Lichens on asbestos-cement roofs: bioweathering and biocovering effects. J. Hazard. Mat. 162(2-3): 1300–1308. https://doi.org/10.1016/j.jhazmat.2008.06.060

Finlay R D, Mahmood S, Rosenstock N, Bolou-Bi E B, Köhler S J, Fahad Z, Rosling A, Wallander H, Belyazid S, Bishop K, Lian B (2019) Biological weathering and its consequences at different spatial levels – from nanoscale to global scale. Biogeosciences Discuss. https://doi.org/10.5194/bg-2019-41

Gadd G M (2017) Fungi, rocks, and minerals. Elements 13(2): 171-176. https://doi.org/10.2113/gselements.13.3.171

Garg KL, Mishra AK, Singh A, Jain K K (1988). Biodeterioration of cultural heritage some case studies. In: Conservation, Preservation and Restoration Traditions, Trends and Techniques (eds) Kamalkar V, Pandit Rao V and Veerender M. Pub. By Birla Archaeological and Cultural Research Institute, Hyderabad. 31–38 pp.

Gayathri P (1980) Effects of lichens on granite statues. Res. Bull. 2: 41–52.

Gazzano C, Favero-Longo S E, Matteucci E, Roccardi A,Piervittori R(2009) Index of Lichen Potential Biodeteriogenic Activity (LPBA): a tentative tool to evaluate the lichen impact on stonework.Int.Biodeter.Biodeg. 63(7): 836–843. https://doi.org/10.1016/j.ibiod.2009.05.006

Gholipour-Shahraki M, Mohammadi P (2017) The study of growth of *Caloplaca* sp. PLM8 on Cyrus the Great's Tomb, UNESCO World Heritage Site in Iran. Int. J. Environ. Res. 11(4): 501–513. https://doi.org/10.1007/s41742-017-0044-0

Herrera LK, Arroyave C, Guiamet P, de Saravia SG, Videla H (2004) Biodeterioration of peridotite and other constructional materials in a building of the Colombian cultural heritage. Int.Biodeterior.Biodeg. 54: 135–141.

Hueck HJ (1986) The biodeterioration of materials – an appraisal. In: Biodeterioration of Materials (eds) Walters AH and Elphick JS. Pub. By Elsevier, London. 6–12 pp.

Marques J, Gonçalves J, Oliveira C,Favero-Longo SE, Paz-Bermúdez G, Almeida R, Prieto B (2016) On the dual nature of lichen-induced rock surface weathering in contrasting micro-environments. Ecology 97(10): 2844–2857. https://doi.org/10.1002/ecy.1525/full

Porter D, Broxton D, Bass A,Neher DA,Weicht TR, Longmire P,Spilde M, Domingue R. (2017) The role of case hardening in the preservation of the Cavates and Petroglyphs of Bandelier. MRS Advances 2(37-38): 1969–2005. https://doi.org/10.1557/adv.2017.277

Salvadori O, Municchia A C (2016) The role of fungi and lichens in the biodeterioration of stone monuments. The Open Conference Journal 7, (suppl.1: M4) 39–54.

Seaward M R D (1997) Urban deserts bloom: a lichen renaissance. Bibl.Lichenol. 67: 297–309.

Seaward M R D (2015) Lichens as agents of biodeterioration. In: Recent Advances in Lichenology (eds) Upreti D K, Divakar P K, Shukla V and Bajpai R (eds.). Pub. By Springer, India. 189–211 pp. https://doi.org/10.1007/978-81-322-2181-4_9

Seward MRD, Giacobini C, Giuliani MR,Roccardi A (1989) The role of lichens in the biodeterioration of ancient monuments with particular reference to Central Italy. Int.Biodeter. 25: 19–55.

Singh A, Chatterjee S, Sinha GP (1999) Lichens of Indian monuments. In: Biology of lichens (eds)Mukerjee KG,Chamola BP, Upreti DK and Upadhyaya RK. Pub. ByAravali, Book International, New Delhi. 115–151 pp.

Tretiach M and Geletti, A (1997) A CO_2 exchange of the endolithic lichen *Verrucaria baldensis* from Karst habitats in northern Italy. Oecologia 111: 515-522.

Uppadhyay V, Ingle KK, Trivedi S, Upreti DK (2016) Diversity and distribution of lichens from the monuments of Gwalior division, Madhya Pradesh with special reference to rock porosity and lichen growth. Tropical Plant Research 3(2): 384–389.

Upreti DK, Bajpai R, Nayaka S (2009) Indian monuments need lichen biodeterioration study. New Horizons7: 64–69.

Upreti DK, Nayaka S, Joshi Y (2004) Lichen activity over rock shelters of Bhimbetka world heritage zone., Madhya Pradesh. Rock Art Research: Changing Paradigms. The 10th Congress of The International Federation of Rock Art Organization (IFRAO) and the Exhibition of Rock Art (28th Nov. 2nd Dec. 2004). Rock Art Society of India (RASI).

Chapter 17

Biodeterioration and Certain Conservation Aspects of Champaner Pavagadh Monuments: A UNESCO World Heritage Site in Gujarat

Disha Mehta

Department of Botany, Faculty of Science, The Maharaja Sayajirao University of Baroda, Vadodara – 390 002, Gujarat, India
e-mail: mehtadisha661@gmail.com

ABSTRACT

Champaner Pavagadh Archaeological complex situated in Panchmahal district of Gujarat is designated as UNESCO World Heritage site since 2004. It comprises of various monuments at Champaner and adjoining Pavagadh hills. Among all these monuments, few monuments are ruined due to the attack of different biological organisms on their facades. Current study addresses the diversity of these biological organisms and damage caused to the structure of selected monuments of Champaner Pavagadh. A total of thirty-one species of biological organisms, which included cyanobacteria, algae, bryophytes and lichens were observed on the different monuments. Monuments like Navlakha Kothar, Makai Kothar, Jain Temple, Saher ki Masjid, Amir Manzil and Mandvi were studied for the presence of biological organisms. Occurrence of 17 species of bryophytes, 8 species of lichens, one alga and five species of cyanobacteria were recorded from different monuments of Champaner Pavagadh complex. Dominant group of organisms like bryophytes and lichens had invaded on the Navlakha Kothar followed by antiquity of Jain Temple and Makai Kothar. Biodeterioration to these monuments might have caused because of intrinsic and external factors. This is due to the elevation, high humidity during rains, anthropogenic activities and disturbances caused by tourists. Owing to which historical monuments are losing their aesthetic value and archeological importance. To preserve these UNESCO World Heritage site buildings, use of different preventive and certain chemical treatment measures are suggested.

***Keywords**: Biodeterioration, Champaner monuments, World heritage, Gujarat, Lichens, Bryophytes.*

17.1 Introduction

Biodeterioration is defined as "the undesirable changes in the properties or qualities of a material or a structure by the vital activities of organisms" (Allsopp *et al.*, 2004). The discoloration of the buildings and the damage to their structure is one of the most common manifestations of biodeterioration. It occurs on humid and illuminated substrates such as old buildings, structures and historical monument walls. This biodeterioration process is mostly initiated by pioneering microorganism cyanobacteria and very few green algae (Gil and Saiz-Jimenez 1992; Adhikary and Kovacik 2010; Ortega-Morales *et al.*, 2013; Keshari and Adhikary 2014). These microorganisms are attached to the surfaces, to each other and embedded in the self-produced matrix resulting in the formation of biofilms. These biofilms extensively cover the exposed surface that caused the disfiguration and discoloration of the structures (Ortega-Morales *et al.*, 2013; Keshari and Adhikary 2014). On these biofilms, propagules of other biological organisms adhere like bryophytes, lichens and allied vascular plants leading to further colonization (Herrera *et al.*, 2004). These biological organisms colonize and deteriorate the surface of the monuments which are made up of different materials, can have different structures like being porous or semi-porous and also varying surface features like being smooth or rough surface (Gil and Saiz-Jimenez 1992; Adhikary and Kovacik 2010; Ortega-Morales *et al.*, 2013; Verma *et al.*, 2014; Joshi *et al.*, 2015; Nayak *et al.*, 2017).

India has a rich diversity of monuments and several ancient monuments are an inspiration for the present and future generations. It is important that such sites are properly preserved and maintained so that citizen can make full use of their social and cultural values. Among the many ancient monuments in India, several such important monuments are protected by the 'Ancient Monuments and Archaeological Sites and Remains Act 1958 (Anon 2010). Monuments are generally endangered due to adverse environmental conditions like typically hot and humid surroundings that are common in a tropical country like India and this facilitates biodeterioration. Among all the protected monuments, based on their structure, arts and cultural past some of these monuments have been designated as World Heritage Sites by UNESCO for appreciation of their importance globally. Current study focus on the renowned sites of Champaner Pavagadh for their architectural beauty and history for tourism and pilgrimage.

17.2 Study Site: Champaner Pavagadh Monuments

Champaner is a historical city in the state of Gujarat and is located in

Panchmahal district. Thakur (1987) first proposed a management plan for preservation of monuments at Champaner – Pavagadh complex. A new study was jointly prepared by researchers from the University of Illinois and the Heritage Trust of Baroda with the objective of proposing it as a World Heritage Site to UNESCO (Ruggles and Sinha 2009). In July, 2004 Champaner-Pavagadh was officially designated a World Heritage Site and was named as the Champaner-Pavagadh Archaeological Park (Ruggles and Sinha 2009). This world heritage site comprises of various monuments at Champaner and the adjoining Pavagadh hills (Sinha *et al.*, 2004; Modi 2008). It has a split identity between Pavagadh hill as the abode of a Hindu goddess and Champaner with remains of a historical Islamic city (Sinha *et al.*, 2004; Ruggles and Sinha 2009). The main Pavagadh hill is the highest point in the district, rising to a height of about 800 m surrounded by several small hillocks ranging from 200 m – 300 m in height. The temple of Kalika Mata has immense value that it adds to the region and is visited by over 2 million pilgrims annually (Ravdandekar 2014).

Few of these historical monuments are vanishing on account of natural and man-made causes of which the major one is the traffic of pilgrims. Some of the monuments, mostly temples at Pavagadh are regularly repainted due to their religious importance while some are subjected to cleaning and preservation measures by the ASI workers (Ruggles and Sinha 2009). However, there are still several monuments of Champaner Pavagadh which are at different stages of degradation. The degradation seems to be due to the growth of biological organisms as well as anthropogenic activity due to the heavy rush of tourists. Hence, there is an urgent need to preserve such monuments which have historical as well as cultural importance. Therefore, six monuments were selected for the current study from the Champaner Pavagadh site. Three monuments namely Makai Kothar, Navlakha Kothar and Jain temple are located on the Pavagadh hills. Two of the other three monuments, Saher ki Masjid and Mandavi are located next to each other within the Champaner fort, while Amir Manzil (not ASI protected but under its supervision) is located outside the fort ramparts. All these selected monuments broadly built by different building materials or substrate such as different mortar, plasters, bricks and rocks. The locations of the six monuments have been given in the map Figure 17.1.

17.3 Weather conditions of Champaner Area

Champaner Pavagadh site has broadly huge difference in temperature ranges between 9° to 44° C. During winters, which are cool, the days having an average temperature of 26 °C while the nights are as cold as 9 °C. In summer season, the temperatures are severe, reaching up to 44 °C. This duration of summer is scorching. The average annual rainfall in the region is 944 mm. The major rain months are July and August. Five years (2016 to 2020) cumulative data of average monthly rainfall and temperature for the Panchmahal district is shown in Figure 17.2.

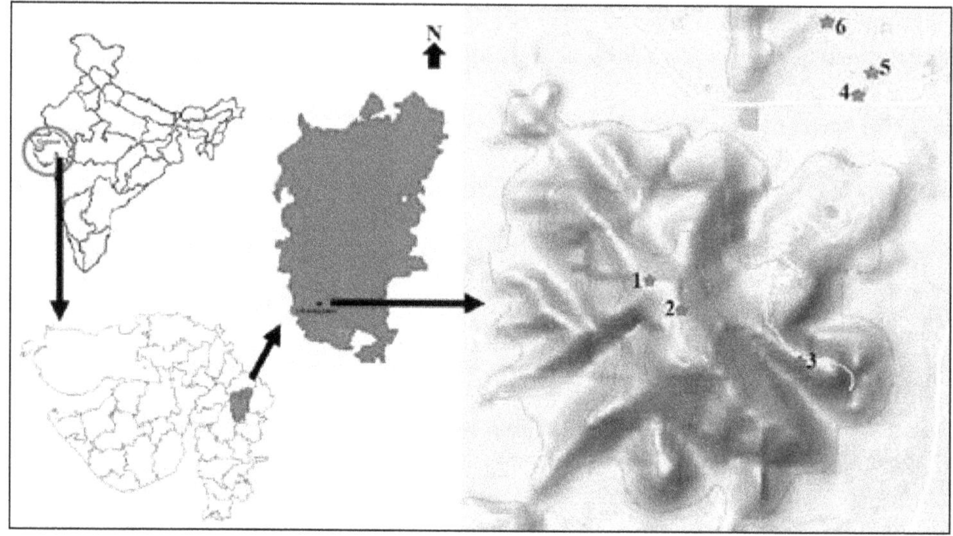

Figure 17.1: Map of Champaner-Pavagadh Complex Showing Location of the Six Selected Sites. 1. Navlakha Kothar, 2. Jain Temple, 3. Makai Kothar, 4. Saher ki Masjid, 5. Mandavi, 6. Amir Manzil.

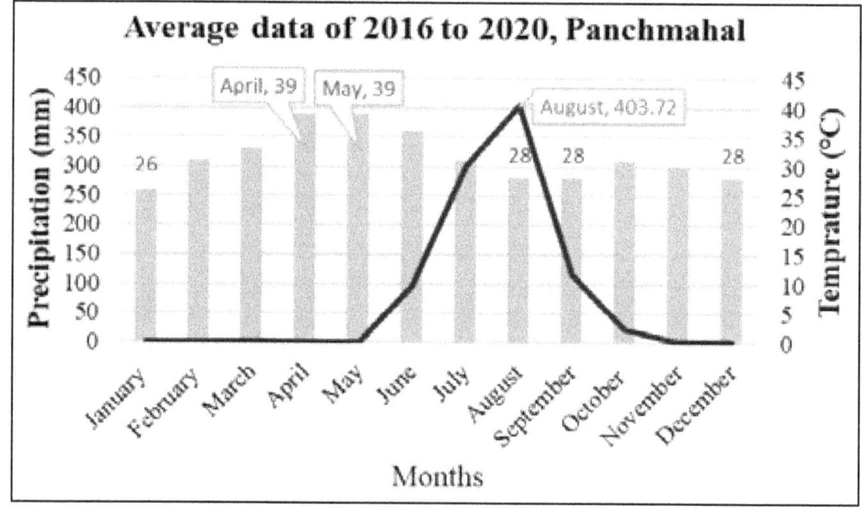

Figure 17.2: Average Monthly Temperature and Rainfall Data for 2016-2020 for Panchmahal District (*Source*: IMD 2021).

17.4 Methodology

Selected sites were visited after taking permission from the ASI, Vadodara circle. Biological samples such bryophytes, lichens and biofilms causing deterioration of the monuments were collected by non-destructive method (La

Cono and Urzi 2003; Glime 2013; Bajpai and Upreti 2014). Morphology and micro-morphology of all the collected samples were studied and identified by standard floras or monographs (Desikachary 1959; Gangulee 1969-72; 1974-78; Chaudhary and Deora 1993; Chaudhary *et al.*, 2006; Awasthi 1991; 2007; Aziz and Vohra 2008; and Bajpai and Upreti 2014; Mehta and Shah, 2021). Limitation in collection of lichen samples was encountered since sites are protected. Hence, in situ chemical spot test were also conducted for identification of lichen species. Figure 17.3 shows the sampling of biological organisms and chemical spot test for lichen.

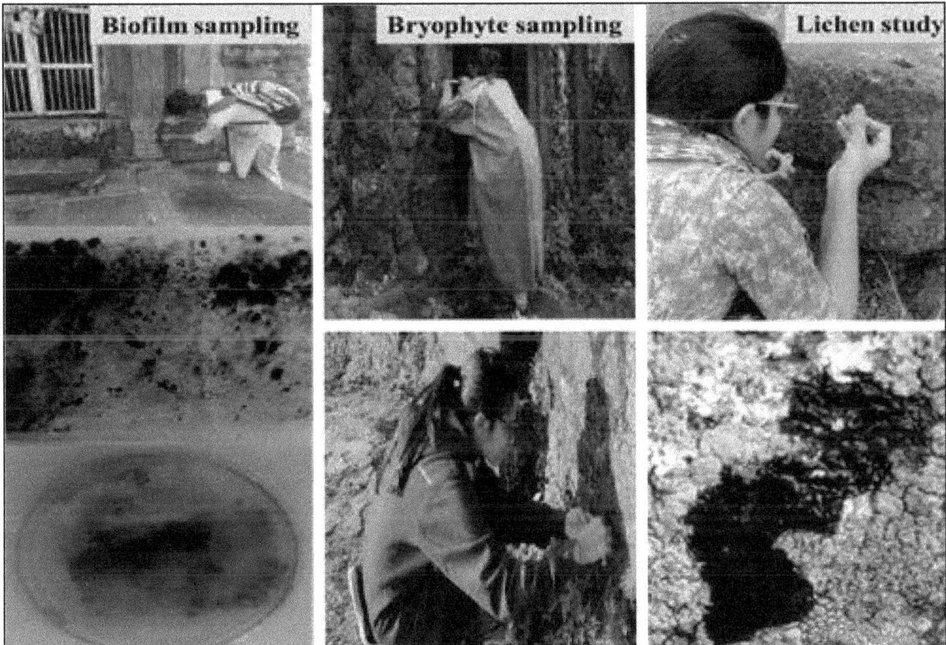

Figuro 17.3: Sampling of Biological Organisms and Chemical Spot Test of Lichen.

17.5 Diversity of Biological Organisms

A total of 31 species of biological organisms from which 17 species of bryophytes, 8 species of lichens, 5 species of cyanobacteria and 1 species of micro green alga were found from the selected sites causing deterioration of the monuments.

17.5.1 Bryophytes

Total 17 bryophytes species, included nine species of liverworts, six species of mosses and two species of hornworts. Among these species, maximum species of liverworts were from the order Marchantiales (*Riccia, Plagiochasma, Asterella* and *Cyathodium*), mosses from the order Pottiales (*Semibarbulla, Hyophila,*

Hydrogonium and *Gymnostomiella*) and hornworts from the order Anthocerotales (*Anthoceros*). Table 17.1 shows distribution of bryophyte species on the specific monuments.

Table 17.1: Diversity and Distribution of Bryophytes on Selected Monuments

Sl.No.	Name of the Bryophytes Species	Selected Monument Sites					
		Champaner Pavagadh					
		1	2	3	4	5	6
	Mosses						
1.	*Anomobryum auratum* (Mitt.) Jaeg.					+	
2.	*Fissidens splachnobryoides* Broth. in Schum. et Lauterb.			+		+	
3.	*Hydrogonium arcuatum* (Griff.) Wijk. et. Marg.	+					
4.	*Hyophila involuta* (Hook.) Jaeg.	+	+	+	+	+	
5.	*Semibarbula orientalis* (Web.) Wijk et Marg.					+	
6.	*Gymnostomiella vernicosa* (Hook.) Fleisch.					+	
	Liverworts						
7.	*Asterella angusta* (Steph.) Kachroo.				+	+	
8.	*Cyathodium cavernarum* Kunze.					+	
9.	*Lejeunea aloba* Sande Lac.					+	
10.	*Riccia gangetica* Ahmad.						+
11.	*Riccia discolor* L.rt L.		+				
12.	*Riccia grollei* Udar.			+			
13.	*Riccia billardieri* Mont. et Nees.		+				+
14.	*Plagiochasma microcephalum* (Steph.) Steph.					+	
15.	*Plagiochasma appendiculatum* L. et L.				+	+	
	Hornworts						
16.	*Anthoceros bharadwajii* Udar et. Asthana						+
17.	*Anthoceros subtilis* Steph.						+

1. Saher ki Masjid, 2. Mandavi, 3. Amir Manzil, 4. Makai Kothar, 5. Navlakha Kothar, 6. Jain Temple.

Note: + indicates presence.

17.5.2 Lichens

Total 8 species of lichens were noticed causing deterioration of the monument structure. They included four crustose lichens (*Caloplaca awasthii, C. cuplifera, Pertusaria multipunta* and *Diploschistes* sp.), two crustose to leprose lichens (*Lepraria coriensis, L. lobificans*), one squamulose lichen (*Endocarpon nanum*) and one foliose lichen (*Phaeophysia hispidula*). All lichen samples were deposited in the National Botanical Research Institute (NBRI), Lucknow herbarium repository and their accession numbers were obtained. Lichens were not found on the monuments of the Champaner at the foothill of Pavagadh. Distribution of lichen species on specific monuments are mentioned in Table 17.2.

Table 17.2: Diversity and Distribution of Lichen on Selected Monuments

Sl.No.	Name of the Lichen species	Selected Monument Sites		
		Champaner Pavagadh		
		1	*2*	*3*
1.	*Caloplaca awasthii* Y. Joshi and Upreti		+	
2.	*Caloplaca cupulifera* (Vain.) Zahlbr.	+		
3.	*Pertusaria multipuncta* (Turner) Nyl.		+	+
4.	*Diploschistes* Norman		+	
5.	*Lepraria coriensis* (Hue) Sipman			+
6.	*Lepraria lobificans* Nyl.		+	
7.	*Endocarpon nanum* Ajay Singh and Upreti		+	
8.	*Phaeophyscia hispidula* (Ach.) Essl.		+	+

1. Makai Kothar, 2. Navlakha Kothar, 3. Antiquety from surrounding the Jain Temple

Note: + indicates presence.

17.5.3 Cyanobacteria and Micro-green Algae Biofilms

Biofilms samples with six species were found from the selected monuments of Champaner-Pavagadh included two species (single genus) from the order

Table 17.3: Diversity and Distribution of Biofilm Species on Selected Monuments

Sl.No.	Name of the Species Forming Biofilms	Selected Monument Sites					
		Champaner Pavagadh					
		1	*2*	*3*	*4*	*5*	*6*
	Cyanobacteria						
1.	*Chroococcidiopsis cubana* Komarek and Hindak	+	+				
2.	*Leptolyngbya foveolarum* (Montagne ex Gomont) Anagnostidis et Komarek				+	+	
3.	*Leptolyngbya crispata* (Playfair) Anagnostidis and Komarek			+			
4.	*Nostoc punctiforme* (Kutz.) Hariot				+	+	
5.	*Desmonostoc muscorum* Agardh ex Bornet and Flahault			+			
	Microalga						
1.	*Asterarcys quadricellulare* (K. Behre) E. Hegewald and A.W.F. Schmidt						+

1. Saher ki Masjid, 2. Mandavi, 3. Amir Manzil, 4. Makai Kothar, 5. Navlakha Kothar, 6. Jain Temple

Note: + indicates presence.

Synechococcales (*Leptolyngbya*) as well as Nostocales (*Nostoc*) and single species from the order Chroococcidiopsidales (*Chroococcidiopsis*) and Chlorococales (*Asterarcys*). The members of Synechococcales and Nostocales were dominated on the monuments of Champaner-Pavagadh.

17.6. Dominant Flora occurring on Pawagad Monuments

Among all the studied plant groups on selected sites, bryophyte group was dominant on the monuments causing the deterioration the structure. Especially in monsoon seasons, Navlkha Kothar looks like heaven for bryophytes. Followed by Bryophyte group, Lichen and Cyanobacterial biofilms covered the large area on the monument walls and causing degradation and deterioration of the structures. As per species wise in each group, *Asterella angusta*, *Plagiochasma appendiculatum*, *Cyathodium cavernarum* and *Hyophilla involuta* from bryophyte in Figure 17.4a-d respectively, in lichen species *Caloplaca awasthii*, *Caloplaca cupulifera*, *Phaeophyscia hispidula* and *Pertusaria multipuncta* in Figure 17.4e-h respectively while in cyanobacterial biofilms species *Nostoc punctiforme*, *Leptolyngbya foveolarum* and *Chroococcidiopsis cubana* (Figure 17.4i-k respectively) were dominant on the selected monuments of Champaner Pavagadh. An alga species *Asterarcys quadricellulare* (Figure 17.4l) biofilms were observed from the single site namely Jain temple, where whole site having discolour with this alga.

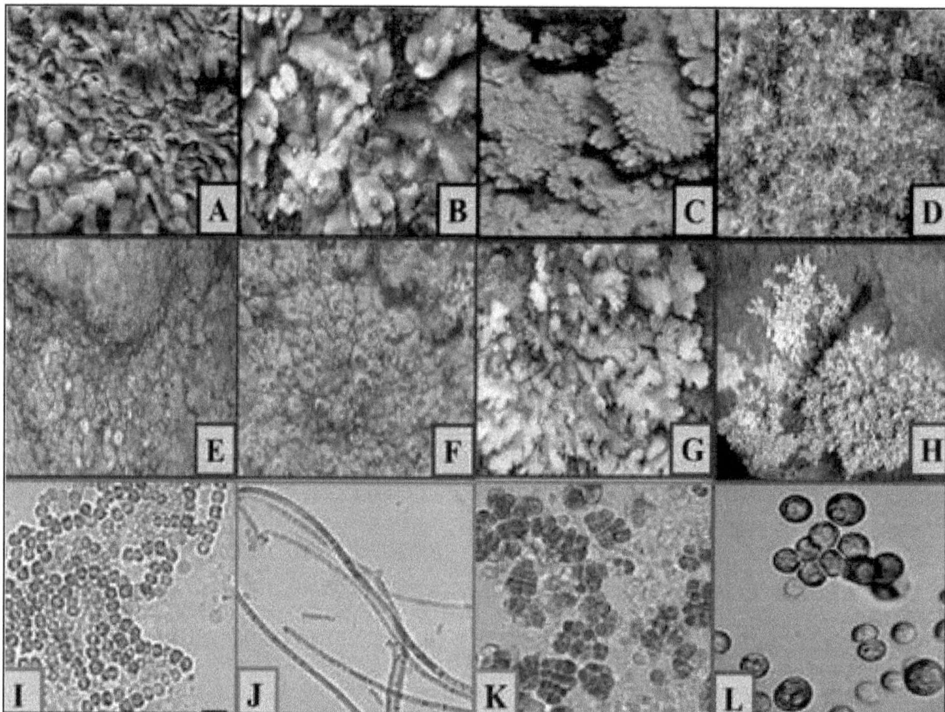

Figure 17.4a-l: Glimpses of Dominant Biological Organisms Causing Deterioration of the Monuments of Champaner Pavagadh (Scale bar for I to L = 10 μm).

17.7. Community Succession and Mechanism of Biodeterioration

Microorganisms like cyanobacteria, micro green algae and lichens are pioneering organisms (Adhikary and Kovacik 2010; Bajpai and Upreti 2014) which colonize such monuments. The environmental conditions in the vicinity and the surface of the material that constitute the monument plays an important role in encouraging the growth of microorganisms on it. Over a period of time these sites were taken over by bryophytes and some allied vascular plants forming intermediate stage of community and ultimately leading to an advanced stage of seral community (Figure 17.5). Plants of advance stage of seral community have root that penetrate and grow in between fissures which causes the structure to crack resulting in the deterioration of the structure (Crispim and Gaylarde 2005). Community succession plays leading role for degradation and deterioration of the monuments. The Angkor Wat temple complex in Cambodia is a classic example of community succession on monuments (Bartoli *et al.*, 2014).

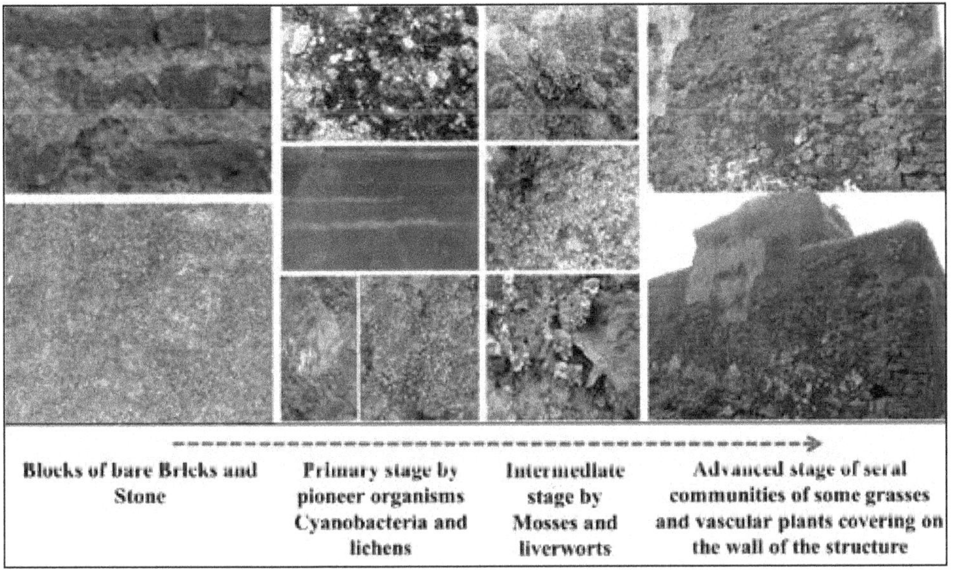

Figure 17.5: Ecological Succession on the Monument wall Causing Deterioration.

17.8. Control Measures

Traditionally, control measure is meant for doing something to the degraded structure like patching it up with mortar or cutting out decayed stone and replacing it with new stone. But as per historical and architectural importance ideally, method employed as control measure should not replace old structure material with new materials owing to which it ruins the aesthetic beauty and lose the architectural importance. To overcome all these problems, proper method and regular maintenance is vitally important for long term preservation. This reflects the growing awareness of the importance of preventive conservation with the principle of minimum intervention (Goncalves *et al.*, 2009). Hence, need to limit the use of materials that might prove harmful to either the stone or to the environment (Albero *et al.*, 2004).

17.8.1 Preventive

Preventive conservation comprehends very wide range of topics such as legislation to protect individual monuments, visitor management, traffic control and disaster planning as well as modelling of interior environments.

Preventive conservation measures are more effective when concern with keeping water out of the structure material and with controlling the relative humidity and temperature of the air around the structure. The main purpose of relative humidity control is to reduce damage from salt and moisture cycles.

17.8.2 Curative

In curative measure, cleaning is the first step after survey has been completed. Cleaning may serve some circumstances to remove harmful materials from the surface. A dirty monument does not look well cared and the dirt may well obscure both fine detail and major architectural features. Ideally cleaning should be done by physical method. In this method, gently rub the surface and remove the dirt and some other unwanted materials. Followed by physical method, apply biocide suitable for the monument materials (Mehta 2022).

17.8.3 Chemical Treatment

Chemical treatment is an active conservation method for the structure. Ideally chemical treatment needs to be reasonably cheap, easy to apply and safe to handle as well as having VOC (Volatile organic compound) regulations. For this treatment, water repellent chemical should be used, due to this structure surface resist the water. Chemicals are usually applied to the surface of the monuments by brush or spray and are drawn into the structure by capillary action (Mehta 2022).

Acknowledgements

Author is grateful to Archaeological Survey of India (ASI), Vadodara circle to cooperate and approve necessary permissions (36/13/MIS/T and A-2016-17-3771 and 36/13/MIS/T and A-2018-19-2701) for the study of the selected sites. The author expresses her gratitude to Dr. Dharmendra Shah her Ph. D. guide, for all support and encouragement during this research and also thanks to the Head, Department of Botany, The M. S. University of Baroda for providing the facilities to carry out this research under DRS programme.

REFERENCES

Adhikary S P, Kovacik L (2010) Comparative analysis of cyanobacteria and micro-algae in the biofilms on the exterior of stone monuments in Bratislava, Slovakia and in Bhubaneswar, India. The Journal of Indian Botanical Society, 89(1 and 2): 19-23.

Albero S, Giavarini C, Santarelli M L, Vodret A (2004) CFD modeling for the conservation of the Gilded Vault Hall in the Domus Aurea. Journal of Cultural Heritage 5 (2): 197–203.

Allsopp D, Seal K J, Gaylarde C C (2004) Introduction to biodeterioration. Cambridge University Press. p.237.

Anon (2010). Ancient monuments and archaeological sites remains act 1958.

Awasthi D D (2007) A Compendium of the Macrolichens from India, Nepal and Shri Lanka. Bishen Singh Mahendra Pal Singh, Dehradun. India. p. 580

Awasthi D D (1991) A Key to the Microlichens of India, Nepal and Shri Lanka. J. Cramer, Berlin. Stuttgart. p.337

Aziz M N, Vohra J N (2008) Pottiaceae (Musci) of India. Bishen Singh Mahendra Pal Singh.p.366

Bajpai R, Upreti D K (2014) Lichens on Indian Monuments Biodeterioration and Biomonitoring. Bishen Singh Mahendra Pal Singh, Dehradun. India. p. 222.

Bartoli F, Municchia A C, Futagami Y, Kashiwadani H, Moon K H, Caneva G (2014) Biological colonization patterns on the ruins of Angkor temples (Cambodia) in the biodeterioration vs bio protection debate. International Biodeterioration and Biodegradation, 96: 157-165.

Chaudhary B L, Deora G S (1993) Moss flora of Rajasthan. Himanshu publications, Udaipur, India. p.127.

Chaudhary B L, Sharma T P, Sanadhya C (2006) Bryophyte Flora of Gujarat. Himanshu Publications Udaipur and New Delhi, p.198.

Crispim C A, Gaylarde C C (2005) Cyanobacteria and biodeterioration of cultural heritage: a review. Microbial Ecology 49: 1–9.

Desikachary TV (1959) Cyanophyta. Indian Council of Agricultural Research, New Delhi p.686

Gangulee H C (1969-72) Mosses of Eastern India and Adjacent Regions. Vol. I. Books and Allied (P) Ltd. Kolkata, India. p.830.

Gangulee H C (1974-78) Mosses of Eastern India and Adjacent Regions. Vol. II. Books and Allied (P) Ltd., Kolkata, India. p.1546.

Gil J A, Sáiz-Jiménez C (1992) Biodeterioration of Roman mosaics by Bryophytes.

Glime J M (2013) Bryophyte ecology. Vol. 1. Physiological ecology. E-book sponsored by Michigan Technological University and the International

Association of Bryologists. Available at http: //digitalcommons.mtu.edu/bryophyte-ecology.

Goncalves T D, Pel L, Delgado Rodrigues., J (2009) Influence of paints on drying and salt distribution processes in porous building materials. Construction and Building Materials 23 (5): 1751–59.

Herrera L K, Arroyave C, Guiamet P, de Saravia S G, Videla H (2004) Biodeterioration of peridotite and other constructional materials in a building of the Colombian cultural heritage, International biodeterioration and biodegaradation. 54(2-3): 135-141.

IMD (2021) Customized Rainfall Information System, Hydromet Division. India Meteorological Department.http: //hydro.imd.gov.in/hydrometweb/(S(ah3z5k55icx21j45v0pygfuo)/DistrictRaifall.aspx

Joshi Y, Nayal S, Tripathi M, Bisht K, Upreti D K (2015) Distribution and diversity of lichenized fungi colonizing Jageshwar group of temples, Almora, Uttarakhand. Proceedings of the National Academy of Sciences, India Section B: Biological Sciences. 85(2): 545-554.

Keshari N, Adhikary S P (2014) Diversity of cyanobacteria on stone monuments and building facades of India and their phylogenetic analysis. International Biodeterioration and Biodegradation, 90: 45–51.

La Cono V, Urzýì C (2003) Fluorescent in situ hybridization applied on samples taken with adhesive tape strips. Journal of Microbiological methods 55: 65–71.

Mehta D (2022) Biodeterioration of selected historical monuments by lower plants and cyanobacteria. Ph. D. Thesis, The Maharaja Sayajirao University of Baroda, Vadodara. Repository at Shodhganga: a reservoir of Indian theses.

Mehta D, Shah D (2021) Cyanobacteria and microalgae growing on monuments of UNESCO World Heritage site Champaner Pavagadh, India: biofilms and their exopolysaccharide composition. Archives of Microbiology. P.1-9

Modi S M (2008) Champaner Pavagadh Managing conflicts A conservation challenge. Structural analysis of historic construction: preserving safety and significance: proceedings of the VI international conference on structural analysis of historic construction, SAHC08, 2-4 July 2008, vol 1. CRC Press, Bath, UK, p 175.

Nayaka S K, Behera, P K, Bajpai R, Upreti D K, Satapathy K B (2017) Lichens growth on Sun Temple of Konark in Odisha, India-A curse or blessing. Cryptogam Biodiversity and Assessment. 2: 48-52.

Ortega-Morales BO, Nakamura S, Montejano-Zurita G, Camacho-Chab J C, Quintana P, De la Rosa, S D C (2013) Implications of colonizing biofilms and microclimate on west stucco masks at North Acropolis, Tikal, Guatemala. Heritage Science, 1(1): 1-8.

Ravdandekar A (2014). Ecologically sustainable development plan for Pavagadh forest area using satellite data. India. Ph.D. Thesis, The Maharaja Sayajirao University of Baroda, Vadodara.

Ruggles D F, Sinha A (2009) Preserving the cultural landscape heritage of Champaner-Pavagadh, Gujarat, India. In Intangible heritage embodied. Springer, New York, p. 79-99.

Sinha A, Kesler G, Ruggles D F, Wescoat J Jr (2004) Champaner Pavagadh, Gujarat, India: challenges and responses in cultural heritage planning and design. Tourism Recreation Research 29: 75–78.

Thakur N (1987) Champaner: Draft Action Plan for Integrated Conservation. Heritage Trust, Baroda

Verma P K, Kumar V, Kaushik P K, Yadav A (2014) Bryophyte invasion on famous archaeological site of Ahom Dynasty 'Talatal Ghar'of Sibsagar, Assam (India). Proceedings of the National Academy of Sciences, India Section B: 84: 71-74.

Chapter 18

Deterioration of Historical Buildings by Bryophytes, and their Possible Role in Ecological Restoration of Urban Spaces

Virendra Nath

Chief Scientist (Sci. 'G'), CSIR-Emeritus Scientist. Head Bryology Group, CSIR-National Botanical Research Institute (NBRI), Rana Pratap Marg Lucknow – 226 001, Uttar Pradesh, India
e-mail: drvirendranath2001@rediffmail.com

ABSTRACT

World cultural heritage and old constructed buildings and historical monuments suffer from deterioration due to ageing of the construction materials exposed to varied climatic conditions over time. Besides physical and chemical, the damage caused by microbiota, bryophytes and plants play a dominant role in deterioration. The bryophytes are unique, they help to improve soil stability, fix basic nutrients like nitrogen (N) and carbon (C), and increase the organic matter content in soils, facilitating other plants to grow roots and serving as a habitat for other organisms. They are also susceptible to rapid dehydration under low relative humidity and quickly resume metabolic activity upon rehydration. Furthermore, mosses can recover quickly after an environmental perturbation, being one of the earlier colonizers among biological soil crust (BSC) components. Moreover, they are totipotent, i.e., any vegetative moss tissue can be a propagule from which a new plant can grow. In addition, due to their poikilohydric nature, they have a great ability to capture and retain atmospheric pollutants and, therefore, are widely used to monitor air quality. The cell walls of the bryophytes can exchange cation with their surroundings, which enables them to uptake essential Ca^+ into the extracellular or intracellular region of the protoplast. The level of exchangeable Ca^+ have been reported to be 16-17 times higher in calcicole mosses in comparison to other mosses. The humus produced by death and decay of these thalli helps the seeds of grasses *and plants to initiate germination and the roots find their way in rock or building crevices and can cause further damage. Although deterioration by microbes may be harmful to monuments in certain cases the growth of bryophytes have helped to increase the aesthetic appearance. Like protecting the huge tree growth along with sand stone temples in famous Angkor the growth of mosses is appreciated with a number of old buildings. The use of such initiatives is also promoted in order to restore the degraded ecosystem and reduce the effects of climate change. The chapter discusses both the issues deterioration as well as restoration caused by tiny photosynthetic organisms of different life forms.*

***Keywords**: Deterioration, Historical buildings, Conservation, Bryophytes, Mosses, Ecological restoration, Urban spaces.*

18.1 Introduction

Cultural heritage assets are exposed to weather and submitted to influence of environmental parameters. Physicals, chemicals and biological factors interact with constitutive materials inducing changes both in its compositional and structural characteristics. The matter transformation may occur due to the metabolic activity connected with the growth of living organisms. This activity is needed to maintain in equilibrium the "matter transformation cycles" and contribute to very important aspect of "life" such as to transform hard rocks in soft soils (pedogenesis) or to reduce the complex biological structures into simpler components. Micro and macroorganisms can find a suitable habitat for their growth either on monumental buildings and archaeological remains. The living species dwelling on these materials are ranging from microscopical bacteria to the higher plants and animals. The intensity of the damage caused is correlated with: type and dimension of the organism involved; kind of material and state of its conservation; environmental conditions, micro-climatic exposure; level and types of air pollutants. Photosynthetic organisms such as bryophytes (mosses and liverwort) and vascular plants abundantly grow on the archaeological area and on the buildings when the environmental conditions or especially the water content is available for such biological growth. Their activities are cause of concern.

Bryophytes are poikilohydric organisms that play a key role in ecosystems. They are called amphibians among plants. Their luxuriant growth can be seen during rains on rocks, old buildings and empty spaces on land. As soon as rains are over the bryophytes lose their green carpet appearance and remain in dormant phase till next season. With their small, yet remarkably varied gametophytes and solitary sporangia, bryophytes exhibit innovative means of flourishing under the ecological constraints of life on land. Next to flowering plants, bryophytes are the most diverse and species-rich lineages of the embryophytes. They are found on all continents and occupy xeric to aquatic niches, with the greatest diversity thriving in specialized mesic microenvironments and the greatest biomass in cool temperate regions where the single genus *Sphagnum* contains roughly 16 per cent of the earth's carbon (Halsey *et al.*, 2000). Bryophytes were among the original colonizers of terrestrial habitats and their status as the oldest living land plants is rarely contested (Mishler and Churchill 1985; Shaw and Renzaglia 2004). Although the fossil record is scant, it is likely that the three lineages of bryophytes (liverworts, hornworts and mosses) diversified during the Upper Ordovician-Silurian phase of the primary radiation of land biota (Bateman *et al.*, 1998). Spore microfossils support the widespread occurrence of spore-bearing bryophyte-like plants by the end of the Ordovician (Wellman *et al.*, 2003). While a growing number of Devonian meso-fossils have more in common with extant bryophytes than with extant tracheophytes, taphonomic comparisons between bryophytes and fossils of problematic affinities are less convincing (Graham *et al.*, 2004). Thus, bryophytes evolved during a pivotal moment in the history of life on earth and have persisted through hundreds of millions of years.

The desert bryophytes require morphologies and life forms that allow them to retain more water for longer periods. Most probably, bryophytes from habitats with higher moisture or even full aquatic ones have shoot morphology and colony life forms more adapted to other limiting factors than water such as nutrient interception. These colony life forms can be classified in cushions, short or tall turfs, mats, pendants, fans, dendroids and streamers (Gimingham and Birse 1957; Glime 2017) formed by shoots with different lengths and ramifications with axis supporting the leaves consisting of a single layer of cells. The shoot morphology and colony life forms are traits in the adaptation of bryophytes to the different habitats that correspond to predicted levels of desiccation. Since we cannot measure the rate of cellular dehydration, we can make use of techniques such as microscopic studies that provide interesting details to identify species and investigate several other aspects, including cell structure and ultrastructure. However, the analysis of complex matrixes such as bryophyte colonies interacting with environmental contexts in 3D qualitative or quantitative forms is hard to assess but is becoming of utmost importance. X-ray computed microtomography (μ-XCT) enables new qualitative or quantitative approaches in the study of shoot or colony morphology and life form (Maurício *et al.*, 2013). The determination of colony water storage locations is important to understand how life form can control, to some extent, dehydration rate.

Abel (1956) did one of the first attempts to classify vegetative desiccation tolerance (DT) in a wide range of bryophytes through vitality tests based on the exposure to a very wide range of relative humidity (RH) atmospheres (0 to 96 per cent RH). Other authors reported similar conclusions based on measurements of photosynthesis (Lee and Stewart 1971), electrolyte leakage (Brown and Buck 1979) or habitat preference (Franks and Bergstrom, 2000). Most studies only followed recovery after rehydration over a few hours (Lee and Stewart, 1971; Brown and Buck, 1979). Abel (1956) had already pointed out that some other aquatic bryophytes could develop DT under certain conditions. However, the recent works on *Fontinalis antipyretica* (Cruz de Carvalho *et al.*, 2015; 2017) demonstrated that this bryophyte can cope with desiccation in certain conditions. Dehydration rate can be classified as fast (less than an hour), slow (a few hours) or very slow (hours to days) (Cruz de Carvalho *et al.*, 2011; Stark *et al.*, 2013). However, cellular dehydration rate is difficult to standardize across different bryophyte species (Alpert and Oliver 2002), due to several parameters that influence dehydration rate such colony size and shoot morphology (Proctor 2001; Elumeeva *et al.*, 2011). It is expected that bryophytes with different morphologies and life forms, dried in the same conditions and maintaining their initial tissue organization, will have different rates of dehydration at the cellular level.

The bryophytes are the only group of land plants that have a dominant gametophyte plant body that is independent and autotrophic. The sporophyte stage is dependent on the gametophyte for nutrition and physical support. It is differentiated into foot, seta and spore producing structure called capsule. All the bryophytes have capacity for regeneration fast and these are homosporous (Rashid 2007). Bryophytes are broadly divided into three major groups *viz.*, liverworts, hornworts and mosses. From a biodeterioration perspective, the liverworts and mosses are found on the walls and other parts of the building or monuments. Hornworts generally grow on soil and do not play a role in biodeterioration of the wall structure. They are however important in monument locations where soil accumulates. The protonema stage of mosses is the initial phase that covers the walls. Mature gametophyte has the leaf, stem and rhizoids. Rhizoids penetrate into the pores and fissures of the substratum (Gil and Saiz-Jimenez, 1992). Due to this, mechanical forces are generated which result in formation of cracks in their surroundings resulting in damage and deterioration of the wall structure. The classification of bryophytes is undergoing constant revisions and this revision is based on several adaptations such as special reproductive structures, alternation of generation, desiccation tolerance capacity *etc.* that help to survive in any harsh environment condition by Renzaglia *et al.* (2000).

The growth and metabolic activity of algae, mosses and higher plants, is regulated by natural parameters such as light and moisture (Monte 1993; Ortega-Calvo *et al.*, 1995). In rural areas, their growth is enhanced by nitrogen-rich air from fertilizers (Cadot-Leroux, 1996; Young, 1997). Phototrophic microorganisms may grow on the stone surface (called epilithicphototrophs) or may penetrate some millimeters into the rock pore system (called endolithicphototrophs) (Friedmann and Ocampo-Friedmann, 1984). Bryophyte life-forms as adaptations that maximize water use efficiency, but many have also recognized relationships between colony organization and light intensity, and with the type of substratum. In some cases bryophyte colonies may trap beneficial above-ambient concentrations of CO_2 from the respiration of microorganisms and invertebrates in the underlying substratum leading to increased photosynthetic rates (Sveinbjdrnsson and Oechel 1992, Tarnawski *et al.*, 1992). Restorers and conservators should consider biodeterioration processes as part of a complete and careful diagnosis of stone decay in cultural objects (Warscheid 1996).

18.1.1 Recent Techniques to Measure Deterioration and Identify the Deteriogens

Evaluation of the biological contribution to stone decay starts with the description of the type of stone material (Warscheid and Braams 2000) and exposure conditions for the entire object or heritage building, including water presence (*e.g.*, rising dampness, damaged water drainage, condensational moisture) and nutrients (*e.g.*, inorganic and organic compounds from natural or anthropogenic sources). The diagnosis should also provide detailed information on the form, intensity, and extent of weathering damages as well as its distribution on the monument (Fitzner *et al.*, 1992). The outgrowth or crust can be subjected to TTC test and change in colour will signify the presence of living organisms. In case of bryophytes soaking the material in water for 6 or more hours was required. Culture-independent methods based on the PCR amplification of DNA genes (DGGE, RISA, SSCP *etc.*) and culture-dependent techniques can be applied to resolve the identity of organisms present in the samples (Urzi *et al.*, 2003). In the latter case, molecular tools for genetic studies, such as FISH, 16S rDNA sequencing, ITS-PCR and fITS, are conducted to gain information on metabolically active biodeteriogens (Pangallo *et al.*, 2009). Since culture-dependent techniques have limits, we should always be aware that a difference of 1:100 fold exists between numbers of viable culturable microorganisms (VCM) and direct total counts (DTC). The vantage of using cultures, however, is that researchers can isolate the organisms in monospecific, axenic and clonal strains and carry out studies on the taxonomic position of isolates, test their metabolic profiles and design probes for the direct in situ detection of specific cells (Albertano and Urzì 1999). Furthermore, while the direct extraction of DNA provides information on the totality of live and dead microorganisms without discriminating if they are metabolically active, cultivation is true evidence about those organisms responsible for biodeterioration.

No single technique is sufficient to measure stone deterioration, since decay takes many different forms. Some techniques, such as 3D laser scanning and fluorescence LIDAR (light detection and ranging), look only at the surface, and they are well suited to decay that consists of a gradual loss of surface, leaving sound stone behind. Other techniques, such as ultrasonic measurements, thermography, or magnetic resonance imaging (MRI) are designed to probe below the surface, and these are useful where decay consists of a loss of cohesion within the stone, or the development of detached layers, blisters, or internal voids. Before using more complex methods, simple visual examination plays an important role in quantifying decay. A single examination can convey the state of the stone at a particular moment, but it does not capture the rate of decay. For this, a series of inspections is required, usually over a period of several years. Photographs are of immense value here, but their objectivity can be abused. Winkler (1975), for example, constructs an alarming graph of exponentially increasing decay on the basis of just two photographs. Even within a series of photographs, a fundamental difficulty is that often they have been shot under differing lighting conditions, making the interpretation of surface loss challenging (Thornbush and Viles 2008). Two improvements in traditional photographic documentation show promise. One is the use of time-lapse method to provide more frequent images (Zehnder and Schoch 2009),and the other is polynomial transform mapping (PTM), a subset of RTI (Relectance Transform Imaging), that is, the use of multiple photographs from different angles to document more comprehensively the texture of stone surfaces. This gives the viewer the ability to control the angle of the light source in a given image using Java-based software (Padield *et al.,* 2005). Bioalterations in rocks can be seen by ultra-thin section techniques, optical and electron microscopy, microanalysis (EDXRA), X-ray diffractry (XRD), infrared (FTIR) and Raman spectroscopy, as well as analytical chemistry investigations (Piervittori 2004).

The IR thermography is a detection technique that allows capturing images, called thermograms, in the infrared thermal range of the electromagnetic spectrum. These images are got by means of a proper detector capable of measuring the emissivity of a substrate, or the radiation of the heat emanating from the body as a function of its temperature. Nowadays, this detector is installed in a commercial thermocamera quite similar to a normal digital photocamera for size and weight. All structures, with a temperature above absolute zero, emit radiations in the infrared range and the camera is able to measure their temperature by measuring the intensity of the emitted radiation (based on the Stefan-Boltzmann law).

The multispectral photographic techniques called imaging are non-invasive and portable methods of analysis of works of art (paintings, canvas, wall paintings and also painted stone artifacts). The imaging techniques present in the ICVBC mobile Laboratory are the following: Ultraviolet reflected photography (UVr); Ultraviolet fluorescence photography (UV); Near Infrared photography (IR); Visible Induced Luminescence photography (VIL); Infrared False Color (IRFC); Ultraviolet False Color (UVFC).

Chlorophyll fluorometers imaging: Chlorophyll fluorometers are highly sensitive research instruments which give quantitative information on the quantum yield of photosynthetic energy conversion. Their measuring method is based on the Pulse Amplitude Modulation (PAM) technique. The essentials features of the PAM fluorometry are the extremely selective detection system to distinguish between the fluorescence excited by the measuring light and the much stronger signals caused by ambient and actinic light. The imaging portable instrument (PSI Photon Systems Instruments) is specialized for the study of two dimensional heterogeneities of the photosynthetic activity. PAM fluorometer provides not only images of chlorophyll fluorescence, but also images of all relevant chlorophyll fluorescence parameters using the Saturation Pulse method.

18.2 Deterioration of Heritage Buildings

Biophysical deterioration is mechanical damage exacted upon the substrate due to surface detachment resulting in superficial losses, or penetration and exerted pressure during growth resulting in increased porosity. Biochemical deterioration is the direct action by biological organisms through metabolic processes on the substrate. This involves the exudation of organic acids which can *etc*.h or solubilize stone, the exudation of organic chelating agents which sequester metallic cations from stone, or the conversion of inorganic substances by redox reactions which form inorganic acids that *etc*.h stone and contribute to salt formation. Aerobic organisms produce respiratory, carbon dioxide which becomes carbonic acid and contributes to dissolution of the stone and soluble salt formation. The biodeterioration of building materials results in the uptake of calcium or other ions, leaving the surface eroded and exposed to water and frost attacks. The interrelationship between biodegradation and other environmental degradative processes has been noted Krumbein (1968), Krumbein and Altmann (1973), and Krumbein and Lange (1978). Biodeterioration when coupled with other environmentally induced degradation is usually synergistic: the presence of the one makes deterioration by the other all the more effective, Biodeterioration is typically a secondary degradation process which begins after some degree of deterioration has occurred as the result of other causes. This damage can take the form of a rough surface, a soil-like powder on the surface, or the deposition of inorganic or organic matter. Biodeterioration may be enhanced or subdued by other environmental conditions. A deteriorated monument usually exhibits several types of deterioration processes operating in tandem, making it difficult to attribute damage specifically to a single cause (Hueck-van derPlas 1968; Eckhardt 1978; Caneva and Salvadori 1989).

Mosses and higher plants exhibit chemical degradative effects on stone similar to lichens. Mechanical damage by mosses is less threatening than for higher plants: they possess rhizoids rather than roots, and require a surface layer of soil before they can grow. Growth at mortar joints, however, can cause problems for the overall structure. Bech-Andersen (1985) and Tiano (1987) discussed the deterioration mechanism of mosses and higher plants. Mosses and liverworts chiefly deteriorate stone aesthetically. Humus deposits formed by death and decay of bryophyte thalli support the growth of higher plants. Biochemical disintegration of stone surfaces (Saiz-Jimenez 1995) occur owing to higher acidity of their rhizoids and have an elevated ability of extracting mineral cations from the stones (Bech-Anderson 1985; Saiz-Jimenez 1995). Though, mechanical action is less threatening (Shah and Shah 1992-93; Jain *et al.,* 1993) because these organisms possess rhizoids rather than real roots. The presence of clay in the stone favors their growth (Hyvert 1972).

The source of nutrients for bryophytes are precipitation, dust and substrate (Glime 2017). From the substrate the bryophytes uptake some nutrient for their growth due to which it causes degradation of the substrate. The cell walls of the bryophytes can exchange cation with their surroundings. This enables bryophytes to uptake essential cations like Ca+ into the extracellular or intracellular region of the protoplast. Calcium held on the exchange sites is thought to influence the permeability of the adjacent plasmalemma. The level of exchangeable Ca+ have been reported to be 16-17 times higher in calcicole mosses compared to calcifuges mosses (Bates 1998; Mehta 2021).

Moisture relations A relatively dense colony of shoots or thalli (cushions, turfs) has two obvious advantages over widelyspaced individuals. It can store water in the capillary spaces created between the components and its relatively solid form is likely to become enveloped in a laminar boundary layer (Proctor 1981). Single shoots generate turbulence in a moving air stream which induces rapid evaporative water loss. In contrast, the laminar layer surrounding an aerodynamically smooth colony becomes saturated with water vapour, this reduces the concentration gradient for molecular diffusion and thus resists further moisture transfer from leaf to free air. Measurements of water loss from intact bryophyte colonies held in a wind tunnel (Proctor 1981) have confirmed the considerable advantage of colonial organization in increasing aerodynamic resistance. With increasing windspeeds the relatively smooth colonies of *Ceratodon purpureus* (short turf) and *Grimmia pulvinata* (small cushion) gave evaporation rates close to those predicted for the standard relations of a smooth plane, whereas rougher cushions (*e.g. Mnium hornum, Dicranum majus*) underwent greater than predicted evaporation. In the latter cases the divergence from standard laminar flow commenced at a wind speed where the laminar boundary layer had thinned approximately to the dimensions of surface irregularities of the colony. Leaf hair-points, responsible for the hoary appearance of colonies of species like *G. pulvinata*, Although densely-packed cushions and short turfs predominate in the most xeric habitats, even these life-forms have strong limitations when compared with the cuticularized and stoma-regulated leaf of a vascular plant. Thus Proctor and Smith (1995) found that cushions of the xerophytic moss *G. pulvinata* remained moist for only an hour or two after precipitation in summer, but in the cooler and cloudier autumn and early winter they were continuously hydrated for long periods. The contrasting arrangements of the shoots, radial in cushions and vertical in turfs, may have a functional significance. A plausible interpretation is that the vertical arrangement characterizing turfs on soil produces the shortest distances for conduction of moisture from the soil to the growing apices. Many short-turf bryophytes are endohydric or mixohydric which supports this view. Alternatively, radial expansion of protonema may be inhibited earlier on hard surfaces by drought so that increase in density is mainly by production of offsets from the few colonizing individuals. It is also possible that the radial shoot arrangement in small cushions has advantages in colony water storage capacity and in providing an aerodynamically smooth cushion.

Heavy growth of bryophytes can obstruct the passage of rainwater from gutters causing stagnation, and consequently increase damage through freezing. Higher plants with a real root system cause physical damage, through the pressure exerted by roots growth (up to about 15 atm.) and chemical through the production of acidity and exudates from their rootlets.

18.3 Management Strategies for Biological Growth on Monuments

The growth of microorganisms and photosynthetic organisms like cyanobacteria, algae and bryophytes on monuments is a phenomenon that results in loss of the original rock material and the irreversible transformation of the substrata (Urzi 2004). To counteract biodeterioration, the common conservation practice for monuments applies direct as well as indirect methods that are generally based on treatments of the affected object sometimes without an appropriate management of the surrounding environment (Caneva and Simona 2009). However, the nature of the artefact is highly dynamic and open to the continuously changing physical, chemical and biological variables that influence the environment in which it stands. Even small shifts of the microclimatic parameters can compromise the delicate equilibrium existing between the environment and the item/surface, and initiate deleterious processes or, on the contrary, create conditions no longer suitable for the development of biodeteriogens and stop further corrosion of materials. Various non-invasive (NIT) and non-destructive (NDT) techniques may be for the diagnosis, monitoring and control of microorganisms.

The treatment of cultural property: the practical aspects of the protection of building materials has also been reviewed (Hugo and Russell 1982). Biodeterioration is typically seen as only a cosmetic problem: it is noted chiefly as a difference in appearance from unaffected stone. This approach reflects a misunderstanding of the nature of biodeterioration and its synergistic effect on other degradative processes. Treatment involves solving problems posed by the deterioration caused by other physical processes. This approach does not address the problem of biodeterioration, and may indeed exacerbate it. Aqueous treatments used to clean stone or remove soluble salts are not necessarily effective in removing biological growth, and could result in an increase in growth as a result of the wetted stone (Warscheid *et al.*, 1988). Consolidation treatments may provide new nutrients for biological growth, ultimately rendering the consolidation ineffective, and leading to further deterioration (Nugari and Priori 1985; Salvadori and Nugari 1988). Indirect treatment involves the alteration of environmental factors to make growth unfavorable, such as reducing contact with water, and lowering ambient humidity or temperature, or light levels. Effecting environmental changes has been established as effective means to control biological growth on many substrates (Hopton, 1988: Van der Molen *et al.*, 1980). Although not all recommended practices are applicable to exterior stone monuments, biological growths on stone have been controlled in part by changing environmental conditions like access to moisture (Charola *et al.*, 1986; Del Monte *et al.*, 1987). Direct treatments with a biostat or biocide can be done to inhibit and or eradicate future growth, and further application of a moisture barrier or consolidant.

Efforts have been made to use 2-5 per cent ammonia to clean, zinc flurosilicate to inhibit growth, and polyvinyl acetate (PVA) as a moisture barrier (Sharma *et al.*, 1985). At Borobadur (Siswowiyanto 1981), biological growths were treated using AC 322 (a variation of AB 57, the "Mora" poultice, Mora *et al.*, 1984) to clean, and a quaternary ammonium salt compound to inhibit growth. Treatment provided temporary inhibition, and periodic maintenance was required. A mixture of borates and boric acid, has been used in the field by Richardson without observed damage. Sodium borates such as Borax have also often been used for cleaning as well as biocidal properties, but can lead to soluble salt formation in urban atmospheres. Borax and Chlorox were used in tandem to control growths on Mayan ruins by Hale (1975). Studies indicate that no biocide is uniformly effective on all organisms and on all stone substrates (Emmel *et al.*, 1988). A review of available biocides and recommended usage was compiled by Allsopp and Allsopp (1983). Quaternary ammonium compounds have been shown to contribute to the degadation of lime mortars and hardened Portland cement (Fearn 1978). Other potential problems are treatments including biocides or chemical cleaning solutions that introduce materials which can form soluble salts, or consolidation or water-repellent treatments which utilize materials, which provide nutrients for biological growth. Several recent studies indicate that many materials commonly used to waterproof or consolidate stone increased the potential for biological growth (Koestler and Santoro 1988).

Biocides affect the metabolic activity of organisms, thus cause severe damage and even death of the organisms. Spraying and brushing of diluted biocide solutions are the most common systems of treatment. Prior to their use on monuments, their effectiveness against target organisms, resistance of target organisms, toxicity to humans, risks of environmental pollution, compatibility with stone, and effects of interactions with other chemical conservation treatments, is to be assayed (Krumbein *et al.*, 1993). Various biocides as polybor (a mixture of polyborates and boric acid), borax and clorox could be used to control microbial growth. Surface-active quaternary compounds have strong biocidal activity (Richardson 1988). Quaternary ammonium compounds are reported to effectively inhibit microbial growth on sandstone monuments (Sadirin 1988). Lichens can be temporarily inhibited by aqueous solutions of benzalkonium chloride (20 per cent), sodium hypochlorite (13 per cent), and formaldehyde (5 per cent), using soaked cotton strips for about 16 hours, followed by scrubbing with a brush and water (Nishiura and Ebisawa 1992). The treatment can kill all organisms and microorganisms but does not give any kind of protection for future recolonization, thus biocides are to be applied regularly to control the biological growth.

Other mitigation strategies such as UV light treatments for biological greening have been well-studied, and are recommended as part of short-term conservation management strategies for cultural heritage, but may have negative long-term effects on the substrate and environment (Silva *et al.*, 2017). The problems caused by moss growth on roofs, such as blocked gutters, pitting, water retention, and chemical deterioration caused by water runoff on slate roofs (Austin-Smith: Lord LLP, 2007). Finally, the Edinburgh World Heritage Historic Home Guide for Roofs recommends brushing moss off of clay roof tiles as they hold moisture which causes the tiles to deteriorate more quickly and cause frost damage (Mayhew 2018).

18.4 Bioreceptivity of Substrate Materials

The concept of bioreceptivity was focused by Guillitte (1995) as the aptitude of a material to be colonized by one or several groups of living organisms. Otherwise, the mere occurrence of organisms on stone surfaces does not automatically imply destructive action but in some cases could be considered: as only aesthetic detrimental appearance (Warscheid 1988); perceived aesthetically pleasing or credit these with a protective role against weather induced aggression (Nimis *et al.*, 1992). The bioreceptivity must be related with the totality of material properties that contribute to the establishment, anchorage and development of flora and/or fauna. In stony material it relates mainly with surface roughness, moisture content, chemical composition of the outer layers and with the structure-texture of the rock. The "Primary or Intrinsic Bioreceptivity" is connected with the initial potential of colonization of a sound stone. Then, following the evolution over time of surface characteristics, under the action of colonizing organisms and environmental factors, it becomes the more important "Secondary Bioreceptivity". The conservative treatments applied on stone induce a "Tertiary Bioreceptivity". Moreover, we can have "Extrinsic Bioreceptivity" when the stone colonization is essentially due to the presence of settled matter not related with the beneath stone (Guillitte 1995; Miller *et al.*, 2012). One can assess the bioreceptivity of a material to an organism by artificially inoculating the material with the diaspores of the organism itself and placing them under optimal environmental conditions (growth chamber). Thus a specific "Bioreceptivity Index" could be determined for a stony material related to its susceptibility to biodeterioration. However, these studies are always indicatives, as many types of colonization are part of a synecological mechanism and not all the organisms present on exposed stone surfaces can develop on artificial conditions, even if very close to those present in nature.

Hence the test made with a single organism or few isolated strains can become either impossible or completely atypical. In fact, the results reported in few lab experiments can be considered pertinent only for the stones used, the organisms tested and the incubation conditions applied (Guillitte and Dreesen 1995).

18.5 Biological Growth adds to the beauty of Buildings

Mosses are the ideal organisms for ecological restoration projects in disturbed ecosystems (Cruz de Carvalho *et al.*, 2018) or even for use on roofs and green walls in urban environments (Garabito *et al.*, 2017). However, it is not acceptable to harvest mosses in the field and transplant them elsewhere, as their growth rate in the natural environment is very slow. For example, Tian *et al.* (12006) report that moss *Bryum argenteum* is able to cover 70 per cent of 10 × 10 cm squares in 3–4 years in desert dunes, so another source of mosses is needed, which is why laboratory cultures were started. Studies aimed at restoring or rehabilitating dry land areas have been artificially growing moss for years (Zhao *et al.*, 2016). The procedure followed by most of these works is practically the same: they collect bryophytes from the study area and grow them in greenhouses or growing chambers inside plastic containers using autoclaved sand from such areas as a substrate and adding culture medium to speed up their growth. However, these moss culture methods have been poorly described and no information is available concerning their performance. It is unknown which photoperiod or temperature they use to cultivate the different species or if the moss samples were irrigated for most studies. It is only noted that high rates of moss growth are reached in two and six months.

Bryophytes are poikilohydric organisms that play a key role in ecosystems, while some of them are also resistant to drought and environmental disturbances but present a slow growth rate. Recent researches have uncovered possible benefits from biological colonization such as protection from mechanical weathering (Chen *et al.*, 2000), and has defined these under the umbrella term of 'bioprotection' (Carter and Viles 2005). Moss culture in the laboratory can be a very useful tool for ecological restoration or the development of urban green spaces. Particularly roofs and walls in the Mediterranean region. Therefore, a study was undertaken to: (i) determine the optimal culture conditions for the growth of four moss species present in the Mediterranean climate, such as *Bryum argenteum, Hypnum cupressiforme, Tortella nitida, and Tortella squarrosa*; (ii) study the optimal growth conditions of the invasive moss *Campylopus introflexus* to find out if it can be a threat to native species. Photoperiod does not seem to cause any recognizable pattern in moss growth. However, temperature produces more linear but slower growth at 15 °C than at 20 and 25 °C. In addition, the lower temperature produced faster maximum cover values within 5–8 weeks, with at least 60 per cent of the culture area covered. The study concludes that the culture of moss artificially in the organic gardening substrate without fertilizers is feasible and could be of great help for further use in environmental projects to restore degraded ecosystems or to facilitate urban green spaces in the Mediterranean area.

As biological growth and patinas can change the viewer's perception and experience of a cultural heritage site, it is important to determine how the presence of mosses will affect values such as aesthetics, authenticity, and perceived damage as defined in the original Burra Charter (Australia ICOMOS 1979), and more recent heritage-based values typologies (Fredheim and Khalaf 2016). In certain cases, the presence of mosses may positively impact the aesthetic value of a site by contributing to its historic or authentic values. As examined in previous research, the greening of historic sites can add important contextual information regarding the age of the site. For example, in the case study of Dryburgh Abbey, Scotland, biological growth on the ruined abbey is seen as intrinsic to the beauty and 'romantic' aesthetic of the site (Douglas-Jones *et al.*, 2016). Established c. 1150 by the Premonstratensians, Dryburgh Abbey was a Cistercian monastery destroyed during the Scottish Reformation in 1544. Today the site is designated as a scheduled monument in the care of Historic Scotland (HS). The surviving ruins are typically medieval gothic, constructed of local sandstone faced with coursed ashlar. House has been noted as a cause for concern due to the building having "national significance as the largest area of medieval polychromatic wall decoration in Scotland" (Douglas-Jones *et al.*, 2016). However, when visitors were asked directly about the removal or cleaning of biological growth many visitors expressed an appreciation for the biological growth. One visitor expressed the desire for the abbey to have "death in beauty," while another visitor noted that the abbey should "be able to decay slowly, without us preventing it" (Douglas-Jones *et al.*, 2016).

Perhaps in the case of sites like Dryburgh Abbeywhere the natural values and context of the site are intrinsically intertwined, the impact of cleaning biological growth needs to be carefully evaluated. This can justify the value reasoning behind the expert and visitor surveys. If both built heritage experts and visitors value the presence of biological growth (and indeed mosses) on built heritage, then perhaps cultural heritage management strategies need to be modified to align with these insights.

The presence of biological growth on cultural heritage is well documented, with primarily negative consequences for both the physical and chemical deterioration of built heritage materials often outlined, as well as effects on the visual aesthetics (Bartoli *et al.*, 2014). The term biodeterioration describes the "physical or chemical damage caused by biological agents on structural materials and occurs through the formation of biofilms" and can refer to multiple types of biological growths including vascular plants, lichens, fungi, bryophytes, algae, and cyanobacteria (Herrera *et al.*, 2009). In particular, moisture uptake and surface saturation are concerns when discussing moss colonization (Garcia-Rowe and Saiz-Jiminez 1991). However, the source of moisture encouraging such growth must be addressed in order to prevent re-growth. The biocide is unlikely to have a long-lasting effect and may damage the masonry, hence should not form part of a regular maintenance regime" (Historic Environment Scotland 2014).

However, strict cleaning schedules and the portrayal of pristine heritage sites oppose the aesthetic values appreciated by heritage visitors as seen in previous case studies (Douglas-Jones *et al.*, 2016). Exploration of the values perceived by biological greening on cultural heritage as well as the advice from heritage organizations regarding cleaning and management regimes helped to shape and inform the questions provided in the survey. Overall, experts are more likely to enjoy the look of mosses on historic buildings than visitors, with 90 per cent of respondents marking 3/5 or higher when asked if they agree with the statement 'I enjoy the look of mosses on historic buildings' compared to visitors, with 77 per cent of respondents marking 3/5 or higher. Results from the Pearson Correlation Coefficient gives a p-value of < 0.1164, at $p = 0.05$, the result is not statistically significant. The results suggest that both visitors and experts mildly enjoy the appearance of mosses on historic buildings.

18.6 Conclusion

The potential for damage to stone substrates by biological organisms has been demonstrated by observations in the field and experiments in the laboratory. It is difficult, however, to quantify and attribute observed deterioration on monuments specifically to biological organisms rather than to other environmental factors, The type of organism present may affect observed damage and treatment options. Indirect and direct treatments have been used with success to reduce the deleterious effects of biological organisms. If direct treatment is used, the chosen materials should be evaluated to determine possible chemical changes to stone and potential reactivity with other materials from previous, current, or future treatments. In addition, materials used in the treatment should be evaluated to determine their prohibitory or promotional effect on the biological organisms which are present

REFERENCES

Abel W O (1956). Die Austrocknungsresistenz der Laubmoose. S B Wien Akad Wiss Math.-naturw Kl. Abt. I 165: 619–707.

Albertano P, Urzì C (1999) Structural interactions among epilithic cyanobacteria and heterotrophic microorganisms in Roman hypogea. Microbial Ecology 38: 244-252.

lpert P, Oliver M J (2002). "Drying without dying," in Desiccation and Survival in Plants: Drying Without Dying. Eds. Black, M., and Pritchard, H. W. (Wallingford, UK: CABI Publishing), 3–43.

Allsopp C, Allsopp D (1983) An updated survey of commercial products used to protect materials against biodeterioration. Biodet. Bull., 19: 99-146

Altieri A, Ricci S (1994) Il ruolo delle briofite nel biodeterioramento di materia lapidei. In Proceedings of the 3rd International Symposium "The Conservation of Monuments in the Mediterranean Basin", Venezia, V. Fassina, H. Off, F. Zezza, Eds. 329-333.

Austin-Smith: Lord LLP (2007). Conservation Area Maintenance Guide. East Ayrshire Council. Retrieved from: east-ayrshire.gov.uk/Resources/ PDF/K/Kilmarnock Conservation Area Maintenance Guide.

Australia ICOMOS (1979) The Australia ICOMOS guidelines for the conservation of places of cultural significance " Burra Chapter"

Bartoli F, Municcchia AC, Futagami Y, Kashiwadani H, Moon K H, Caneva G (2014) Biological colonization patterns on the ruins of Angkor temples (Cambodia) in the biodeterioration vs bioprotection debate. Int. Biodeterior. Biodegrad. 96: 157-165

Bates J W (1998) Is 'life-form' a useful concept in bryophyte ecology? Oikos 82, 223–237. doi: 10.2307/3546962

Bateman R M, Crane P R, DiMichele WA, KenrickP R, Rowe. N P (1998) Early evolution of land plants: phylogeny, physiology and ecology of the primary terrestrial radiation. Annual Review of Ecology and Systematics 29: 263–292

Bech-Andersen J (1985) Biodeterioration of natural and artifcial stone caused by algae, lichens, mosses and higher plants. In: Barry, S., Houghton, D.R., Llewellyn, G.C., O'Rear, C.E. (Eds.), Biodeterioration, Vol. 6. CAB International Mycological Institute and The Biodeterioration Society. Slough, UK, pp. 126–131.

Bettini C, Villa A (1976) Il Problema della vegetazione infestante nelle aree archeologiche. In The Conservation of Stone I, R.Rossi-Manaresi Ed., Bologna, 191-204.

Brown D H, Buck G W (1979) Desiccation effects and cation distribution in bryophytes. New Phytol. 82, 115–125. doi: 10.1111/j.1469-8137.1979.tb07565.

Cadot-Leroux L (1996) Rural nitrogenous pollution and deterioration of granitic monuments: results of a statistical study. In: Riederer, J. (Ed.), Proceedings of the Eighth International Congress on Deterioration and Conservation of Stone, Vol. 1. Rathgen-Forschungslabor, Berlin, Germany, pp. 301–309

Caneva G, Galotta G (1994) Floristic and structural changes of plant communities of the Domus Aurea (Rome) related to a different weed control. In Proceedings of the 3rd International Symposium "The Conservation of Monuments in the Mediterranean Basin", Venezia, V. Fassina, H. Off, F. Zezza,(eds.) 317-322.

Caneva G, Nugari M P, Salvadori O (2009) Plant Biology for Cultural Heritage, Getty Conservation Institute, Los Angeles CA (USA): 408 pp.

Caneva G, Salvadori O (1989) Biodeterioration of stone. In: Larraini, L., Pieper, R. (eds.), The Deterioration and Conservation of Stone. UNESCO, Paris, pp. 182–234.

Caneva G, Simona C (2009) Ecology of biodeterioration. Plant Biology for Cultural Heritage. Biodeterioration and Conservation. 35-58.

Carter NEA, Viles HA (2005) Bioprotection explored: the story of a little known earth surface process. Geomorphol 67: 273–281

Charola A., Lazzarini E, Wheeler G E, Koestler R J (1986) The Spanish apse from San Martin de Fuentiduena at the Cloisters, Metropolitan Museum of Art, New York. In Case Studies in the Conservation of Stone and Wall Paintings: Preprints of the Contributions to the Bologna Congress. ed. N. S. Brommelle and P. Smith, International Institute for Conservation, London, pp. 18-21.

Chen J, Blume H P, Beyer L (2000) Weathering of rocks induced by lichen colonization – a review. *Catena* 39: 121–146

Cruz de Carvalho R, Branquinho C, Marques da Silva J (2011) Physiological consequences of desiccation in the aquatic bryophyte Fontinalis antipyretica. Planta 234, 195–205. doi: 10.1007/s00425-011-1388-x

Cruz de Carvalho R, Silva A B, Branquinho C, Marques da Silva J (2015) Influence of dehydration rate on cell sucrose and water relations parameters in an inducible desiccation tolerant aquatic bryophyte. Environ. Exp. Bot. 120, 18–22. doi: 10.1016/j.envexpbot.2015.07.002

Cruz de Carvalho R, Catalá M, Branquinho C, Marques da Silva J, Barreno E (2017) Dehydration rate determines the degree of membrane damage and desiccation tolerance in bryophytes. Physiol. Plantarum. 159, 277–289. doi: 10.1111/ppl.12511

Cruz de Carvalho R, dos Santos P, Branquinho C (2018) Production of moss-dominated biocrusts to enhance the stability and function of the margins of artificial water bodies. Restor. Ecol. 2018, 26, 419–421.

Del Monte M, Sabbioni C, Zappia G (1987) The origin of calcium oxalates on historical buildings, monuments and natural outcrops. The Science of the Total Environment. 6'7, 17-39.

Douglas-Jones, R., Hughes, J.J., Jones, S., and Yarrow, T. (2016) Science, value and material decay in the conservation of historic environments. J Cult Heritage 21: 823–833.

Eckhardt, F.E.W. (1978). Microorganisms and weathering of a sandstone monument. In: Krumbein, W.E. (Ed.), Environmental Biogeochemistry and Geomicrobiology. The Terrestrial Environment. Ann Arbor Science Publishers, Michigan, pp. 675–686.

Elumeeva, T. G., Soudzilovskaia, N. A., During, H. J., and Cornelissen, J. H. C. (2011). The importance of colony structure versus shoot morphology for the water balance of 22 subarctic bryophyte species. J. Veg. Sci. 22, 152–164. doi: 10.1111/j.1654-1103.2010.01237.

Emmel, T., Brill, H., Sand, W. and Bock, E. (1988) Screening for biocides to inhibit biogenic suiphuric acid corrosion in sewage pipeline. In Biodeterioration 7. ed. D. R. Houghton. R. N. Smith. and H. O. W. Eggins. Elsevier, New York, pp. 118-22.

Fitzner, B., Heinrichs, K., Kownatzki, R. (1992) Classi cation and mapping of weathering forms. In: Rodrigues, J.D., Henriques, F., Jeremias, F.T. (Eds.) Proceedings of the Seventh International Congress on Deterioration and Conservation of Stone. Laboratorio Nacional de Engenharia Civil, Lisbon, Portugal, 15.-18.06.1992, pp. 957–968.

Fearn, J. E. (1978) The Effects of Herbicides on Masons. National Bureau of Standards. Washington. D.C.

Fitzner B, Heinrichs K, La Bouchardiere D (2003) Weathering damage on Pharaonic sandstone monuments in Luxor-Egypt. Build. Environ. 38: 1089–1103.

Franks, A. J., and Bergstrom, D. M. (2000). Corticolous bryophytes in microphyll fern forests of south-east Queensland: distribution on Antarctic beech (Nothofagus moorei). Austral. Ecol. 25, 386–393. doi: 10.1046/j.1442-9993.2000.01048.x

Fredheim H L, Khalaf M (2016) The significance of values: heritage value typologies re-examined. International Journal of Heritage Studies. 22: 466 – 481

Friedmann, E.I., Ocampo-Friedmann, R (1984) Endolithic microorganisms in extreme dry environments: analysis of a lithobiontic microbial habitat. In: Klug, M.J., Reddy, C.A. (Eds.), Microbial Ecology. American Society for Microbiology, Washington DC, pp. 177–185.

Gadd G M, Dyer TD (2017) Bioprotection of the built environment and cultural heritage. Microbial Biotechnology.10(5): 1152–1156 doi: 10.1111/1751-7915.12750

Garabito D, Vallejo R, Montero E, Garabito J, Martínez-Abaigar J (2017) Green buildings envelopes with bryophytes. A review of the state of the art. Boletín de la Soc. Espanola de Briología 48–49.

Garcia Rowe, J. and Saiz-Jimenez, C. (1991) Lichens and bryophytes as agents of deterioration of building materials in Spanish cathedrals, International Biodeterioration, 28, 151-163.

Gil J A, Saiz- Jimenez C (1992) Biodeterioratlon of Roman mosaics by bryophytes. Instiuto de Recursos Naturales y Agrobiologia, Apartado 1052.

Gimingham C H, Birse E M (1957) Ecological studies on growth-form in bryophytes I Correlations between growth-form and habitat. Journal of Ecology 45(2): 545

Glime J M (2017) Bryophyte Ecology. In: Physiological Ecology. (ed.) Glime J. M. Ebook sponsored by Michigan Technological University and the International Association of Bryologists. http: //digitalcommons.mtu.edu/bryophyte-ecology/

Graham L E, Wilcox L W, Cook M E, Gensel.P G (2004) Resistant tissues of modern marchantioid liverworts resemble enigmatic early Paleozoic microfossils. Proceedings of the National Academy of Sciences, U.S.A. 101: 11025–11029

Guillitte O (1995) Bioreceptivity: a new concept for building ecology studies. The Science of the Total Environment. 167: 215-220

Guillitte O, Dreesen R (1995) Laboratory chamber studies and petrographical analysis as bioreceptivity assessment tool of building materials. The Science of the Total Environment. 167: 365-374.

Hale M E (1975) Control of biological growths on Mayan archaeological ruins in Guatemala and Honduras. In National Geographic Research Reports. The National Geographic Society. Washington, DC, pp. 305-21.

Halsey L A, Vitt D H, Gignac L D (2000) *Sphagnum* dominated peat lands in North America since the last glacial maximum: their occurrence and extent. The Bryologist 103: 334–352.

Herrera L K, Le Borgne S, Videla H A (2009) Modern methods for materials characterization and surface analysis to study the effects of biodeterioration and weathering on buildings of cultural heritage. International Journal of Architectural Heritage, 3(1): 74–91. https: //doi.org/10.1080/15583050802149995

Historic Environment Scotland (2014). Advisory Standards of Conservation and Repair for the Historic Building Environment in Scotland. Retrieved from: historicenvironment.scot/media/5118/advisory-standards-repair.pdf

Hopton J W (1988) Physical conditions and microbial growth: Some implications for biodeterioration. In Biodeterioration 7. (ed.) D. R. Houghton, R. N. Smith. and H. O. W. Eggins. Elsevier, London. pp. 441-8.

Hugo W B, Russell A D (1982) T.vpes of Antimicrobial Agents. Principles and Practices of Disinfection. Preservation and Sterilisation. (eds.) A. D. Russell. W. B. Hugo and G. A. J. Ayliffe. Blackweli Scientific. St. Louis. pp. 8-106.

Hyvert G (1972) The Conservation of Borobudur Temple. UNESCO document no. RMO. RD/2646/CLP. Paris: UNESCO.

Hueck-van der Plas E H (1968) The microbiological deterioration of porous building materials. International Biodeterioration Bulletin 4: 11–28.

Jain K K, Mishra A K, Singh T (1993) Biodeterioration of stone: A review of mechanism involved. In: Recent Advances in Biodeterioration and Biodegradation, vol. 1, (ed.) K.L. Garg, N. Garg, and K.G. Mukerji, Calcutta: Naya Prokash. 323- 54.

Keller N D, Frederickson A F (1952) The role of plants and colloid acids in the mechanisms of weathering. Am. Journ. Sci. 250: 594-608.

Koestler R J, Santoro E D (1988) Assessment of the susceptibility to biodeterioration of selected polymers and resins. GCI Scientific Program Report, The Getty Conservation Institute, Marina del Rey, CA.

Krumbein W E (1968) Geomicrobiology and geochemistry of the "Nari-Lime Crust' (Israel). In Recent Developments in Carbonate Sedimentology in Central Europe, (ed.) G. Muller and G. M. Friedman. Springer. New York. pp. 138-47.

Krumbein W E, Altmann H J (1973) A new method for the detection and enumeration of manganese oxidizing and reducing microorganisms. Helgolander Wiss. Meeresunters. 25: 347-56.

Krumbein W E, Lange C (1978) Decay of plaster, paintings and wall material of the interior of buildings via microbial activity. In Environmental Biogeochemistry, and Geomicrobiologv, Vol. 2: The Terrestrial Environment, (ed.) W. E. Krumbein. Ann Arbor Science, Ann Arbor, MI. pp. 687-97.

Krumbein W E, Diakumaku E, Petersen K, Warscheid Th, Urzi C E (1993) Interactions of microbes with consolidants and biocides used in conservation of rocks and mural paintings. In: Thiel, M.-J. (Ed.), Conservation of stone and other materials, Vol. 2. E and FN Spon, London, pp. 589–596.

Lee J A, Stewart G R (1971) Desiccation injury in mosses: I. Intra-specific differences in the effect of moisture stress on photosynthesis. New Phytol. 70: 1061–1068. doi: 10.1111/j.1469-8137.1971.tb04588.x

Lewis C C, Eisenmenger W S (1948) Relationship of plant development to the capacity to utilize potassium in orthoclase feldspar. Soil Sci. 65: 495-500.

Mayhew R (2018) Historic Home Guide: Roofs. Edinburgh World Heritage. Retrieved from: ewh.org.uk/wp-content/uploads/2018/02/EWH-Roofs-Guide-1.pdf

Mehta D N (2021) Biodeterioration of selected historical monuments by lower plants and cyanobacteria. Ph.D. Thesis The M.S. University of Baroda, Vadodara 390002

Miller A Z, Sanmartín P, Pereira-Pardo L, Dionísio A, Saiz-Jimenez C, Macedo M F, Prieto B (2012) Bioreceptivity of building stones: A review. Science of The Total Environment 426: 1-12

Mishler B D, Churchill S P (1984) A cladistic approach to the phylogeny of the "bryophytes." Brittonia 36: 406–424.

Monte M, Sabbioni C (1987) A study of the patina called scialbatura. Studies in Conservation 32: 114–121

Monte M (1993) The in uence of environmental conditions on the reproduction and distribution of epilithic lichens. Aerobiologia 9: 169– 176.

Mora P, Mora L, Philippot P (1984) Conservation and Museologv Series. Conservation of Wall Paintings. Butterworths. London. 576 pp.

Nimis P L, Pinna D, Salvadori O (1992) Licheni e conservazione dei monumenti C.L.U.E. Ed., Bologna.

Nishiura T, Ebisawa T (1992) Conservation of carved natural stone under extremely severe conditions on the top of an high mountain. In: K. Toishi, H. Arai, T. Kenjo, and K. Yamano [eds.]. Proceedings of the 2nd International Conference on Biodeterioration of Cultural Property, October 5-8. Held at Pacifico Yokohama, International Communications Specialists. Tokyo. pp. 506-511.

Nugari M P, Priori G F (1985) Resistance of acrylic polymers (Paraloid B72. Primal AC33) to microorganisms. First part. In Vth International Congress on Deterioration and Conservation of Stone. Proceedings, Vol. 2, (ed.) G. Felix. Presses Polytechniques Romandes. Lausanne, pp. 685-93.

Ortega-Calvo JJ, Arino X, Hernandez-Marine M, Saiz-Jimenez C (1995) Factors affecting the weathering and colonization of monuments by phototrophic microorganisms. Science of the Total Environment 167: 329–341.

Piervittori R, Salvadori O, Isocrono D (2004) Literature on lichens and biodeterioration of stonework, IV. Lichenologist 36 (2): 145–57

Proctor M C F (1981) "Diffusion resistances in bryophytes," in Plants and their Atmospheric Environment. 21 Symposium of the British Ecological Society. Eds. Grace, J., Ford, E. D., and Jarvis, P. G. (Oxford, UK: Blackwell Scientific Publications), 219–229.

Proctor M C F (2001) Patterns of desiccation tolerance and recovery in bryophytes. Plant Growth Regul. 35, 147–156. doi: 10.1023/a: 1014429720821

Proctor M C F, Smith A J E (1995) Ecological and sytematic implications ofbranching patterns in bryophytes. - In: Hoch, P. C. and Stephenson, A.G. (eds), Experimental and molecular approaches to plant biosystematics. Missouri Botanical Garden, MO, pp. 87-110

Rashid A (2007) An introduction to Bryophyta. Vikas House Publishers, New Delhi 308pp.

Renzaglia K S, Duff R J, Nickrent D L, Garbary D (2000) Vegetative and reproductive innovations of early land plants: implications for a unified phylogeny. Transactions of the Royal Society, London 355: 769–793.

Richardson BA (1988) Control of microbial growth on stone and concrete. In: Houghton, D.R., Smith, R.N., Eggins, H.O.W. (Eds.), Biodeterioration, Vol. 7. Elsevier Applied Science, London, New York, pp. 101–106.

Sadirin H (1988) The deterioration and conservation of stone historical monuments in Indonesia. In: J. Ciabach (comp), 6th International Congress on Deterioration and Conservation of Stone. Vol. 1, Nicholas Copernicus University Press Department, Torun, Poland. pp. 722-731.

Salvadori O, Nugari M P (1988) The effect of microbial growth on synthetic polymers used on works of art. In Biodeterioration 7. ed. D. R. Houghton. R. N. Smith and H. O. W. Eggins. Elsevier. New York. 424-7, Sand. W. and Bock. E.

Saiz-Jimenez C (1995) Microbial melanins in stone monuments. The Science of the Total Environment, 167: 273-286.

Shah R P, Shah N R (1992-93) Growth of plants on monuments. Studies in Museology 26: 29-34.

Sharma B R N, Chaturvedi K, Samadhia N K, Tailor P N (1985) Biological growth removal and comparative effectiveness of fungicides from central India temples for a decade in situ. In Vth International Congress on Deterioration and Conservation of Stone. Proceedings, ed. G. Felix. Presses Polytechniques Romandes, Lausanne, pp. 675-83.

Saiz-Jimenez C (1993) Deposition of airborne organic pollutants on historic buildings. Atmos Environ 27B: 77-85

Silva M (2017) Novel biocides for cultural heritage. Doctoral dissertation, Institute for Advanced Studies and Research, University of Évora. http://hdl.handle.net/10174/21001

Silva M, Rosado T, Teixeira D, Candeias A, Caldeira A T (2017) Green mitigation strategy for cultural heritage: bacterial potential for biocide production. Environmental Science and Pollution Research, 24(5). 4871-4881.

Siswowiyanto S (1981) How to control the organic growth on Borobudu r stones after the restoration. Conservation of Stone. IL ed. R. Rossi-Manaresi, Bologna. pp. 759-68.

Shaw J, Renzaglia K S (2004) Phylogeny and diversification of bryophytes. American Journal of Botany 9: 1557-1581.

Stark L R, Greenwood J L, Brinda J C, Oliver M J (2013) The desert moss Pterygoneurum lamellatum (Pottiaceae) exhibits an inducible ecological strategy of desiccation tolerance: Effects of rate of drying on shoot damage and regeneration. Am. J. Bot. 100, 1522–1531. doi: 10.3732/ajb.1200648

Sveinbjbrnsson B, Oechel W C (1992) Controls on growth and productivity of bryophytes: environmental limitations under current and anticipated conditions. - In: Bates, J. W. and Farmer, A. M. (eds), Bryophytes and lichens in a changing environment. Clarendon Press, Oxford, pp. 77-102.

Tarnawski M, Melick D, Roser D, Adamson E, Adamson H, Seppelt R (1992) In situ carbon dioxide levels in cushion and turf forms of Grimmia antarctici at Casey Station, East Antarctica. - J. Bryol. 17: 241- 249.

Tian G Q, Bai XL, Xu J, Wang X D (2006) Experimental Studies on the Natural Restoration and the Artificial Culture of the Moss Crusts on Fixed Dunes in the Tengger Desert, China. Front. Biol. China 1: 13

Tiano P (1986) Problemi biologici nella conservazione del materiale lapido. La Prefabbricazione 22: 261-272.

Tiano P (1987) Biological deterioration of exposed works of art made of stone. In Biodeterioration of Constructional Materials. The Biodeterioration Society, Publication Service. Lancashire Pol~echnic, Kew, UK, pp. 37-44.

Tiano P, Riminesi C (2017) State of arts of monumental stones diagnosis and monitoring.The International Archives of the Photogrammetry, Remote Sensing and Spatial Information Sciences, Volume XLII-2/W5, 26th International CIPA Symposium, 28 August–01 September 2017, Ottawa, Canada

Tiano P, Accolla P, Tomaselli L (1995) Phototrophic Biodeteriogens on lithoid surfaces: an ecological study. Microbial Ecology, vol 29 n° 3: 299-309

Tomaselli L, Lamenti G, Tiano P (2017) Diagnostic tools for monitoring phototrophic biodeteriogens. 10.1201/9780203746578-33.

Thornbush M J, Viles H A (2008) Photographic monitoring of soiling and decay of roadside walls in central Oxford, England. Environmental Geology 56 (3–4): 777–87.

Urzì C (2004) Microbial deterioration of rocks and marble monuments of the Mediterranean basin: a review. Corrosion Review 22: 441- 457.

Urzì C, De Leo F, Schumann P (2008) *Kribbella catacumbae* sp. nov. and *Kribbella sancticallisti* sp. nov., isolated from whitish-grey patinas in the Catacombs of St Callistus in Rome, Italy. International Journal of Systematic and Evolutionary Microbiology 58: 2090-2097.

Urzi C, De Leo F (2001) Sampling with adhesive tape strips: an easy and rapid method to monitor microbial colonization on monument surfaces. Journal of Microbiological Methods 44: 1-11.

Urzì C, Albertano P (2001) Studying phototrophic and heterotrophic microbial communities on stone monuments. Methods in Enzymology 336: 340-355

Urzì C, De Leo F, Donato., La Cono V (2003) Study of microbial communities colonizing hypogean monument surfaces using non destructive and destructive sampling methods.

Urzì C, Realini M (1996) Bioreceptivity of Rock Surfaces and Its implication in Colour Changes and Alteration of Monuments, Study Case of Noto's Calcareous Sandstones. Dechema Monographs vol 133, VCH Ed.151-159.

Van Der Molen, L, Garty J, Aardema BW, Krumbein W (1980) Growth control of algae and cyanobacteria on historical monuments by a mobile UV unit (MUVU), Studies in Conservation, 25 (2): 71-77.

Varela Z, Real C, Branquinho C, Paço T A, Carvalho R C D (2021) Optimising Artificial Moss Growth for Environmental Studies in the Mediterranean Area. Plants 10: 2523. https://doi.org/10.3390/plants10112523

Warscheid Th (1996) Biodeterioration of stones: analysis, quanti cation and evaluation. Proceedings of the 10th International Biodeterioration and Biodegradation Symposium, Dechema-Monograph No. 133, Dechema, Frankfurt, pp. 115 –120.

Warscheid T, Braams J (2000) Biodeterioration of stone: a review. International Biodeterioration and Biodegradation 46: 343–368

Warscheid T, Petersen K, Krumbein W (1988) Effect of cleaning on the distribution of microorganisms on rock surfaces. In : Biodeterioration 7, D.R.Houghton, R.N. Smith and H.O.W.Eggings Eds., Elsevier Applied Science Publ. 455-460.

Wellman C, Osterloff P, Mohluddin U (2003) Fragments of the earliest land plants. Nature 425: 282–285.

Winkler E M (1975) Stone decay by plants and animals. Stone properties, durabilities in man's environment. Springer Verlag Ed. 154-163.

Young M E (1997) Biological growth and their relationship to the physical and chemical characteristics of sandstones before and after cleaning. Ph.D. Thesis, The Robert Gordon University, Aberdeen

Zehnder K, Schoch O (2009) Eflorescence of mirabilite, epsomite and gypsum traced by automated monitoring on-site. Journal of Cultural Heritage 10 (3): 319–30.

Zhao Y, Bowker M A, Zhang Y, Zaady E (2016) Enhanced recovery of biological soil crusts after disturbance. In Biological Soil Crusts: An Organizing Principle in Drylands; Springer International Publishing: Berlin, pp. 499–523.

Chapter 19

Certain Archaeological Features and Deterioration Aspects of Two Buddhist Sites: Devnimori and Sanchi with Respect to Lichens

Arun Arya

Department of Environmental Studies, Faculty of Science, The Maharaja Sayajirao University of Baroda, Vadodara – 390 002, Gujarat, India
e-mail: sarojarun10arya@rediffmail.com

ABSTRACT

Buddhism is one of the world's major religions. It originated in India in 563–483 B.C.E. with Siddhartha Gautama, and over the next millennia it spread across Asia and the rest of the world. Buddhists believe that human life is a cycle of suffering and rebirth, but that if one achieves a state of enlightenment (nirvana), it is possible to escape this cycle forever. Siddhartha Gautama was the first person to reach this state of enlightenment and was, and is still today, known as the Buddha. Buddhists do not believe in any kind of deity or god, although there are supernatural figures who can help or hinder people on the path towards enlightenment. According to legend, King Ashoka, who was the first king to embrace Buddhism (he ruled over most of the Indian subcontinent from c. 269 - 232 B.C.E.), created 84,000 stupas and divided the Buddha's ashes among them all. Prominent Buddhist sites in the country include Bodh Gaya, Devnimori, Sanchi, Sarnath, Sravasti, Kaushambi and Kapilvastu.

Deterioration of stone monuments is a complex process and is brought about by a number of factors working together. Besides weathering microbes and lichens can cause deterioration. They may be crustose, filamentous, thallose or much branched structure, and are inhabitants on old as well as modern construction made of cement or lime plaster which helps in their favourable growth. Lime loving or calicicolous lichens have inbuilt tolerance to overcome the air pollution and dry condition of the substratum. The chapter discuss archaeological details of one Buddhist site Devnimori situated in Gujarat and another world heritage site Sanchi present in Madhya Paradesh. Deterioration caused to the stone structure at Sanchi by microorganisms and few lichens is described, no such occurrence is reported in literature from the site.

Keywords: Archaeological details, Stupa, Deterioration, Lichens, Buddhist sites, Devnimori, Sanchi

19.1 Introduction

Deterioration of the cultural heritage and historic buildings is a result of both natural and anthropological contribution involving flora, fauna and microorganisms. While the world travels towards urbanisation, the nature is getting degraded in the pursuit. Though we attempt to foresee the major environmental impacts of pollution, the complete scenario of sustainability remains uncared. Apart from the health issues and ecological drawbacks, there are multitudes of socio-economic problems due to pollution which are not fully addressed in the popular scientific investigations. There is no exaggeration in stating that air pollutants has made a massive impact on the ancient monuments, which are standing as proofs of ancient cultural heritage and legacy of every nation. As they are inevitably waiting for the ultimate dismissal, no modern technique can bring them back at any cost. But we can reduce the rate of damage by adopting scientific conservation methods or restrict the microbial damage. Comite *et al.* (2021) studied the impact of atmospheric pollution on outdoor cultural heritage. The major sources of human induced air pollutants include industries, power plants, mining, and automobiles. There are multiple pathways in which the age-old artefacts and structures are damaged due to continuous exposure to the prevailing air pollutants. The environmental factors like high temperature, humidity, rainfall and sunlight interacts with the surface of the stone and make it susceptible for biological damage (Singh and Bhadauria 2014).

The Indian culture, often labelled as an amalgamation of various cultures, spans across the Indian subcontinent and has been influenced and shaped by a history that is several thousand years old. India is the birthplace of Hinduism, Vedic, Buddhism, Jainism, Sikhism, and other religions. Some of the important Buddhist sites include Sanchi, Sarnath, Kapilvastu and Bodh Gaya. After Nirvana the Buddha remains were cremated under eight mounds with two further mounds encasing the urn and the embers. The relics of the Buddha were spread between eight stupas, in Rajagriha, Vaishali, Kapilvastu, Allakappa, Ramagrama, Pava, Kushinagar and Vethapida. The Pipraha stupa was built first. King Ashoka embraced Buddhism and recovered relics from stupas (except Ramagrama) and created 84,000 stupas throughout the country. Bodh Gaya contains one of the holiest of Buddhist sites: the location where, under the sacred peepal (*Ficus religiosa* L.), or Bo tree, Gautama Buddha (Prince

Siddhartha) attained enlightenment and became the Buddha. A simple shrine was built by the emperor Ashoka in 3rd century BC to mark the spot, and this was later enclosed by a stone railing (1st century BCE), part of which still remains. The uprights have representations of the Vedic Gods Indra and Surya, and the railing medallions are carved with imaginary beasts. The shrine was replaced in the Kushan period *i.e.* 2nd Century CE by the present Mahabodhi temple designated a UNESCO World Heritage site in 2002.

Devnimori, is a Buddhist archaeological site in northern Gujarat, about 2 km from the city of Shamlaji, in the Aravalli District of Northern Gujarat, India. The site is variously dated to the 3rd century or 4th century CE. Its location was associated with trade routes and caravans in the area of Gujarat. Site excavations have yielded Buddhist artefacts dated prior to 8th -century in the lowest layer, Buddhist artwork from the Gurjara-Pratihara period. The site was excavated between 1960 and 1963. It is thought that this architectural pattern then became the prototype for the later development of monasteries with shrines in Devnimori, Ajanta, Aurangabad, Ellora, Nalanda, Ratnagiri, Odisha and others. The viharas in Devnimori were built from fired bricks. It has also has residential caves with water cisterns, as at Uparkot in Junagadh. Devnimori also has a stupa where stacked relic deposits were found. This is the only case of a free-standing stupa in the area of Gujarat. Nine images of the Buddha were found inside the stupa. The Buddha images clearly show the influence of the Greco-Buddhist art of Gandhara and have been described as examples of the Western Indian art of the Western Satraps.

About 10 km from Vidisha, the Buddhist monuments at Sanchi are located on a serene and picturesque forested plateau, are also considered to be the sacrosanct Cetiyagiri in the Sri Lankan Buddhist chronicles, where Mahindra, the son of Emperor Aœoka, stopped prior to undertaking his journey as a missionary to Sri Lanka. The stupas, temples, viharas, and stambha at Sanchi in central India are among the oldest and most mature examples of aniconic arts and free-standing architecture that comprehensively document the history of Buddhism from the 3rd century BCE to the 12th century CE. The enshrined remains of Sariputra and Maudgalyayana (chief disciples of Buddha) in Sanchi were venerated by Theravadins, and continue to be revered to the present day. Sanchi remained an important seat of Buddhism until the 13th century CE.

All over the world, the deterioration of rocks and stones have become matter of great concern. Airborne fungi, algae and lichens and angiospermic plants were responsible for decay of rock cut images at Gwalior fort (Jain 2001), cyanobacterial forms like *Tolypothrix, Chrococcidiopsis, Xenococcus* and *Nostoc* were associated with small rock pieces from Konark and Lingraj temples of Bhubaneswar (Adhikary 2001), lichens were also reported causing damage in Jokaomyo-ji temple in Kamakura, Japan (Imai *et al.*2005). Blackening of plastered walls of Himeji castle and Oura Roman Catholic Church in Nagasaki, Japan was caused by the fungus *Cladosporium* (Arai and Yamagishi 2005). Heritage monument like Buddhist Stupa of Sanchi is made up of bricks, stones and mortar, plaster *etc.* These stone structures suffer damage caused by bideteriogens like microbes, lichens and plants. Lichens biodiversity, their biodeteriogenic action on historical monuments and the possible preservation aspects for the substrata have been investigated by many authors (Nimis *et al.*, 1992; Romão and Rattazzi, 1996; Rigamonti, 2008; Nascimbene *et al.*, 2009) The lichens are of great ecological importance because of their pedogenetic action on the lithic surfaces (Schatz 1963; Lounamaa 1965; Syers 1969; Iskandar and Syers 1972; Williams and Rudolph 1974; Ascaso *et al.*, 1982; Wilson and Jones 1983). Lichens a unique combination of Myco and photobionts are considered as pioneer on stones, plants and on soil. Through respiration they produce CO_2, which is transformed, within the thallus into carbonic acid. With their metabolism release organic acids with chelating properties, which complexes with mineral cations present in the rock substratum. These metabolic substances cause compositional change in the stone surface layer immediately beneath the lichen structure, with depletion of its main chemical elements (Al, Mg, Mn, Zn, Si, Ca, K, Fe) and the accumulation of some of these, especially Ca, inside the thallus (Jones and Wilson 1985). Moreover they can cause physical damage (Jones *et al.*, 1987) as a result of the hyphal deep penetration (up to 15 mm) and periodical thallus contraction and swelling (up to 35 time its dry weight; Gehrmann *et al.*, 1989) as a consequence of climatic changes. With their slow constant growth (about 1 mm per year Ciarallo *et al.*, 1985), they can cover large stone areas with undesired chromatic patches.

19.2 About certain Buddhist Sites in India

Buddhists believe in a wheel of rebirth, where souls are born again into different bodies depending on how they conducted themselves in their previous lives. Buddhism is connected to concept of karma. The Buddha taught about Four Noble Truths. The first truth is called "Suffering (*dukkha*)," second is "Origin of suffering (*samudâya*) due to desire (*tanhâ*), third is "Cessation of suffering (*nirodha*), and the fourth is "Path to the cessation of suffering (*magga*)" is about the way to achieve enlightenment. There are two main groups of Buddhism: Mahayana Buddhism is common in Tibet, China, Taiwan, Japan, Korea, and Mongolia. It emphasizes the role models of *bodhisattvas*. Theravada Buddhism is common in Sri Lanka, Cambodia, Thailand, Laos, and Myanmar *etc.* It emphasizes a monastic lifestyle and meditation as the way to enlightenment.

The Lord Buddha was born in 623 BC in the sacred area of Lumbini located in the Terai plains of southern Nepal, testified by the inscription on the pillar erected by the Mauryan Emperor Asoka in 249 BC. Lumbini is one of the holiest places of one of the world's great religions, and its remains contain important evidence about the nature of Buddhist pilgrimage centres from as early as the 3rd century BC. Prominent Buddhist sites in the country include Bodh Gaya, Devnimori, Sanchi, Sarnath, Sravasti, Kaushambi and Kapilvastu. Bodh Gaya in Bihar, contains one of the holiest of Buddhist sites: the location where, under the sacred pipal (*Ficus religiosa*), tree, Prince Siddhartha attained enlightenment and became the Buddha. A simple shrine was built by the emperor Ashoka (3rd century BCE) to mark the spot, and this was later enclosed by a stone railing (1st century BCE), part of which still remains. The uprights have representations of the Vedic Gods Indra and Surya, and the railing medallions are carved with imaginary beasts. The shrine was replaced in the Kushan period (2nd century CE) by the present Mahabodhi temple designated a UNESCO World Heritage site in 2002. Sarnath is 10 km away from Varanasi. Lord Buddha himself selected Sarnath for deliverance of his first sermon to his five disciples. Chaukhandi and Dhamek stupa, Ashokan Pillar, Chinese Buddha temples and museum are situated here. Another site Kapilvastu is 20 km from Siddarthnagar and 105 km from Gorakhpur city in U.P. Sakya ruler and father of Siddharth king Shuddhodhana ruled here and childhood was spent here. Nirwan stupa and Mahanirvana temple are situated in Kushinagar 51 km from Gorakhpur. Lord Buddha breathed his last here and achieved Mahanirvana. Kaushambi is 60 km from Prayagraj Buddha gave several sermons in this city of Vatsa janpad. Ruins of old fort, Ghosshitaram Monastery and Ashokan pillar can be found here.

The "stupa" in Sanskrit is for heap. It is an important form of Buddhist architecture –a place of burial or a receptacle for religious objects. At its simplest, a stupa is a dirt burial mound faced with stone. Archaeologists in India have observed that a number of early Buddhist stupas or burials are found in the vicinity of much older, pre-historic burials, including megalithic sites. Associated with the Indus Valley as broken Indus-era pottery was incorporated into later Buddhist burials. Structural features of the stupa- including its general shape and the practice of surrounding stupas with a stone or wooden railing. The Buddha had left instructions about how to pay homage to the stupas, "And whoever lays wreaths or puts sweet perfumes and colours there with a devout heart, will reap benefits for a long time". This practice would lead to the decoration of the stupas with stone sculptures of flower garlands in the Classical period.

19.3 Buddhist Site in Gujarat: Devni Mori

The existence of Buddhists was present centuries ago in Gujarat. Devni Mori is located two kilometers away from Shamlaji and 20 km away from Bhiloda of Arvalli district. There were many hills on the banks of the river Meshwo on that time. One of them named 'Bharajaraja's Tekara', and the surrounding land were the area of Vihara for the Buddhist Monks of the time. The height of the stupa was 85 inches, and there were 36 rooms for the adjoining Buddhist monks. Two bridges of jewellery were found during the research, on one of them, it was written in a Sanskrit language that 'this construction was designed by Agnivarma Sudarshan and it was constructed by a king named Rudrasen. At that time, this supreme stupa was in Kshatriya time. Such stupas were also found in Sindh (Eastern India before India) and Takshashila.

Figure 19.1

(a) Buddha relics placed in the Museum of Archaeology Department of The M.S. University of Baroda; (b) Chancellor of the University Rajmata Shrimati Subhanginiraje Gaekwad during her visit to the museum.

These stupas appear to be shown above that the time was a science progressive and great. Around 272 iron pieces of Buddhist times were found around the stupa. It is believed to be of the third century. The site of Devni Mori included numerous terracotta Buddhist sculptures (but no stone sculptures), Pottery *etc.* were obtained from here. The site is variously dated to the 3rd century or 4th century CE, or circa 400 CE. Its location was associated with trade routes and caravans in the area of Gujarat. Site excavations have yielded Buddhist artefacts dated prior to 8th-century in the lowest layer, Buddhist artwork from the Gurjara-Pratihara period.

The site was excavated between 1960 and 1963 under the guidance of Mehta and Chowdhary (1966). The site became flooded by a water reservoir, a project started in 1959 and completed over 1971-1972 over the nearby Meshwo River. The site of Devni Mori included numerous terracotta Buddhist sculptures (but no stone sculptures), also dated to the 3rd-4th century CE, and which are among the earliest sculptures that can be found in Gujarat. The remains are located in the Shamlaji Museum and Baroda Museum and Picture Gallery. Devni Mori has a specific construction pattern for a monastery, with an image shrine built opposite the entrance. This kind of arrangement was initiated in northwestern sites such as Kalawan (in the Taxila area) or Dharmarajika. It is thought that this architectural pattern then became the prototype for the later development of monasteries with shrines in Devni Mori, Ajanta, Aurangabad, Ellora, Nalanda, Ratnagiri, Odisha and others. The viharas in Devni Mori were built from fired bricks. It also had residential caves with water cisterns, as at Uparkot in Junagadh.

Devnimori also has a stupa where stacked relic deposits were found. This is the only case of a free-standing stupa in the area of Gujarat. Nine images of the Buddha were found inside the stupa. The Buddha images clearly show the influence of the Greco-Buddhist art of Gandhara and have been described as examples of the Western Indian art of the Western Satraps. Three relics caskets were retrieved from the stupa. One of these caskets bears an inscription which mentions a date: the 127th year in the reign of Western Satrap ruler Rudrasena. In the year 127 of the Kathika kings, when king Rudrasena was ruling, the erection of this stupa, which was banner of this earth, was done. It was the 5th day of Bhadrapada – *Anonymous (2014).* According to Mehta and Chowdhary (1966), the art of Devnimori prove the existence of a pre-Gupta era Western Indian artistic tradition. This tradition, they suggest may have influenced the art of the Ajanta Caves, Sarnath and other places from the 5th century onward. As a matter of fact, Devnimori represents the extension of Gandharan influence to the subcontinent, which persisted locally with the sites of Mirpur Khas, Úâmalâjî or Dhânk, a century before this influence would further extend to Ajanta and Sarnath.

Figure 19.2: Stone Relic Casket Recovered from Devnimori.

19.4 Sânchi : Buddhist site in Madhya Pradesh

The town of Sanchi is synonymous with Buddhist Stupas - hemispherical structures typically containing relics of the Buddha or his followers. The Stupas of Sanchi were constructed on the orders of Emperor Ashoka in the 1st century BC to preserve and spread the Buddhist philosophy. Sanchi has been protecting these beautiful and sacred architectural wonders, just the way these wonders have been safeguarding ancient history and art of the Mauryan period. The region of Sânchi, however, like the great centres at Sârnâth and Mathura, had a continuous artistic history from the 3rd century BC to the 11th century AD. The city declined in importance in the 12th century. The numerous stupas, temples, monasteries and an Ashokan pillar have been the focus of interest and awe for global audiences as well. In fact, UNESCO has given the status of 'World Heritage Site' to the Mahastupa in 1989.

Sânchi is the site of three stupas: stupa No.1, an Aúhokan foundation enlarged in succeeding centuries; No. 2, with railing decorations of the late ŒuEga period (*c.* 1st century BC); and No. 3, with its single toran (ceremonial gateway) of the late 1st century BC–1st century AD. Other features of interest include a commemorative pillar erected by the emperor Aœhoka (*c.* 265–238 BC); an early Gupta temple (temple No. 17), early 5th century, with a flat roof and pillared portico; and monastic buildings ranging over several centuries.

The four torans of the Great Stupam which were added in the 1st century BC are the crowning achievements of Sânchi. Each gateway is made up of two square posts topped by capitals of sculptured animals or dwarfs, surmounted by three architraves, which end in spirals not unlike the rolled ends of scrolls.

On the topmost crossbar were placed originally the trident-like symbol of the triratna and the wheel of the law. The crossbars and the intervening square dies between them are covered with relief sculpture depicting the events of the Buddha's life, legends of his previous births (*Jâtaka* stories), and other scenes important to early Buddhism such as the emperor Aœoka's visit to the Bo tree, as well as auspicious symbols. Inscriptions give the names of the donors of the relief; one commemorates the gift of the ivory workers of Vidisha and has given rise to the suggestion that the tradition of working in ivory may have been translated into stone. The reliefs are deeply carved, so that the figures seem to swim against a sea of dark shadow cast by the strong Indian sun. The panels, which employ the device of continuous narration, are crowded, rich, and brimming with life. The Buddha is depicted throughout in symbolic form, by a wheel, an empty throne, or a pair of footprints.

The gateways are having magnificent figures of female yakshas (earthly spirits). They serve no true architectural purpose, yet their pose, with leg thrusting against the post and arms entwined in the branches of a tree, is appropriate to the space they fill. A damaged torso of a Sânchi yaksha is preserved at the Museum of Fine Arts, Boston. The sculptural treatment shows considerable advance over similar yaksha figures at Bhârhut (stupa of the mid-2nd century BC, in Madhya Pradesh). There is much smoother movement in the flexed bodies and more attention given to the open space around the figure. The fertility aspect of the union of maiden and tree is emphasized in the heavy breasts and hips and the transparent draperies. The smooth modelling and roundness of forms combine to give the 'yaksha' figures a wonderful vitality and a sense of "swelling from within" characteristic of the finest Indian sculpture of all periods.

Buddhist monuments at Sanchi contain an appreciable concentration of early Indian artistic techniques and Buddhist art, referred to as its Anionic School. Depicting Buddha through symbols, the sculpted art shows the evolution in sculpting techniques and the elaboration of icons. Stories and facts of great religious and historical significance, enlivened with bas-relief and high-relief techniques, are also depicted. The quality of craftsmanship in representing the gamut of symbolism through plants, animals, human beings, and Jataka stories shows the development of art though the integration of indigenous and non-indigenous sculpting traditions.

Why this Buddhist Site is Unique in its Architecture?

Criterion (i): The perfection of its proportions and the richness of the sculpted decorative work on its four gateways make Stupa 1 an incomparable artistic achievement. The group of Buddhist monuments at Sanchi – stupas, temples and monasteries – is unique in India because of its age and quality.

Criterion (ii): From the time that the oldest preserved monument on the site was erected, *i.e.*, Aœhoka's column with its projecting capital of lions inspired by Achaemenid art, Sanchi's role as intermediary for the spread of cultures and their peripheral arts throughout the Mauryan Empire, and later in India of the Sunga, Shatavahana, Kushan and Gupta dynasties, was confirmed.

Figure 19.3: Picture of Buddha on Folder and Postage Stamp of
2 Anna Depicting Leaf of Bodhi Tree (Peepal) Released by Postal Dept. on
24/5/1956 to Commemorate 2500th year of Buddha Birth.

Criterion (iii): Having remained a principal centre of Buddhism up to early medieval India following the spread of Hinduism, Sanchi bears unique witness as a major Buddhist sanctuary in the period from the 3rd century BCE to the 1st century CE.

Criterion (iv): The stupas at Sanchi, represent the most accomplished form of this type of monument. The hemispherical, egg-shaped dome (anda), topped with a cubical relic chamber (harmika), is built on a circular terrace (medhi); it has one or two ambulatories for the faithful to use (pradakshina patha). Representing a transition from wood structures to stone, the railings (vedika) and the gateways (torana) also bear witness to the continued use of the primitive forms of megalithic tumuli covered with an outer layer and surrounded by a palisade.

Criterion (vi): Sanchi is one of the oldest extant Buddhist sanctuaries. Although Buddha never visited the site during his earthly existence, the religious nature of this shrine is obvious. The chamber of relics of Stupa 3 contained the remains of Sariputra, a disciple of Shakyamuni who died six months before his master; he is especially venerated by the occupants of the "small vehicle" or Hinayana.

Integrity

Within the boundaries of the property are all the known elements necessary to express its Outstanding Universal Value, including the sculpted monolithic pillars, sanctuaries, temples, and viharas atop and along the slopes of the hillock of Sanchi. These elements demonstrate the complete vocabulary of mature Buddhist aniconic art and free-standing architecture. The property, which also encompasses its near natural setting, is thus of adequate size to ensure the complete representation of the features and processes that convey the significance of the Buddhist Monuments at Sanchi. The property is in a good state of conservation.

Authenticity

The archaeological remains of the Buddhist Monuments at Sanchi are authentic in terms of their locations and setting, forms and design, and materials and substance, as well as, to a degree, their spirit. These representations of mature Buddhist free-standing architecture and aniconic sculpted art remain at their original locations and in a setting that is sympathetic. The Sanchi stupas were restored in the early 20th century and demonstrate all the original features characteristic of mature Indian stupas. Though abandoned for about 600 years, Sanchi has witnessed the revival of a pilgrimage from all over the Buddhist world, and in particular from Sri Lanka, thus testifying to the religious significance of this place. The site is alive with chants and prayers to immortalize the remains of Sariputra and Maudgalyayana, two of the foremost disciples of Lord Buddha.

Figure 19.4: Picture of Great Stupa of Sanchi during Sunset (Photographed on 15/3/22).

Figure 19.5: Beautiful Gate and Hemispherical Structure of Great Stupa Containing Relics of the Buddha.

Temple 45 is a ruined Latina temple set within the eastern walls of Monastery 45, built, according to this study, in at least two different stages between the mid 9th century and the beginning of the 10th century AD. Indian temple types are distinguished by the shape of their spires, and Latina temples

Figure 19.6a-c

(a) Architraves of the north gateway (Toran) to the Great Stupa (stupa No. 1) at Sânchi; (b) Stupa 2; (c) Sculpture of Lord Buddha placed in front of a Toran gate.

have smoothly curving edifices with quadrangular plans, each face faceted by projections made up of piled courses. The layered eaves of Latina temples' central projections are covered in 'creepers' or 'lata' of interlocking gavaksas (stylised gable and dormer window forms), lending the temple type its name. The Latina spire from Temple 45 has now fallen away, leaving in place only the central sanctum, the rough inner core of the lowest part of the spire, and the base of its entrance hall. In addition to the standing structures, however, about 500 of the temple's fragmented remains are stacked around Monastery 45's neighbouring areas, many of which come from the Latina superstructure.

19.5 Protection and Management by the State Government

The long-term challenges for the protection and management of the property are to control the impact of visitors, and natural impacts including humidity and the industrial development in the region. A management plan is in the process of being developed to ensure the long-term safeguarding of the archaeological vestiges of the property while allowing for the property to continue being visited by pilgrims and tourists from around the world. The rural landscape surrounding the property is managed by the Nagar panchayat and is governed by the M. P. Bhumi Vikas Rules (1984), which can regulate and protect heritage sites. In addition, clause 17 of section 49 of the Madhya Pradesh Panchayati Rajya Adhiniyam (1993) provides heritage protection. Governed by the aforementioned legislative instruments, including the AMASR Act 2010, the Sanchi Vikas Yojna Praroop (2001) and a plan under Nagar tatha gram nivesh Adhiniyam (1971), prepared by the Madhya Pradesh Town and Rural Planning Department, Bhopal, are being implemented to manage areas beyond the protected and prohibited area.

19.6 Deterioration Caused by Cyanobacteria, Algae and Fungi

Bacteria, cyanobacteria and other microbes cause biological weathering (Dorn 1998). Although fungi and algae may colonize surfaces faster than lichens but the damage is lesser (Dorn 1998; Hale 1975). Brock (1987) defined microbial ecology as the study of microorganisms in their natural environment, and emphasized that the researcher should examine the microorganisms *in situ*. Genera *Chroococcus* and *Nostoc* were forming brilliant green patina, red powdery granules and black biofilm on wall paintings of Holy Saviour's cave in Vallerano, Italy. *Gloeocapsa* caused discoloration of architectural acrylic paints in Sao Paulo and Ubatuba, Brazil, as well as formation of black biofilm. Biodeterioration of mortar was prevented by addition of anatase to the formulation that was used on two external walls of Palacio Nacional da Pena in Sintra, Portugal (Falkiewicz-Dulik *et al.*, 2015).

Early investigations on biodeterioration were mainly focused on the isolation and identification of the cultural microorganisms for a description of them without specific function to the biodeterioration (May 2000). Such practice had been persisted before the application of polymerase chain reaction (PCR) to cultural heritage microbiology research to reveal the complex microbial community without culturing and isolation (Rölleke *et al.*, 1998). The latest high-throughput sequencing and metagenomics provide a much deeper description of the microbial community and composition without culturing or isolation (May 2000; Sterflinger and Pinar 2013; Zhang *et al.*, 2019; Ma *et*

Figure 19.7

(a) Beautiful Toran gate in front of a Sanchi Stupa 1 showing blackening; (b and c) Two stone pillars near Stupa showing black coloured growth caused by Cyanobacteria and fungi; (d) Stone pillar showing growth of Lichen.

al., 2020; Meng *et al.*, 2020). This new approach also has some major weakness in its inability to identify the active microbes from the genomic DNA-based community analysis and, in addition, the active deteriorating ones are not identified from the community. Because of these, the advances made on knowing the community better have not translated to useful results on the deteriorating processes for a more effective management and prevention (Liu *et al.*, 2020). Amman *et al.* (1995) found that many fungi inhabiting rock substrate cannot be isolated or cultured. These microbes were viable but non-culturable. This study presented by Gutarowska and Zakowska (2010) showed an estimation of mould contamination of 31 samples, as well as a comparison of two methods of mould estimation: dilution plate method and UV-determination of ergosterol. Many fungi which fail to be cultured can identified by this method. Cecchi *et al.* (1996) proposed fluorescence lidar technique for the monitoring of biodeteriogens by remote sensing method.

Preliminary examination of stone pillar as shown in Figure 7 B and C shows the blackening caused by Cyanobacteria like *Nostoc*, *Scytonema* and certain Dematiaceous fungi like *Cladosporium* and *Drechslera*. In certain places the pillars are too much dark due to growth of microbes. Microorganisms colonizing surfaces of cultural heritage can destruct the underlying materials through their influences on the physical, chemical, and bio-receptibility of the substratum materials, especially metabolic activities and biochemical reaction. Biofilms as a physical layer on the outer surface of cultural heritage alter the thermal property and moisture contents, which in turn have their impact on the biofilms and the activities of the microorganisms. The mechanisms involved in biodeterioration of natural sandstone and man-made materials include both abiotic and biological ones (Mitchell and Gu 2000; Liu *et al.*, 2020). The pioneering microorganisms form biofilms on rock and stone surfaces and cause the discoloration in appearance, alteration of porosity and moisture diffusivity in and out of the stone. Adhikary and Satapathy DP (1996) reported *Tolypothrix byssoidea* from temple rock surfaces of coastal Orissa. In only two of the six sites sampled in Angkor Wat were filamentous cyanobacteria, *Microcoleus*, *Leptolyngbya*, and *Scytonema*, found; the first two detected by sequencing of 16S rRNA gene library clones from samples of a moist green biofilm on internal walls in Preah Khan, where *Lyngbya* (possibly synonymous with *Microcoleus*) was seen by direct microscopy as major colonizer. *Scytonema* was detected also by microscopy on an internal wall in the Bayon. This suggests that filamentous cyanobacteria are more prevalent in internal (high moisture) areas (Gaylarde *et al.*, 2012).

Agrawal (2007) suggested use of 10 per cent aqueous ammonium hydroxide solution in which small amount of neutral detergent and hydrogen peroxide were added. Application of CMC paste and peeling off of the dried layers was suggested to remove microbial growth from red stones. Presence of dirt and dirt is always associated with such spots. Detergent can help to clean it. There is conspicuous growth of microbial flora on these standing stone pillars. The efforts should be made to restore them in a sequential treatment to minimize the damage. Stone stabilization studies were performed by Ginell and Kumar (2004), they concluded that cycloaliphatic epoxy and bisphenol- an epoxy resin when applied as dilute solution in water can consolidate soft lime stone and increase resistance. A polyurethane dispersion resin in an organic solvent can increase erosion resistance. Biocides are effective before or after consolidation. Copper compounds as $CuCl_2$ are effective against existing microflora. Silver nitrate and Polybar are useful biocides

19.7 Association of Lichens with Heritage Monuments of Sanchi

The role of lichens in stone weathering is strictly related to understanding the damage and removal treatments. Ascaso *et al.* (1976; 1982) reported pedogenic action caused by parmelia and other lichens. The chelating properties of lichens and lichenic acids were studied by Schatz (1963). A few recent articles are a plea for greater consideration when treating exterior stone covered with lichens. Fry (1927) suggested mechanical action for removal of crustose lichens. Although their removal from tombstones, sculptures, and monuments is widely practiced, it can damage the stone, and, in the case of extensive and repeated use of biocides, the environment. The mechanical removal of crustose lichens is particularly difficult because the thallus forms an intimate association with the substrate. Hence, its removal leads to severe structural damage (Scheerer *et al.*, 2009). Rather than resort to mechanical, or biocidal cleaning, all of which have major disadvantages. Sheppard (2007) favours minimal intervention proposing non-destructive documenting/recording of the monuments and letting the lichens contribute to the aesthetics of churchyards and cemeteries. Moreover, certain forms of biological growth can have a scarcity or rarity that must be taken into account when planning conservation work (Watt 2006). The metabolic substances produced by lichens, cause compositional change in the stone surface layer immediately beneath the lichen structure, with depletion of its main chemical elements (Al, Mg, Mn, Zn, Si, Ca, K, Fe) and the accumulation of some of these, especially Ca, inside the thallus (Jones, and Wilson1985). Moreover they can cause physical damage as a result of the hyphal deep penetration and periodical thallus contraction and swelling (up to 35 time its dry weight Gehrmann *et al.*, 1989) as a consequence of climatic changes. With their slow constant growth about 1 mm per year (Ciarallo *et al.*, 1985), they can cover large stone areas with undesired chromatic patches. With respect to their morphology, crustose lichens are closely attached to the substratum and are more significant in stone weathering than foliose and fruticose lichens.

Among crustose lichens the ephi and endolithic strains are very important for stone biodeterioration (Richardson1973). In fact, these organisms have been proposed and used as biomonitors (Monte 1993). Different lichenic species

Figure 19.8a-b

(a) Part of a stone pillar showing growth of lichen patches of *Lecanora*; (b) Part of beautiful Toran gate near Stupa 1, note the blackening and repair work done.

develop selectively upon specific lithotypes (Piervittori and Laccisaglia1993) and their colonization can be supported by the contribution of the organic nitrogen present in bird excrements such as crows and pigeons (Jones and Wilson1985). The genera of ephi and endolithic crustose lichens, frequently isolated from stone monuments, are *Protoblastenia, Verrucaria, Caloplaca, Aspicilia, Lecanora and Xanthoria* (Garcia-Rowe and Saiz-Jimenez1991; Pinna 2014). It is evident that it is this ability which helps the lichens to survive in a desert environment. Measurements of CO_2 exchange in natural habitats in the Negev desert (Lange *et al.*1970) demonstrated that *Ramalina maciformis, Teloschistes lacunosus* and six additional species of *Caloplaca, Diploschistes, Squamaria,* and *Xanthoria* were able to photosynthesize by means of dew-water condensation in the thallus and by water-vapour intake.

The saxicolous and terricolous lichens also grow luxuriantly on exposed rocks in moist humid places (Thakur *et al.*, 2020). The common saxicolous lichens of the temperate regions were described by Mishra and Upreti (2015). Many authors showed that lichens are indeed generally defacing and intrinsically damaging. Moreover, the decay of the lichen thallus, which occurs on the centre of the colonies of some species, can open the underlying area to further weathering, resulting in cratered mounts on the rock surface (Mottershead and Lucas 2000). Although the protective effects of lichens deserve further research, this aspect cannot be generalized, and each case should be examined on its own merits (Pinna 2014).

Following lichens were observed on different structures and loose stones present in Sanchi during Feb- March 2022.

19.7.1 *Buellia chloroleuca* (Figure 19.9.a1)

Buellia is a crustose lichen given this name due to fungus known as *Buellia*. The algal partner is always *Trebouxia*. The lichen is also termed as disc lichen or button lichen. The lichen is wide spread and has more than 450 spp.

19.7. 2 *Diplotomma alboatrum* (Figure 19.9.b3)

Diplotomma alboatrum is a species of lichen of the family Caliciaceae. It is a crustose lichen with grey thallus and black apothecia. Fungi belong to Ascomycota. It was first described as Lichen *alboater* by German lichenologist Hoffmann (1784). Lichenologist Flotow (1849) transferred it to the genus *Diplotomma* The lichen is widely distributed, and has been recorded in Europe, North America, and Australasia.

19.7. 3 *Verrucaria nigrescens* Pers. (Figure 19.9.b4)

The species is known from numerous localities in Poland (Fa³tynowicz 2003). It is one of the most frequently reported *Verrucaria* species, but probably it is less common; it has been confused with other species such as *V. tectorum* (A.) Massal.) Körb. or *V. nigroumbrina*

It has drawn little interest from lichenologists and are especially poorly known in the area.

It is a noteworthy record of pyrenocarpous crustose lichens from Sanchi.

19.7. 4 *Verrucaria amphibia* Cle. (Figure 19.9.a2)

It is crustose lichen thallus is blackish- grey. It belongs to Pezizomycotina group of Ascolichen. The photobiont is *Nostoc* like cyanobacteria. Occurring on stones or marine rocks. Orange species *Verrucaria* was rare in Europe and was reported from central Poland by Krzewicka and Hachu³ka (2008).

19.7. 5 *Caloplaca cinnabarina* (Figure 19.9.c6)

Pyrenocarpous lichen have crustose habit, areolate, margin abrupt at edge or slightly lobed or notched, without elongated lobes; surface orange, smooth, without asexual propagules; cortex cellular, 14-28 µm thick, granules absent; medulla dense; Apothecia: apothecia immersed, lecanorine; disc flat, epruinose; pycnidia present, totally immersed, ostiole orange; Spot tests: apothecial margin K+ red; thallus K+ red, H-, 10 per cent N-, CN-, C-; present on rocks; worldwide distribution; Another sp. with similar colour is *C. rubelliana*. The colour of the apothecial discs. *C. rubelliana* has a scarlet red disc.

19.7.6 *Lecanora conizaeoides* (Figures 19.8a; 19.9d7)

The name *Lecanora* comes from the Greek lekanon (a small bowl) and ora (beauty), in reference to the appearance of the apothecia. *Lecanora* is characterized by asci of the Lecanora-type, simple hyaline ascospores, and crustose thalli. The apothecial margin usually contains algal cells. It is a heterogeneous assemblage of

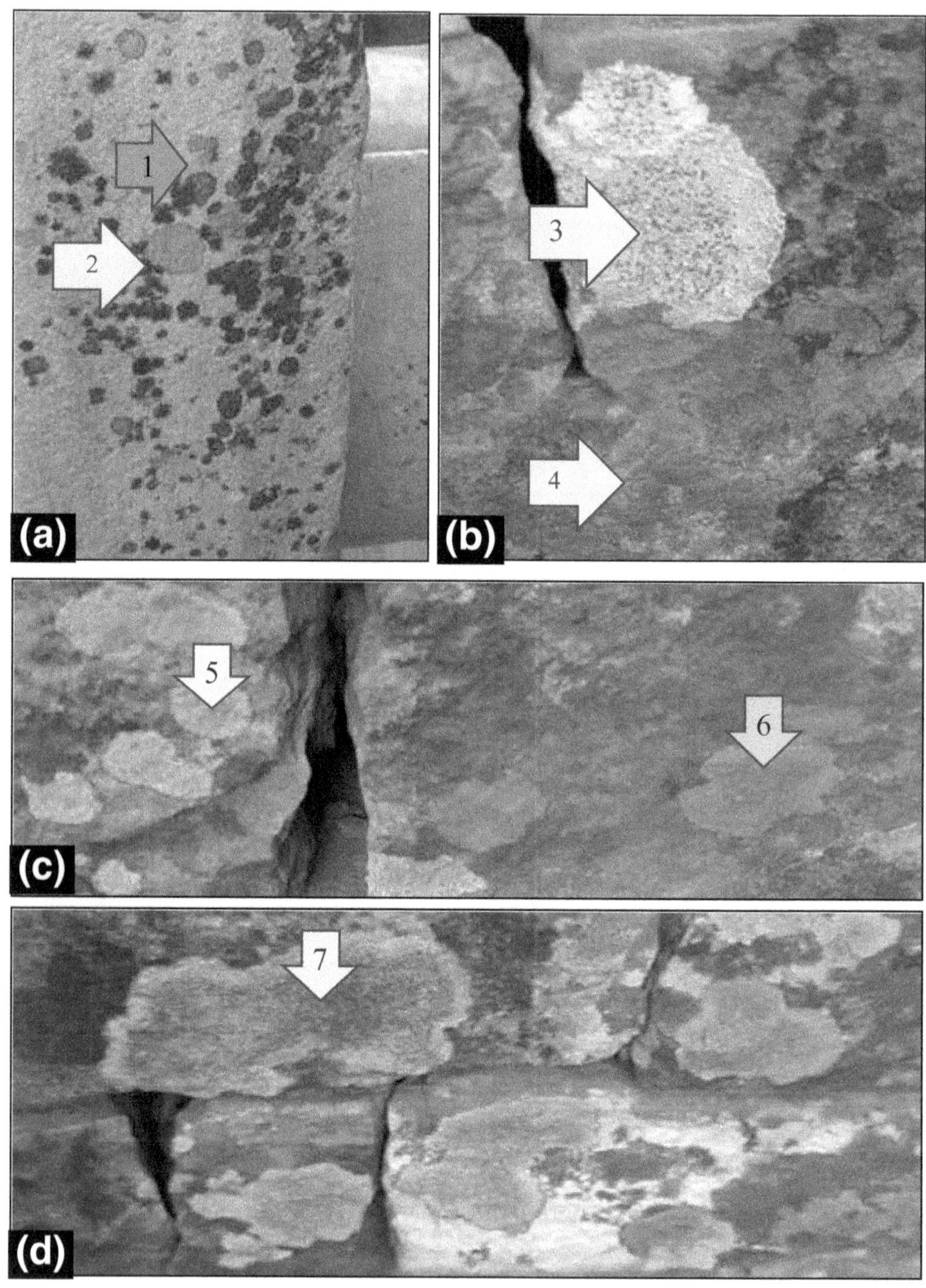

Figure 19.9a-d

(a) Part of stone pillar showing growth of lichens *Buellia chloroleuca* (a1) **Verrucaria amphibia** (a2); (b) Growth of light grey coloured lichen **Diplotomma alboatrum** (b3); (c) Pyrenocarpous reddish lichen **Caloplaca cinnabarina** (c6); (d) Blackish- brown thalli of **Lacanora** (d7).

different groups, several of which deserve generic rank. *Lecanora* is characterized by the presence of atranorin and oxalate crystals in the amphithecium. lichen thallus is leprose, adnate, granular, areolate, placodioid or peltate, rarely immersed in the substrate; prothallus: blackish brown, white to greenish-grey or not visible; surface is white or various shades of grey, yellow or brown, soredia absent or present, isidia and cephalodia absent; cortex often false and composed partly of dead algal cells), or absent; photobiont is a trebouxioid green alga, secondary one absent; The fungal component has clavate asci, Lecanora-type, 8-spored (or in some non-Sonoran species multispored); ascospores are simple, narrowly to broadly ellipsoid, smooth-walled, hyaline; Distribution: cosmopolitan, Upreti (2002) reported *Lecanora* from Khajuraho temple near Parvathi temple.

Scott (1960) pointed out that spore discharge is affected by the moistening of apothecia by rain, dew, and mist. In temperate regions the annual distribution of these types of precipitation will, he believes, result in maximum sporulation during spring and autumn. Verseghy (1965) agrees that the amount of rainfall undoubtedly influences the spore production of lichens and found that in dry years the asci are sterile in both spring and autumn. Also she noted that in rainy years the spore production is intensive in autumn. The adaptation of lichens to desert conditions was demonstrated by Pearson and Skye (1965) Similarly, Antarctic lichens that exist also under drought conditions can regain their full photosynthetic capacity after several weeks of desiccation (Lange and Kappen 1972).

Upreti (2002) suggested use of 5 per cent solution of morpholine in methylated spirit, in place of use of ammonia vapours for neutralizing the acids. Then an application of citrimide 0.1 per cent in alcohol is given to provide a long term protection. The final coat of Polyvinylacetate (PVA) in solution form has been found useful by Awasthi (1991). Lichen encrustations on the basaltic stone were removed by UV laser pulses (355nm) at a fluence of 0.35J/cm ^{2}. Laser at 1064nm was not successful as the high thermal energy caused the stone minerals to melt (Cappitelli 2020). In vitro experiments carried out by Speranza *et al.* (2013) showed sever damage of cellular structure of *Verrucaria nigrescense* and fungi and algae grown endolithically after Nd:YAG laser treatment with an infrared wavelength (1064nm,5ns). Both lasers Nd:YAG and an Er:YAG, damaged the Asco-lichen thalli of *Circinaria hoffmanniana* from archaeological park in Portugal. Pozo- Antonio *et al.* (2019) recommended use of a Nd:YAG laser treatment because it induced less intense physical changes than the Er:YAG laser.

19.8 Conclusion

Buddha lived 563-483 BC., the paper discusses two prominent sites one in Gujarat Denimori and another in Madhya Pradesh – Sanchi. Flourishing of Buddhism in Gujarat is indicated by increased activity during the time of the Kshatrapa kings (2nd- 4th cent. AD). Elaborate structural monuments such as stupas and vihars of Devnimori, Vadnagar and Junagarh, Sana, Talaja. The unique relic stone casket containing bodily relics were unearthed from 13 feet deep stupa. The holy ashes were placed in a small gold bottle and is preserved in Museum of archaeological department. The association of 6 different lichens and occurrence of cyanobacteria and fungi is reported from the stone pillars and loose stones from Sanchi. The architectural details of two sites are presented. The microbial deterioration of cultural heritage includes physical and chemical damage as well as aesthetic alteration. With the technological advancement, a plethora of techniques for removing unwanted microorganisms have opened up new opportunities for microbiologists and conservators. These techniques include chemical methods, *i.e.*, traditional biocides and nanoparticles, physical methods, such as mechanical removal, UV irradiation, gamma radiation, laser cleaning *etc.* Certain treatments for restricting the growth of microbes and lichens are proposed.

REFERENCES

Adhikary S P (2001) Survival strategies of lithophytic cyanobacteria on temples and monuments. In: Studies in biodeterioration of materials -1. O.P. Agrawal, S. Dhawan, R. Pathak (eds.). Pub. by INTACH and ICBCP, Lucknow. 29-47.

Adhikary SP, Satapathy DP (1996) *Tolypothrix byssoidea* (Cyanophyceae/ Cyanobacteria) from temple rock surfaces of coastal Orissa, India. Nova Hedwigia 62: 419–423

Agrawal O P (2007) Essentials of Conservation and Museology. Sundeep Prakashan, Delhi, 185pp.

Anonymous (2014) "Dev ni Mori". *19 November 2014.* Retrieved 4 November 2017.

Amman R I, Ludwig W, Schleifer KH (1995) Phylogenetic identification and in situ detection of individual microbial cells without cultivation. Microbiological Reviews 59: 143-169.

Arai H, Wada K (2005) Blackening on monuments and its control: Long-term prevention of biodeterioration. In: Biodeterioration of Cultural Property-7. S. Dhawan, Abduraheem K, and V. Nath (eds.) Pub. by ICBCP, Lucknow. 89-94.

Ascaso C, Galvan J, Ortega C (1976) The pedogenic action of *Parmelia conspersa, Rhizocarpon geographicum* and *Umbilicaria pustulata*. Lichenologist 19: 151-171.

Ascaso C, Galvan J, Rodriguez C (1982) The Weathering of Calcareous Rocks by Lichens. Pedobiologia 24: 219-229.

Awasthi D D (1991) Lichens and monuments. In: Biodeteriorationof Cultural Property. O.P. Agrawal, S. Dhawan (eds.) Mac Millan, India, New Delhi 207-211.

Brock TD (1987) The study of the microorganisms in situ. Progress and problems. In: Ecology of microbial communities (M Fletcher TRG Gray and J G Jones eds.) Society for General Microbiology, Cambridge. 1-17.

Cappitelli F, Cattò C, Villa F (2020) The control of cultural heritage microbial deterioration. Microorganisms. 8(10): 1542. doi: 10.3390/microorganisms8101542.

Cecchi G, Pantani L, Raimondi V, Tirelli D, Tomaselli L, Lamenti G, Bosco M, Tiano P, (1996) Fluorescence lidar technique for the monitoring of biodeteriogens on the cultural heritage in Remote Sensing for Geography, Geology, Land Planning and Cultural Heritage. D.Arroyo-Bishop *et al.* (eds.), SPIE, Bellingham, 2960, 137-147.

Ciarallo A, Festa L, Piccioli C, Raniello M (1985). Microflora action in the decay of stone monuments. In Vth International Congress on Deterioration and Conservation of Stone, Press Polytechniques Romandes, Lausanne, 2: 607-616.

Comite V, Miani A, Ricca M, La Russa M, Pulimeno M, Fermo P. (2021) The impact of atmospheric pollution on outdoor cultural heritage: an analytic methodology for the characterization of the carbonaceous fraction in black crusts present on stone surfaces. Environ Res. 201: 111565. doi: 10.1016/j.envres.2021.111565.

Dorn R I (1998) Rock Coatings. Elsevier, Amsterdam.

Falkiewicz-Dulik M, Janda K, George W (2015) Handbook of Microorganism Involved in Biodegradation of Materials. 2nd ed. Pub. by Chem Tec. Pub. Toronto, Ontario M1E 1C6, Canada ISBN 978-1-895198-87-4

Fa³tynowicz W (2003) The lichens, lichenicolous and allied fungi of Poland – an annotated checklist. (In:) Z. Mirek (ed.) Biodiversity of Poland 6. W. Szafer Institute of Botany, Polish Academy of Sciences, Kraków: 435 pp.

Flotow J V (1849) "Lichenes florae silesiae". *Jahresbericht der Schlesischen Gesellschaft für Vaterländische Kultur (in German). 27: 130.*

Fry E J (1927) The mechanical action of crustaceous lichens on substrata of shale, schist, gneiss, limestone and obsidian. Ann. of Botany 41: 437-460.

Garcia-Rowe J, Saiz-Jimenez C (1991) Lichens and bryophytes as agents of deterioration of building materials in Spanish Cathedrals. International Biodeterioration 28: 151-163.

Gaylarde C C, Rodríguez C I I, Navarro-Noya Y E, Hernández C, Ortega-Morales B O (2012) Microbial Biofilms on the Sandstone Monuments of the Angkor Wat Complex, Cambodia. Current Microbiology 64: 85–92.

Gehrmann C K, Petersen K, Krumbein WE (1989) Silicole and calcicole lichens on jewish tombstones - Interactions with environment and bio corrosion,

VI International Congress on Deterioration and Conservation of Stone, Torun, Suppl. Vol. 33-38.

Ginell W, Kumar R (2004)Limestone stabilization studies at a Maya site in Belize. In: biodeterioration of stone surfaces. L.L. St. Clair and M.R.D. Seaward (Eds.) Springer Science, Dordrecht. DOi 10.1007/978-1-4020-2845-8

Gutarowska B, Zakowska Z (2010) Estimation of fungal contamination of various plant materials with UV-determination of fungal ergosterol. Annals of Microbiology. 60: 415-422.

Hale ME (1975) Control of biological growths on Mayan Archaeological ruins in Guatemala and Honduras. National Geographic Society Research Reports 1975 Projects; 305-321.

Hoffmann G.F. (1784) Enumeratio Lichenum (in Latin). W. Waltheri. p. 30.

Imai N, Kashiwadani H, Arai H (2005) A novel method for the control and removal of lichens and algae on stone cultural heritage sites. In: Biodeterioration of Cultural Property-4. O.P. Agrawal, S. Dhawan (eds.) Pub. by ICBCP, Lucknow. 111-131.

Iskandar I K, Syers J K (1972) Metal-complex formation by lichen compounds. J. Soil. Sci. 23: 255-265.

Jain A K (2001) Biodeterioration of rock cut images at Gwalior Fort. In: Studies in biodeterioration of materials -1. O.P. Agrawal, S. Dhawan, R. Pathak (eds.). Pub. by INTACH and ICBCP, Lucknow. 29-47.

Jones D, Wilson, M J (1985) Chemical Activity of Lichens on Mineral Surfaces - A Review. International Biodeterioration Bulletin. 21: 99-105.

Jones D, Wilson MJ, An Mc Hardy WJ (1987) Effects of lichens on mineral surfaces. In Biodeterioration 7. D.R. Houghton, R.N. Smith and H.O.W. Eggins Eds., Elsevier Applied Science, 129-134.

Krzewicka B, Mariusz Hachu³ka M (2008) New and interesting records of freshwater *Verrucaria* in Central Poland. Acta Mycologica 43 (1): 91–98.

Lounamaa, K.J. (1965) Studies on the content of iron, manganese and zinc in macrolichens. Ann. Bot. Fenn. 2: 127-137.

Lange O L, Schulze ED, Koch W (1970) Experimentell-okologische Untersuchungen an Flechten der Negev-Wuste. II. CO_2 -Gaswechsel und Wasserhaushalt von *Ramalina maciformis* (Del.) Bory, am naturlichen Standort wahrend der sommerlichen Trockenperiode. Flora (Jena) 159: 38-62

Lange O L, Kappen L (1972) Photosynthesis of lichens from Antarctica.//! "Antarctic Research Series" (G. A. Llano, ed.) 20: 83-95

Liu X, Koestler RJ, Warscheid T, Katayama Y, Gu J-D (2020) Microbial biodeterioration and sustainable conservation of stone monuments and buildings. Nature Sustainability.

Ma W, Wu F, Tian T, He D, Zhang Q, Gu J-D, Duan Y, Wang W, Feng H (2020) Fungal diversity and potential biodeterioration of mural paintings on bricks in two 1700-year-old tombs of China. Int Biodeterior Biodegrad 152: 104972. https://doi.org/10.1016/j.ibiod.2020.104972

Mirek Z (2002) Biodiversity of Poland 6. W. Szafer Institute of Botany, Polish Academy of Sciences,

May E (2000) Stone biodeterioration. In: Mitchell R, McNamara CJ (eds) Cultural heritage microbiology: fundamental studies in conservation science. American Society for Microbiology, Washington, DC, pp 221–234.

Mehta R N, Chowdhary SN (1966) Excavation at Devnimori. A Report by Dept. of Archaeology and ancient history, M.S. University of Baroda. 197pp.

Monte M (1993) The influence of environmental conditions on the reproduction and distribution of ephilitic lichens. Aereobiologia vol 9, n° 2, 169-180.

Mottershead D, Lucas G (2000). The role of lichens in inhibiting erosion of a soluble rock. *Lichenologist* 32, 601–609 10.1006/lich.2000.0300

Meng H, Zhang X, Katayama Y, Ge Q, Gu J-D (2020) Microbial diversity and composition of the Preah Vihear temple in Cambodia by high-throughput sequencing based on both genomic DNA and RNA. Int Biodeterior Biodegrad 149: 104936

Mishra G K, Upreti D K (2015) Lichen flora of Kumaun Himalaya. Lap Lambert Academic, Deutschland, Germany.

Mitchell R, Gu J D (2000) Changes in the biofilm microflora of limestone caused by atmospheric pollutants. Int Biodeterior Biodegrad 46: 299–303

Nascimbene J, Salvadori O, Nimis P L (2009) Monitoring lichen recolonization on a restored calcareous statue. Science of the Total Environment 407: 2420-2426.

Nimis P L, Monte M, Tretiach M (1987) Flora e vegetazione lichenica di aree archeologiche del Lazio. Studia Geobotanica 7: 3-161.

Nimis P L, Pinna D, Salvadori O (1992) Licheni e conservazione dei monumenti. Cooperativa Libraria Universitaria Press, Bologna. 7e164.

Pearson, L., and Skye, E. (1965). Air pollution affects pattern of photosynthesis in Parmelia sulcata a corticulous lichen. Science 148: 1600

Pinna D (2014) Biofilms and lichens on stone monuments: do they damage or protect? doi: 10.3389/fmicb.2014.00133

Pozo- Antonio J S, Barreiro P, Gonzalez P, Paz-Bermudez G (2019) Nd: YAG and Er: YAG laser cleaning to remove *Circinaria hoffmanniana* (Lichens, Ascomycota) from schist located in the Coa valley archaeological park. Int. Biodeter. Biodegr. 144: 104748.

Richardson B A (1973) Control of Biological growths. Stone Industries 8: 22-26.

Rigamonti M (2008) Lichens and monuments: searching for the balance between the necessity of restoration and preservation, and the improvement of the naturalistic-environmental aspects of the historical-architectural heritage. Scientifica Acta 2: 93-96.

Rölleke S, Witte A, Wanner G, Lubitz W (1998) Medieval wall paintings-a habitat for archaea: identification of archaea by denaturing gradient gel electrophoresis (DGGE) of PCR-amplified gene fragments coding for 16S rRNA in a medieval wall painting. Int Biodeterior Biodegradation 41: 85–92

Romão P M S, Rattazzi A (1996) Biodeterioration on megalithic monuments. Studies of lichens' colonization on Tapadão and Zambujeiro Dolmens (southern Portugal). International Biodeterioration and Biodegradation 95: 23-35.

Schatz A (1963) The importance of metal-binding phenomena in the chemistry and microbiology of the soil. Part I: The chelating properties of lichens and lichens acids. Advancing Frontiers Pl. Sci. 6: 113-134.

Scheerer S, Ortega-Morales O, Gaylarde C (2009) Microbial deterioration of stone monuments-an updated overview. Adv. Appl. Microbiol. 66: 97-139.

Scott, G. D. (1960). Studies of the lichen symbiosis. I. The relationship between nutrition and moisture content in the maintenance of the symbiotic state. New Phytol. 59,374-384

Sheppard M (2007) A liking for lichen. ICON News: The Magazine of the Institute of Conservation 13, 22–26

Speranza M, Sanz M, Oujja M, de los Rios a, Wierzchos J, Perez- Ortega S, Castillejo M. Asco C (2013) nd-YAG laser irradiation damages to *Verrucaria nigrescense*. Int. Biodeter. Biodegr. 84: 281-290.

Sterflinger K, Pinar G (2013) Microbial deterioration of cultural heritage and works of art – tilting at windmills? Appl Microbiol Biotechnol 97: 9637–9646

Schatz A (1963) Soil microorganisms and Soil Chelation. The pedogenic Action of lichens and Lichen Acids. Agric. Food Chem. 11: 112-118.

Syers J K (1969) Chelating ability of Fumarprotocetraric a cid and *Parmelia conspersa*. Plant Soil 31: 205-208.

Singh D, Bhadauria S (2014) Effects of growth of angiosperms on the Awagarh Fort (A historical monument): a case study. In: Biodeterioration of Cultural Property- 7. S. Dhawan, Abduraheem K, and V. Nath (eds.) Pub. by ICBCP, Lucknow. 149-156.

Thakur M, Mishra G, Nayaka S, Chander H (2020). An Assessment of Lichens Diversity from Mandi District, Himachal Pradesh, India. International Journal of Plant and Environment. 6: 277-282. 10.18811/ijpen.v6i04.6.

Upreti D K (2002) Lichens on Khajuraho temple and nearby area of Mahoba and Chattarpur district. In: Biodeterioration of Materials- 2. R.B. Srivastava, G.N. Mathur, O.P. Agrawal (eds) Pub. by DMSRDE, ICBCP and INTACH, Lucknow. 123-127.

Williams M E, Rudolph ED (1974) The Role of Lichens and Associated Fungi in the Chemical Weathering of Rock. Mycologia 66: 648-660.

Verseghy K (1965) Acta Biol. (Budapest) 16: 85.

Watt D (2006) Managing biological growth on buildings. *Historic Churches*. The Building Conservation Directory: Special Report magazine. 13: 36–38

Wilson M J, Jones D (1983) Lichen weathering of minerals: implications for pedogenesis. In Wilson, R.C. Ed., Blackwell, London, 5-12.

Zhang G, Gong C, Gu J, Katamaya Y, Ji-Dong G (2019) Biodeterioration and the mechanisms involved of sandstone monuments of World Cultural Heritage sites in tropical regions. Int. Biodeter. Biodegr. 143: 104723

Chapter 20

Termite Menace Management in Heritage Buildings

S.P. Singh

Former Director Conservation, National Museum, Janpath, New Delhi – 110 011, India
e-mail: spsingh7401@gmail.com

ABSTRACT

Termites have been around for over 200 million years, making them one of the oldest insects living on almost every continent. Management deterioration of monuments and heritage structures caused by termites in the tropics is a major issue. All relevant precautions are necessary in the fabrication, design and building specifications in both pre- and post-construction. The termites can enter houses, buildings and other structures through a variety of pathways. If unprotected, the floors, walls and roofs of structures offer easy entry to termites. By combining chemical, physical and natural termite prevention strategies a structure can be made resistant to termite attacks. To prevent termite entry into buildings multiple approaches must be used. Before construction begins, all nests and palatable wood must be removed from the construction area and the wood should be selected based on its natural repellency. In case naturally repellent timber is costly, not available, then wood should be impregnated with appropriate chemical preservative. A chemical barrier must be created by treating all exposed wood in the structure with recommended termiticides and all wood should be placed above a concrete footing. Once materials have been selected added constructional preventative measures must be adopted. Trace, Target and Treatment may be the technique to deal with termite menace.

Keywords: Termite, Deterioration, Management, Heritage buildings, Wooden objects, Pesticides.

20.1 Introduction

Termite is derive from the Latinword *termes* (woodworm, white ant), altered by the influence of Latin *terere* "to rub, wear, erode" from the earlier word *tarmes*. A termite nest is referred as a *termitary* or *termitarium*. In earlier English, termites were known as "wood ants" or "white ants" and the modern term was first used in 1781(Anonymous 2015). They are common in tropics and subtropics and often hollow out wood completely leaving a thin outer shell of undamaged wood (Bhatnagar *et al.*, 2010). Termite attack often remains undetected until the whole structure collapses. Certain insects destroy wood by living inside it and feeding on it. The infestation caused by insects of Coleoptera group or beetles may be detected by small flight holes or by the formation of little bore dust deposits. Anobiidae insects are found in deterioration of furniture, sculpture and other wooden objects. High RH (relative humidity) and moderate temperature help to increase the growth of termites. Unlike the Anobiidae, the powder post beetles of Lyctidae can even thrive in dry conditions and attack mainly on sap wood. Beetles of Cerambycidae can cause serious damage, mainly to the structural timber such as roofs or floors (Kirk and Shimada 1985; Allsopp and Seal 1986; Bravery *et al.*, 1987). In 16th Century Chinese used 800 chemicals for its control containing mercury, thallium, zinc and selenium *etc.* For last three decades Organochlorine compound such as Dialdrin was used. But now these are banned. A number of new methods and technologies have been introduced such as Borate, bait and barrier methods against termites (Esenther and Gray 1968). Use of pheromones is practiced as chemical barriers. Central Insecticide Board has provided methods of treatment within BIS 6313 part III 2011. In India a number of thermal resistance techniques have also been used extensively in historical buildings and for controlling molding *etc.* in Darbar Sahib and Sri Akal Takht Sahib.

Wood being an organic material consisting mainly of cellulose, hemicellulose, lignin and extractives (a variety of chemical substances responsible for properties such as color, odor, taste, resistance to decay, flammability and hygroscopicity); there are also very small quantities of inorganic substances. The elemental composition of dry wood is approximately 50 per cent carbon, 44 per cent oxygen, 6 per cent hydrogen and 0.1 per cent nitrogen. Its general chemical composition is $C_6H_9O_4$ (Mindess 2007). Wood is used to refer to the basic material trunk from trees. Timber or lumber, refers to the sawn structural members (beams, planks or boards), which are used in construction. Timber usually contains a large number of macroscopic defects, such as knots and cracks; the properties of timber are governed both by these macroscopic features and the underlying structure of the wood itself.

Termite, a group of cellulose-eating insects belonging to order Isoptera, has the social system which shows remarkable parallels with those of ants and bees, although it has evolved independently. Even though termites are not closely related to ants, they are sometimes referred to as white ants. Phylogenetic studies have shown that the closest relative to the termite is the cockroach; for this reason termites are sometimes placed in the order Dictyoptera. Termites feed primarily on cellulose and Lignin cell wall. Termites cannot digest cellulose directly so they rely on symbiotic bacteria and protozoa living with in their intestine to supply most of the enzymes needed for cellulose digestion. Economically termites play an important role in environment by helping to break down and recycle dead plant species. They are known for their damaging property as they cause extensive loss to the agriculture or horticulture, Forestry garden plants and vegetable *etc.* They are more famous for their damage to wood work in buildings however, termites are beneficial also and are doing a great service to the mankind (Edwards and Mill 1986). Less than 10 per cent of world's termite species are pests and other 90 per cent form a keystone role in bioconversion of lignin and cellulosic organic matter in biosphere. Besides there is an intimate relationship between termites and soil quality which is least studied. Termites are known to ameliorate the soil conditions and a large quantity is brought to the surface. Garbage disposal has become a major problem in the recent times. The termites, along with Queen are highly proteinaceous in nature, are being consumed as food be men in many parts of the world and are also used as food for the poultry (Mahapatro and Chatterjee 2017). In short, perhaps the beneficial role of termites is much more than the destructive parts to ponder upon. A well defined strategy is needed to analyze this natural resource. The cultural property in the buildings are to be managed as it's rare. I will record wooden antique textile objects canvas painting monuments materials used in the storage infection departments issue the damages surface is deceptively sound but inside it is allowed in kitchen in Central part of Kerala temple are made up of Murali and it's each about 7.8kg. wood that is week. Coastal regions play a great role in the destruction of the material in Museum Archives and other historical collection.

Termites usually prefer soft woods compared to hard woods (Pearce 1997) because the latter contains a larger amount of lignin and is not easily digested by termites (Ghaly and Edwards 2011; Scholz *et al.*, 2010). Undigested lignin is excreted by termites as feces and used for nest building (Amelung *et al.*, 2002). One of the oldest timber structures is situated in China: a Liao dynasty temple constructed in 984 A.D. (Figure 20.1). The oldest existing wood-frame structure in the United States is a home built in the year 1636 (Mindess 2007). On the other hand, there are all too many cases in which wood-frame structures have had to be rebuilt after only three to five years in service. Traditional Acehnese houses were usually made with preferred hardwood species, often with a combination of wood and concrete (Figure 20.2). It was found that the age of traditional houses are affected by the intensity of the termite attacks.

20.2 Termites in Ancient Texts and Modern Researches

Termites secured their place in the human life and culture of India since the

Vedic period. In ancient Indian literature they were referred as "Kashtaharika" (wood feeders) (Rao 1957).Varah Mihir (505 AD- 587 AD), the famous astronomer, mathematician, and astrologer of India in his Brihat Samhita, refers to Dacragals meaning the science of understanding water exploration, wherein the role of termite as an indicator of underground water has been explained. Inverse 54.9 of the Samhita it is stated that the sweet ground water would be found near termite mound located of the Jamun tree (botanical name *Syzigium cumini* (L) Skeels/*Eugenia jambolana*) at a specific depth and at a distance of 15 feet at the south of the tree.

"Without exception the water requirement of the insects are generally very high, and they need to protect themselves against fatal desiccation by living and working within the climatically sealed environment of their nest or within covered galleries. According to present level of research the atmosphere within the nest has to be maintained practically at saturation moisture level (99-100 per cent humidity). It is a matter of common observation that whenever a termite nest or runway is damaged. The insect immediately rush to the breach and repair it with wet soil brought up from within the nest. From an overall consideration of the evidence it seems to be safe to conclude, that, while normally the insects use every readily available source of water close to the underground surface, under conditions of severe climatic stress, they can and they probably do descend to the water table, no matter how deep it may be. Hence a well developed, active, permanent colony of mound building termites can be taken as an indication of underground spring in proximity."

Figure 20.1: Liao Dynasty Temple at Dule, China (984AD).

Figure 20.2: Traditional Acehnese House (still- style, with a single story) in Pidie Regency. Aceh. Indonesia. Lower right: semi-modern additional construction (combination of concrete and wood material: pink wall), not part of tradition.

Three examples mentioned in the publication are (a)Termites seen in the Katanga province (Congo, Kinshasa) right up to the hill slopes where spring emerge,(b) in the dry jungle upland of coastal zone of Karnataka state, and (c) in the Deccan Plateau area.

It is also asserted in the verse 54.85 that among a group of termite mounds a water vein is sure to be found below the smaller of the mounds. Verse 52 mentions that in a desert reason if a group of mounds are found, and if the middle one is in white colour, then water will be found within the depth of 55 Purushas (in Sanskrit one Purusha is equivalent to 2.28m) or 125.7m.

As a common observation of a combination of different symptoms termite mounds are said to be close to the trees and Hindus exploited this knowledge in the exploration of underground springs.

Termites live in colonies that can consist of millions of termites. They're organized, social insects that have very specific roles within their colonies and use pheromones to communicate with one another. Queens and kings reproduce and care for nymphs (babies), workers provide food, and soldiers protect the colony. They are found in the field of conservation of cultural property near ancient forts, palaces, our heritage buildings housed in heritage buildings in the form of museums. These monuments contain organic materials such as manuscripts, antiquities, textile objects *etc.* which act as food for termites, this results in deterioration or loss of such valuable records. Since this causes irreversible damage the losses are incalculable.

Along with the modernization, approaches to manage termite menace, researches also followed new pathways. Modern scientific research on termites kicked off by Konig (1779) in peninsular India. At the onset of twentieth century termite research was not in limelight other than listing species. Lefroy (1909) listed as many as 20 species. With advancement of science in the twentieth century, evidences of active research on termite came into public by various scientist *viz.* Snyder, Light, Roonwal (1979). This period enriched termite research in India by publication of various reports, books and other scientific scripts. In the year of 1960, the "International Symposium on Termites in The Humid Tropics" was organized in New Delhi under patronage of UNESCO and Zoological Survey of India. Along with other aspects of termites, their damaging impact was also discussed and scripted in that symposium proceedings. In this period scientists and researchers were engaged in termite damage and their sustainable management in India. Sen-Sarma *et al.* (1975) listed many wood destroying termites in their report entitled 'Wood Destroying Termites of India.' Roonwal (1979) described life cycle in his book entitled 'Termite Life and Termite Control in Tropical South Asia'. Under patronage of Zoological Survey of India, another landmark book series was published in 1989 under the title 'The Fauna of India and The Adjunct Countries: Isoptera' (Roonwal and Chhotani 1989). Such scientific literature surely laid the foundation stone of 'Termite R and D' in India. The first report of termites to the humankind dates back to the publication of Systema Naturae by Carolus Linneus in eighteenth century. He listed three species of termites in his zoological landmark book including Odontotermes. Subterranean termites belonging to Macrotermitinae and Rhinotermitinae attained pest status owing to their capacity of damaging buildings and other constructions in both rural and urban areas (Sornnuwat 1996; Kirton and Azmi 2005). Many of subterranean termite species manage their dispersal from native to the distant countries through Shipboard infestations (Scheffrahn and Su 2005). Initial records of termite attack to manmade structures dates back to the start period of twentieth century(Seabra 1907).United Nation Environment Program (UNEP) suggested use of alternatives for Persistent Organic Pollutants (POPs) for termite management (Anonymous 2014).

Termites help keep our planet clean by eating or decomposing old, moldy, decaying, wet trees, wood, debris and other plant material. This process is vital to our eco-system and produces new soil which then grows new life – trees, that provide food sources and homes to other animals and insects. Termites provide an excellent food source for the organisms like bugs, lizards, birds, anteaters, other small animals, and even people in some parts of the world. Termites eat poop, to prepare their digestive systems or gut flora, for eating wood and other plant material – a process called trophallaxis. Swarming termites sometimes mistaken for flying ants don't actually cause structural damage. These winged termites (reproductive kings and queens) fly from the colony in search of mates to reproduce with. Termites look similar to ants, but they're actually more closely related to cockroach species (Singh 2015).

20.3 Termites: Nests and Life Cycle

Termites build mounds or nests where the colony resides – usually made of digested wood, soil, mud, and feces. Sometimes these nests are extremely large and intricate, with the one of the largest found being 12.8 m tall!There are more than 3000 species of termites which fall broadly in three categories: subterranean, damp wood and dry wood termites. One lives inside the soil making monumental mounds, while the other two live and feast on wood. The colony comprises of the queen, king, soldiers, and workers. Led without a leader, together they make sure their empire functions efficiently, going about responsibly carrying on their assigned tasks.

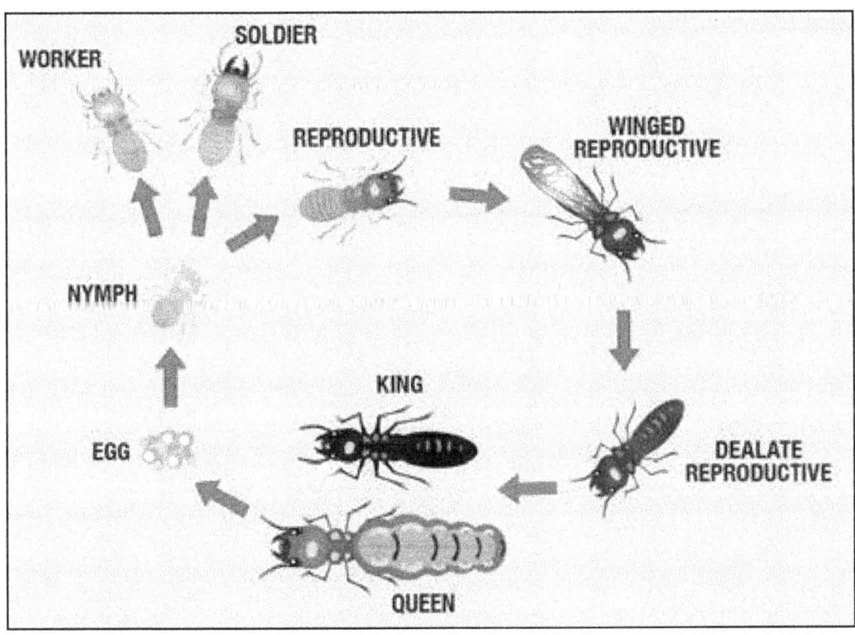

Figure 20.3: The Members of Colony Comprising of Queen, Soldier, Worker and Reproductive Termites.

They have been evolved here for millions of years ago. Our agricultural fields and buildings when fall in their habitat are attacked, hence our rampant construction activities needs to be checked! These detrivores are responsible for decomposing most of the dead matter in a forest. They make tunnels and tubes in order to stay protected while foraging food, in the process making the soil porous and allowing water percolation.

Termites are revered in cultures that understood their values. Presence of termite mounds indicates presence of groundwater and mineral deposits. These natures' recyclers enrich the soil and improve crop yield. Inside the termite gut, there is a whole ecosystem of microbes that are capable of breaking down the

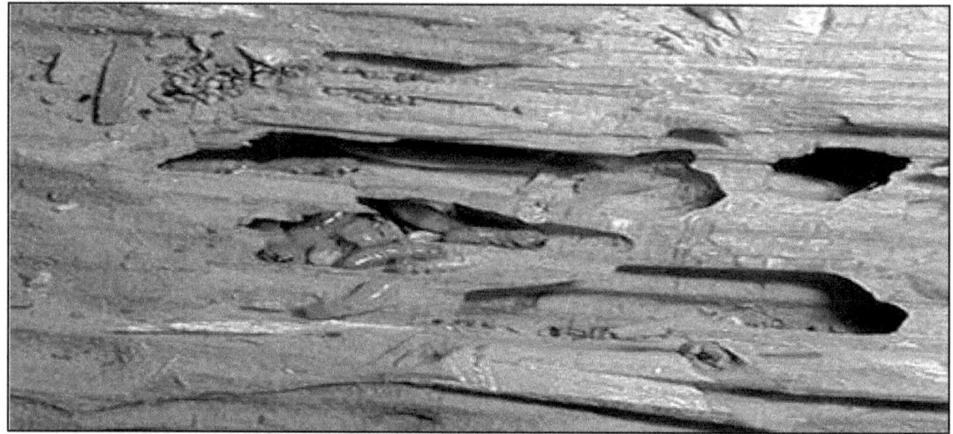

Figure 20.4: Tunnels Seen in Wood.

Figure 20.5: A Blue Colored Housing Colony Near Mehran Garh Fort, Jodhpur.

most complex dead matter. Termites feast on the cellulose and have a huge insatiable appetite, chewing 24 hours in a day! They are undoubtedly one of the most sustainable creatures, consuming and transforming everything dead helping our planet bit by bit. Termite swarming acts as an indicator of the presence of termite infestation. Swarming termites generally do not cause harm to the structures because they aim at reproduction. But the wingless workers of the colony tend to cast absolute harm to the buildings and structures (Cooper Pest Solution 2005).

The subterranean termites create citadels of literally epic proportions. Made with soil, cellulose, saliva and feces these are hard to break. They have an intricate system of ventilation to regulate temperature and humidity for them to survive and carry on their activities. The mound material is of great value. It can be utilized with our mud in construction, and in medicine. The mound is in itself a self-sustaining town with systems designed for water storage, breeding chambers, farming and air conditioning. Fungus *Termitomyces* is a mushroom forming fungi associated with termite nests (Arya and Katerina 2022).

20.4 Division of Termites on the Basis of their Habitat

1. **Subterranean or ground nesting termites,** build nests in soil in humid climates.

2. **Non-subterranean Dry wood nesting,** termites with no contact to soil.

3. **Non-subterranean Humid wood nesting termite,** which grows in humid conditions.

The moisture creates a favourable condition for existance the life of the termites. They have access to ground and other floors. Termites build tunnels between their nest and search for food through covered runways, which provide humid conditions, desiccation and protection. The darkness is favourable for the movement of these white ants. They always maintain contact with earth. They have nocturnal habits.

The subterranean termites enter into buildings from ground level, which are not pre treated. Such buildings are to be chemically treated with specified pesticides in a specific concentration per unit area and in linear following the provisions of BIS Code 6313(Part III-2011).

The chemical barriers, which prevent the termites in reaching to the super structure of the building and the foundation with a soil are created. Pesticide treatment is an ideal solution. The chemical barrier between soil and wood work *i.e.* cellulosic material and other contents of the building, which act as a food for the pest. Subterranean termites may be of two types, dry wood termite and humid or moist wood termite. Although *Nasutitermes* spp. are rarely reported to be destructive pests of wood in service (when compared to subterranean termite species, such as *Coptotermes* spp. in Southeast Asia (Kirton and Azmi 2005; Siswanto *et al.*, 2015), this genus exhibits a relatively wide adaptation to environmental conditions and was reported to be a successful pioneering insect species in the recolonization of the Krakatau Islands (Indonesia) after the catastrophic eruption in 1883 (Gathorne *et al.*, 2002).

20.5 Termites: Systematic Position

Termites belong to order Isoptera of super order Dictyoptera of Subclass Pterygota.

Class Insecta in Arthropoda.

Classification of the termites is as follows:,

Termites subterranean white ants =*Termes flevipes*

Termitidae-(family) Odontotermes obesus Capriteres. (Mound Builders)

Microtermes, Nasutermes, (Carton Nest Builders)

Rhinotermitidae -(family) *Coptoterme ceylonicus, Coptotermes parvulus, Rhinotermes, Reticulotermis, Stylotermes fletcheri* (Mound Builders).

Hodotermitidae-(family) Hodotermopsis, Zootermopsis, Archotermopsis **Mastotermitide** (family) Darwiniensis (primitive).

Kalotermitidae- (family) Kalotermes, Glyptotermes, Neotermes (primitive). (Damp wood termites).

Cryptotermes and powder post termites (Dry wood termites) (serious pest menace in furniture)

Termopsidae (family) Archotermopsis.

20.6 Diversity of Termite Species in India

India is bestowed with a wide range of species diversity in both plant and animals, out of which, entomofauna stands out for its extreme diverse interaction with humankind. Thirteen species belonging to three major families of termites were reported to damage structures from various parts of India (Roonwal and Chhotani 1989;Thakur 2007).

Out of 337 species of termite fauna reported from Indian region, 92 species belonging to 22 genera of five families *viz.* Termopsidae Holmgren, Kalotermitidae Froggatt, Rhinotermitidae Froggatt, Stylotermitidae Holmgren and Termitidae Westwood were found to be wood destroying, attacking standing trees in forest and timber-in-service as well (Shanbhag and Sundararaj (2013). Sen-Sarma *et al.* (1975) in their studies reported 35–40 species of termite damaging wooden structures in various parts of the country. In India*Coptotermes heimi* and *Heterotermes indicola* are the prominent wood destroying pests. Other subterranean termites found around household premises were outdoor may cause damage to buildings (*Odontotermes* and *Microtermes* species) and indoor as well. Close association of *Odontotermes feae* with other termites was reported by Roonwal and Verma (1991). According to Assmuth (1915), *Microtermes obesi* was found closely associated to the nest of *O. feae* in eastern part of India. Interestingly in case of structures like wooden beams and other wood works, *M.obesi* was reported to attack always after *O. feae*.

20.7 Field Survey and Collection of the Data on Structural Damage

Field survey and collection of data on structural damage caused by Termites was done in places like Haridwar, Rishikesh, Dehradun (Uttarakhand),

Muzzaffarnagar, and Varanasi (UP), Jaipur (Rajasthan), Bilaspur and Raipur (Chhattisgarh), Jammu (Jand K), Sambalpur and Puri (Odisha), Bengaluru (Karnataka), Patiala and Jalandhar (Punjab), Nagpur (Maharastra) and Chennai (Tamil Nadu). Termite infestation was documented in various Museums, National Parks and Sanctuaries, and Botanical gardens of India *viz.*, Indian Museum, Kolkata and Eco-park (WB), Ahar National Museum, Udaipur and Gulab-bag (Rajasthan), Panipat District Museum (Haryana), Bhitarkanika National Park, Nandankanan Botanical Garden (Odisha), Lalbagh Botanical Garden (Karnataka), and IARICentral Museum. Based on the present survey to major places of the country, pest prevalence mapping was done for major termite-pests *Coptotermes heimi* and *Heterotermes indicola*. Four major structural termite pests were reported from India *viz. Heterotermes indicola, Coptotermes sp, Odontotermes sp and Microtermes sp.* Media headlines in Indian media, one notably munching of millions of Indian currency note by termites inside the safety steel bank lockers in Uttar Pradesh (Annonymous 2011). Archives and libraries were reported to be infested by subterranean termites from India since long time (Roonwal 1979). Drywood and subterranean termites were reported attacking various archives, libraries, and museums in mostly tropical countries where books and furniture were included as chief items of display (Pinniger 2012). Drywood termites including genera Cryptotermes and Kalotermes were found to make tunnels in wooden structures and were also reported to inhabit in the stacks of papers and books. Subterranean termites including Reticulitermes, Coptotermes and Macrotermes were found to attack such constructions through soil and trees. Use of hygienic storage for books and other termitophilic articles inside the libraries, archives and museums and keeping such vulnerable items away from floor and walls were recommended strongly (Roonwal 1979; Mahapatro 2014).

According to Novita (2020) age of traditional houses affected the intensity of the termite attacks. All the traditional houses, that were more than 200 years old (19 houses) were attacked by termites at a serious (32 per cent) or moderate level (68 per cent). Many wood species contained chemical compounds that can suppress the attacks of xylophageous insects (Maia and Moore 2011). Age of Traditional Houses According to interviews with home owners, 72 houses (90 per cent) were built using selected hard wood species *e.g., Shorea* sp., *Artocarpus* sp., and *Vitex* sp.) depending on their culture, art, beliefs, and traditional knowledge of biological resources, as reported in Saudi Arabia (Abu-Ghazzeh 2001), and Africa (Debelo and Degaga 2014). Such wood species were common around the villages at the time of construction. However, with increasing deforestation, these wood species have become scarce and expensive. The local people have been forced to use low-quality wood for building or repairing their houses. These circumstances make such houses vulnerable to termite disturbances. Termites usually prefer soft woods compared to hard woods (Pearce 1990) because the latter contains a larger amount of lignin and is not easily digested by termites (Ghaly and Edwards 2011; Scholz *et al.*2010).

Undigested lignin is excreted by termites as feces and used for nest building (Amelung 2002). Traditional Acehnese houses were usually made with preferred hardwood species, often with a combination of wood and concrete. We found that the ages of traditional houses affected the intensity of the termite attacks.

Some wood species are preferred by termite species (Ribera, *et al.*, 2017). However, given the long time since their construction (>200 years), the quality of the wood had deteriorated due to environmental factors. Basidiomycete fungi can change the wood's physical–mechanical properties and reduce the durability of wooden materials (Hyde *et al.*, 2019). Fungal mycelium can trigger and accelerate termite activity and consequently increase the level of wood destruction (Rouland-Lefevre 2000; Morales-Ramos and Rojas 2001). The annual local climate conditions in Aceh (24–29 °C and 80 per cent humidity) are ideal for fungal growth and termite activity (Novita *et al.*, 2020).

20.8 Termite Management

Management measures included treatment of wooden structures and poles with recommended termiticides and destruction of termite galleries detected anywhere in or around the buildings. Use of hygienic storage for books and other termitophilic articles inside the libraries, archives and museums and keeping such vulnerable items away from floor and walls have been recommended strongly (Roonwal 1979; Mahapatro *et al.*, 2014).

20.8.1 Preventive Method

Prevention of decay is the main problem here. Subterranean termites cause the majority of the economically important damage to wood products throughout the world. They typically require some connection to moist soil through a system of galleries. In order to cross inert substrates to access wood, they will build shelter tubes from earth, wood fragments, faecal excretions and salivary excretions. Formosan subterranean termites have, however, been known to start colonies out of ground contact, for example, around water tanks in high-rises in Hawaii. This particularly voracious species has also been introduced into the continental United States via several Gulf Coast ports. It is also prevalent in southern Japan.Drywood termites attack dry healthy wood and donot inhabit the soil. A mated pair of winged termite can fly in and start a new colony directly in wood in buildings. In North America, drywood termites are confined to the extreme southern United States and Mexico. In Canada, subterranean termites are only considered a serious pest in some southern Ontario city, specially downtown Toronto and in some of the drier areas of British Columbia including Eastern Vancouver Island, the Gulf Islands, the Sunshine Coast and the Okanagan.

Apart from chemical treatments, which are discussed here, the best protection against termites is to prevent them from gaining access to the structure. This may involve ensuring that the timber parts of a structure are not in contact with damp soil, preventing cracks in concrete slabs, providing proper ventilation, *etc.* Proper painting will generally prevent the ingress of dry wood termites (Mindess 2007).

20.8.2 Chemical Control Methods

To fill up the cracks treatment with coal tar and creosote mixture (1:2 ratio) or phenyl was recommended. All over brush coating of wooden articles with either chlophenol-napthalene-petrolium mixture (1:1: 40, 2 coats) or hot coal tar was recommended by Roonwal (1979) Landscapes, national parks and even the avenue trees could not manage to escape from the paws of these tiny giants and were subjected to annihilation. Old oak trees adding beauty to the streets of New Orleans in United States of America were reported to be severely attacked by Formosan subterranean termite, *Coptotermes formosanus* (Gilberg and Su 2012). The present research endeavors through extensive field surveys revealed that termite infestation is prevalent in a number of important national parks, gardens from various parts of India, in addition to the roadside avenue trees across the nation.

In North America, the most common of the chemicals is chromated copper arsenate (CCA) that is available in three different combinations of chromium trioxide, copper oxide and arsenic pentoxide, depending on the specific application. It is now often used for poles, posts, pilings and foundation timbers as it provides protection against fungi and termites. Some environmental concerns have, however, been raised about its use. It has been found that in some areas, the CCA may drain off freshly treated wood, or leach out of treated wood into the soil, groundwater or nearby rivers and streams, leading to elevated concentrations of arsenic. Several other water soluble preservatives are also recommended by ASTM. Acid copper chromate (ACC) is a combination of copper sulphate and sodium dichromate with some chromic acid, and makes wood more resistant to termites and decay. Ammoniacal copper arsenite (ACA) is often used for timber and plywood intended for house foundations. It consists of copper and arsenic salts in an aqueous ammonia solution. Chromated zinc chloride (CZC) can provide protection not only against insects and decay, but also against fire. It consists of a combination of zinc chloride and sodium dichromate (Mindess 2007). By far the most widely used of the oil-borne preservatives is creosote, produced mostly from the distillation of coal tar.

The Apex agency of Central Insecticide Board and Registration Committee under the provision of Insecticides act of 1968 and the Rules of Insecticide 1971 has approved the chemicals to be used for the treatment of the structures at pre and post construction stages. Lindane is (bannedd vide Gazette Notification number S.O. 637 (E) Dated 25/03/2011) banned for manufacturing, selling and formulate. Proper safety precautions should be taken while the use of chemicals. The water bodies near building should not be contaminated. If applied to skin or body the part must be washed with care to remove toxic traces. Keep the food items away from pesticides (Agrawal 2007).

A) Basic Principle

Chemicals toxic to the subterranean termites are used to effectively control termite infestations. This treatment which is applied to the infected buildings and is used to control both existing infrastructure and to prevent reinfestation in the building. The residual activity depends upon the choice of the chemicals, doses adopted and administration. Essential requirement as barrier and methods of applications should be completed and continuous monitoring should be done.

Time of application of treatment must commence as the appearance of termite in the building premises without any further delay, for avoiding the present and future damage. Once a chemical barrier is found it should not be distributed at chance or emergency, it should be treated to restore the continuity and completeness of the barrier systems. For post anti termite treatment along outside the foundations drill is used digging 12mm holes, which are laid as close as possible to the plant. The water emulsion pumped into the holes at a rate of 2.25 liter in furniture, in absence of percolation to the tune for the dispersal of the chemical of few mm should be obtained by sustainable, and suitable methodology. The emulsion is to be applied at a rate of 2.25 running meter operations followed inside the walls as well.

Treatment at point of contact of woodwork all lessons in wood work in the building which is in contact with the process of valves and is infected with termites should be treated by spraying or injecting the termiticide in 6mm holes at downward angle 45 degree refusal for the maximum of 500ml for hole which are sealed off.woodwork what work may be classified as a friend and brother world which has been damaged by termites beyond repair shall be replaced after track paints white washes. Termites can cause damage due to cracks in concrete on the floor surface owing to the constructional defects. At the construction joints in the concrete floor 12mm holes are made at the junction of floor and constructed e walls and in expansion joints at 300ml/mm at intervals and treat the soil below 1 per cent RO.

Light switches, boxes involved in case of infestation in electrical fixtures in buildings are treated effectively with 5 per cent powder per switch box and shall be fixed after nesting the treating mixture of 1 per cent Cholopyriphos. In contrast with the floor the best protective treatment shall be provided by drilling holes of about 3 mm diameter, towards the 45° downwards to the core of wood work. The holes are made with both sides of the plinth wall at 300 mm intervals. The treatment should start from the internal walls and foundation. In the soil operation it will be executed at wall corners and doors as well as windows frames are embedded in the ground. The insecticide is sequestered through the holes till refusal or maximum 1 liter for home and these are to be sealed by paint carried as per modus operandi provisions of BIS code 6313 part III 2011, against guarantee of 5 years. Onwards periodically inspection and vigilance should be observed after treatment to the existing building during subsequent humid/hot season. The rectification is executed in case of re-infestation to the treatment points during the guarantee period *etc.*

B) Termite Baits

Baits are an important option in termite prevention and control programs. In some instances, baits are used when a liquid treatment is considered unacceptable for whatever reason or they may be used in conjunction with a liquid treatment. With new construction, baiting systems are not actually installed until after final grading is completed so that the stations are not accidentally damaged or covered with soil.

Baiting technologies such as the Sentricon system described in this study and by others (Su *et al.*, 1998; 2002; Su and Hsu 2003) are nonintrusive and non-interruptive and may provide important tools for managing subterranean termite infestations in historic sites. Baits, however, may not be effective against other termite species such as *Nasutitermes* spp. or drywood termites. For such species, heat or fumigation with inert gases, or localized treatment with liquid insecticides may have to be used. One seeming disadvantage of using termite baits to protect historic properties is the lengthy time (months–years) required to eliminate termite populations from a site, during which additional damage may occur. However, because termite infestations at many historic sites have been on-going for decades, and in the case of Fort Christiansvaern, centuries, damage potential during the baiting period is probably smaller than damage that has already occurred. Although the conventional use of insecticide spraying or injection may provide a faster result by killing a small portion of termites at the point of treatments, it only drives termites from one section to other sections of the property without affecting the overall population.

20.8.3 Traditional Knowledge for Termite Control

Various plant based chemicals have been tried. These chemicals are eco-friendly and most effective against termite treatment to the Heritage buildings which remain free from termite infestation for 4 to 5 year.

Indigenous Traditional Knowledge (ITK) for termite management in India is a lesser known but it is an eco-friendly way-out of pest management. These practices are in function since ages in various tribes and village communities in India, which pass them as legacy from generation to generation. Even though most of these practices are aimed at agricultural practices, many of them are being used to protect agricultural constructions against termites. In some areas of Gujarat state, the oil wastes from ONGC oil wells are being applied to the bamboo strips which are used for making various supporting structures in agricultural fields. This practice was successfully reported to keep termites away from bamboo structures in field.

In order to prevent termite attack on Aangali, a special arrangement of bundles of pearl millet and sorghum straws is made, which is structurally specialized to prevent percolation of rain water. Ash and salt are mixed and layered in different layers on the storage place before arrangement of fodder. In Jodhpur city of Rajasthan, the buildings were painted with white and blue in order to prevent termite attack on wall and interior. Likewise, many other applications of indigenous knowledge were documented under Inventory of Indigenous Technical Knowledge (ITK) in Agriculture by Mission Unit, Division of Agricultural Extension, Indian Council of Agricultural Research, New Delhi where examples of using ITK against termite are available.

REFERENCES

Agrawal O P (2007) Termite control in Museums. Essentials of Conservation and museology. Sandeep Prakshan Delhi. 23-31pp.

Annonymous (2011) Termites eat up Rs 1 crore at SBI branch in Uttar Pradesh. http: //indiatoday.intoday.in/story/termites-eatsbis-rs-1-crore-in-uttar-pradesh/1/135890.

Anonymous (2015) Termite - Merriam-Webster Online Dictionary. Retrieved 5 January 2015.

Arya A, Katerina R (2022) Biology cultivation and Application of Mushrooms. Pub. By Springer. 663 pp.

Assmuth J (1915) Indian wood-destroying white ants. J Bombay Nat Hist Soc 23: 690–694

Abu-Ghazzeh, T.M. The art of architectural decoration in the traditional house of Al-Alkhalaf, Saudi Arabia. J. Archite. Plan. Res. 2001, 18, 156–177.

Amelung W. Martius C, Bandeira A G, Garcia M V B, Zech, W (2002) Lignin characteristics and density fractions of termite nests in an Amazonian rain forest-indicators of termite feeding Guilds? Soil. Biol. Biochem. 2002, 34, 367–372. [CrossRef]

Becker G (1972) Protection of wood particle board against termites. Wood Sci Technol 6: 239–248

Cooper Pest Solution (2005) Termite swarm fact sheet http: //www.cooperpest. com/FactSheets/F_05_Termiteswarmer. pdf. Accessed 3 March 2016, 61.

Debelo, D.G.; Degaga, E.G. (2014) Preliminary studies on termite damage on rural houses in the Central Rift Valley of Ethiopia. Afr. J. Agric. Res. 2014, 9, 2901–2910.

Edwards R, Mill AE (1986) Termites in buildings. Their biology and control. Rentokil Ltd., West Sussex, pp 54–67

Esenther G R, Gray D E (1968) Subterranean termite studies in southern Ontario. Can. Entomol. 100: 827-834.

Gathorne-Hardy, F.J., Jones, D.T., Mawdsley, N.A. (2002) The recolonization of the Krakatau islands by termites (Isoptera), and their biogeographical origins. Biol. J. Linn. Soc. 2002, 71, 251–267.

Ghaly A, Edwards S (2011)Termite damage to buildings: Nature of attacks and preventive construction methods. Ame. J. Eng. Appl. Sci. 2011, 4, 187–200.

Gilberg M, Su N Y (2012) New termite baiting technologies for the preservation of cultural resources: results of field trials in the national park system. Park Sci 2: 16–22

Hyde K D, Xu J, Rapior S (2019) The amazing potential of fungi: 50 ways we can exploit fungi industrially. Fung. Diver. 2019, 97, 1–136.

Kirton, L.G.; Azmi, M. (2005) Patterns in the relative incidence of subterranean termite species infesting buildings in Peninsular Malaysia. Sociobiology 2005, 46, 11–15.

Konig (1779) Naturgeschicte der sogenannten weissen Ameisen. Beschr. Berlin. Ges. Naturf. Freunde: 4: 1–28. English translation and comments, by T B Fletcher, In: Proceedings, 4th entomological meeting (Pusa 1921), 4: 312–333, 6 pls., 1921

Lefroy H M (1909) Indian insect life. A Manual of the Insects of the Plains (Tropical India). Xii ? 786 pp. pls. Calcutta and London. (Thacker Spink and Co.) (Reprinted 1971), New Delhi, Today and Tomorrow's Printers, *etc.*

Mahapatro G K, Chatterjee D (2017) Termites as Structural Pest: Status in Indian Scenario. Proc. Natl. Acad. Sci., India, Sect. B Biol. Sci. DOI 10.1007/s40011-016-0837-5

Mahapatro G K, Kumar S, Chakraborty S (2014) Krishi nirman me deemak niyantran–kuchh vyabaharik sujhaw. Prasar Doot, February, 48-49 71.

Maia M F, Moore S J (2011) Plant-based insect repellents: A review of their efficacy, development and testing. Malar. J. 2011, 10, S11.

Manzoor F, Mir N (2010) Survey of termite infested houses, indigenous building materials and construction techniques in Pakistan. Pak J Zool 42: 693–696

Mindess S (2007) Environmental deterioration of timber. WIT Transactions on State of the Art in Science and Engineering,28: 297-305. WIT Press doi: 10.2495/978-1-84564-032-3/09 288 Environmental Deterioration of Material, www.witpress.com, ISSN 1755 8336 (on linc)

Morales-Ramos, J.A., Rojas, M.D. (2001) Nutritional ecology of the Formosan subterranean termite (Isoptera: Rhinotermitidae)-feeding response to commercial wood species. J. Econ. Entomol. 94: 516–523.

Novita N, Hasbi A, Husaini I, Teuku M J, Syaukani S, Emiko O, Katsuyuki E (2020) Investigation of Termite Attack on Cultural Heritage Buildings: A Case Study in Aceh Province, Indonesia, Insects 2020 (11)385; doi: 10.3390/insects11060385

Pearce M (1997)Termites: Biology and Pest Management, 1st ed.; CAB International: Chatham, UK, 1997; p. 172.

Pinniger D B (2012) Managing pests in paper-based collections. Preservation Advisory Centre. The British Library, London

Rao H S (1957) History of our knowledge of the Indian fauna through ages. J Bombay Nat Hist Soc 54: 251–280

Ravan S (2010) Ecological distribution and feeding preferences of Iran termites. Afr J Plant Sci 4: 360–367

Ribera, J., Schubert, M., Fink, S., Cartabia, M., Schwarze, F.W.M.R. (2016) Premature failure of utility poles in Switzerland and Germany related to wood decay basidiomycetes. Holzforschung 2016, 71, 241–247.

Roonwal M L (1979) Termite life and termite control in tropical south Asia. Scientific Publishers, Jodhpur

Roonwal M L, Chhotani O B (1989) The fauna of India and the adjunct countries, Isoptera (Termites). Zoological Survey of India, Calcutta

Roonwal M L, Verma S C (1991) The South Asian wood-destroying termite, Odontotermes feae (synonym O. indicus). Identity, biology and economic importance (Termitidae, Macrotermitinae). Rec Zool Surv India 129: 1–33

Rouland-Lefevre, C. (2000) Symbiosis with fungi. In Termites: Evolution, Sociality, Symbiosis, Ecology; Abe, T., Bignell, D.E., Higashi, M., Eds.; Kluwer Academic Publishers: Dordrecht, The Netherlands, pp. 289–306.

Scholz, G.; Militz, H.; Gascon-Garrido, P.; Ibiza-Palacios, M.S.; Oliver-Villanueva, J.V.; Peter, B.C.; Fitzgerald, C.J. (2010) Improved termite resistance of wood by wax impregnation. Int. Biodete. Biodegra 64: 688–693.

Shanbhag R, Sundararaj R (2013) Host range, pest status and distribution of wood destroying termites of India. J Trop Asian Entomol 2: 12–27

Singh S. P. (2015) Anti-termite treatment of historic buildings. In: A bouquet of Indian heritage Research and Management: Dr. Agam Prasad Felicitation Volume (2 Vols-Set) (Eds) Prashant Srivastava P and Sanjaya Kumar Mahapatra SK, By Swati Publications, 447-456, ISBN : 978938184315

Scheffrahn RH, Su NY (2005) Distribution of the termite genus Coptotermes (Isoptera: Rhinotermitidae) in Florida. Florida Entomol 88: 201–203

Sen-Sarma PK, Thakur ML, Misra SC, Gupta BK (1975) Studies on wood destroying termites in relation to natural termite resistance of timber. Project Report published by Forest Research Institute 21.

Seabra AF (1907) Some observations on the *Calotermes flavicollis* (Fab.) and the *Termes lucifugus* Rossi. Bull Portuguese Soc Nat Sci 1: 122–123

Siswanto, E.; Ahmad, I.; Dungani, R. (2015) Treat of subterranean termite attack in the Asia Countries and their control: A review. Asia J. Appl. Sci. 2015, 8, 227–239.

Sornnuwat Y (1996) Studies on damage of constructions caused by subterranean termites and its control in Thailand. Wood Research No. 83

Su, N.-Y., P. M. Ban, and R. H. Scheffrahn. (2002) Control of subterranean termite populations at San Cristóbal and El Morro, San Juan National Historic Site. J. Cult. Heritage 3: 217–225

Su NY, Hillis-Starr Z, Ban PM, Scheffrahn RH(2003) Protecting historic properties from terrestrial termites. American Entomologist. 20-32pp.

Su, N.-Y, Thomas J D, Scheffrahn R (1998) Elimination of subterranean termite populations from the Statue of Liberty National Monument using a bait matrix containing an insect growth regulator, hexaflumuron. J. Amer. Inst. Conserv. 37: 282–292.

Thakur R K (2007) Termites from Delhi (Insecta: Isoptera), with new distributional records. Indian J Forestry 30: 505–508

United Nations Environment Programme chemicals finding alternatives to persistent organic pollutants (pops) for termite management. http://www.unep.org/chemicalsandwaste/Portals/9/Pesticides/Alternatives-termite-fulldocument.pdf. Assessed 14 Jan 2014

Zhong JH, Liu LL (2002) Termite fauna in China and their economic importance. Sociobiology 40: 25–32

Chapter 21

Impact of Atmospheric Pollution and other Environmental Factors on Deterioration of Heritage Monuments

S.P. Singh

Former Director Conservation, National Museum, Janpath, New Delhi – 110 011, India
e-mail: spsingh7401@gmail.com

ABSTRACT

The monuments made up of stones and mortar, not only show the architectural beauty but are good examples of cultural heritage prevailing at that time. One can know various things by peculiar designs of these monuments. Monuments and old buildings in India with natural stone and baked bricks have been exposed to decay for centuries. It is said that vast wealth of ancient Indian culture is still preserved in temples. The numerous temples have their own museums and collections of artifacts including paintings, utensils and armaments. The building materials are decayed by adverse environmental conditions and the extent of damage depends on both the materials and the conditions. Growth of microorganisms and plants on monument, accelerated due to favorable environmental conditions is not only dangerous but even fatal to them; the biochemical damage is due to extraction of organic acids which can erode the surface. If plants grow inside fissures or cracks present in the surface, the gap may increase further.

As an Indian we inherit priceless cultural wealth in the form of monuments, archival materials, archaeological sites, artifacts and vast variety of museum materials. To protect this cumulative heritage for posterity's unspoiled as possible is indeed a challenging task. Environment *is changing fast, the content of SO_2, CO_2, Oxides of nitrogen, dust and water vapors in atmosphere are instrumental in causing physical and chemical alterations, Environmental pollution are causing direct and indirect effect by influencing activity of microbes on heritage monuments. Problems of decay due to pollution in some national and international monuments are discussed.*

Keywords: *Deterioration, Impact, Atmospheric pollution, Environmental factors, Monuments, Marble, Taj Mahal, Dust, SO_2.*

21.1 Introduction

Conservation science helps preserve the memory of monuments which have been shown to be in such a declining state of neglect and decay that time is running out. Monuments that very probably cannot be restored as part of a conservation project risk disappearing without a trace. An analytical campaign of what is still possible to characterize them and can therefore reveal the materials, artistic techniques, decorations and effect of anthropogenic and environmental factors which can be effectively further utilized to restore many disappearing monuments (Rumpazzi *et al.*, 2021). Safeguarding our cultural heritage is a widespread problem of global relevance; preservation of our past is a way to improve human progress. In recent decades, weathering processes have greatly increased, especially in urban areas, owing to the higher incidence of environmental– anthropogenic pollution. The deleterious effects of these processes on cultural heritage are well known (Gomez-Alarc on *et al.*, 1995; Ortega-Calvo *et al.*, 1992). Stone monuments like pyramids of Egypt, Basilica of Venice and Taj Mahal of Agra and many more represent a considerable part of the multiform expressions of art, and owing to their prevalently outdoor location they are strongly subjected to weathering and decay processes (Albertano 1995; Caneva *et al.*, 1992; Sharma and Sharma 1982). In addition, humidity and other environmental factors accelerate the deterioration process of stone monuments caused by microorganisms.

The monuments are made up of stones and mortar, they not only show the beauty but also are good examples of cultural heritage prevailing that time. Different kinds of microorganisms may colonize art works: chemoheterotrophic bacteria, chemolithotrophic bacteria, phototrophic bacteria, algae, were reported from marble statues placed in Boboli Gardens in Florence, Italy (Lamenti *et al.*, 2000; Tomselli *et al.*, 2000).On one hand the natural disasters have been posing a big question on our heritages, while on the other the change in climate, pollutants in environment and biological agents have been playing their major role in promoting the decay of our ancestral bounties, the improper maintenance has pushed the issues far more ahead than it ought to be. Conservation experts will have to work together and seek the solutions to the problems related to deterioration. The reason that cultural property distributed in Asian district has received serious damage by various biological agents in the environment condition in high humidity and temperature.

Monuments and old buildings in India with natural stone have been exposed to decay for centuries. Upreti *et al.* (2002) reported 22 spp. of lichens from cultivated trees and old walls in Lucknow. Due to increasing pollution none of these are hardly visible now inside the city. The higher pollution levels of heavy metals were recorded in Lucknow by the use of lichens- a biological indicator of pollution (Bajpai *et al.*, 2009; Upreti *et al.*, 2009). Arya and Gupta (2016) found deposition of pollutants on Lehripura gate in Vadodara promoting growth of bacteria and fungi forming biofilm. The pollutants like SO_2 and $PM_{2.5}$ were the cause of concern in the city of Taj and necessary steps are needed to reduce the atmospheric pollution there (Goyal and Singh 1990). Growth of microorganisms and plants on monuments is not only dangerous but even fatal to them. After chemical weathering and microbial decay the plants may find cracks, fissures and drainage openings on the buildings and can cause serious damage. Blackening of stone monuments in Angkor Wat had been caused by growth of Cyanobacteria (Arai 2014). A similar study was conducted on the Japan's Himeji castle where *Cladosporium* species was isolated. Its control was sought under the application of hydrotherm on plastered walls in 1991 (Arai and Wada 2014).The Cyanobacteria and weeds were found growing on the top of the famous ancient and renovated brick temple of Vishnu in Bhitargaon, Kanpur (Figure 21.4).

Morelia belongs to the UNESCO Cultural Heritage and is located at the crossing of the Mexican Volcanic Belt, the Sierra Madre Mountains, the Coconuts and the Rivera plaques. The Cathedral and a beautiful Aqueduct are the main monuments built with blocks of ignimbrites. This study was performed on blocks of ignimbrites removed from the cathedral during a restoration campaign. The damage was assessed on ignimbrite blocks used to build major architectonic monuments in Mexico (Alonso and Martinez 2003). A clear correlation was found between the contents of environmental sulfur and the degradation of the mechanical strength of the ignimbrites. Since the main sources of environmental sulfur in the city are motor vehicles and a paper factory in the southwest from where the main winds come from, it is conclude that the damage is from anthropogenic origin. Air pollution is a key factor in the degradation of surfaces of historical buildings and monuments. The impact of pollutants emitted into the atmosphere on materials is enormous and often irreversible. Corrosion caused by chemicals and soiling caused by particles can lead to economic losses but, more importantly, to the destruction of our cultural heritage, an important component of our individual and collective identity. A recent study led by the Italian Institute for Environmental Protection and Research (ISPRA: 2015) and the Institute for Conservation and Restoration of Heritage (ISCR) shows that in Rome about 3600 cultural heritage made of calcareous stone (limestone) and 60 cultural heritage objects made of bronze are at risk of deterioration. As a response to this threat, Italy has been engaged in the development of strategies and technologies to safeguard cultural heritage assets for many years. The recent study finds that loss of material as a result of air pollution in Rome is estimated to be between 5.2 and 5.9 microns per year for marble and between 0.30 and 0.35 microns per year for bronze. Air pollution is more harmful than land pollution, water pollution and noise pollution *etc.* The latter's effects are yet to be studied with reference to Indian monuments whereas the effect of air pollution on monument has received considerable attention from various Indian experts. Some such studies are discussed in this article

21.2 Role of Pollutants in Changing Microclimate and Accelerating Biodeterioration

The biodeterioration process affects monumental stones, statues, historical buildings, wall paintings, archaeological remains, and, to a lesser extent, glasses and metals (Giustetto *et al.*, 2015; Biswas *et al.*, 2013; Gorbushina and Palinska 1999; Gorbushina *et al.*, 2004; Piñar and Sterflinger 2009). Numerous parameters influence the succession of microorganisms on stone; firstly the properties of the stone itself determine the colonization pattern. The mineral composition, structure-texture, porosity, and permeability of stone may influence the distribution of such organisms in the monuments (Miller and Macedo 2006). Favorable environmental conditions and the presence of nourishment sources allow the biological colonization of an exposed stone surface (Miller *et al.*, 2000; Palla *et al.*, 2003).

It is known that biological growth on stone is highly dependent on climatic and microclimatic conditions, such as humidity, temperature, light, and atmospheric pollutants (Moroni and Pitzurra 2008; Mansch and Bock 1998; Zanardini *et al.*, 2000; Nuhoglu *et al.*, 2006). To avoid microbial proliferation, it is necessary to control environmental factors, which becomes difficult in archaeological sites or in urban spaces, but is more easily achieved in closed environments where the control of climatic and microclimatic conditions is easier (Salvadori and Charola 2011). Occasionally, the metabolic activities of microorganisms (autotrophic and heterotrophic) induce different types of damage: physical, when pressure is exerted by the growth of vegetative structures (*e.g.*, lichenic and fungal thalli); chemical, when the excretion of enzymes, the production of inorganic and organic acids, and the liberation of chelating compounds occur; and aesthetical, as in the effect of acid rain yellowing of Taj mahal was observed (Sharma and Sharma 1982).

The Palazzo Pitti remained the official residence of the Medici Rulers of Florence through the 17th century. The Medici constructed an 11 acre complex of gardens, sculptures, and fountains behind the Palazzo Pitti, known as the Boboli Gardens (Figure 21.1). These gardens are in a similar style to many Italian Gardens from the Baroque Age. Several different grottos were built within the Boboli Gardens, including the Grotto of Adam and Eve. The Boboli Gardens Amphitheater contains dozens of marble sculptures all orientated around a central fountain. Its called the *"Amphitheater"* because the statues are arranged around a large oval-shaped walkway, in the same shape as an ancient

Roman Amphitheatre. The sculptures themselves are also in the same style as many marble statues from classical Greece and Rome. Borrowing elements from the architecture of antiquity is an important characteristic of the heavily decorated style in building art that was popular in Europe in the 17th century, and Renaissance Architecture.

Figure 21.1: Marble Statues Placed in Boboli Gardens, Italy.

According to Warscheid (2003), moderate climates with regular rainfall tend to give rise to a mixed consortium of microorganisms on exposed stone surfaces. The highest degree of damage to monuments may occurs in the tropics, because of high humidity and temperatures. The stone microflora here is considered to be very aggressive, with a high capacity for "biocorrosion" and biofouling (Warscheid 2003). These two terms are defined by Warscheid (2003) as: (1) microbially induced or influenced corrosion of materials, altering the structure and stability of the substrate, and (2) the presence of colloidal microbial biofilms on or inside materials, leading to visual impairment and potentially altering the physiochemical characteristics of the substrate. The production of pigments, thick walls, help in protection of microorganisms from adverse climates. It may cause severe aesthetic damage. Deeply-colored coccoid and filamentous cyanobacteria, which predominate in biofilms on buildings in the hot and humid climates of Latin America (Gaylarde and Gaylarde 2005), are more frequently present on surfaces of buildings at high altitude in the tropics and subtropics than at lower altitudes (Gaylarde and Englert 2006; Gaylarde and Gaylarde 2005).

Table 21.1 presents a selected list of monuments, country and their observed damages due to pollutants. Different materials of stone determine the damage caused. A thick black crust was formed on dolomite, while sulphate deposition was common on monuments made of marble.

Table 21.1: Damage Caused by Pollutants in Certain International Monuments

Monument	Country	Stone	Damage	Reference
Boboli Gardens Florence	Italy	Marble Sand stone	Pollution Photosynthetic bacteria due to humidity	Tomselli *et al.*, 2000
CA Granda Milan Building	Italy	Dolomite	Thick black crust Solidity loss Efforescence	Peruzzi *et al.*, 1978
Cathedral of Sarville	Spain	Calcite	Carbonaceous matter gypsum formed between 1-17 per cent	Alcaeda and Martin 1988
Cleopetra's needle N.Y. city	USA	Limestone	Weathered up to 3 mm	Bugess and Schaffer 1952
Cologne Cathedral	Germany	Sandstone	Pollution affected	Graue *et al.*, 2012
Flanders monument	Belgium	Sandy limestone	Gypsum layer formation of calcite matrix	Nijs 1985
Florentine Architecture, Florence	Italy	Marble	Sulphate deposition	Frediani *et al.*, 1976
Herten Castle Westfalia, Racklinghausen	Germany	Sandstone	Complete destruction due to pollution	Winkler 1975
Lincon cathedral	U.K.	Limestone	Blisters of stone, blackened crust	Butin *et al.*, 1985
Plazzo Deiquiurcconsult, Milan	Italy	Sandstone/ Limestone	Gypsum formation, 70 per cent black due to unburnt material deposition	Bertolaccini *et al.*, a975
St Marks' Bascilica, Venice	Italy	Marble	Increase of sulphate layers. Black carbonaceous matter	Fassina 1986

21.3 Air pollution and Ancient temples

Bhargav *et al.* (1999) assessed the effect of major air pollutants and meteorological parameters with respect to the degradation of the stone monuments during the seventh to tenth centuries A.D. in the city of Bhubaneswar, Orissa, India. The four monuments such as Bharateswara temple, Vaital temple, Parasurameswara temple and Mukteswara temple are located in a region where contamination through vehicular emission was relatively high. Crusts of biomass were observed on the portico which is not explicitly visible to the sun during the tropical climate. The biomass crusts and lichens were grown on the fine carvings, thereby deteriorating the artistic value of the sculptures. Microphotographs showing chemical leaching and opaque minerals are posted by Bhargav *et al.* (1999).

In general, the climatological parameters, relative humidity and wind direction, along with the pollutants, were found to be responsible for the deterioration of the monument. The concentrations of sulphur oxides (SOx), nitrogen oxides (NOx) and suspended particulate matter (SPM) were found to be high during the winter season compared to spring, summer and post-summer seasons at all the four stations considered for monitoring. The rates of deposition and erosion were high at all the four directions of the four temples due to heavy traffic volumes on the highways neighbouring to the temples (around 200–300 heavy vehicles per day near the Bharateswara and Vaital temples and 1200–1300 vehicles per day near the Mukteswara and Parasurameswara temples) during the busiest hours of the day. Calcium chloride ($CaCl_2$) was the main salt present on the stone surface, carried by the salt-laden air from the neighbouring sea. This compound gets converted to calcium hydrate due to temperature variations. Furthermore, crystallization of this salt also results in scaling of the stone surface. The occurrence of sulphate (SO4 2–) ions on the surface of the stone monuments was due to the sulphuric acid formed by the surface catalysed reaction between SO_2 and the iron minerals in the stone.

Goel *et al.* (2017) assessed the magnitude and size aggregated distribution of particulate matter (PM) in temples of Kanpur city in India. Active sampling was conducted in three temples (nine sites), and PM_{10} mass concentration was observed to be as high as 2184 µg m–3 inside the premises of the temple. The concentration of the pollutants surpassed the Central Pollution Control Board (CPCB) norms of 100 µg m–3. They found that three types of incenses used, namely, Agarbatti, Dhoop and Diya may add to particulate pollution. This concentration further increased during festival days. The $PM_{2.5}$ concentrations was found to be 99.9 per cent of the total particle load and contributed 75–92 per cent in terms of the mass. The particles were poly-dispersed, and coagulation mechanism was observed to be dominant in the winter time. They also concluded that ventilation and seasonal weather during which measurements were taken played a major role in the particle size distribution measurement

Pollution negatively impacts historical monuments and ancient temples around the world, from the Acropolis in Greece to the America's own Lincoln Memorial. The threat is in the risk of losing these irreplaceable structures forever. Many of these monuments have cultural and aesthetic value that is beyond price. The Figures 21.2–21.4 depicts the oldest surviving 5th Century brick temple situated in Bhitargaon area near village Umari in Kanpur, U.P. It was built in the 5th century A.D during the Gupta Empire. According to Cunningham, because of the Varaha incarnation at the back of the temple, it was probably a Vishnu temple. The old temple survived because the temple is situated away from city in clean environment. But due to climatic effect the biological weathering took place and ASI has restored the beauty of temple. No deity is placed here. There is no evidence of the existence of temple architecture during the Vedic period. Worshipping was systematized and paved the way for the evolution of

Figure 21.2: Oldest 5th Century Brick Temple of Vishnu Situated in Bhitargaon, Kanpur.

Figure 21.3: Brick Temple of Vishnu after Restoration by ASI in 2011.

Figure 21.4: Oldest 5th Century Brick Temple of Vishnu Situated in Bhitargaon, Kanpur.

temple structures. Rock-cut architecture began to develop from the 3rd century BCE. Though the earliest rock-cut architecture is from the Mauryan dynasty, the Ajanta caves which belong to post Mauryan period are among the earliest rock-cut temples. Rock-cut temples gave way to stone temples and as stone was not easily available everywhere, it gave way to brick temples. The entrance into the sanctum shows one of the first uses of a semi-circular doorway. Alexander Cunningham (First Director General of the Archaeological Survey of India, 1871) called this as the 'Hindu arch' which was peculiar to India. The temple has a tall pyramidical spire (shikhara) above the inner sanctum (garbha griha). This shikhara became the standard feature of the Nagara temple architecture of India. The walls of the temple are decorated with terracotta sculptures of God and Goddesses like Shiva, Parvati, Ganesha, Vishnu *etc.* on panels separated by bold ornamental pilasters (rectangular column that projects slightly beyond the wall).

21.4 Effects of Acid Rain on Buildings and Monuments

Some damage, such as from wind or rain, is unavoidable. However, pollution contributes additional risk factors that can increase the level of destruction. The effects may be minor, such as a blackening of the surface of monuments due to dust. Other impacts can have permanent consequences. Acid rain, first recognized in Sweden in 1872, was considered a local problem for a long time. But in the 1950s recognition that acid rain in Scandinavia originated in Britain and northern Europe showed instead that acid rain was not only a regional, but global problem. Normal rain generally ranges from about 6.5 to about 5.6 on the pH scale. Acid rain, however, measures below 5.5. Acid rain has been measured at the bottoms of clouds at pH 2.6, and in fog in Los Angeles, as low as 2.0. Common naturally occurring materials used for buildings and monuments include sandstone, limestone, marble and granite. Acid rain corrodes all these materials to some degree and accelerates natural decomposition. Limestone and marble dissolve in acids. The sand particles forming sandstone often are held together by calcium carbonate, which dissolves in acid.

Granite, while much more resistant to acid, still can be *etc.*hed and stained by acid rain and the pollutants it carries. Cement also reacts to acid rain. Cement is calcium carbonate, which dissolves in acid. Concrete buildings, sidewalks and artwork made with cement show the effects of acid rain. In addition, slabs of granite and other decorative materials are often held in place using Portland cement. Acid rain damage to concrete buildings in heavily polluted cities like Hangzhou, China, can be extensive. Copper, bronze and other metals react with acids as well. Corrosion of the bronze sheeting on the Ulysses S. Grant Memorial, for example, shows as green streaks down the pedestal. Copper dissolved from the bronze has washed down the base and oxidized into green stains. The Thomas Jefferson Memorial in Washington, D.C., is one of many monuments affected by acid rain. The dissolving calcite releases the silicate minerals contained within the marble. The loss of material weakened the structure enough that reinforcing straps were added during the 2004 restoration. In addition, a black crust left by dirt caught in the *etc.*hed marble must be gently washed away. Many sculptures throughout the United States and Europe are carved from marble or limestone. When sulfuric acid rain strikes these statues, the reaction of the sulfuric acid with the calcium carbonate yields calcium sulfate and carbonic acid. The carbonic acid further breaks down into water and carbon dioxide. Calcium sulfate is water-soluble so washes away from the statue or sculpture.

The effect of acid rain on Taj Mahal structures serves as one example of how acid rain impacts buildings. Air pollution from a local refinery has caused acid rain to form, turning the white marble yellow. Although some have argued that the yellowing is natural, or caused by iron supports in the marble, the local courts agreed that air pollution has impacted the Taj Mahal. In response, the Indian government has established local strict emission controls to help protect the Taj Mahal. Sharma and Sharma (1982) analyzed the water-soluble samples taken from the marble and sandstone parts of the monuments located in Agra. They also analyzed the impact of combustion, manufacturing and other polluting operations on the monuments located in Taj Mahal, Red Fort, Itmad-ud Daula, Sikandra and Fatehpur Sikri. The annual average SO_2 ranged between 16 and 20 $\mu g\ m^3$ causing corrosion of the building stones. The SO_2 content present in the atmosphere with high relative humidity aids in the formation of sulphurous acid.

Figure 21.5: Beautiful Taj Mahal in 1890.

Figure 21.6L Beautiful Taj Mahal during Restoration by ASI in 2015.

Efflorescence's on the sandstone surfaces around the Taj Mahal consist of whitish, patchy encrustations of gypsum, which was deposited contemporaneously with the formation of the sandstone. Initially, the gypsum was uniformly dispersed in the sandstone, but centuries of wetting followed by evaporation of water at the surface has concentrated these salts at the surface and in the subsurface regions (Gauri and Holdren Jr.1981). The efflorescence's, being water soluble, repeatedly dissolve and recrystallize in the alternating episodes of wetness and dryness of the stone. Thermodynamics suggest that enormous pressures are generated in such processes. For instance, gypsum crystallizing from an IOX supersaturated solution at 50° C generates a pressure of 334 atmospheres (Winkler and Singer 1972). While such super-saturation conditions do not exist in the pore space of sandstone, the repeated crystallization over long periods of time has resulted in failure of the stone' and creation of hollow patches.

The most damaging of all forms of pollution on the marble of Taj have been air and water pollution (IL and FS 2009). Effluents were released from over 1700 factories that existed in and around Agra, in the form of gases such as CO, CO_2, SO_2, chlorofluoro carbons, nitrogen oxides, particulate matter *etc.* Out of these, SO_2 in particular causes yellowing and intensive damage to marble (Ramchandran 2007). Further, emission of particulate matter from refinery emissions will make these potentially hazardous salts even more dangerous by facilitating their migration and their consequent accumulation near the stone surfaces. Existing technology lacks proven methods for preserving stone structures containing gypsum in the zone of weathering (UNESCO 1977). Cleaning the Taj Mahal is a relatively simple matter. The brownish discolorations under arches at the Taj are presumably only environmental dust adhering with recrystallized calcite formed by CO_2 reaction with marble; these encrustations should be mechanically removed. Fine abrasives should be used to polish the marble surface. Such an operation should, in addition to removing dirt, give a polish to the marble increasing its water repellency. The cleaned marble surfaces then should be washed regularly with water to prevent crusting. After cleaning, the marble may be given a surface treatment with polymeric materials, which act as semi-permeable membranes (they do not absorb atmospheric gases) and are resistant to UV radiation (Gauri 1978a). Calcite grains mixed in such polymers may be forced into cracks caused by the expansion of iron attachments to inhibit the movement of water into the stone. The sandstone structures, however, need optimal removal of efflorescence's and consolidation improve strength and water-repellency (Gauri 1978b).

21.5 Chemical Weathering

Acid rain will seriously corrode monuments over time and completely different building materials are subjected to different types of chemical weathering. Metals like bronze or copper, are at risk of corrosion by the interaction with changing weather conditions. They create a skinny layer of "patina," a by-product of chemical weathering that includes inexperienced hue, and might be seen on the sculpture of Liberty. As shortly because the sculpture was erected in 1886, it began to bear the natural processes of physical and chemical weathering. The salt water of New York's harbor began to weather it through the processes of ice and salt crystallization, and salt water also acted as a chemical agent that exacerbated the corrosion of Lady Liberty's copper skin. After nearly 100 years, the statue needed extensive repairs, which were undertaken from 1984 to 1986 and cost millions of dollars. If the statue is to exist for future generations, it will need periodic restoration, because the processes of weathering are relentless. All monuments similarly must be cared for if they are to survive. Sulfuric acid aerosols will readily attack building materials and will erode them.

When a microbial colonization is evident, its relevance to degradation and weathering of inorganic materials should be carefully evaluated as biotic and abiotic agents interact in quantitatively variable relations (Siegesmund and Snethlage 2014). Any organisms on cultural objects may not necessarily modifying the chemical composition or physical properties of the materials (Pinna 2017). Only in particular conditions and in combination with other factors they can initiate, facilitate, or accelerate deterioration processes. Moreover, the growth of some organisms is very slow, and the damage becomes visible only after years or even decades. At present, the importance of biodeterioration processes on historical objects of art has reached growing attention of people in charge of the conservation of cultural heritage. A large set of relevant studies have documented and discussed the interaction between biological colonization and cultural heritage objects. Despite considerable research efforts, there are still general issues that need to be addressed. Regarding stones, many aspects of the interaction between microbial communities, lichens and these materials are still unknown (Di Martino 2016). Not surprisingly, in recent times many papers report that the biological colonization of outdoor stones may act as a protective layer shielding the materials from other factors that cause decay, such as wind and rainwater. In addition, the species and their amount within biofilm-forming microbial communities can change over time. Progress in microbial ecology and genomics, in parallel with developments in biological imaging and analytical surface techniques, can promote a comprehensive insight into the dynamics of the structured microbial community within biofilms (Pinna 2021).

21.6 Effect of Pollutants on different Stone Monuments

Pollutants affect different stone monuments on the basis of their chemical interactions. An important part of this process is the oxidation of sulfur dioxide, either in the gas phase or in moisture films on building stones. This oxidation is assisted by atmospheric oxidants, such as ozone and hydrogen peroxide, as well as by catalysts, such as soot or black carbon and smoke. Sulfates can attack carbonate rocks through dissolution by sulfuric or sulfurous acid and by converting calcium and magnesium carbonates to more soluble sulfates or sulfites. The sulfate attack on silicate rocks, in contrast, is not easily measured. Leaching of alkali metals and iron may occur, which often forms black crusts on the stone surface, causing discoloration. Török *et al.* (2011) investigated the effect of the environment on the formation of these crusts in urban and rural areas and Farkas *et al.* (2018) studied the crusts in different countries. Gibeaux *et al.* (2018) determined the pollution rates from the speed of color change measured by fixed-point observations. Efficient automotive combustion is an important contributor of nitrogen oxides, which are readily converted to corrosive nitric acid in the presence of oxidants. Although it is also a strong acid, nitric acid is less damaging to carbonate rock than sulfuric acid and sulfates due to the greater reactivity of sulfates with stone. Hydrochloric acid, which readily dissolves carbonate rock, is an important pollutant source of chloride ions, in addition to sea spray and desert dust in many areas. Coal combustion is the major non-natural source of hydrochloric acid. All these atmospheric pollutants, after undergoing emission from natural and anthropogenic sources, transmission (over short or long distances), emission (reaching certain local environments), and deposition (on dry and wet surfaces with variable reactions), can cause devastating salt weathering hazards. Table 21.2 shows effect of different pollutants and weather conditions on deterioration of Indian monuments of heritage importance.

21.6.1 Limestone, Sand, Marbles

The particulate pollution makes building dirty and the tarry matter occasionally causes staining. The acidic pollutants greatly enhance the rate of acid based decay of the limestones. However, even if there is no man made pollutants carbon dioxide present in the air and Sulphur dioxide decay of seaweeds would be sufficient to cause deterioration of these building materials. The main acid attack is due to SO_2. It is highly soluble in water and it reacts with water to forms sulfuric acid. Two reactions may occur as follows

1. Sulphurous acid + O_2 from the air could produce sulphuric acid (H_2SO_4), which will then attack limestone ($CaCO_3$) to give calcium sulphate ($CaSO_4$) and water.

   The Calcium sulphate then takes up the water as it crystallizes in the form of mineral gypsum ($CaSO_4 2 H_2O$).

2. Sulfuric acid can directly attack the limestone to give calcium sulfate ($CaSO_4$). This also crystallizes as gypsum.

Table 21.2: Effect of Environmental Conditions on Certain Indian Monuments

Sl.No.	Monument	Stone Type	Env. Factors and Effect	References
1.	Ajanta	Volcanic rock	Heavy rainfall 720mm Temp.19-36 Paint affected by moist condition/Microbial damage	Tilak *et al.*, 1970
2.	Dwarkadhis Temple	Limestone	Humid weather/temp. 15-35 C Salt deposition/growth of moss and lichens	Lal 1978
3.	Fatehpur Sikri	Sandstone/ marble	Dry Temp. 5-45C Dust high/growth of lichens	Lal 1978
4.	Khajuraho	Sandstone	Dry climate Dusty blackened atmosphere	Lal 1985
5.	Konark	Khandolite	Humid/Temp. 12-35C Salt laden winds/Moss/lichens	Lal 1978
6.	Lotus temple	Marble	Vehicular pollution Yellowing of marble	Kaur 2015; Narain and Bell 2005
7.	Red Fort Agra	Sandstone	Dry climate Temp. 5-45C Higher SO_2 levels	Lal 1978
8.	Mahabalipuram Temple	Granite and chronokite	Humid Salt affected	Bahadur 1994
9.	Taj Mahal	Marble	Dry climate/20-75 per cent humidity Temp. 5-40C Pollution due to SO and particulate matter	Sharma and Sharma 1982; Goyal and Singh 1990

The first of these parts is probably more prevalent party under the dam conditions. The second one can also apply where calcium sulphite is present in the gypsum coating of the exposed limestone surface. In any event the gypsum coating slows down the attack. Further action depends on how often the affected stone is washed by the rain.

The slightly soluble gypsum is steadily removed from those parts of limestone faced building that are frequently washed by the rain. When there is no rain water to keep these parts clean droplets of acids in the polluted air continue to condense on them under foggy conditions. The acid will react with any uncharged limestone surface and bind any available particulate pollutants to that surface. Thus these areas become darker and the skin on them becomes less and less permeable in urban districts, where the particulate pollution is high, at the surface often becomes black. The fate of the Impermeable skin depends on the resitance of the limestone to weathering. The most durable limestone appears to be able to retain a dirty inert skin more or less indefinitely.

21.6.2 Marble

Marbles consist essentially of calcium carbonate, they initially undergo same chemical reactions as limestone, when they are in moist air containing sulphur based acids. A layer is usually formed that can incorporate some dirt particles. As with limestone further action depends whether the marble is well washed by rain. In well washed areas the stone patch is dissolved, no dirt accumulates and the marble surface is gradually weathered. But where the surfaces nearly free from pores, more crystallization of gypsum occurs. This type of corrosion is normally considered less than with the limestone in the same environment, though polished marble will lose its smooth surface quickly (Singh 2007).

The atmospheric corrosion of marble was evaluated in terms of SO_2 concentration as air pollution and climatic factors such as rainfall, relative humidity, temperature and so on under the field exposure. Marble of calcite type ($CaCO_3$) was exposed to outdoor atmospheric environment with and without a rain shelter at four test sites in the southern part of Vietnam for 3-month, 1- and 2-year periods from July 2001 to September 2003. The thickness loss of marble was investigated gravimetrically. X-ray diffraction and X-ray fluorescent methods were applied to study corrosion products on marble. The corrosion product of marble was only gypsum ($CaSO_4\,2H_2O$) and was washed out by rain under the unsheltered exposure condition. It was found that the most substantial factors influencing the corrosion of marble were rainfall, SO_2 concentration in the air and relative humidity (Lan *et al.*, 2005).

21.6.3 Sandstones, Slates and Granite

The majority of sandstones consists of grains of Quartz a crystalline form of silica SiO_3 deposited together by silica in a less well crystallized form. Iron oxides or hydroxides are sometimes accompanied by grains of feldspars and micas.

Quartz based sandstones are very resistant to sulphate based acids in the air, but they can become very dirty. They tend to be dirtier in the rain-sorbed areas than in the sheltered part of the building. In this sense their behavior is quite different from that of line stones or marbles. Very occasionally, sulphur based acids in the air will attack the sandstones and then converted it to soluble form. This can then migrate to the surface of the stone, where lime, derived perhaps from mortar reconverts it to the dirty looking in -soluble form. Such deposits can remain unnoticed beneath the soot layer until the sandstone is cleared by mechanical or chemical process.

Some sandstones are cemented with dolomite. In general, these withstand the acid polluted atmosphere better than calcareous sandstone, particularly when ample cementing material is present. This is probably because dolomite is much less readily attacked by acids than calcite. When a dolomite sandstone seems to respond like a calcareous sandstone, the cement probably contains calcite as well as dolomite. It is usually the calcite that is attacked by the acid. Closely allied to the attack by acidic gases on calcareous sandstones is the attack by search gases on certain roofing slate that contain up to 13 per cent calcite. Sulphur based acidic gases in the rain water and arched by capillary, between the lap, that is in the overlap between adjacent in a roof. They form acids which attack the calcite in slates, thus weakening it. The gypsum formed by the reaction causes further weakening by crystallization of the substratum.

In complete contrast, acidic pollutants in the air are unlikely to cause any significant decay of granite used for building. However, natural decay of stones get accelerated by many reasons. Acid rains accelerates the leaching rate of soils, change the growth rate of the vegetation. According to 1972 claim, the pH of rainwater of Rewa industrial area of Germany changed from about 5 to less than 4 between the years 1956 and 1966. The decrease of rainwater pH in the last 20 years in England and the world is due to the increased use of fuel oil and natural gas, and also to treating shoot by mandatory industrial precipitation.

One of the most aggressive atmospheric pollutant affecting the building materials is sulphur dioxide (SO_2) which is very reactive and corrosive. Acid deposition is a more precise acid rain term, which has two main parts: dry and wet (Olaru *et al.*, 2010). Dry deposition refers to the deposition of pollutant gases and particles in the absence of rain. Wet deposition is concerned with the incorporation of pollutant substances in cloud droplets (occult deposition or rainout), or in regular precipitation (wet-only precipitation, acid rain or wash-out). Dry deposition, which is more important than wet deposition for highly polluted areas (Furlan and Girardet 1983), results from the transfer of pollutant gases and/or particles, including aerosols, from the atmosphere to a surface in the absence of rain (short range deposition). The main effects of sulfur dioxide on limestones are the formation of crusts and the loss of material due to solubilization, which can represent the 30–50 per cent of material loss (Pérez and Bello 2003). The loss of material can also be produced where weathering crusts reach certain thicknesses and then drop from the stone surface (Camuffo *et al.*, 1983). The stone surface where the crust is detached usually presents disaggregation and higher porosity and surface area than the original stone, becoming weaker to further weathering processes (McGee and Mossotti 1992).

21.7 Microbial Biodeterioration

The colonization of external surfaces of buildings by microorganisms causes the well-known aesthetically unacceptable appearance of staining of the stone surfaces by biogenic pigments (Urzi *et al.*, 1992; 1993) and the production of extracellular polymeric substances (EPS) that cause mechanical stresses to the mineral structure due to shrinking and swelling cycles of these colloidal biogenic slimes inside the pore system (Dornieden *et al.*, 2000; Warscheid 1996). This can lead to the alteration of pore size and distribution, together with changes in moisture circulation patterns and temperature response (Warscheid 1996; Warscheidand Krumbein 1996). Microorganisms may also alter the water permeability of the minerals by the deposition of surfactants (Gaylarde and Morton 1999). Last but not least, it has been shown that the early presence of biofilms on exposed stone surfaces accelerates the accumulation of atmospheric pollutants. Thus, microbial contamination acts as a precursor of the formation of detrimental crusts on rock surfaces caused by acidolytic and oxidoreductive (bio-) erosion of the mineral structure (Blaschke 1987; Ortega-Calvo *et al.*, 1994). Organisms present on stone monuments can include photolithoautotrophs, such as algae, cyanobacteria, mosses, and higher plants. Also present are chemo-lithoautotrophic bacteria that can release acids such as nitrous acid (*Nitrosomonas* spp.), nitric acid (*Nitrobacter* spp.), or sulfuric acid (*Acidothiobacillus* spp.), thus changing the local pH (Wagner and Schwartz 1967), and chemoorganotrophic bacteria and fungi that may release chelating organic compounds (Ascaso *et al.*, 1990; Palmer 1994), or weaken the mineral lattice by the oxidation of metal cations such as $Fe2+$ or $Mn2+$ (Vourinen 1981; Wagner and Schwartz 1967). Literature data show that the phototrophs dwelling on stone monuments are represented by a rather large number of genera and species in microbial consortia (Pantazidou and Theoulakis 1997). Although chemolithotrophic microorganisms have often been described in association with damaged inorganic materials (Wolters *et al.*, 1988) and were first suggested to play a role in stone deterioration in the 19th century (Mu"ntz 1890), more recent studies have emphasized the significance of chemoorganotrophic bacteria and fungi, together with photoautotrophs (May *et al.*, 1993). Even in the absence of a primary colonizing film of phototrophs, heterotrophic bacteria and fungi can grow on painted surfaces, using organic compounds from the paint as substrates, and produce acids that cause degradation of the paint and the underlying surface (Shirakawa *et al.*, 2002). A high number of taxa found on a substratum does not necessarily imply high bioreceptivity of that substratum, since many other environmental parameters play an important role in a successful colonization (solar radiation, temperature, water regime, climate, *etc.*). A study was undertaken by (Macedo *et al.,* 2009) to study deterioration in selected monuments subjected to similar climatic conditions (the Mediterranean climate); never the less, specific microclimatic parameters *i.e.* orientation, exposure to shadow, permanent capillary humidity, *etc.* are generally missing in the available literature. The biological colonization showed a characteristic trend with the microclimate (orientation and presence or absence of trees.

In the literature, many of black fungi are reported as RIF (rock inhabiting fungi) to emphasize that the "rock" is their preferred or exclusive habitat. However, this terminology does not include their main features such as melanin production, pleomorphism, or meristematic development; for this reason, we do not use it in this context. In the frame of cultural heritage the acronym MCF (Micro Colonial Fungi) as first employed by Staley *et al.* (1982) is widely used for their description. It refers to the typical black cauliform-like colonies visible on the rocks and stones. Humidity may affect the settlement of MCF on the stone artifacts as unique inhabitants or as associated with other stone colonizers. In fact, in lower or sheltered parts near the ground, where there is a sufficient availability of water, MCF are strictly associated with phototrophic microorganisms with whom, however, they do not establish a symbiotic relationship; in harsh, dry micro-environmental conditions, MCF become the unique colonizers (Marvasi *et al.,* 2012; De Leo *et al.,* 2019; Santo *et al.,* 2021).

21.8 Need for Stone Conservation

Everything weathers and erodes and will eventually disappear. To some geomorphologists, the act of "conserving" may seem to be an ill-advised attempt at arresting nature. It is as if we wish to arbitrarily freeze a snapshot of the building or monument in question at a point in its ruination. This notion was recognized in nineteenth century European landscape architecture when "ruins"–real or created– were incorporated into the designs of gardens and courtyards. Geomorphologist Emery's (1960) last statement on the weathering of the Great Pyramid of Giza was that the ancient structure should "remain as the last of the seven ancient wonders of the world for 100,000 years to come." Weathering crusts, case hardening, and even biotic colonies act to bind and indurate surfaces. Cleaning and resurfacing can destroy this natural protection. The applications to the stone (binders, sealants, biocides, repellants, *etc.*) have unknown long-term effects on weathering processes. We have witnessed the disastrous consequences of inappropriate conservation efforts from the past. Cleopatra's Needle, the Egyptian obelisk in New York City, was once treated with wax to seal it from the elements, but this also sealed in saline moisture. This treatment exacerbated an already deleterious action of moving the monument from an arid to a humid environment (Winkler 1978). Binders derived from mortars or cements may exert new pressure on masonry when they solidify or crystallize. Price (1996) points out that we know little about the microscale structure and interactive effects of sealants and binders on rock.

Geomorphologists can lend their expertise in this research (Young *et al.*, 2000). Fourth, there is a growing trend in the application of "artificial weathering" substances to mask fresh, cleaned, or repaired surfaces (Elvidge and Moore, 1980; Griswold, 1999). The best recommendation for slowing the rates of deterioration is to limit human contact. This goes for tourists–touching, vibrating, evapo-transpiring, breathing–as well as scholars, doing the same but also prodding, measuring, and conserving. All of our research points to human impact as the greatest cause of stone deterioration. Appreciation of our cultural treasures, scientifically and aesthetically, unfortunately contributes to the human impact on stone. Geomorphologists can lend their comprehension of interacting physical, biological, and human processes toward the study of cultural stone while adopting the "don't touch" attitude of the art conservators (Pop *et al.*, 2002).

REFERENCES

Ascaso C, Sancho L, Rodriguez-Pascual C (1990) Weathering action of saxicolous lichens in maritime Antartica. Polar Biol 11: 33–39

Albertano P (1995) Deterioration of Roman hypogea by epilithic cyanobacteria and microalgae. In: First International Congress on Science and Technology for the Safeguard of Cultural Heritage in the Mediterranean Basin, Catania, Siracusa, Italy, pp. 1303–1308.

Alcaeda M, Martin A (1988) Macroscopical study of stone alteration of the Cathedral of Seville, Paper presented at the VI International Congress on Deterioration and Conservation of Stone, Tarun, Sept. 12-14, pp. 216-224

Alonso E, Martýìnez L (2003) The role of environmental sulfur on degradation of ignimbrites of the Cathedral in Morelia, Mexico.Building and Environment 38(6): 861-867

Arai H (2014) Microbial problems in Biodeterioration of Cultural property: Forty years of Study. In : Biodeterioration of Cultural Property- 7 (eds) Dhawan S. Abduraheem K. and Nath V Pub By Sukriti Nikunj, Lucknow 1-10pp

Arai H, Wada K (2014) Blackening on monuments and its control: Long term prevention of biodeterioration. In : Biodeterioration of Cultural Property- 7 (eds) Dhawan S. Abduraheem K. and Nath V. Pub. By Sukriti Nikunj, Lucknow 1-10pp 89-94

Arya A, Gupta S P (2016)In: Anthropogenic pollution causes and concern paperback – 1 (eds) Arya A, Basu S.K.The Readers Paradise; 1st edition (1 January 2016) 116pp

Ascaso C, Sancho L, Rodriguez-Pascual C (1990) Weathering action of saxicolous lichens in maritime Antartica. Polar Biol 11: 33–39

Bahadur A K (1994) 'Impact of environment on Shore Temple, Mahabalipuram (Tamil Nadu), India', Conservation of Cultural Property in India, Vol. 27. pp. 62-73.

Bajpai R, Upreti D K, Dwivedi S K (2009) Arsenic accumulation in lichens of Mandav monuments, Dhar district, Madhya Pradesh, India. Environ Monit Assess 159: 437–442. https://doi.org/10.1007/s10661-008-0641-7

Bertolaccini M A, Cerquiglini S, Fassina V, Torraca G (1975) 'Study of some gaseous and particulate pollutants in the atmosphere of Venice 1972-73 and their effect on deterioration of Istrian stone', Report to UNESCO, ICCROM Rome

Bhargav J S, Mishra R C, Das C R (1999) Environmental deterioration of stone monuments of Bhubaneswar, the temple city of India. Stud Conserv 44: 1–11. https://doi.org/10.1179/sic.1999.44.1.1

Blaschke R (1987) "Natural building stone damaged by slime and acid producing microbes." In: Proceedings of the Ninth International Conference on Cement Microscopy, Reno, NV. USA, pp 70– 81

Bugess, Schaffer R J (1952) 'Cleopatra's needle', Chemistry and Industry, pp. 1026

Butlin R N, Cooke R U, Jaynes S M, Sharp A S (1985) 'Research on limestone decay in the U.K.', Proceedings of the Vth International Congress on Deterioration and Conservation of Stone, Lausanne, pp. 537- 546

Camuffo D, del Monte M, Sabbioni C (1983) Origin and growth mechanisms of the sulfated crusts on urban limestone. Water Air Soil Pollut. 1983, 19: 351–359

Caneva G, Nugari M P, Ricci S, Salvadori O (1992) Pitting of marble roman monuments and the related micro ora. In: Delgado, J., Enriques, F., Telmo, F. (Eds.), Seventh International Congress on Deterioration and Conservation of Stone. LNNA, Lisbon, pp. 521–530.

De Leo F, Antonelli F, Pietrini A M, Ricci S, Urzì C (2019) Study of the euendolithic activity of black meristematic fungi isolated from a marble statue in the Quirinale Palace's Gardens in Rome, Italy. Facies. 65: 18.

Di Martino P (2016) What about biofilms on the surface of stone monuments? Open Confer Proc J 6: 14–28

Dornieden T H, Gorbushina A A, Krumbein W E (2000) Patina–physical and chemical interactions of sub-aerial biofilms with objects of art. In: Ciferri, O, Tiano, P, Mastromei, G (Eds.) Of Microbes and Art: The Role of Microbial Communities in the Degradation and Protection of Cultural Heritage, Kluwer Academic, Dordrecht, pp 105–11

Elvidge C D, Moore C (1980) Restoration of petroglyphs with artificial desert varnish. Studies in Conservation 25, 108 – 117.

Emery K O (1960) Weathering of the Great Pyramid. Journal of Sedimentary Petrology 30, 140 – 143

Farkas O, Siegesmund S, Licha T, Török Á (2018) Geochemical and mineralogical composition of black weathering crusts on limestones from seven different European countries. Environ Earth Sci 77: 211. https: //doi.org/10.1007/s12665-018-7384-8

Fassina V (1986b) The stone decay of main portal of Saint Marks Basilica in relation to natural weathering agents and air pollution, Paper presented at the VIth International Congress on Air Pollution and Conservation, Rome

Frediani P, Malesani P G, Vannucci S (1976) Weathering of Florentine stone, Athens, pp. 117-119

Furlan V, Girardet F (1983) Considerations on the rate of accumulation and distribution of sulphorous pollutants in exposed stones. In Materials Science and Restoration; Wittmann, F.H., Eds.; Lack und Chemie: Filderstadt, Germany, pp. 285–290.

Gauri K L, Holdren G C (1981) Pollutant effects on stone monuments Environ. Sci.Technol. 15 (4): 386–390. https: //doi.org/10.1021/cs00086a001

Gauri K L. (1978a) Sci. Am. June. 126- 136.

Gauri K L. (1978b) In "Proc. International Sympsium on Delcrioration and Protection of Stone Monuments": UNESCO RII.EM: Paris. Junc 5-9. pp. 1-20.

Gaylarde C C, Morton L H G (1999) Deteriogenic biofilms on buildings and their control: a review. Biofouling 14: 59–74

Gibeaux S, Vázquez P, De Kock T, Cnudde V, Thomachot-Schneider C (2018) Weathering assessment under x-ray tomography of building stones exposed to acid atmospheres at current pollution rate. Constr Build Mater 168: 187–198. https: //doi.org/10.1016/j.conbuildmat.2018.02.120

Gomez-Alarc G, Munoz M, Arino X, Ortega-Calvo J J (1995) Microbial communities in weathered sandstones: the case of Carrascosa del Campo church, Spain. The Science of the Total Environment 167, 249–254.

Goyal P, Singh M P (1990) The long-term concentration of sulphur dioxide at Taj Mahal due to the Mathura Refinery.Atmospheric Environment. Part B. Urban Atmosphere. 24 (3): 407-411https: //doi.org/10.1016/0957-1272(90)90048-Y

Graue B, Siegfried S, Tobias L, Klaus S, Pedro O, Bernhard M (2012) The effect of air pollution on the stone decay of the Cologne Cathedral. 9893

Griswold J, (1999) Camouflaging graffiti: the problem of outdoor inpainting. In: Dean, J.C. (Ed.), Images Past, Images Present: The Conservation and Preservation of Rock Art. IRAC Proceedings, vol. 2, pp. 41 – 46

IL and FS Ecosmart Ltd. (2009) "Technical EIA Guidance Manual for Common Effluent Treatment Plants" Hyderabad, September 2009

Kaur R (2015) Vehicular Pollution in Delhi and Its Impact on Lotus Temple. My India (on line) April 21. 2015

Krumbein WE, Diakumaku E, Gehrmann C, Gorbushina AA, Grote G, Heyn C, Kuroczkin J, Schostak V, Sterflinger K, Warscheid Th, Wolf B, Wollenzien U, Yun-Kyung Y, Petersen K (1996) Chemoorganotrophic microorganisms as agents in the destruction of objects of art–a summary. In: Riederer, J (Ed.) Proc Eighth International Congress on Deterioration and Conservation of Stone, vol. 2, Rathgen-Forschungslabor, Berlin, pp 631–636 67

Lal B B (1978) 'Weathering and preservation of stone monuments under tropical conditions', International Symposium on Deterioration and Protection of Stone Monuments, Paris, pp. 1-9

Lamenti G, Tiano P, Tomaselli L (2000) Microbial communities dwelling on marble statues. In: Monte, M. (Ed.), Eighth Workshop EUROCARE–EUROMARBLE, EU 496, CNR, Roma, pp. 83–87

Lan T T N, Nishimura R, Tsujino Y, Satoh Y, Thoa NTP, Yokoi M, Maeda Y(2005) The effects of air pollution and climatic factors on atmospheric corrosion of marble under field exposure. Corrosion ScienceV 47 (4) : 1023-1038https://doi.org/10.1016/j.corsci.2004.06.013

Macedo M F, Miller A Z, Dioný´sio A, Saiz-Jimenez C (2009) Biodiversity of cyanobacteria and green algae on monuments in the Mediterranean Basin: an overview. Microbiology, 155: 3476–3490 DOI 10.1099/mic.0.032508-0

Marvasi M, Donnarumma F, Frandi A, Mastromei G, Sterflinger K, Tiano P, Perito B (2012) Black microcolonial fungi as deteriogens of two famous marble statues in Florence, Italy. Int. Biodeterior. Biodegrad. 68: 36–44.

May E, Lewis FJ, Pereira S, Tayler S, Seaward MRD, Allsopp D (1993) Microbial deterioration of building stone–a review. Biodet Abs 72: 109–123

McGee E S, Mossotti VG (1992) Gypsum accumulation on carbonate stone. Atmos. Environ. (26B): 249–253.

Mihaela O, Aflori M, Simionescu B, Doroftei F, Stratulat L (2010) Effect of SO$_2$ Dry Deposition on Porous Dolomitic Limestones. Materials 3: 216-231; doi: 10.3390/ma3010216 materials ISSN 1996-1944 www.mdpi.com/journal/materials Article

Mu¨ntz A (1890) Compt Rend Acad Sci 110: 1370–1372. Cited in Waksman, SA (1932) Principles of Soil Microbiology. The Williams and Wilkins Co., Baltimore, p 567

Narain U, Bell RG (2005) Who Changed Delhi's Air? The Roles of the Court and the Executive in Environmental Policymaking. Resources for the Future. Dec. 2005. 1616 P St. NW Washington, DC 20036 202-328-5000 www.rff.org

Nijs R (1985) 'Petrographical characterization of calcareous building stones in Northern Belgium' in G. Felix (ed.) Proceeding Vth Internaional Congress on Deterioration and Conservation of Stone. Ecole, Polytechnique, Federale, Lausanne, Italy, pp. 13-21

Ortega-Calvo J J, Arino X, Stal L J, Saiz-Jimenez C (1994) Cyanobacterial sulfate accumulation from black crusts of a historic building. J Geomicrobiol 12: 15–22

Palmer Jr, R J (1994) Mikrobielle Aktivita¨ten in verwitterndem Gesteinsmaterial: Biomasse, Gesellschaftsstruktur und Na¨hrstoffe. Werkst Korros 45: 114–116

Pantazidou A, Theoulakis P (1997) Cyanophytes and associated flora at the neoclassical palace of St. George and Michael in Corfu (Greece). Aspects of cleaning procedures. In: Moropoulou, A, Zezza, F, Kollias, E, Papachistodoulou, I (Eds.) 4th Int Symp Conservation of Monuments, vol. 4, Technical Chamber of Greece, Rhodes, pp 355–368

Pérez B J L, Bello M A (2003) Modeling sulfur dioxide deposition on calcium carbonate. Ind. Eng. Chem. Res. 42: 1028–1034.

Peruzzi R, Alessandrini |G, Liboriola G, Giambelli L, De Captione (1978) Characteristics of decay of materials used in Ca Granda (Milan Building), Paper presented at the International Symposium Rilem, Deterioration and Protection of Stone Monuments, Paris

Pinna D (2017) Coping with biological growth on stone heritage objects. Methods, products, applications, and perspectives. Apple Academic Press, Palm Bay

Pinna D (2021) Microbial growth and its effects on inorganic heritage materials. Joseph E. (ed.), Microorganisms in the Deterioration and Preservation of Cultural Heritage, https://doi.org/10.1007/978-3-030-69411-1_1

Price C A (1996) Stone Conservation: An Overview of Current Research. Getty Conservation Institute, Los Angeles.

Pope G A, Meierding T C, Paradise T R (2002) Geomorphology's role in the study of weathering of cultural stone. Geomorphology. 47(2): 211-225.DOI: 10.1016/S0169-555X(02)00098-3

Ramachandran S (2007) "A touch up for the Taj", Online Asia Times, South Asia Nov 9, 2007

Rampazzi L, Corti C, Geminiani L, Recchia S (2021) Unexpected Findings in 16th Century Wall Paintings: Identification of Aragonite and Unusual Pigments. Heritage 2021, 4: 2431–2448. https://doi.org/10.3390/heritage4030137 Academic Editor: Diego Tam

Santo A P, Cuzman O A, Petrocchi D, Pinna D, Salvatici T, Perito B (2021) Black on white: Microbial growth darkens the external marble of Florence cathedral. Appl. Sci. 2021, 11, 6163.

Sharma J S, Sharma D N (1982) Atmospheric contamination of archaeological monuments in the Agra Region (India). Sci Total Environ 23: 31–40. https://doi.org/10.1016/S0166-1116(08)70988-9

Shirakawa M A, Gaylarde C C, Gaylarde P M, John V, Gambale V (2002) Fungal colonization and succession on newly painted buildings and the effect of biocide. FEMS Microbiol Ecol 39: 165– 173

Siegesmund S, Snethlage R E (2014) Stone in architecture. Properties, durability. Springer, Berlin Heidelberg, Germany

Singh S P (2007) Effect of Pollution and other Environmental Factors on Monuments. Context, built living natural.4 (1): 55-60

Tomasellia L, Lamentia G, Boscob M, Tianoc P (2000) Biodiversity of photosynthetic micro-organisms dwelling on stone monuments. International Biodeterioration and Biodegradation 46: 251–258

Török A, Licha T, Simon K, Siegesmund S (2011) Urban and rural limestone weathering; the contribution of dust to black crust formation. Environ Earth Sci 63: 675–693. https://doi.org/10.1007/s12665-010-0737-6

UNESCO (1977) Intergovernmental Conference on Environmental Education, Tbilisi, USSR, 14-26 October 1977: final report 101pp Document code: ED/MD/49

Upreti D K, Yadav V, Nayaka S (2002) A note on lichens from environs of Lucknow district, U.P. India. In Proc. 89th Indian Science Congress. Lucknow University, Lucknow

Upreti D K, Bajpai R, Nayak S (2009) Indian monuments need lichen biodeterioration study. New Horizons. 7: 64-69

Urzi C E, Krumbein W E, Warscheid T H (1992) On the question of biogenic colour changes of mediterranean monuments (coatingcrust-microstromatolite-patina-scialbatura-skin-rock varnish). In: Decrouez, D, Chamay, J, Zezza, F (Eds.) Proc Second Int Symp Conservation of Monuments in Mediterranean, Basins, Geneva, pp 397–420

Urzi C E, Criseo G, Krumbein W E, Wollenzien U, Gorbushina A A (1993) Are colour changes of rocks caused by climate pollution, biological growth, or by interactions of the three? In: Thiel, M-J (Ed.) Conservation of Stone and Other Materials, vol. 1, E and FN Spon, London, 1: 279–286

Venkata Rao N, Rajasekhar M, Rao C G (2014) Detrimental effect of air pollution, corrosion on building materials and historical structures. Am J Eng Res 3(3): 359–364

Vourinen A, Manterc-Alhonen S, Uusinoka R, Alhonen P (1981) Bacterial weathering of Rapakivi granite. J Geomicrobiol 2: 317– 325

Wagner M, Schwartz W (1967) Geomikrobiologische Untersunchungen. VIII. u¨ber das Verhalten von Bakterien auf der Oberfla¨che von Gesteinen und

Mineralien und ihre Rolle bei der Verwittenrung. Z Allgem Mikrobiol 7: 33–52

Warscheid T (1996) Impacts of microbial biofilms in the deterioration of inorganic building materials and their relevance for the conservation practice. Int Zeitsch Bauinstandsetzen 2: 493–504

Wagner M, Schwartz W (1967) Geomikrobiologische Untersunchungen. VIII. u¨ber das Verhalten von Bakterien auf der Oberfla¨che von Gesteinen und Mineralien und ihre Rolle bei der Verwittenrung. Z Allgem Mikrobiol 7: 33–52.

Warscheid T, Krumbein WE (1996) Biodeterioration of inorganic nonmetallic materials–general aspects and selected cases. In: Heitz, H, Sand, W, Flemming, HC (Eds.) Microbially Induced Corrosion of Materials, Springer, Berlin, pp 273–295

Winkler E M (1975) 'Weathering rates of stone in urban atmosphere', Proceedings of an International Symposium, Bologona, pp. 27-36.

Winkler EM (1978) The complex history of salt weathering observed on Cleopatra's Needle in New York Central Park. Geological Society of America Abstracts with Programs 10: 578.

Winker E M, Singer P C (1972) Crystallization pressure of salts in stone and concrete. Bull. Geo. Soc. Am. Bull. 83: 3509-3513

Wolters B, Sand W, Ahlers B, Sameluck F, Meinke M, Meyer C, Krause-Kupsch T, Bock E (1988) "Nitrification: the main source for nitrate deposition in building stones." Proc Sixth Int Cong Deterioration and Conservation of Stones. Nicolaus Copernicus University, Torun, Poland, 12–14 September 1988, pp 24–31

Young M E, Ball J, Laing R A (2000) Quantification of the long term effects of stone cleaning on decay of building sandstones. 9th International Congress on Deterioration and Conservation of Stone, Venice, June 19 – 24, pp. 179 – 186.

Chapter 22

Microbial Deterioration and Occurrence of Bryophytes and Angiospermic Plants on Certain Heritage Buildings of Vadodara

Niharika Nema Baijal[1] and Arun Arya[2]

[1]Food Technology Division, Homi Bhabha National Institute, Bhabha Atomic Research Centre, Mumbai – 400 094, Maharashtra, India
[2]Department of Environmental Studies, Faculty of Science, The Maharaja Sayajirao University of Baroda, Vadodara – 390 002, Gujarat, India
e-mail: niharikanema@gmail.com; sarojarun10arya@rediffmail.com

ABSTRACT

Monuments and old buildings in India with natural stone and baked bricks have been exposed to decay for centuries. The monuments made up of stones and mortar, not only show the architectural beauty but are good examples of cultural heritage prevailing that time. One can know various things by peculiar designs of these monuments. It is said that vast wealth of ancient Indian culture is still preserved in temples. The numerous temples have their own museums and collections of artifacts including paintings, utensils and armaments. The building materials are decayed by adverse environmental conditions and the extent of damage depends on both the materials and the conditions. Growth of microorganisms and plants on monument is not only dangerous but even fatal to them; the biochemical damage is due to extraction of organic acids which can erode the surface. If plants grow inside fissures or cracks present in the surface, the gap may increase further.

It has been recorded that roots of many flowering plants by penetrating and enlarging disorganize the structure of monuments. Indian Peepal and banyan tree in particular are the most destructive. Their root penetrates into crevices,loosens and dislodges stones. Various plants growing in stone monuments and old buildings produce complex organic acids like gluconic acid, gallic acid, itaconic acid, kojic acid, aspergillic acid, gyrophoric acid, salicylic acid, and usnic acid. These organic acids produced by the plants on the monuments, thus weakening the structure and pitting the smooth and polished surface. A study was undertaken during July-August months of 2014 and assessment of deterioration caused by bryophytes and phanerogams was studied. The survey of faculty of Performing arts, near Sur Sagar revealed the presence of two bryophytes on walls and six flowering plants. The growth of Ficus religiosa and F.tsiela was more prominent. In the terrace of old Faculty of Science building the presence of sixteen different flowering plants and one moss was observed. The presence of these plants was more prominent on joints of walls and near drainage-pipes. The Angiospermic plants were not present in the large number as expected, this may be due to the renovated terrace, it was having water-proofing sheets all around, but on the walls and corners of the buildings many flowering plants were seen. A detailed account of these plants is presented in paper. There is a need for regular survey of Hajira tomb and the two old buildings of The M. S. University of Baroda. Certain preventive and control measures are proposed so that we can preserve the beauty of these heritage buildings for a longer time.

Keywords: Microbes, Deterioration, Bryophytes, Angiospermic plants, Heritage buildings, Hajira, Vadodara

22.1 Introduction

Buildings consisting of bricks and stones, are rapidly colonized by microbial communities forming subaerial biofilms. These biofilms are composed of different microorganisms (mainly algae, Cyanobacteria, bacteria and fungi) immobilized on the stone surface and embedded in a matrix formed by extracellular polymeric substances (EPS), which spreads onto the substratum and serves both to keep the microbes together and to facilitate further adhesion onto and incursion into the substratum (Gorbushina 2007; Vázquez-Nion *et al.*, 2016). Formation of such biofilms are common on heritage buildings, as regular monitoring and repair is not a routine feature. Lichens are considered pioneers. According to several authors (Ortega-Calvo *et al.*, 1991; Tiano *et al.*, 1995; Crispim andGaylarde 2005), Cyanobacteria and members of Chlorophyta are considered first to colonize the stones, providing an excellent organic nutrient base for subsequent heterotrophic microflora. Heterotrophs, lithotrophs. In thin films the association of organisms is symbioses type. Once vital organic substratum is produced the higher plants start appearing in the buildings where proper humidity and light is available.

Although any warfare, earthquake or volcanic eruption can destroy a heritage monument within seconds the threats from Pollution, favorable environmental conditions and changing climate are equally important to deteriorate the aesthetic beauty of any monument and always need a disaster preparedness management to reduce the damage. Such monuments are historically important and serve as centers of tourist attraction. Archaeological Survey of India helps to maintain valuable heritage of the nation.

Conservation of old monuments is possible only when we understand clearly the cause of deterioration. The reason that cultural property distributed in Asian countries has received serious damage by various biological agents in the favorable environmental conditions like high humidity and temperature. The monuments are made up of stones and mortar not only show the with beauty but also are good examples of cultural heritage prevailing that time and its been a serious world-wide concern to protect this treasure for learning the past history and for future generations. On one hand the natural disasters are the problem, while on the other hand pollution, climate change and global warming are posing severe threat to old monuments. The improper maintenance and lack of knowledge about conservation has pushed the issues far more ahead than it ought to be. Monuments and old buildings in India with natural stone have been exposed to decay for centuries. Growth of microorganisms and plants on monuments is not only dangerous but even fatal to them. Mainly the plants are found in cracks, fissures and drainage openings on the buildings (Agrawal *et al.*, 1994). Some researchers (Shah and Shah 1992; Agrawal *et al.*, 1994; Shah 1999; Arya and Shah 2009)have tried to study the issue in different buildings in the city of Vadodara. Salient features and structural details of few buildings under discussion is as follows:

22.1.1 The Hajira Tomb

The Mausoleum of Qutubuddin Mohammad Khan, situated in Pratapnagar area of Vadodara is popularly known as the Hajira. It has the distinction of being possibly the first monument of the Baburi dynasty and was built around 1586. Mr. Khan was the tutor of Salim, son and successor of Akbar and also that his son Naurang Khan, who held important office in Gujarat under Akbar. Qutubuddin was Governor of Gujarat thrice between 1573 and 1605. The mausoleum houses about a dozen tombs in and around the impressive makbara. It has a cenotaph, a feature common to most of the Mughal tomb architecture. Also are the surrounding gardens laid out as precise the geometric in spirit as the architecture (Parimoo 1989).

It is built on a high octagonal platform with smaller gates on the cardinal directions and five arches one on each side. As in the other tombs, the real grave is in an underground chamber and the false grave in the tomb chamber (Commissariat 1868).

22.1.2 Faculty of Science

The Maharaja Sayajirao University of Baroda was established in April 1949, under the patronage of Maharaja Pratapsinghrao Gaekwad. The visionary ruler of Baroda state Srimant Maharaja Sayajirao Gaekwad III after whom it is named started initially a Baroda college in 1881 (Figure 22.1). Although Maharaja made primary education compulsory for the children, he was not of the opinion to start a university considering high expenditure required for generating infrastructure, decent labs and continuous running cost. It was Maharishi Aurobindo Ghosh, who convinced him later on to start a university. The university has total of 94 departments out of these 50 departments are in Science, Social Science and Humanities.

Figure 22.1: Special Cover Showing Ancient Building of Baroda college (Arts Faculty).

The faculty of Science is located in the main university campus near railway station in Fatehgunj. The faculty of science is a constituent institution of M.S. University of Baroda under the direct management and control of the university. This faculty started its independent existence in March 1951with Dr.C.S Patel as its first Dean. Its foundation was laid on 23rd January 1926. The old building which houses the faculty at present was completed in about 1934 in the reign of Srimant Sayajirao Gaekwad III. It houses departments of Chemistry and Zoology, a big statue of Maharaja is placed in front of the building (Figure 22.3a). Since building is old time to time repairs have been undertaken. The construction is made up of bricks and red stone. A telescope is installed at the top of building in Astronomical observatory. A gift from the Maharaja it houses an 8-inch refractor telescope. The observatory has been visited by several well-known scientists. It organizes shows on important astronomical events for general public and school children. The observatory has a copper dome which can be opened during observations. The copper dome showed erosion due to pollutant gases.

22.1.3 Sangeet Shala of MSU Baroda

The city of Baroda considers this faculty as the cultural hub of the city. Located opp. Sursagar, Nyaymandir, the Faculty of Performing Arts, formerly known as college of Indian Music, Dance and Dramatics was established as "Chookraoni Gayan Shala-Singing School for children" in February 1886 by great visionary Maharaja Sayajirao Gaekwad – III, with learned musician Prof. Mawlabaksh as its founder principal. It is a foremost constituent of the Maharaja

Sayajirao University of Baroda, Vadodara – established in 1950 (Figures 22.4a and 22.5a). It is India's premier institution imparting training at the University level. It houses five vibrant departments *viz*, Vocal music, Instrumental music (Sitar/Violin), Tabla, Dance (Kathak/Bharat Natyam) and Dramatics offering different programme. The potential training is towards developing and fostering both creative and intuitive abilities in the student whereby attaining a systematic understanding of the art and science of performing arts and its aesthetic applications and to enable them to respond to changing times of the contemporary society.The Faculty enjoys a status of unique centre of Performing Arts in the entire western region engaged in its mission of defining cultural dynamism of Indian classical arts in the rapidly evolving global economy. It helps in preserving and practicing ancient tradition of Indian Performing Art. Practicing Ancient Indian Tradition of "Gharanas and Guru-Shishya Parmpara" blended with Modern Pedagogy. Faculties are also relates with Traditional Schools of Music (Gharana-Sampraday) with their required **Prominent Gurus and Alumni:** Prof. Mawla Bux, Sufi Inayat Khan, Ut. Faiyaz Khan, Ut. Ata Hussain Khan, Ut.Nissar Hussain Khan, Shri HirjibhaiPatrawala, Prof. C. C. Mehta, Prof. R. C. Mehta, Pt. Shivkumar Shukla, Prof. C. V. Chandrashekhar, Pt. Madhusudan Joshi, Ut. Shameem Ahmed, Smt. Swapna Waghmare, Shri Kirtidan Gadhvi, Shri Nitiranjan Biswas.

22.2 Materials and Methods

A) The sites were visited and photographs were taken.

B) Isolation of fungi, and algae/cyanobacteria was tried.

C) Microorganisms were identified on the basis of their cultural and morphological characters.

D) Collection of Bryophytes and plants was done from the two buildings of the M.S. University in two different seasons (during rains and winter 2018) and identification was done.

E) Main purpose for identification is to know the occurrence of weeds and suggest measures to destroy these deteriorating agents.

22.2.1 Biodeterioration in the Hajira Tomb

In the presence of moisture roof top of dome and walls showed the occurrence of three Cyanobacteria, one alga, four fungi, two bryophytes and 18 Angiospermic plants belonging to 14 families. Data are recorded in Tables 22.1 and 22.2. Out of the 18 angiospermic plants Shah (1999) reported that 12 were new reports from monuments in India. According to Agrawal *et al.* (1994) both herbaceous and woody plants may grow on old monuments if enough moisture and light is available. Three spp. of *Ficus* (*F. religiosa, F. benghalensis* and *F.glomerata*), margosa or neem (*Azadirachta indica*) and Manila tamarind or goras amli (*Pithecelobium dulce*) were occurring on this monument. All these plants are of woody nature. *Chlorella* sp. was only alga and *Funaria* and *Cyathodium*were

Figure 22.2a-c

(a) Heritage monument of Hajira Tomb in Vadodara; (b) Growth of peepal (*Ficus religiosa*) visible on the tomb; (c) Hajira Tomb after conservation done by ASI (plants removed).

two bryophytes reported (Shah 1999). Continued growth of these life forms on building is detrimental for its health. They may cause erosion and deterioration by exerting physical pressure and biochemical weathering. Archaeological survey of India later on removed these plants and beauty was restored.

22.2.2 Biodeterioration in the Faculty of Science

Monuments and old buildings in India with natural stone have been exposed to decay for centuries. Growth of microorganisms, lichens, bryophytes and phanerogams on monuments is not only dangerous but also fatal to them. Mainly the plants are found in cracks, fissures, and drainage openings on the building which tend to widen the masonry gaps. Growth of plants may be useful from economical or ecological point of view, providing clean air, O_2 and reducing the CO_2, a major greenhouse gas, and pollutants but their occurrence on old buildings can damage them and are extremely harmful for their survival. University has to spent lot of money to tackle the problem. In a survey conducted a large number of herbs were identified (Figures 22.3c). The occurrence of sacred fig (*Ficus religiosa*) can be prominently seen in picture (Figure 22.3b). This is in front of main building of Faculty of Science just above the statue of Maharaja. Removal of such plants is most essential. Since renovation was done at different places above the Department of Zoology, weeds were removed but at one place lot of grass was observed (Figure 22.3d), this growth was promoted by excreta of monkeys which serve as substrate for growth of common grass (*Cyanodon dactylon* Figure 22.e) and other weeds. Total 17 angiospermic plants were recorded (Table 22.2).

22.2.3 Biodeterioration in the Sangeet Shala of MSU Baroda

A study was undertaken during July- August during rains and another during November months to assess the occurrence of flora and the deterioration caused by bryophytes and phanerogams. The study revealed the presence of two bryophytes and six flowering plants on the walls at the faculty of performing arts. The beauty of city can be visualized from the tower of building. The old structure has terracotta tiles and lot of accumulated dustpromotes luxuriant growth of annual plants. The seeds may be brought by birds, bats, rodents and monkeys. Total 16and fourteen different types of flowering plants were found at the faculty of Performing Arts. Today the sight is enough to make anyone sore. The splendid tower is covered with algae, plant growth and bird droppings.

The broken railings are to be restored as at certain places growth of *Ficus* sp. was observed (Figure 22.5a, c). It is very important that in main campus wherever the wooden floor and stair cases or railing is present should not be replaced by other structural material but restored by wood only. This will intact its beauty and add to architectural and heritage value.

Figures 22.3a-e

(a) Front view of Faculty of Science; (b) Growth of *Ficus religiosa* on the top of Faculty of Science; (c) Growth of *Lindenbergia urticifolia* on Faculty of Science building; (d) *Ficus* sp. growing in old building of Chemistry Dept.; (e) Growth of *Cynodondactylon* on the top of terrae of Zoology Dept. Note the presence of excreta of monkeys.

Table 22.1: Occurrence of different Biodeteriogens at 3 different Locations

Sl.No.	Group	Organism	Hajira1	Science	Sangeetshala
1.	Bacteria	Gram positive	++	++	++
2.	Algae	*Chlorela*	+	-	-
3.	Cyanobacteria	*Gloeocapsa*	++	-	-
4.		*Nostoc*	++	+	++
5.		*Oscillatoria*	-	+	-
6.	Fungi	*Aspergillus niger*	++	+	+
7.		*A. flavipes*	++	+	-
8.		*Fusarium sp.*	+	-	-
9.	Bryophyta	*Cyathodium barodae*	++	++	+++
10.		*Funaria hygrometrica*	+++	++	+
11.		*Funaria* sp.	-	++	-

Data Source: Shah, 2000.

-: Absence. +: Presence; ++L More occurrence; +++: Present in large numbers.

Table 22.2: Occurrence of different Biodeteriogens at 3 different Heritage Buildings

Sl. No.	Plant	Family	Common	Hajira[1]	Science Faculty	Sangeet-shala
1.	*Acalypha indica*	Euphorbiaceae		+	+	-
2.	*Achyranthes aspera*	Amaranthaceae		-	+	+
3.	*Alternanthera sessilis*	Asteraceae		**+**	-	-
4.	*Amaranthus tricolor*	Amaranthaceae		+	+	-
5.	*Amaranthus viridis*	"	Chaulai/ Tandalja	+	+	+
6.	*Azadirachta indica*	Meliaceae	Neem/ margosa	+	-	-
7.	*Boerhaavia diffusa*	Nyctaginaeae	Patahrchata	+	+	-
8.	*Calotropis procera*	Asclepediaceae	Akdo	-	+	-
9.	*Cyperus rotundus*	Cyperaceae		+	+	+
10.	*Cynodon dactylon*	Gramineae	Green grass	+	+	+
11.	*Eclipta alba*	Asteraceae	Bhangra	+	+	-
12.	*Ficus bonghalensis*	Moraceae	Vad/ Banyan	+	+	+
13.	*F. glomerata*	"	Umaro	+	+	-
14.	*F. religiosa*	"	Peepal	+	+	+
15.	*F. teisela*	"		-	-	+
16.	*Indigofera enneaphylla*	Fabaceae		-	-	+

Sl. No.	Plant	Family	Common	Hajira[1]	Science Faculty	Sangeet-shala
17.	*Lindenbergia urticifolia*	Scrophulariaceae		+	+	+
18.	*Lycopersicon lycopersicum*	Solanaceae	Tomato	+		-
19.	*Pentatropis capensis*	Apocynaceae		+	-	-
20.	*Pithecellobium dulci*	Caesalpinaceae	Gorasamli		+	-
21.	*Physalis peruviana*	Solanaceae	Rsabhari	+	+	+
22.	*Ruellia tuberosa*	Acanthaceae		+	-	+
23.	*Sesamum indicum*	Pedaliaceae	Til	_	-	+
24.	*Solanum nigrum*	Solanceae	Black nightshade	+	+	+
25.	*Trichodesma amplexicaule*	Boraginaceae		-	+	+

Data Source: Shah, 1999;

-: Absence. +: Presence of plant species.

22.3 Damage to the Buildings by Pollution and Environmental Factors

Attempts have been made to understand the phenomenon of weathering of old buildings and monuments with an objective to develop better methods of preservation. Besides the nature of stone or plaster used on brick work, a host of climatic factors working constantly on the building play a vital role in bringing about the destruction. Winkler (1975) reported Physical, Chemical and Biological factors responsible for decay. Importance of lichen and plant growth on degradation of monuments was highlighted by several workers (Riederer 1981, Agarwal *et al.*, 1994; Mottershead and Lucas 2000; Lisci *et al.*, 2003).

Urbanization has caused air,water and soil pollution. The complex physical and chemical interaction of the atmospheric pollutants with the mineral material and microorganisms have accelerated the decay of stones (Warscheid and Braams 2000; Tecer and Cerit 2002). Studies conducted by Arya and Gupta(2016) and Prasad and Arya (2020) on aeromycoflora and aeroallergens in the city showed formation ofa film composed by hydrocarbons generated by vehicular exhaust and dust, which promoted settlement and accelerated growth of bacteria and fungi on the walls of Lehripura gate in busy market area of Vadodara. This gate is 15[th]century sand stone structure up to a height, then bricks and plaster is used. The deterioration was reported by occurrence of fungi like, *Trichoderma viride, Rhizopus stolonifer, Aspergillus niger, A. terreus, Alternaria alternata* and *Curvularia lunata*. These fungi were responsible for the blackening of walls (Prasad and Arya 2020). Patel and Choksi (2012) reported that ecofriendly materials such as wood, clay, stones, clay *etc.* are used. Arai and

Figure 22.4a-c

(a) Main building of Faculty of Performing Arts or Sangeet Shala of MSU Baroda; (b) Plants growing along drain pipe in Sangeet Shala; (c) A variety of plats growing in terrace of Sangeetshala.

Yamagishi (2005) reported that traditional castles in Japan are having plaster or stucco on the surface. They observed blackening due to *Cladosporium* spp.

22.4 Biodeterioration: Causes and Concern

Because of the complexity of the microbial composition of most subaerial biofilms, understanding their development requires knowledge of all the

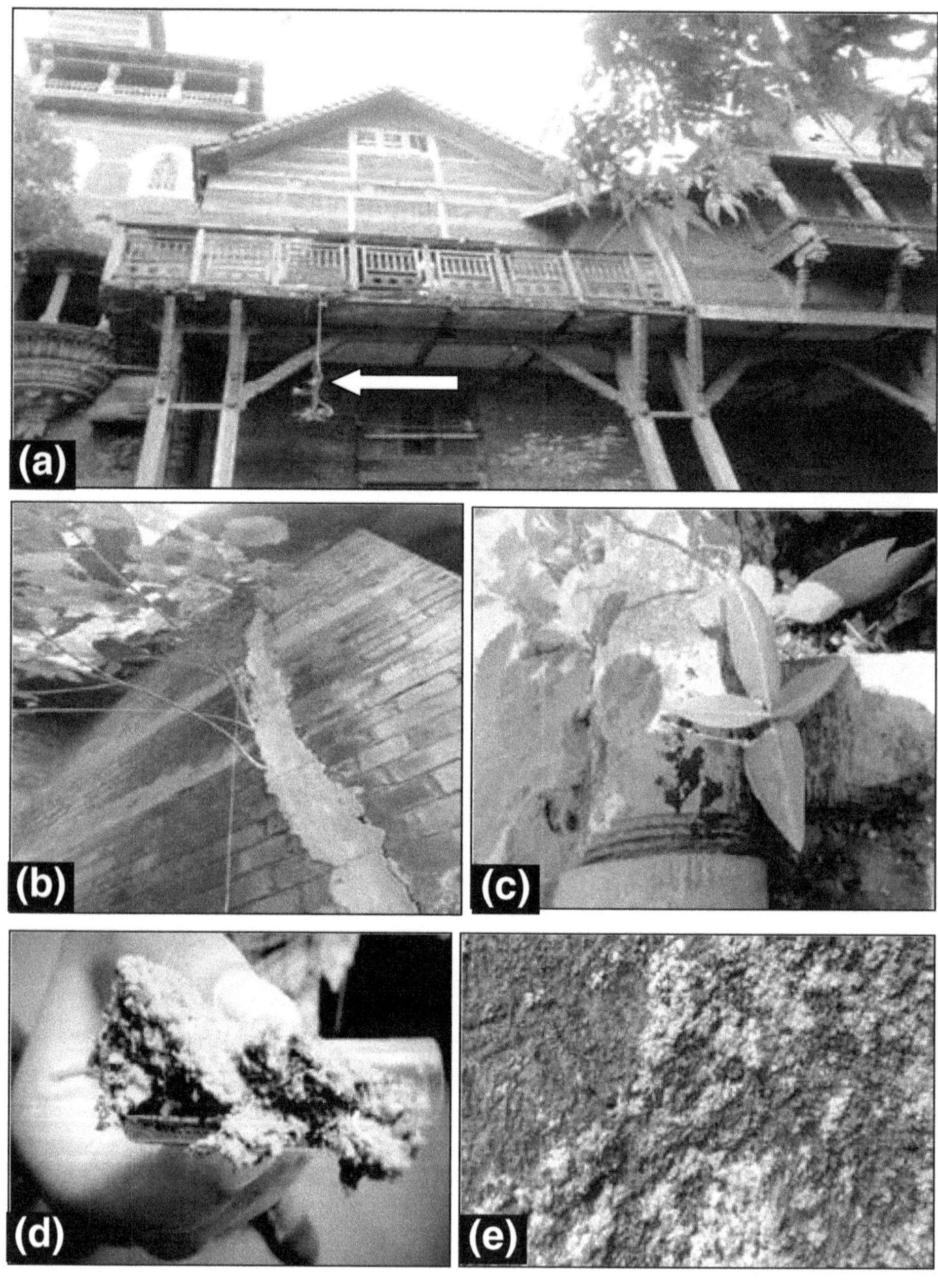

Figure 22.5a-e

(a) Hanging *Ficus* plants from railing of building; (b) Plants finding space for growth from a big crevice, repaired in main building of Faculty of Performing Arts; (c) From a joint of drainage pipe a big plant of *Ficus teisela* can be seen growing; (d) A bryophyte recovered from a wall during rainy season; (e) Luxuriant growth of a Bryophyte – *Cyathodium barodae* from a wall of Performing Arts.

components (aerobic heterotrophs, lithotrophs, oxygenic phototrophs, and even heterotrophic phagotrophs) (Gorbushina 2007). Some researches has been focused on characterizing the microbial composition of biofilms growing on historic buildings, mainly from the point of view of their biodeterioration potential. Thus, for example, Gaylarde and Gaylarde (2000) analyzed the major microbial biomass of 230 biofilms developed on building surfaces, concluding that Cyanobacteria, followed by fungi, were most common in Latin America, whereas green algae, followed by Cyanobacteria, were most frequently present as major biomass in Europe (. Moreover,

Microorganisms are able to utilize different elements by bio-solubilization of building materials like stone, brick or plastered surface. Recent researches have revealed that fungi can utilize elements from airborne organic compounds too and inorganic particulate matter settled on stone surfaces, They mainly arise from incomplete combustion of fossil fuels and dusts (Saiz-Jimenez 1997; Zanardini *et al.*, 2000). They further reported that heterotrophic organisms act as first colonizers in the area with high level of organic pollutants.

In 2001 the dome of the faculty of arts of M. S. University was de-weeded to preserve its architectural beauty. And the growth of Cynobacteria, algae and fungi was removed. The original structure as well as its color could not last longer and again growth of bacteria can be seen. The top of dome has become black. With the help of ASI experts now again University has undertaken its restoration work. Cultural heritage of monuments and temples is in constant exchange with the surrounding environmental conditions, both water and biological attacks may take place simultaneously (Gu and Katayama 2021). Initially oligotrophic niche on sandstone becomes enriched through phototrophic growth to sequestrate organic substances onto and into the surface layer of sandstone, in doing so the local conditions are altered from the oligotrophic to organic carbon rich one. After this transformation, heterotrophic microorganisms become dominant including both bacteria and fungi. The change in the color of sandstone from its original gray or light brown to dark and black occurs. Fresh microbial biofilm consisting of algae and cyanobacteria was reported (Lan *et al.*, 2010) and the different colors and pigmentations on sandstone wall are associated with different microbial groups at Bayon and Angkor Wat (Kusumi *et al.*, 2013). These microorganisms need to be further evaluated for their specific role in thebiodeterioration at different development stages considered to advance our comprehensive understanding of their specific role over time to provide more accurate information for future management of cultural heritage effectively. Subsequent results indicate the acid attack initiated by microbial oxidation of S from reduction product of H_2S or SO_4 (Li *et al.*, 2007; 2010). A research report showed that ammonia oxidation is an active biochemical reaction mechanism carried out mainly by AOA at several temples in Cambodia (Meng *et al.*, 2017). It is certain that more contributors are involved in the biodeterioration of sandstone monuments and destruction of valuable galleries and based on that, key factors should be identified and ranked according to the extent of destruction by each of them so that management can be focused on a few most important ones first to monitor the effectiveness over time to verify the conservation results.

The organic acids produced by microbes and lichens, to name a few, usnic acid, gallic acid, aspergillic acid *etc.*, loosen and dislodge the stones, but when they grow in such gaps, they increase the masonry space. Indian peepal (*Ficus religiosa*) and banyan (*Ficus benghalensis*) are the trees causative of the most harm by pitting the smooth and polished surface. Chemically the root's surface is acidic in nature. Various plants growing in stone monuments and old buildings produce complex organic acids. They are gluconic acid,itaconic acid, gyrophoric acid, and salicylic acid. These organic acids produced by the plants on the monuments thus weakening the structure and pitting the smooth and polished surface. The roots tend to secrete even more Hz ions to the surroundings, when under the stress of low water availability. Earlier growth of plants have been reported on stone monuments like Konark Temple of Puri, Temple of Khajuraho, Fatehpur Sikri and Purana Kila of Delhi. The weeds on the top of the Faculty of Arts of M.S. University were found reducing its architectural beauty. Mishra *et al.* (1995) studied the growth of higher plants over monuments. They reported 30 plant species on Indian monuments in tropical environment.

Monuments show irreversible deterioration and destruction under the tropical climate conditions which support a colonizing microbial community and flora, and, more importantly, the active growth of various microorganisms under favorable conditions of the seasonality (Keshari and Adhikary 2014; Meng *et al.*, 2020). Due to anthropogenic influences of pollutants microbial biofilms are formed on stone surfaces, which alter the physical properties, thermal and water uptake and loss from the materials, and also the dynamics of soluble salt uptake and transport, into and out of the stone block, to initiate any physical stress and damage internally, *e.g.*, crystallization of minerals to induce the pressure and delamination (Zhang *et al.*, 2019).

Knowledge and management of the flora on cultural heritage is a phenomenon affecting the entire globe. Botanical studies on archaeological areas should be not only limited to the protection of monuments but have to give the guidelines for the introduction of new plant elements to support the current fruition of the areas compatible with the original landscape (Caneva 1999). The study of flora occurring in the archaeological area and in its surroundings allow to select which species can be maintained or planted without excessive management costs. With the exception of the single Flavian Amphitheatre in Rome (the Colosseum), which has become a case study with 8 published contributions from the XVII Century to nowadays, there are no studies published on most of the sites and monuments in Italy (Domina 2018). A study of the present flora of Flavian Amphitheatre (better known as *Colosseum)* was carried out to complete the investigations of its natural flora, already analyzed during the last four centuries. The present flora is richer than it was expected despite the vegetation on the site is periodically cleared for conservation purposes. As a result 242 species were listed, showing an uneven distribution: the majority of species is found in the hypogeous (168 species) whereas towards the upper levels of the monument drastic drop in species number is observed. Several species encountered are rare or even very rare for the flora of Rome, or now disappeared, as in particular *Asphodelus fistolosus* and *Sedum dasyphyllum*. The distribution of life forms shows a clear dominance of therophytes (Grapow *et al.*, 2001).

22.5 Management of Biodeterioration: Recent Researches

Plant based biocides or green biocides are tried to control microbial growth from monuments. Aqueous leaf extract of *Calotropis gigantea* showed inhibition of biodeteriogens like *Aspergillus niger* and *Curvularia lunata*, extract from *Pongamia pinnata* was effective against *Cladosporium cladosporioides* (Gupta 2017). Dhawan (2001) while studying the biodeterioration of Hukuru Mosque of Maldives, suggested use of 0.50 per cent polycide, jet of hot water, and 12 ppm solution of Zinc dimethyldithiocrbamate, to control algal growth, which is having high algicidal property (Maloney and Palmer 1956).

Biochemical deterioration is the most complex form of biodeterioration and occurs by the direct action by microorganisms' metabolic processes on the material substrate. This may involve a bio-corrosion process whereby acidic and pigmented organic and inorganic products are excreted which can *etc.*h and stain the object, weakening the matrix of the material, leading to more favorable conditions for further attachment and growth and therefore continuing to add to the degradative process(Dakal and Cameotra 2012). A great variety of microorganisms have been found on culturally significant materials. However research into the mechanisms of biodeterioration on these types of materials indicates that bacterial microorganisms – such as prototrophs, which do not require organic material for growth – play a significant role as primary colonizers. They do so by modifying substrates and providing through their growth, movement, metabolic processes and cell mass, organic nutrients for successive colonizers such as fungi (Ciferri 1999). The proliferation and persistence of these organisms and their potential to cause biodeterioration, either directly or through successive colonization, is the establishment of surface-associated assemblages of microorganisms, known as biofilms which provide enhanced resistance to physical and chemical stress through a multicellular existence (Richards and Melander 2009). The formation and survival of bacterial biofilms on various surfaces is well documented. Although still not fully understood, through microscopic and molecular techniques, it has been revealed that biofilm formation is a complex process regulated by both genetic and environmental factors (Stoodley *et al.*, 2002). While many species of bacteria are able to form biofilms, further discussions on structure, function and treatment of biofilms will focus primarily on the ubiquitous gram-negative bacterium, *Pseudomonas aeruginosa*. Identified in association with biodeteriorated cultural material, *P. aeruginosa* remains one of the most extensively studied model organisms for bacterial biofilm formation (Cappitelli and Sorlini 2008).Encouraging biofilms to disperse by beneficially interrupting these chemical signaling processes lies at the heart of solving biofilm-related problems in areas such as materials science, energy efficiency, hygiene and cultural materials conservation (Alexander and Schiesser 2017).

Sandstone is a porous material of natural origin and is not strong enough compared to other stone and rock from igneous formation. This is susceptible to attack by weathering. Water management is an important issue in the protection of world natural heritage for an effective and long term plan (Liu *et al.*, 2019; Gu and Katayama 2021).The lowering of light intensity (Zelinka 1996; Faimon 2003) together with a source light spectrum not supporting the chlorophillianprocess (Kartalis and Mais 2000) was found effective. Unfortunately, a certain amount of lampflora nevertheless appears in these caves from time to time (Kartalis and Mais 2000; Rajczy *et al.*, 2000). Hence, such lamp-flora must be eliminated chemically. Generally a solution of sodium hypochlorite is used (Zelinka 1996). On contact, this solution effectively eradicates any presence of cyanobacteria, algae, and mosses. During these operations, however, measurable amounts of Cl_2 was released into the atmosphere and polluted the cave environment. Some bats died in the Javor¡ý´c¡ko caves (Czech Republic) in 1981 as consequence of hypochlorite cleansing. The disappearance of some insect species in some of the show caves in Slovakia could also be connected with hypochlorite applications. The preliminary interaction period should not be prolonged over 24 h because of a catalytic effect of mineral surface on peroxide decomposition. A similar study was conducted on Japan's castle where fungal *Cladosporium* spp. was isolated and identified. Its control was sought under the application of Hydrotherm on plastered walls.

Competent bacteria were selected according to the ability to metabolize the deposits to be removed, without damaging the original material of the artifact. For instance, to clean paper, bacteria must not show cellulolytic activity as well as to clean limestone they must not have the ability to solubilize carbonates (Sprocati *et al.*, 2021). In this way the cleaning is selective and safe for the artifact. The bacteria used in the case studies belong to the laboratory in-house collection ENEA (De Vero *et al.*, 2019). The collection gathers over 800 strains of biotechnogical interest mainly isolated from harsh environments, hypogea, artworks, *etc.* All strains are wild type and non-pathogenic (mainly belonging to the risk group 1), their gene sequences (rDNA16S) are deposited in the Gen-Bank database. The support agent must be compatible with the survival of the cells.

Ranalli and Zanardini (2021) offered a comprehensive view on the role and potentiality of virtuous microorganisms in the bio-cleaning and bio-removal of black crusts and salts altering CH stonework's. A careful selection of the most performant microorganism is needed in the removal of the undesired substances present on stone surfaces as deposits and crusts (in this case mainly represented by salts such as nitrates and sulfates) is the first step in the biorestoration strategies. The application of the selected microorganism is also important since the carrier, used to apply the viable cells on the altered stone surfaces, must have specific properties and provide them an adequate microenvironment to optimize their activity. The ideal delivery system should have the following characteristics: (i) able to guarantee the appropriate water content for the microbial activity; (ii) not toxic both for the microorganisms and for the environment; (iii) do not cause any color variation to the stone surface; (iv) applicable to all types of surfaces and finally, (v) easy to prepare and also to apply and to remove at the end of the treatment. All these aspects about the carrier properties and characteristics were evaluated forthe different situations (Bosch-Roig *et al.,* 2015) The bio-cleaning process also requires a deep diagnostic study of the work of art to be cleaned, an assessment of any associated risks, an evaluation of the effectiveness and efficiency of the process, and last but not least, an evaluation in terms of economic and environmental sustainability (Bosch-Roig and Ranalli 2014).

To conclude it can be said that there's a need of regular surveys and cleaning. The angiospermic plants, and dicot weeds such as *Ficus* spp. can be controlled by spraying a weedicide on the plants like Gramoxone, 2,4-D and then physically removing the dried roots and sealing the gap if any. Chemicals are applied by means of hand sprayers with multi-nozzles or power mist applicator. Water is generally used as carrier. Phenoxy herbicides such as 2,4 D, 2,45-T and Sivex are used for foliage spraying. Aminotriazole and ammonium sulphamate are also effective in foliage treatment (Agrawal *et al.,* 1994).

22.6 Conclusion

Biological factors cause deterioration through biochemical effects and intrusion of organisms. The former is a crucial or essential factor in biodeterioration of building structures as the metabolites (enzymes, excrements or feces) of micro and macro organisms, plant and animals living in a material can cause chemical damage of the material. Fungal hyphae, lichens and plant root systems which spread through structures can induce a mechanical damage. Also, boring insects may destroy structure cohesion which can encourage water penetration more quickly and deeply facilitating other deteriorating processes. Findings represent deterioration of Hajira tomb and old buildings of Faculty of Science and Performing Arts called Sangeet Shala of The Maharaja Sayajirao University of Baroda. Certain methods are suggested to remove the plants and arrest the growth of different microorganisms.

REFERENCES

Agrawal OP, Mishra AK, Jain KK (1994) Removal of plants and trees from historic buildings. INTAC, Indian Conservation Institute. 98 pp.

Alexander SA, Carl H. Schiesser CH (2017) Heteroorganic molecules and bacterial biofilms: Controlling biodeterioration of cultural heritage. Archive for Organic Chemistry Arkivoc part ii, 180-222.

Arai H, Yamagishi T, Kashiwadani H, Imai N (2005) Blackening effect on monuments and their control. In: Biodeterioration of Cultural Property-4. O.P. Agrawal and S. Dhawan. Pub. by ICBCP, Lucknow.80-93.

Bosch-Roig P, RanalliG (2014). The Safety of biocleaning technologies for Cultural Heritage. Frontiers in microbiology. 5. 155. 10.3389/fmicb.2014.00155.

Caneva G (1999)A botanical approach to the planning of archaelogical parks in Italy. – Conserv. Management Archaeolog. Sites 3: 127-134. doi: 10.1179/135050399793138590

Cappitelli F. Sorlini C. Appl. Environ. Microbiol. (2008) 74, 564-569. http: // dx.doi.org/10.1128/AEM.01768-07

Ciferri O (1999) Microbial degradation of paintings. Appl. Environ. Microbiol. 65: 879-885.

Crispim CA, Gaylarde CC (2005) Cyanobacteria and biodeterioration of cultural heritage: a review. Microb Ecol. 49: 1–9. doi: http: //dx.doi.org/10.1007/ s00248-003-1052-5.

Dakal TC, Cameotra SS (2012) Environ. Sci. Eur. 2012, 24, 1-13. http: //dx.doi. org/10.1186/2190-4715-24-36

Dhawan S. (2001) Biodeterioration of Hukuru mosque of male(Maldives) and its preservation treatment. In: Studies in Biodeterioration of materials-1. (eds.)O.P. Agrawal, S. Dhawan,R. Pathak. Pub. by INTACH and ICBCP, Lucknow.48-58

De Vero L. Boniotti MB. Budroni M. Buzzini P. Cassanelli S. Comunian R. Gullo M. Logrieco AF. Mannazzu I. Musumeci R. Perugini I. Perrone G. Pulvirenti A. Romano P. Turchetti, B. Varese GC (2019) Preservation, Characterization and Exploitation of Microbial Biodiversity: The Perspective of the Italian Network of Culture Collections. *Microorganisms, 7*, 685. https: //doi. org/10.3390/microorganisms7120685

Domina G (2018) The floristic research in Italian archaeological sites. – Fl. Medit. 28: 377-383

Faimona J, Sₗtelcl J, Kubesₗova´ S, Zima´k J (2003) Environmentally acceptable effect of hydrogen peroxide on cave "lamp-flora", calcite speleothems and limestones. Environmental Pollution 122: 417–422

Gaylarde CC, Gaylarde PM (2000) "Biodeterioration of external painted walls and its control." In: Rilem Workshop on Microbial Impacts on Building Materials, 2000, Sa~o Paulo, Brazil, paper 4 41.

Gorbushina AA (2007) Life on the rocks. Environ Microbiol. 9: 1613–1631. doi: http://dx.doi.org/10.1111/j.1462-2920.2007. 01301.x.

Grapow CL, Caneva G, Pacini A (2001) The Flora of Colosseum (Rome), Webbia, 56: 2, 321-342, DOI: 10.1080/00837792.2001.10670719

Gu JD, Katayama Y (2021) In: Microorganisms in the Deterioration and Preservation of Cultural Heritage, E. Joseph (ed.), https://doi.org/10.1007/978-3-030-69411-1_2

Gupta SP (2017) Mycological studies and scientific conservation of historical monuments. Pub by Aayu Publications, New Delhi. 204pp.

Kartalis N, Mais K (2000) The influence of illumination in the Akistrati Cave, Serron, Greece. International Conference on Cave Lighting, 15–17, November, Budapest, Hungary, Abstracts, pp. 12– 13

Keshari N, Adhikary SP (2014) Diversity of cyanobacteria on stone monuments and building facades of India and their phylogenetic analysis. IntBiodeterior Biodegradation 90: 45–51

Kusumi A, Li X, Osuga Y, Kawashima A, Gu J-D, Nasu M, Katayama Y (2013) Bacterial communities in pigmented biofilms formed on the sandstone bas-relief walls of the Bayon Temple, Angkor Thom, Cambodia. Microbes Environ 28: 422

Lan W, Li H, Wang WD, Katayama Y, Gu J-D (2010) Microbial community analysis of fresh and old microbial biofilms on Bayon temple sandstone of Angkor Thom, Cambodia. MicrobEcol 60: 105–115

Lisci M, Monte M, Pacini E (2003) Lichens and higher plants on stone: a review. IntBiodeterior Biodegradation 51: 1–17

Li X, Arai H, Shimoda I, Kuraishi H, Katayama Y (2007) Enumeration of sulfur-oxidizing microorganisms on deteriorating stone of the Angkor monuments, Cambodia. Microbes Environ 23: 293–298

Li XS, Sato T, Ooiwa Y, Kusumi A, Gu J-D, Katayama Y (2010) Oxidation of elemental sulfur by Fusarium solani strain THIF01 harboring endobacteriumBradyrhizobium sp. MicrobEcol 60: 96–104

Liu YF, Qi ZZ, Shou LB, Liu JF, Yang SZ, Gu JD, Mu BZ (2019) Anaerobic hydrocarbon degradation in candidate phylum 'Antibacteria' (JSI)inferred from genomics. ISME J 13: 2377-2390

Maloney TE. Palmer CM (1956) Toxicity of six chemical compounds to 30 cultures of algae. Water and Sewage Works. 509-513.

Meng H, Katayama Y, Gu JD (2017) Wide occurrence and dominance of ammonia-oxidizing archaea than bacteria at three Angkor sandstone temples Bayon, Phnom Krom and wat Athvea in Cambodia. IntBiodeterior Biodegradation 117: 78–88

Meng H, Zhang X, Katayama Y, Ge Q, Gu JD (2020) Microbial diversity and composition of the PreahVihear temple in Cambodia by high-throughput sequencing based on both genomic DNA and RNA. IntBiodeteriorBiodegrad 149: 104936

Mishra AK., Jain KK., Garg KL(1995) Role of higher plants in the deterioration of historic buildings. The Science of the Total Environment. 167: 375-392

Mitchell R, Gu JD (2000) Changes in the biofilm microflora of limestone caused by atmospheric pollutants. IntBiodeteriorBiodegrad 46: 299–303

Mottershead D, Lucas G (2000) The role of lichens in inhibiting erosion of a soluble rock. Lichenologist. 32: 601–609

Ortega-Calvo JJ, Hernández-Mariné M, Saiz-Jimenez C (1991)Biodeterioration of building materials by cyanobacteria and algae. IntBiodeterior. 28: 165–185. doi: http: //dx.doi. org/10.1016/0265-3036 (91)90041-O.

Patel S. Choksi N (2012) Eco-friendly building materials: A positive step towards environment protection. National Seminar on Environment Education: Concerns, sensitization and action 24-25 Janary, 2012. M.S. University of Baroda. 21

Rajczy M, Buczko´ K, Hazslinszky T (2000) Success and failure– removal of the lamp flora of Abaliget Cave in a perspective of 8 years. International Conference on Cave Lighting, 15–17 November, Budapest, Hungary, Abstracts, pp. 22–23.

Ranalli G, Zanardini E (2021) The Role of Microorganisms in the Removal of Nitrates and Sulfates on Artistic Stonework. In: Microorganisms in the Deterioration and Preservation of Cultural Heritage, E. Joseph (ed.), Pub. Springer 263-280. doi.org/10.1007/978-3-030-69411-1_12

Richards JJ. Melander C (2009) Controlling bacterial biofilms. *Chembiochem : a European Journal of Chemical Biology.* 10: 2287-94. PMID 19681090 DOI: 10.1002/Cbic.200900317

Sprocati AR, Alisi C, Migliore G, Marconi P, Tasso F (2021) Sustainable Restoration Through Biotechnological Processes: A Proof of ConceptMicroorganisms in the Deterioration and Preservation of Cultural Heritage pp 235–261

Stoodley P. Sauer K. Davies DG. Costerton JW (2002) Biofilms as complex differentiated communitiesAnnu. Rev. Microbiol. 56: 187-209. http: // dx.doi.org/10.1146/annurev.micro.56.012302.160705

Tiano P, Accolla P, Tomaselli L (1995) Phototrophic biodeteriogens on lithoid surfaces: an ecological study. Microb Ecol. 29: 299–309. doi: http://dx.doi.org/10.1007/BF00164892.

Vázquez-Niona D, Rodríguez-Castrob J, López-Rodríguezc MC, Fernández-Silvad I, Prietoa B (2016) Subaerial biofilms on granitic historic buildings: microbial diversity and development of phototrophic multi-species cultures. Biofouling, 32 (6): 657–669 http://dx.doi.org/10.1080/08927014.2016.1183121

Zelinka J (1996) Erfahrungenaus der liquidation der unerwu"nschten vegetation in den shauho"hlen der Slowakei. Proceedings of the International Meeting on Show Caves, 10–15 October, 1996, Aggtelek, Hungary, pp. 39–4

Zhang G, Gong C, Gu J, Katamaya Y, Ji-Dong G (2019) Biodeterioration and the mechanisms involved of sandstone monuments of World Cultural Heritage sites in tropical regions. Int Biodeter Biodegr 143: 104723

Chapter 23

Conservation Challenges in Adaptive Reuse of Heritage Buildings as Museums: A Case Study of Rabindra Bharati Museum

Baisakhi Mitra

Department of Museology and Rabindra Bharati Museum, Rabindra Bharati University, Salt Lake City, Near Karunamayee, Kolkata – 700 091, West Bengal, India
e-mail: baisakhimitra@gmail.com

ABSTRACT

The heritage significance of a building is assessed by carrying out historical research, oral history research and research on the fabric of the place and its physical context. While historical research requires a thorough investigation into written and printed records such as newspapers, journals, maps, photographs and illustrations, oral history research involves community and other stakeholder consultation. Fabric research requires a thorough examination of the place for evidence of pre-existing structures, physical changes such as newly constructed or deconstructed walls and partitions and the building's intactness and integrity. This research is then compiled into a historical summary to give a full understanding of the place. The place is then compared to other similar places to determine whether it is significant at local, state, national or international level. These statements of significance provide a basis to formulate policy or strategy about the most appropriate way to keep the place's heritage value.

The Government of West Bengal, reclaimed the Tagore House from the tenants and in 1959 a museum was planned on the premises. Later on the Rabindra Bharati University was established on the same campus for the advancement of learning in arts and culture. The museum is a stunning reminder of the legacy of the Tagore's family and their life and artistic creations, as well as of 19th century Renaissance of Bengal. It stands testimony to Tagore's philosophical leanings as much as it does to the colonial architectural styles of the time. Author discusses conservation challenges of the heritage building since its inception and during COVID.

Keywords: Conservation, Adaptive reuse, Heritage building, Jorasanko Thakurbari, Tagore's house, Rabindra Bharati, Museum.

23.1 Introduction

Heritage buildings are important building blocks of our cultural identity that document the past and, given a sustainable function, may contribute to the present by carrying the history and heritage of a place well into the future. The bricks and mortar not only show us who we were but are also evidences of material technology that were used in the past, the socio-economic changes that occurred over time, the way their spaces were planned and the life and contributions of the people who inhabited them. "Adaptive Reuse" is the process of reusing an old site or building for a purpose other than which it was built or designed for. If a building can no longer function in its original capacity, a new use may be assigned to it to preserve it from destruction (Cunnington 1963; ICOMOS 1964). Sometimes, adaptive reuse is the only way by which a building's fabric may be successfully cared for, preserved or interpreted, while putting the building to alternative use. It is also seen as an effective way of reducing urban problems and adverse environmental impact.

Figure 23.1: Front View of Rabindra Bharati Museum and University at Kolkata.

However, the adaptive reuse of a historic building should have minimal impact on the heritage significance of the building and its setting (Diamonstein

1978). A number of important issues need to be discussed when adaptive reuse of a heritage building is being considered, like:

☆ The existing conservation and heritage policies at the local and national level;

☆ The value of the heritage site and the potential of adaptive reuse of the building;

☆ The social, economic and environmental impact the adaptive reuse will have on the community;

☆ The purpose/s for which it would be reused;

☆ The logistics behind the conversion of the heritage building through adaptive reuse.

23.2 International Conventions: Adaptive Reuse of Heritage Buildings

Article 5 of the Venice Charter for the Conservation and Restoration of Monuments and Sites, a milestone of international theory and practice in heritage building restoration, says: *"The conservation of monuments is always facilitated by making use of them for some socially useful purpose. Such use is therefore desirable but it must not change the lay-out or decoration of the building. It is within these limits only that modifications demanded by a change of function should be envisaged and may be permitted."*

The Venice Charter further states the responsibility of architects and technicians as conservation of historic monuments "in the full richness of their authenticity" (ICOMOS 1964). The concept of authenticity was first mentioned in the Venice Charter and was expanded in the Nara Document on Authenticity in 1994 by stating the indicators of authenticity as "form and design, materials and substance, use and function, traditions and techniques, location and setting, and spirit and feeling, and other internal and external factors" (EIIC 1993).

The other criterion was mentioned as integrity which is a "measure of the wholeness and intactness of the natural and/or cultural heritage and its attributes" (WHC 2005). For the selected museum the principles of authenticity and integrity are evaluated based on the existence and legibility of the style of buildings, their plan of organization, presence of their original architectural elements and building materials and finishes after adaptive reuse projects.

The Burra Charter also states that compatible uses should have no or minimal impact on cultural heritage (ICOMOS Australia 1990). With regards to museum collections the New Orleans Charter recognizes that historic structures and the contents placed within them "deserve equal consideration in planning for their care" (AIC and APT 1992) thus recognizing not only the importance and integrity of the heritage building but also that of the objects housed inside.

In addition, since the artifacts in a museum bear a definitive relationship with the space (Maroevic and Edson 1998) the major challenge of adaptive reuse of historic buildings as museums is the conservation and presentation of the both historic building and the museum collection. The impact of the exhibition space on museum collection is identified as neutral, subordinate, dominant while evaluating the selected case. Also the classification of museum objects *i.e.* typological, chronological or thematic, and their relation with the spatial layout of the building influence the display elements on the museum space, illumination of the space and newly introduced technical installations are also evaluated since these aspects affect the legibility of both historic buildings and the museum collections (ICOMOS New Zealand 1992).

One of the most useful ways of maintaining a heritage building without large scale permanent "modifications" is to set up a museum on a relevant topic. One such example of adaptive reuse was the transformation of the illustrious Jorasanko *Thakurbari* (Tagore's House) to the vibrant Rabindra Bharati Museum, Kolkata, which draws thousands of visitors from all across the globe each year. The old buildings that can be divided into three main types depending on their heritage significance:

1. Buildings of low historical significance.
2. Heritage building: A heritage building may be regarded as having some aspects that are considered important or significant by the community.
3. Heritage-listed building: A heritage-listed building has been assessed as having aspects that are so important or significant that it is considered essential to preserve it. This does not mean that the building cannot be temporarily changed, but it does mean that there are parameters guiding the way the building can be treated. Based on these guidelines the heritage consultant prepares a detailed report called a Conservation Management Plan. This document explains the history and significance of the building and what can be done to reveal and highlight its interesting features. It also analyses the condition of the building fabric and outlines works that the building needs to remain structurally sound and to preserve its significance.

The Kolkata Municipal Corporation Act, 1980 defines a 'heritage building' as any building of one or more premises, or any part thereof, which requires preservation and conservation for historical, architectural, environmental or ecological purpose, and includes such portion of the land adjoining such building or any part thereof as may be required for fencing or covering or otherwise preserving such building, and also includes the areas and buildings requiring preservation and conservation for the purpose as aforesaid under sub-clause (ii) of clause (a) of sub-section (4) of section 31 of the West Bengal Town and Country (Planning and Development) Act, 1979 (West Ben. Act XIII of 1979). According to this definition the *Thakurbari* is a listed heritage building (Lal 1998).

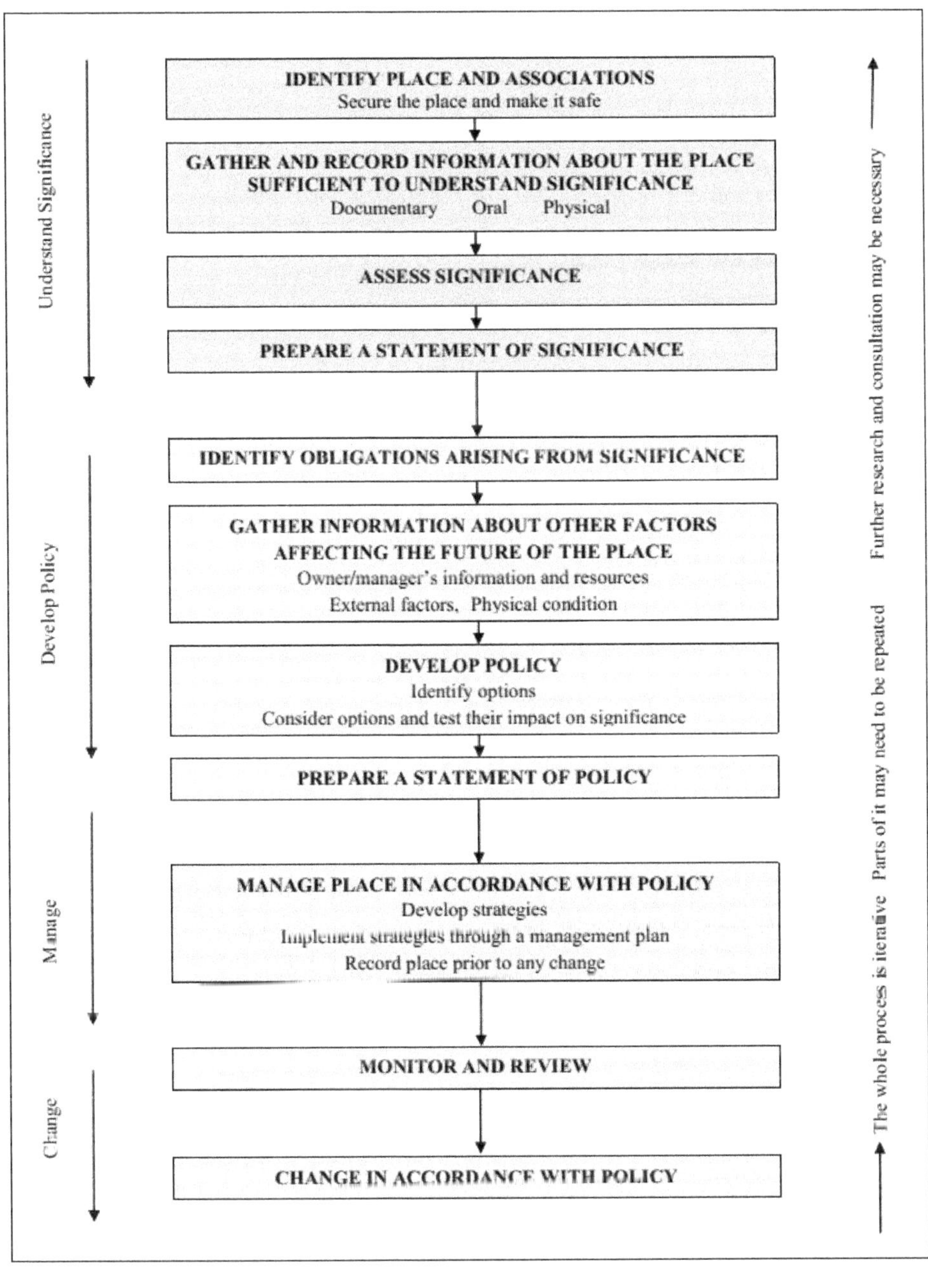

Source: Adopted from the Burra Charter (1999).

23.3 History of the Jorasanko Thakurbari

The mansion, a small house was built by Rabindranath Tagore's great-great grandfather Nilmoni Tagore in 1758. Slowly the different wings and sections were added. The house reached its grand proportion under the poet's grandfather Dwarkanath Tagore (1794-1846), a rich *dewan* working for the East India Company. The architects that worked towards the latter part of the building were all British and thus the architecture was largely based upon the Indo-European concept that exuded the charm of the colonial architecture. The house itself is a three storied structure with thick walls made of bricks that also caters to the need of the local climate. Long corridors give way on either side of the rooms, allowing the summer winds to blow freely, keeping the interiors cool and also host a number of household activities. Stout pillars support the corridors forming extravagant arches near the ceiling about 11-12 feet high. Dark-green wooden Venetian screens hang from the ceiling to bar the rain or scorching sun. An eminent architect Walter Gropius' stated that if buildings were designed following the local climate, then each region would have different form of architectural design is amply exemplified by this house as it combines European style with local needs (Banerjee 1963; Feildene 2003).

In 1823 Dwarkanath constructed another building adjacent to this house and named it *Boithakhana Bari* (Drawing Room House) where he entertained his western guests. This house number 5 was later occupied by his sons Girindranath and his family including his grandsons the celebrated artists Gaganendranath and Abanindranath, nephews of Rabindranath. This house was later pulled down, probably in 1943; thus a priceless piece of Bengal's heritage was lost and the integrity of the main structure was compromised.

The life of a house is infused by its residents. The Jorasanko Tagore House had a long list of illustrious residents like Dwarkanath Tagore, the pioneering industrialist of India, his son Debendranath who spearheaded the Brahma Samaj movement and his children Satyendranath, the first Indian civil servant, the playwright-musician Jyotirindranath, and of course Noble laureate Rabindranath fondly called *Gurudev*. The later generations of the Tagore, like Gaganendranath and Abanindranath, continued to have profound effect on the cultural scenario of India through their paintings. Abanindranath became the Father of the Bengal School of Art and with students like Nandalal Bose, Asit Kumar Haldar and Jamini Ray created a new Indian style of painting inspired by Indian traditions and mythology. The women of the Tagore House were equally deft at creating new styles be it in wearing the *saree* or in new culinary art. Many of them like Swarnakumari Debi and Sunayani Debi, went on to become leading women writers and painters of the country. While the architecture of the grand house gave it an iconic status, the kaleidoscope of talents residing within gave it a special position in the country.

Following Abanindranath's time the house gradually got fragmented as individual family members sold off their shares to new non-Bengali emigrants to the area. Gradually, the sun set on the glorious days of the Tagore House as the living family members moved to other parts of the city. The house became a bastion of numerous small scale industrial units. The walls that once hummed to the tunes of piano and *esraj* and witnessed great intellectual discourses now stood silent as sounds of anvil and hammer filled its corridors.

However, the house was resuscitated. Heritage buildings located near the center of a town have a better prospect if turned into museums. This is an advantage both in terms of attracting visitors to the site and providing ease of access. The Jorasanko Tagore House is currently located at the heart of the city which in those days formed the crucible of Renaissance Bengal being in close proximity to the residences of other stalwarts like the educator and social reformer Iswar Chandra Vidyasagar, scientist Jagadis Chandra Bose, theatre personality Girish Ghosh and Swami Vivekanada, *etc.* Since many of these residential houses have been turned into museums, it gives Jorasanko Tagore House the distinction of being at the center of this Bengal Renaissance hub thus facilitating a tourism circuit.

23.4 Reclaiming the Jorasanko Tagore House

The Government of West Bengal, under the leadership of then Chief Minister Dr Bidhan Chandra Roy, started the process of reclaiming the Tagore House from the tenants and in 1959 a museum was planned on the premises. On 8 May 1961 the then Prime Minister of India Jawaharlal Nehru inaugurated the Rabindra Bharati Museum. In the following year the Rabindra Bharati University was set up on the same campus for the advancement of learning in art and culture (Ganguly 2012).

23.5 The Museum Building

The museum is a stunning reminder of the legacy of the Tagore family and their life and artistic creations, as well as of 19th century Renaissance of Bengal. In fact, the representation of artistic and historical creations transcends the museum and resonates in the entire physical space of *Thakurbari*. The existing structure consists of three sections – the oldest part Ram Bhavan at the back of the museum, Maharshi Bhavan, and, Bichitra Bhavan constructed by *Gurudeb* himself in 1897 on the west of Maharshi Bhavan. His love for trees and densely growing foliage can be well seen in the south side of the house, which consists of several trees.

The first phase of the museum was spread across the Maharshi Bhavan and Bichitra Bhavan mostly. Only the family maternity room at Ram Bhavan, where Rabindranath too was born, was located in the oldest part. In the later years a number of galleries on the foreign sojourns of Rabindranath Tagore were set up in the Ram Bhavan section thus helping in maintaining the most fragile part of the structure.

23.6. Redefining Domestic Spaces into Public Galleries

To convert a heritage building into a museum it is necessary to propose new meaning to the spaces which starts with planning the exhibition layout after understanding of the exhibition environment. The aim in the Jorasanko Tagore Museum has been to freeze the living quarters of the poet in a time capsule so that the public can experience the space that the great poet lived in. So the décor of these rooms have been retained as it was during the time of Rabindranath with minor alterations mostly in the form of framed photographs and quotes from the works of Tagore. Additional galleries like the ones on paintings and foreign visits were constructed by adopting modern display techniques. Currently, the Museum is fairly big consisting of more than 20 galleries spread over the three sections of building. The galleries include the family maternity room, dining room, dressing room, bedroom, Bichitra Bhavan, Tagore's study room, gallery on Dinendranath Tagore, Mrinalini Devi's room, Mrinalini Devi's kitchen, Western art Gallery, Indian art gallery, Nobel gallery, Renaissance gallery, Dwarkanath Tagore's room, Debendranath's room, Japan Gallery, China Gallery, American Gallery, Hungary Gallery, Thailand Gallery. More galleries are proposed, on completion of which, this would probably be one of the largest biographical or period museum in the country.

Museum functions are divided into three sections: primary spaces which are exhibitions and gallery areas, audio-visual units, library, archive and museum shop intended for visitors; service areas and administrative spaces like workshops, conservation labs, strong room, *etc.*, secondary spaces like restaurants, cafes and lounges for social interaction of visitors and staff. Huang argues that what defines the museum as a spatial type is the organization of spaces in a visitable sequence and the gathering space, the recurrent space in the sequence expressed in the arrangement of objects and among the visitors (Tzortzi 2007). Therefore, the exhibition galleries and administrative spaces in a museum should be separated and the gathering spaces should include the secondary museum spaces such as bookstore, library and café. The coexistence of exhibition and administrative spaces and service units disturbs the museum sequence. The Rabindra Bharati Museum was planned to keep each of these areas distinct from the other. The office space is independent of the galleries and the galleries independent of the service areas.

23.7 The Collection and Interpretation

When an entire building is considered as heritage site, the building and its contents become the display. The collection of the Museum consists of both – organic and inorganic objects. Some of the objects displayed in the museum consist of:

☆ Clothes and furniture used by Rabindranath, his paintings, first editions of some of his publications, exhibition of the poet's life and activities through display of documents.

☆ The contribution of the men and women members of the Jorasanko and Pathuriaghata Tagore family.

☆ Living quarters of the family representing life during 19th century Renaissance Bengal.

☆ Two galleries exhibiting paintings of Bengal and the Anglo-Indian school in watercolor and oil respectively. And four galleries on foreign visits of Rabindranath.

The notable collections include:

☆ Documents : Letters of the members of Tagore family and other contemporary personalities;

☆ Painting and sketches: About 45 drawings and colored sketches of various kinds by Rabindranath, including a self portrait, paintings by Abanindranath, Gaganendranath, and other painters belonging to the Bengal school and European school;

☆ Numerous photographs of the Tagore family;

☆ Songs and music in records and discs;

☆ Miscellaneous objects like furniture, clothes and utensils.

Interpretation is the *raison d'etre* of museums. Regardless of the nature of the collection, the kind of building it is in, or where it is located, the museum exists to communicate to visitors. To successfully interpret a collection or a place, one must consider what needs to be said and how it should be conveyed. It is also important to understand the kinds of spaces available and the collection to be displayed. Though the focus of the museum is on Tagore, the museum takes care to highlight other illustrious members of the family as well like Dwarkanath, Debendranath, Dinendranath and stalwarts of Bengal Renaissance movement like Iswar Chandra Vidyasagar. The interpretation is a judicious mixture of the traditional and the modern.

While the main living quarters of the Tagore household has been kept intact, areas like the painting galleries and foreign galleries have modern display technique accompanied by air conditioning.

23.8 Challenges Faced in Converting the Heritage Building into a Museum

Heritage buildings may be an obvious choice for a museum because the collections and displays relate directly to the building's original use as is the case of the Rabindra Bharati Museum. Factors like the availability of the building (publicly or privately owned, donated or disputed) and availability of funds are important factors in drawing up a feasible museum plan. Although the integrity and historical importance of the Tagore House was beyond doubt it faced a number of practical challenges while being set up. Some issues are discussed as follows:

☆ **No permanent modifications**: Being a heritage-listed building no permanent modifications could be made in the building. The Tagore's House was converted to a museum with no permanent modifications. External improvements, often reversible and usually intrusive (for example ramps, wheelchair platform stair lifts, platform lifts *etc.*) were not possible. Internal improvements, which can often be integrated unobtrusively, was made, especially while constructing the galleries for Indian and Western paintings as these spaces had to be air-conditioned for housing fragile art objects.

Figure 23.2: The Bedroom of *Gurudev* Rabindranath Tagore: Retaining the Aura of the Original Space without any Modern Artificial Intrusions.

☆ **Modification of entry and exit to museum**: The large rambling building which had been built up over two centuries had a number of entrances and exits. For the purpose of the museum entry the steps via the newest Bichitra Bhavan was chosen. The old wooden entrance to the Maharshi Bhavan and other wooden staircases within the building were closed except for emergency visit by the museum staff or for any VIP visits. However, installations of lifts, elevators or ramps were not possible because of the orientation of the building. To protect the oldest part of the building under the Ram Bhavan section it was decided to keep them closed during the huge surge of visitors on the occasion of the birthday celebrations and commemoration of death anniversary of the Poet.

Figure 23.3: The Bengal School of Art Gallery: A modern air-conditioned space within the heritage building with Artificial lighting with lowered ceiling.

☆ A number of ancillary buildings were set up later for facilitating the functioning of the Rabindra Bharati University which is housed on the same campus.

☆ Cost: In this case, other than renovation of the building structure, no other major cost was entailed. Since funding was provided by state govt. it was not a problem.

☆ Loss of integrity: Care was taken not to risk downgrading the building's integrity and heritage value in the process. A part of the Jorasanko Tagore House (No.5) had already been destroyed before the government takeover.

☆ Community cooperation: A building that has a firmly established role in the community may mean that people do not respond well to its new use. This was not the case here. The Jorasanko Tagore House was not only established in the psyche of the Bengalis across the world but Rabindranath Tagore had become the epitome of Bengali sentiments. So the community was only too happy to rescue the house and convert it into a museum.

☆ General pattern: Room sizes and layouts, access and circulation through the building, and light and climate control may not be always conducive to the setting up of a museum. The architects were lucky here too. The overall pattern of the Jorasanko Tagore House combined traditional architectural elements with large rooms, plenty of doors and windows that allowed easy play of light and breeze. The natural lighting available inside the building made it ideal for setting up of galleries showcasing the rooms where Rabindranath and his family lived.

☆ Condition of the structure: Heritage buildings in many cases are in fragile conditions and require immense resources to make it withstand high visitor footfall. The Jorasanko Tagore House when salvaged was more or less in a stable condition. But a number of restorative works was necessary including the consolidation of the building structure to allow it to take the pressure of huge visitor traffic on special occasions.

☆ Non-availability of traditional building materials: A major challenge for building architects, as in most of these cases, are the lack of availability of traditional materials for the building. The materials used for building the Jorasanko Tagore House, the oldest part of which is more than 250 years old, consisted of *khoya, surki* and lime in the ratio 7:2:2. Not only were the materials different, the technique as well as the construction of the bricks was archaic. While the foundation bricks in the oldest Ram Bhavan was large at 10"x 6"x 6", others were 9"x 4 ½"x 3" – sizes not available in the market in the 20th century.

23.9 Climate Adaptability of Heritage Buildings and the Collection

Environmental quality in museum buildings depends on achieving the right balance between several complex and frequently contradictory aspects, such as public enjoyment, touristic entertainment, communication, indoor climate control, users' comfort, display, preservation, energy savings and safety precautions (Holmberg and Burt 2000). Most old buildings are climate responsive and attenuate well with the prevailing external conditions. Hence the notion of an ideal museum environment is relative depending upon the country and climate zone. It is advisable to keep the interior climatic conditions as close as possible to the exterior environment. Minimal intervention is advised with only just enough intrusion to keep limit values and prevent material deterioration. An indoor environment that gradually changes in keeping with the outdoor conditions is less harmful to the building fabric and collections. In most cases climate tempering provided by the building itself is sufficient to keep the structure stable. In case of extreme situation some mechanical devices can provided to moderate to extreme climate conditions (Holmberg and Burt 2000). Changes in buildings can be made to both the interior and the exterior. Interior changes are mainly due to new uses, and they often occur along with replacement of old materials, devices, or design details considered obsolete, applications of new finishes, reduction of ceiling heights, wherever necessary, removal or introduction of partitions, blockage of circulations and openings, and, depending on the new function, increases in the number of users. Changes in the surroundings might be due to urban growth and regulations, reduction of vegetation, alteration of ventilation and solar incidence paths, and, mainly, air pollution, alteration of ventilation and solar incidence paths. All the above issues were relevant in the case of setting up of the Rabindra Bharati Museum.

In such cases, the original building's climatic performance is therefore hindered and in need of improvement. In the case of museums, these improvements should aim at climatic stability and material safety. In warm, humid regions like Kolkata, improvements should also aim at reducing heat and moisture indoors. Instead of paying attention to a building's original design and use, conservation architects tend to address museum building pathologies only, when they result either from natural aging or from lack of maintenance. When dealing with preventive conservation policies, architects tend to overlook the building's original architecture and use, even though they might be potential tools for a better understanding and control of the indoor climate. The Jorasanko Tagore Museum located in warm, humid regions is housed in old, naturally ventilated building, made up of traditional materials and with traditional techniques, and, just by enhancing its original architectural features or by manually controlling the ventilation, climate control was achieved. Humidity and temperature are high most of the time in a year. Daytime temperatures varies from 40°C in summer to 12°C in winter, relative humidity (RH) values range from 58 per cent (winter) to 83 per cent (monsoon), the 69 per cent the mean

It is known that museum collections acclimatize to their immediate surrounding environment (Feilden and Jokiletho 2017), as long as it is stable. Climatic stability is more important than the internationally known and still-recommended standard climatic values. The shapes of rooms, the heights of ceilings, the types and locations of openings, roof forms, and so on –they all play a part in the conditioning of the indoor climate, and these factors should be placed in context when an environmental management plan is to be produced. In case of the present museum a combination of shade and ventilation has played a key role in the process.

However, the most challenging issue is to use natural ventilation and to avoid, at the same time, insects and air pollution indoors. This goal has been accomplished by a careful control of natural ventilation and by the maintaining the natural environment in most of the museum spaces. Attempts to run air-conditioning systems is avoided in most part of the museum as data has shown that when these artificial climate control systems fail, the resultant environment is more damaging than the natural conditions. It results in condensation and fungal outbreaks on surfaces of museum objects. In warm, humid regions a mixed control–*i.e.*, a combination of passive design and air dehumidification and recirculation–seems to be the only solution so far. The Rabindra Bharati Museum has followed this theory and has kept the storage spaces under full air-conditioning. Dehumidifiers have also been used in many places. In exhibition spaces where the collections remain encased (in microclimates) in showcases passive climate control were possible. Natural ventilation was used in the rest of the galleries. In some of the galleries, like the dining hall and bedroom, where the collections were directly exposed to visitors, no artificial environment could be created. Shading devices are quite efficient for reducing heat inside buildings. Thus verandas in the museum have used traditional bamboo curtains (*chic*) to cut off the sun light. They have the advantage of covering most of the facades, protecting doors and windows from both sun and rain.

A major problem with the museum building is that the thick walls have weathered over time, and now some are damp, showing water stains and contributing to increased moisture problems indoors. Sometimes this problem of water seepage affects the museum galleries. It is compounded by the rain water that gets soaked into the building fabric by capillary action. This has been a major problem in the museum, especially in the oldest part of the building for which the foreign galleries need constant monitoring. Fortunately the main museum building on Maharshi and Bichitra Bhavan has remained relatively moisture free and stable. The 24-hour air-conditioned strong room of the museum has been located in the newest part of the Jorasanko *Thakurbari* to avoid any problems related to moisture and termite. The entire building is protected by monthly pest treatment procedure and a dedicated group of civil contractors.

23.10 Heritage Building under COVID-19 Lockdown

The COVID-19 situation put the museum in the face of the biggest challenge since its inception. With the museum shutting down completely the regular maintenance work stopped. It is very essential to air old buildings by keeping the doors and windows open during substantial part of each day. However, even this simple exercise was not possible as most security personnel stayed back home and with high value rare objects in the museum collection it was not prudent to keep the numerous rooms unlocked and unattended. The core museum team was put on emergency duty and with special permission from the administration attended to the maintenance of the objects a few times a month. Though most of the collections could be protected, the building itself suffered with overgrowth of vegetation, spread of termite infestation and accumulation of moisture inside the building with frequent flaking of paint and plaster layers on the walls. The museum reopened to the public in 2021 and the civil contractors and pest treatment regimen swung back to action and within a few months the museum was as good as before. The Museum gradually came back to its normal routine.

Museumizing a house preserves the heritage value of the building, its collections, as well as abstract things like the lifestyle and culture of the residents. This is especially true for writers' museums where the house may have inspired many creations of the author as was the case with *Gurudev* and the Jorasanko Tagore House. Like his literary texts, the heritage building is really and abstract, tangible yet filled with intangible memories of the past. Through adaptive reuse the Tagore House has been turned into the vibrant Rabindra Bharati Museum that not only keeps the memories of the Tagore family alive but also pays tribute to the genius of *Gurudev* in particular. Unlike many adaptive reuse in other parts of the world where heritage buildings have been turned into shopping malls or hotels their conversion to museums seems to give them the right purpose and this has been amply proven in the case of the Rabindra Bharati Museum.

REFERENCES

AIC and APT (1992) New Orleans Charter for Joint Preservation of Historic structures and Artifacts, http://cool.conservationus.org/bytopic/ethics/neworlea.html

Banerjee, H. (1963) The House of Tagore's, Gyan Publishing House

Chakravarti A (2016) Daughters of Jorasanko, Harper Collins

Cunnington P (1988) Change of Use: The Conservation of Old Buildings, London, Alpha Books

Diamonstein B (1978) Buildings Reborn: New Uses, Old Places, New York, Harper and Low Publishers

Feilden B M, Jokiletho J (2017) Management Guidelines for World Cultural Heritage Sites, Rome, ICCROM

Feildene B M (2003) Conservation of Historic Buildings, Oxford, Architectural Press

Ganguly T M (2012) A Pilgrimage to Jorasanko Thakurbari

Hay F (2008) *Intervention*, Centre for the Recycling and Reuse of Buildings. www.recyclingandreuseof buildings.com

Holmberg J G, T Burt (2000) Preventive conservation methods for historic buildings: Improving the indoor climate by changes to the building envelope in Conservation Without Limits: IIC Nordic Group XV Congress, Helsinki, Finland

ICOMOS (1964) International Charter for the Conservation and Restoration of Monuments and Sites www.international.icomos.org/charters/venice_e.htm

ICOMOS, New Zealand (1992) Charter for the Conservation of Places of Cultural Heritage Value http://www.icomos.org/docs/nz_92charter.html

ICOMOS Australia (1999) The Australia ICOMOS Charter for Places of Cultural Significance (Burra Charter) http://australia.icomos.org/wp-content/uploads/BURRA_CHARTER.pdf

Lal ND (1998) *Building Calcutta: Construction Trends in the Making of the Capital of British India, 1880 1911*, Ph.D. thesis, The University of Edinburgh,

Maroevic I, Edson G (1998) Introduction to Museology. The European approach. Munich: Müller Straten.

Stovel H (2008) Origins and Influence of the Nara Document on Authenticity, APT Bulletin, Vol 39(2/3)

Tzortzi, K (2007) *Museum Building Design and Exhibition Layout: Patterns of Interaction*, Proceedings. 6th International Space Syntax Symposium, Istanbul

World Heritage Committee (1993) The Nara Document on Authenticity, http://whc.unesco.org/up;oads/events/documets/event-443-1.pdf

World Heritage Committee (2005) Operational Guidelines for the Implementation of the World Heritage Convention, whc.unesco.org/archive/opguide05-en.pdf

Chapter 24

Conservation of Tangible Cultural Heritage in the Context of Sustainable Development: A Case Study of Jammu and Kashmir

Poonam Chaudhary and Malay Dey

Centre for Studies in Museology, First Floor Gen Zorawar Singh Auditorium, University of Jammu, Jammu – 180 006, J&K

ABSTRACT

Historical buildings, old monuments, heritage objects, antiquities and works of art are the interesting documents of our ancient history and culture. These are included among tangible cultural heritage. According to the International Institute of Conservation, London, the conservation is "an action taken to determine the nature or properties of materials used in any kind of cultural holdings or in their housing, handling or treatment, any action taken to better the condition of such holdings."

The historical places in Jammu are ideal spots to get glimpses of the rich cultural heritage of the bygone era. One can get in-depth information about the rich history of the region by visiting these popular destinations. The ancient sculptures and edifices of the state hold special significance. Built in the 19th century on the right bank of the Tawi River, Amar Mahal Palace is dedicated to Raja Amar Singh. It was designed by the French architect on the lines of the French Chateau. This majestic palace is an erstwhile residence of the Dogra Dynasty, which now has been converted into a museum.

Sustainable goal 11 talks about sustainable cities, the cultural and economic development is possible with developing new facilities and technologies at the same time heritage monuments and objects help to learn about our glorious past. World Heritage properties and heritage in general, help to contribute the sustainable development and therefore increase the effectiveness and relevance of the convention whilst respecting its primary purpose and mandate of protecting the Outstanding Universal value of World Heritage properties. Younger generation can learn from tangible objects and simultaneously tourism can be promoted in the cities with heritage monuments. Paper describes many such heritage buildings and material pertaining to Jammu and Kashmir.

Keywords: *Conservation, Tangible cultural heritage, Sustainable development, Jammu and Kashmir.*

24.1 Introduction

Since the publication of the World Commission on Environment and Development (WCED) report in 1987, the concept of sustainable development has become a burning topic of discussion in numerous studies and publications in diverse fields. Cultural heritage has long been absent from mainstream sustainable development in spite of having a crucial role in the development of society, economy and environment. 'Based on a strong appeal from national and local stakeholders, the 2030 Agenda adopted by the UN General Assembly integrates, for the role of culture, through cultural heritage and creativity, as an enabler of sustainable development across the Sustainable Development Goals'. World Heritage may provide a platform to develop and test new approaches that demonstrate the relevance of heritage for sustainable development' (UNESCO 2022). The 20[th]General Assembly of the State Parties to the World Heritage Convention adopted a Policy on the integration of a sustainable development perspective into the processes of the World Heritage Convention held on 19[th]November 2015. The aim of the policy was to help 'State Parties, practitioners, institutions, communities and networks through appropriate guidance, to harness the potential of World Heritage properties and heritage in general. It contributes to sustainable development and therefore increases the effectiveness and relevance of the Convention whilst respecting its primary purpose and mandate of protecting the Outstanding Universal value of World Heritage properties' (UNESCO 2022). Framework Convention on the Value of Cultural Heritage for Society (FCVCHS) held in 2005 internationally acknowledged the significance of cultural heritage in striving toward sustainable development (Vileniske 2008).

The term culture is used in a wide variety of senses both in general parlance and in social science literature. The concept of culture in anthropological writings covers the total way of life: mental, social, and material. However, culture is the principal concept of the branch of social anthropology and cultural anthropology. The anthropologists define the realist and the idealist approaches to culture. The realist approach considers culture as an attribute of the social life of man. This view advocates that the culture is intimate with the real human behaviour in society. On the other hand, the idealist conceptualization of culture views it to be of the nature of ideas.

British Anthropologist Edward Burnett Tylor (1870) defines 'culture as that complex whole which includes knowledge, belief, art, morals, law, custom, and any other capabilities and habits acquired by man as a member of society' (Avruch 1998). According to David Matsumoto culture is 'a dynamic system of rules, both explicit and implicit, established by groups in order to ensure their survival. This involves attitudes, values, beliefs, norms and behaviours. In other words, culture is dynamic and it changes over time in response to environmental and social changes' (Matsumoto *et al.*, 1999)

If we look at the various definitions of culture, we can realize the numerous variations in the conceptualization of culture. Most anthropologists and sociologists consider culture as a social heritage, a body of human achievements which is transmitted through tradition. The United Nations Educational, Scientific and Cultural Organization (UNESCO) seeks to encourage the identification, protection and preservation of cultural and natural heritage around the world considered to be of outstanding value to humanity.

24.2 Conservation of Tangible Heritage and Sustainable Development of Jammu and Kashmir

Jammu and Kashmir is a haven of vibrant, rich and colourful culture. It takes pride in demonstrating both natural and cultural heritage, which has long attracted national and international tourists to view its beauty and pilgrimage centres. Therefore, the tourism industry in this Union Territory (UT) is one of main sources of income. The cultural heritage of the two regions Jammu and Kashmir cuts across all the regional, religious and ethnic barriers and despite being different both physiographically and climatically it has much in common, acquired through age-old association amongst its people. The cultural similarity, as well as the diversity of the two different regions of the UT, is reflected in its language, literature, religion, arts, crafts, music and religious centres. The years of cultural exchange amongst the people of this region has resulted in cultural amalgamation to such an extent that efforts to study and analyse the people and their culture is stupendous.

The faith and tradition of preaching and respecting each other's religion and pilgrimage centres have been an unique and glorious tradition of the people of the Union Territory. The famous Hindu shrines, mosques, gurudwaras and pilgrimage centres are held in the highest esteem by the people of every faith. Buddhism, which is still followed in the newly formed Union Territory of Ladakh which once was a part of the erstwhile state of Jammu and Kashmir, has its origin in the valley.

The study becomes all the more relevant in view of the fact that since 2010, three sites from the UT of J and K have made it to the UNESCO's "tentative list" to obtain the World Heritage tag, which are:

1. Mughal gardens in Kashmir made it to the tentative list in 2010.

2. Ancient monastery and Stupa in Harwan in 2010. Harwan is also amongst the Silk Route sites in India.

3. Neolithic Settlement of Burzahom in 2014.

All the three sites are located in the valley of Kashmir.

Efforts have been made by government and non-government organisations to protect the age-old heritage of the UT for the cultural as well as socio-economic development of the region. This in turn has led to adoption of the Jammu and Kashmir Heritage Conservation and Preservation Act 2010, which clearly defines the conservation of cultural property. According to the Act 'Conservation means protection, preservation and restoration of heritage sites/ areas, heritage precincts, building, artefacts, handicrafts, paintings, fabrics *etc.* and shall include only such development activity that will enhance the heritage significance of the heritage site within the framework of this Act'. The Act thus covers all tangible movable, immovable and intangible heritage of Jammu and Kashmir for the purposes of conservation.

24.3 Immovable Tangible Heritage of the UT of Jammu and Kashmir

Tangible heritage refers to the physical objects that can be physically touched. From tiny beads of Harappan civilization to the Great Wall of China fall in the category of tangible heritage. The 1954 Hague Convention for the Protection of Cultural Property in the Event of Armed Conflict classifies cultural property according to the identification of three typological categories:

1. Movable or immovable property of great importance to the cultural heritage of every people;

2. Buildings that contain cultural objects, such as museums, libraries and archives; and

3. Historical centres containing monuments.

UNESCO broadens this classification of cultural heritage through the World Heritage Convention. The World Convention reconciles previous definitions of cultural heritage and presents immovable cultural heritage within three categories:

1. Monuments,

2. Groups of buildings and

3. Sites

According to the Jammu and Kashmir Heritage Conservation and Preservation Act 2010, heritage areas denote those areas which are of historical or archaeological or architectural or aesthetic or scientific or environmental or cultural significance. The Act also refers to heritage buildings as buildings or structures which have historical or aesthetic or architectural or environmental significance.

The study of the architectural heritage of the UT of Jammu and Kashmir shows that its tangible immovable heritage dates back to the Neolithic period and its early art and architecture were mostly inspired by Buddhism as revealed from the remains unearthed from various sites specially in Kashmir. Along with stupas, chaityas and viharas, a number of temples representative of great artistic skill have been reported. Some of which are discussed below.

24.4 Some Archaeological Sites of Jammu and Kashmir

There are a lot of archaeological sites in the Jammu and Kashmir which need to be conserved for sustainable development. The use of archaeological and heritage sites as resources for socio-economic development has become increasingly important within the Union Territories. If we study the architecture, the structure and the historical importance of the cultural sites, we can find that the demands of these cultural resources are gradually increasing, prompted by the local communities and the government.

A) **Burzahom:** The importance of the site is reflected in the fact that it was made to the UNESCO Tentative World Heritage List in 2014 under the criteria ii, iii and iv under the category 'culture.' It is the northernmost excavated Neolithic site of India, in the form of circular or oval pits, dug into the Karewas dating between 3000 BC and 1000 BC from Burzahom, Kashmir (Ray1970). The outstanding universal value of the site lies in the fact that it brings to light transitions in human habitation patterns from Neolithic Period to Megalithic period to the early Historic period, from transition in architecture to development in tool-making techniques to introduction and diffusion of lentil in north-western India. The site is considered to be a unique comprehensive storyteller of life between 3000 BCE to 1000 BCE.

B) **Gufkral:** As the name suggests *'guf'* means a cave and *'kral'* means a potter, the site was inhabited by potters who utilized these caves. The site represents five periods from the Aceramic Neolithic to megalithic periods of the history of the UT of Jammu and Kashmir and is located at Banmir village, Pulwama, Kashmir (Ray 1970). In contrast to the pit dwellers of Burzahom, the site of Gufkral was in the form of a cave. The cave of Gufkral is one of the oldest reported caves in Kashmir and some estimates trace their origin to 2000-3000 BCE.

Sites at Akhnoor: (Manda and Ambaran)-Akhnoor is an important site for the region of Jammu. It has been home to two important sites: Manda and Ambaran located on the right bank of river Chenab. It is a place where the first evidence of the Harappan settlement in Jammu has been reported.

C) **Manda:** The significance of the site lies in the fact that it is the only Harappan site reported so far in the UT of Jammu and Kashmir. It is considered to be the northernmost Harappan outpost of India. The excavations at the site which were carried out by the Archaeological Survey of India in 1977 yielded Harappan and black-slipped ware as well as late Harappan red ware and plain grey ware. Bangles and triangular tiles both made of terracotta, potshards with Harappan writings and bone arrowheads have also been found. The continuity of the site to the protohistoric period is proved by the fact that the site has also yielded pottery remains belonging to the Kushan period both incised and plain including terracotta figurines, bone arrowheads, iron daggers and copper antimony rods (ASI). In close proximity to Manda, there are some more sites like Tikri, Guru Baba Ka Tibba, Jhiri, and Jaffarchak, on the other side of the river Chenab, which highlight the cultural development and evolution of Akhnoor from the Harappan age to the early Christian era.

D) **Ambaran- Pambaran:** The site is situated 28 km away from the city of Jammu on the right bank of river Chenab. It is the earliest recorded Buddhist site of the UT of Jammu and Kashmir. The site Ambaran has yielded an eight-spoked stupa base in bricks dated between 1st to 2ndCentury CE, the first of its kind to be reported from any Buddhist site in the entire UT of J and K. The monastic complex is a witness to the vibrant Buddhist phase in Jammu that perhaps led to spread of Buddhism in Kashmir and from thereon to Central Asia. Historians believe that between the 1stand 7thcentury B.C. this place might have been an important place for the followers of Buddhism. It has been suggested that Ambaran is the eighth place in the world where relics of Buddha have been found in a stupa.

Excavations at the site have also unearthed objects, small sculptures, terracotta figures, burnt brick structures and pottery dating back to the pre-Kushan and post-Gupta periods. World-renowned terracotta heads, popularly known as 'Akhnur Terracotta' identical in style to the heads in the Lahore Museum and in private collections in Lahore owe their provenience to the hamlet Pambarvan (Ambaran) (Ray 1970).

The relics of Buddha found at this site has not got the recognition it deserves. Even the visit of Dalai Lama, the Tibetan Buddhist spiritual leader on 16thNovember 2011, has not given much importance to the site to be promoted at an international Buddhist forum. At present, a site museum has been established and opened to visitors by the Archaeological Survey of India.

E) **Harwan:** The archaeological site of Harwan near Kashmir valley has revealed the earliest structures built in pebbles and mud mortar in the form of an apsidal temple with a courtyard, a stupa and a chapel

(Ray1970). Though in ruins, the site is considered to be a 'relic of Kashmir's Buddhist past'. The site is also known for its moulded brick tiles, in various shapes and patterns, which covered the floor of the courtyard of the temple. The subject matter of the tiles deals with the life of humans and the nature surrounding them.

According to UNESCO, 'The site represents a unique tradition of Kashmir which was connected with the Silk Road in its architecture and art, particularly in the decorated patterns of terracotta tile-pavements in the apsidal stupa which became a symbol of the art of Kashmir'. The ancient monastery and stupa in Harwan, is in the UNESCO World Heritage tentative list in 2010. India was well connected with the Silk Road by three probable routes, one of the routes was via Srinagar, Gilgit and Karakoram Range. The route was used by travellers for trade and pilgrimage. The architectural remains at Ushkar and Harwan both in Kashmir were mainly inspired by Gandharan (UNESCO 2022).

F) **Ushkar:** Often referred to as the 'forgotten town of Kushanas', Huvishkapura, which later on called Ushkar, was founded by the Kushana ruler Huvishka in the 2ndcentury A.D. Once a flourishing town because of its prime location on the trade route between Kashmir and north-western India, it emerged as an important Buddhist site, where it has been suggested that Fourth Buddhist Council was held (Ray1970). The place continued to remain important not only from the point of view of Buddhism but also it was patronised by Lalitaditya, the Karkota ruler of Kashmir (695–731A.D) (Ray 1970). Lalitaditya built a shrine of Vishnu named Muktasvamin and a large vihara with a stupa. Hiuan-Tsang, the Chinese pilgrim, who visited Kashmir in 631 A.D. is said to have entered the valley by the Baramulla pass and spent his first night in one of the monasteries at Ushkar. He was accorded a warm welcome by the then reigning ruler (Ray 1970). Except for the ruins of a stupa and its surrounding wall mentioned by the Buddhist traveller Hiuan-Tsang and later by Kalhana, no remains of the monasteries and temples are available now.

G) **Parihaspora:** An important Buddhist site which is now in ruins. Once served as the capital of Lalitaditya and was famous as a 'city of stones'. The site is managed by the Archaeological Survey of India.

24.5 Some Important Historical Monuments of Jammu and Kashmir

A) **The fort of Akhnoor:** Overlooking River Chenab, the fort of Akhnoor, majestically stands on its right bank of an irregular hillock. The northern wall of the fort slopes towards the river, whereas the southern wall follows the irregular edge of the hillock. The fort was planned in a way that its eastern side was protected by the river Chenab. The construction of the fort is attributed to Tegh Singh at a time when a terrible famine broke out in the country. Tegh Singh, in order to save his people from the ravages of famine, started the construction of a fort and palace (Balouria 1930). The fort was constructed within a span of two years. The fort is now under the supervision of ASI and is well maintained. The area around the fort has been beautifully developed making the trip to Akhnoor fulfilling and a memorable one. It is one of the most important historical sites in Jammu and Kashmir. Besides the fort, the place is famous for JioPota tree (*Putranjiva roxburghii*), where Raja Gulab Singh, the Dogra ruler were coronated by the Sikh ruler Ranjit Singh.

B) **Amar Mahal Palace:** Situated on the National Highway towards Srinagar Amar Mahal Palace is the architectural gem of Jammu and Kashmir. Built as a French Chateau on a hill overlooking the river Tawi, is a gorgeous palace made of red sandstone which stands amidst the most picturesque horizons of Jammu city. It was once the residential palace of Raja Amar Singh, a Dogra ruler. The palace was later converted into a museum and is looked after by Hari-Tara Charitable Trust. One of the most attractive exhibits in the museum is the120 kg of pure gold throne. The museum has a paintings gallery and a library containing 25,000 books on various subjects and disciplines are also the centres of attraction. The palace is well maintained and is a good example of a cultural resource which is being used for creating employment (Figure 24.1).

Figure 24.1: Well Maintained Amar Mahal Palace, Jammu.

C) **Bahu Fort-** It is believed that the fort was originally constructed about 3,000 years ago by Raja Bahu Lochan, founder of the state Bahu in Jammu. Hardly anything of the original fort remains and it was only in the 19[th]century that the present fort was rebuilt on the remains of the original fort by the ruling line of the then Dogra rulers. The fort is a living heritage as it houses the temple of Goddess Kali, the presiding deity of the people of Jammu. The temple is popularly known as the *"Baawewali* Mata temple" attracting not only the devotees from Jammu but also those visiting Vaishno Mata shrine.

Bahu Fort is a good example of cultural heritage in the context of sustainable development. Just beneath the fort, the terraced Bagh-e-Bahu Garden has been built in the style of Mughal gardens which gives a panoramic view of the Jammu city. In Bagh-e Bahu Garden. An underground fish aquarium situated here is also a place of attraction. The Light and Sound show is another attraction which has been added recently. The Bahu fort complex and its garden have now become a source of income for the local people.

D) **Mubarak Mandi Palace Complex:** Located at the highest contour of slope overlooking river Tawiat an altitude of about 341m above sea level, the Mubarak Mandi Palace complex occupies an important place not only as the relict feature of the landscape of Jammu but also a symbol of unique feature of heritage. It has played an important role in the development of Jammu and Kashmir township. After the establishment of the Dogra rule over the erstwhile state of Jammu and Kashmir in the second quarter of the 19th century *i.e.* in 1846, the palace complex became a centre of socio-political activities and cultural pluralism (Chaudhary2006). The complex, though a very important cultural asset for the people of Jammu, is deteriorating fast. Natural hazards besides others have played an important role in accelerating the deterioration process of the complex (Figure 24.2).

E) **Mughal Gardens of Kashmir:** The Mughal gardens are generally referred to as the 'Paradise on earth' another significant heritage site of the UT of Jammu and Kashmir which has made it to the UNESCO tentative World Heritage list under the criteria's ii and iv. As the name suggests the Mughal gardens were developed by the Mughal emperors specially Jehangir and Shah Jahan. Nishat Bagh, Shalimar Bagh Achabal Bagh, Chashma Shahi, Pari Mahal and Verinagare are the most well-preserved parts of Mughal gardens. These gardens reflect the engineering skills and aesthetics of the Mughal rulers to achieve the highest potential of exploiting the natural landscape with Natural source of water.

**Figure 24.2: Deteriorated Mubarak Mandi Complex
(Conservation work is going on).**

The Outstanding Universal Value of these gardens lies in the fact that these are "irreplaceable resources for the understanding of garden history in general and the Mughal Period in India." According to the experts "these gardens are an exceptional testimony of the creative and innovative ingenuity demonstrated by the Mughals in taking maximum advantage of the rising slopes of the mountains and the natural setting to fulfil their extraordinary landscape ambitions and needs." There are many other monuments of Jammu and Kashmir which have added to the cultural landscape of Jammu and Kashmir from the point of view of sustainable development.

24.6 Religious Built Heritage

A) **Cave Shrines:** Cave shrines are the oldest form of heritage of Jammu and Kashmir. The region is particularly famous for its cave shrines like Mata Vaishno Devi, Amarnath, Peer Kho, Shiv Khori, *etc.* dedicated to various Indian gods and goddesses. These are viewed as the living places which attract millions of pilgrims from India and abroad. Such heritages fall both in the category of both cultural and natural heritage. These need to be preserved because they not only support socio-cultural activities associated with the religious beliefs of the people but also add to the natural biodiversity, geodiversity, *etc.* of a place.

The region was also famous for its Buddhist heritage and the influence of Buddhism art is reflected clearly on the early art and architectural heritage of early Kashmir. The material used by the Buddhists in the early constructions was grey limestone which could be easily carved and dressed.

B) **Rajvihara Monastery:** Parihaspura is the only surviving example of Buddhist Chaitya or temple in Jammu and Kashmir. The monastery reflects the prevailing Buddhist tradition at that time. The construction of the monastery is attributed to Lalita Muktapida of the Karkota dynasty in the 8th Century (Kak 2002). The statue of Buddha made of copper, which once adorned the monastery is no more present here. The material used for the construction of the monastery was massive blocks of lime stone with lime mortar, which were roughly dressed before their placement for construction (Ray 1970; Kak 2002). The final dressing of the stones was done *in situ* after the embellishments; carvings were covered by a coat of gypsum plaster and perhaps also painted. The floor of the Chaitya was built in single block of stone with no iron or other medium used to support the structure (Ray 1970; Kak 2002). Vestiges of the monastery are an attraction for tourists and forms a part of the Buddhist heritage circuit.

Besides being rich in Buddhist heritage, the UT abounds in temple heritage and fine example of the earliest temples in Kashmir are Loduv, Shankaracharya and Narastan. These were constructed in stone though not as massive as the ones used for the construction of temples at Martand and Avantiswamin (Ray 1970; Kak 2002). Though in ruins now but simplicity with no decoration was the characteristic of these temples.

C) **Shankaracharya Temple:** The temple is dedicated to lord Shiva and has always remained a popular religious and spiritual centre of pilgrimage attracting pilgrims and tourists alike from all over India and abroad as it was also a place where Adi Shankaracharya, a saint and a philosopher performed severe penance (Ray 1970; Kak 2002).

The date of construction of the original temple is a matter of controversy and seems to have been reconstructed several times. Before being named as Shankaracharya, the temple has been identified with the ancient shrine of Jyesthesvara built by a Hindu king named Gopaladitya on the Gopadari hill. Later the name of the temple and the hill was changed to Sankaryacharya and Takht-i-Sulaiman respectively. Though it represents an early temple architectural style of early Kashmir without embellishments, two Persian inscriptions belonging to 17th century have been found on the pillars of the cell (Ray 1970; Kak 2002).

D) **Martand Sun Temple:** Now in ruins, the temple of Martand often referred to as the 'veritable lion of Kashmirian architecture', is one of the exquisitely designed monuments of Kashmir built during the reign of Lalitaditya, the Karkota ruler, who ruled Kashmir in 8th century A.D. (Ray 1970). The temple was built after choosing the most picturesque spot on the 'table land of Matan, overlooking the green plains, intersected by rivers, lakes, villages, surrounded by snow capped mountains (Kak 2002). It was dedicated to the Sun god as it was believed that nature is the manifestation of God.

It is one of the oldest Sun temples of India, which needs to be protected, preserved and made a part of the cultural tourism circuit linking it with the other Sun temples of the country like Konark in Orissa and Modhera temples in Gujarat. Archaeological Survey of India has already declared the Martand temple site as a "site of national importance". Much efforts are being made to restore the past glory of the land one such effort has been holding of religious ceremony on the ruins of the temple. Recently a *"Navagraha ashatamangalm pooja"* was held in the temple, in the presence of saints, members of Kashmiri Pandit community and local residents to protect, promote and develop the site as a centre of cultural and spiritual significance. There is a popular demand to open the temple for religious functions and gatherings.

E) **Avantiswami and Avantisvaratemples (Avantipore):** Located on the bank of river Jehlum with beautiful hills at the backdrop, the temple complex consisting of two temples: Avantiswami, and Avantisvara dedicated to lord Vishnu and lord Shiva respectively were constructed by Avantivarman of Utpala dynasty in the 9th century CE (Ray 1970; Kak 2002). Locals often refer to the temples of Avantiswami and Avantisvara as *Pandav Lari*, meaning "house of the Pandavas".

The temple of Avantiswami though smaller in size was built on the lines of the temple of Martand and is considered to be more perfected and complete in its form.

Material used for the construction of Martand and Avatiswami temples was finely dressed massive blocks of limestone, with little mortar (Ray 1970; Kak 2002). Surfaces of the temples were usually carved with sculpted reliefs, geometrical and floral patterns. Later limestone was replaced by granite. Natural disasters, neglect and iconoclasts are some of the factors responsible for the present condition of the temples.

F) **Krimchi Group of Temples:** Six in number Krimchi group of temples are located in a village, north east of Jammu city on the road which once formed a part of the trade route connecting the valley of Kashmir with Panjab. Belonging to 8th -9th century A.D. The construction of the temples is attributed to the trade guilds which traversed the trade route with their caravans' carrying goods and halted at Kiramchi in course of their journey to Kashmir (Anita1991). Built in Orrian style of architecture the temples are considered to be the most perfect specimens of temple architecture of the early medieval period of

Jammu. Their resemblance to the Bhuvneshwar group of temples, especially Parasura Mesvara temple makes them stand apart from the other temples of Jammu region. Though no icon has been found enshrined in the temples, they were perhaps dedicated to lord Shiva.

24.7 Frescos and Murals

Once a very popular tradition, painting of inner walls of the temples and palaces in Jammu region, is lost to posterity. The skilled painters who painted these walls no longer exist and along with them this art has also vanished forever. The paintings used to be in the form of frescos, and murals depicting the 19th century Jammu society. The subject matter of these paintings on the walls of the temples was based on Hindu mythology narrating various episodes of gods, goddesses and *Avataras* from Puranas, epic, Krishna leela *etc.* Whereas, the royal courts were painted in secular themes. The wall of the temples of Sui Simbli and Burj in Jammu were painted with the pictures of religious theme. A large number of murals have been disfigured by the people who do not have sense to appreciate their historical value (Figure 24.3). No initiative has been taken by the government of Jammu and Kashmir and temple authorities to revive this fading art or to preserve whatever little is left.

**Figure 24.3: Mural Paintings of Sui Simbli Temple Damaged
Due to Lack of Awareness.**

24.8 Movable Tangible Heritage of Jammu

The protection of movable cultural heritage is foregrounded through key legislation such as the 1970 UNESCO Convention on the means of prohibiting and preventing the illicit import, export and transfer of ownership of cultural property and at a later point, the 1995 UNIDROIT Convention on stolen or illegally exported cultural objects.

24.8.1 Epigraphical Records

Number of inscriptions, copper plates, sanads, grants, firmans in Brahmi, Sharda, Takri, Persian and Devnagari scripts have been reported from Jammu and Kashmir from the areas like Bhadarwah, Kishtwar, Rajouri, Poonch, *etc.* (Kaul 2001). These inscriptions have been found engraved on the temples and inside caves. Number of Sarda epigraphic records have been reported from different parts of Kashmir also.

Manuscripts in Sharda in large numbers have been found from Kashmir but there are many more which are still lying with the descendants of royal families and families of royal priests which need to be collected and documented. Many are lying unattended in the libraries of Jammu and Kashmir.

Takri was patronised by the Dogra rulers and was made the official language in combination with Persian and Hindi by the Dogra ruler, Ranbir Singh (Kaul 2001). Though not many earlier coins have been reported from Jammu and Kashmir but coins belonging to the Kushana period have been reported in plenty from the valley (Ray 1970). Coins ascribed to the Hun rulers have also been found in Kashmir. The rulers of Kashmir followed a strong tradition of minting coins, hence a number of gold, silver and copper coins are reported to be housed in the Museums of Jammu and Kashmir.

24.8.2 Miniature Paintings

Miniature paintings in the Pahari style saw their emergence in the hill states of North-western Himalayas including Jammu. From the mid-17th to the 19th century this art was patronised by many rulers and it flourished in a number of small hill principalities like Basholi (Figure 24.4), Jasrota, *etc.* The then hill principalities are now part of Jammu province.

The credit for making the art of miniature paintings popular goes to Pandit Seu and his sons Nainsukh and Manak. Jammu and Kashmir is particularly famous for its Basohli miniatures which have attained international fame. The distinct feature of the Basohli miniature paintings was the use of beetle wings to depict emeralds in the jewellery worn by the Nayakas. Efforts are being made by the govt. of Jammu and Kashmir to protect this art by helping the artists in different ways.

24.8.3 Sculptures

Rich heritage in the form of sculptures have been found in Jammu and Kashmir. Harwan sculptures are rated very high in sophistication and were inspired by the life of upper-class people. Terracotta sculptures reported from Ushkar centred on Lord Buddha and were religious in nature and on the basis of their artistic style have been placed between the 4th and 5th century A.D. (Ray

1970). Huge similarity has been found between the artistic skills of the terracotta figures at Ushkar in Kashmir and Ambaran (Akhnoor) in Jammu (Ray 1970). These two sites can perhaps, be called the twin Buddhist heritage sites of the UT of Jammu and Kashmir and can be promoted as a part of Buddhist heritage tourism.

Figure 24.4: Famous Basholi Painting
(*Courtesy:* Wikimedia).

24.8.4 Dresses

'Pheran' and 'Poot' are the traditional attire of the men and women of Kashmir and are worn as twin dresses one above the other. 'Pheran' is a long robe made of cotton or wool depending on the weather. It is generally worn during winter season to seek protection against cold. It is made in such a way that both men and women during extreme cold can hold 'Kangri', an earthen pot encased in a wicker basket filled with burning coal inside the 'Pheran'. Men wear plain 'Pherans' whereas the 'Pheran' worn by the women are embroidered either in thread or tilla. It is an essential garment which is given to every Kashmiri bride.

Traditional dress worn by the women of Jammu is 'Kurta' and 'Sutthan'. A Dogri bride wears this dress embroidered with 'Kinnari' work. Men folk wear 'Kurta' and 'Kuttana' accompanied by a turban worn in a typical style.

24.8.5 Jewellery

Jewellery worn by the Kashmiri and a Dogra woman is very different.

Traditionally a Kashmiri woman wears an elaborate headwear called 'Chauk Phool' whereas a Dogra woman wears a *Nama, Jhumka, Tikka, Kada, anklet, nath etc.* Kashmir was famous for silver jewellery whereas, in earlier days the traditional jewellery of Jammu was made of a metal 'Gillet'. Earrings made of the bird feather *Murgabi* were particularly popular amongst the Dogra women. Now a day these can be rarely found.

Some famous jewellery of Jammu and Kashmir is *Balaor Deji-hor* (crucial ornament for the Kashmiri Hindu married women), *Kundalas* (circular earrings), Nupur (anklets worn by women), *Jiggni* (head jewellery), *Choker* (Kashmiri tribal ornament), *Bangar* (wrist ornament) *etc.* This magnificent traditional jewellery of Jammu and Kashmir is not only the identity of the region but can also play a crucial role in the development of the cultural values.

24.9 Museums and other Stakeholders for the Protection of Cultural Heritage

The museums and other cultural institutions of Jammu and Kashmir are committed to protect the Cultural Heritage both tangible and intangible. In this context, the following principal guidelines are to be followed by the cultural institutions:

A) **Adaptation of General Policy:** The museum and other cultural institutions like the Jammu and Kashmir chapter of the Archaeological Survey of India, The Indian National Trust for Art and Cultural Heritage *etc.* will take some policies such as collection management, documentation, exhibition, and conservation for safeguarding Cultural Heritage. The condition of the cultural heritage sites and the building of the UT have already been discussed. There are some heritage sites which need immediate steps for conservation.

B) **Formation of Competent Body:** Protection of Cultural Heritage is not an easy matter, therefore, the museum and other cultural authorities must establish a competent body consisting of experts in the field of museum and heritage specially conservation experts and members of different sections of the society, especially local people.

C) **Territory Selection:** It is not possible for a museum or a cultural institute to cover all the heritage sites and heritage components of UT from a conservation point of view. Therefore, local museums should take the responsibility for some selected areas in which cultural heritage is to be protected. It is better to select the neighbouring areas of the museums or cultural institutions.

D) **Identification:** Identification of both tangible and intangible cultural heritage objects in a particular area is very important for the protection of the same. Identification should be done through experts and with the help of local communities.

E) **Documentation and mapping:** Documentation is a very important part of keeping records of the Tangible Cultural Heritage of Jammu and Kashmir. The museums should document the tangible cultural heritage and should ensure that everyone can access the information. A heritage tourism map needs to be prepared for the tourists.

F) **Formation of Database:** A publicly accessible national cultural heritage database is very important to correlate cultural heritage and sustainable development. Any noteworthy details related to the conceptualisation of heritage can easily be tracked through a perfect database. International heritage charters, conventions and recommendations have encouraged the improvement of inventories and current databases of cultural heritage. This includes the UNESCO World Heritage, Athens Charter Convention, and the UNESCO recommendation concerning the protection, at the national level, of the cultural and natural heritage. The ICOMOS (1996) principles for the recording of monuments, groups of buildings and sites, the UNESCO (2001) convention on the protection of underwater cultural heritage, the UNESCO (2003) convention for the safeguarding of the intangible cultural heritage, the ICOMOS (2008; 2011) charter on cultural routes and the Valletta principles are very important for the safeguarding and management of historic cities, towns and urban areas.

G) **Research Work:** Museum can be a research institute for conducting scientific, technical and artistic studies, as well as research methodologies for the purpose of effective safeguarding the tangible cultural heritage of the Union Territory of Jammu and Kashmir, in particular, the cultural heritage which is getting lost. In this context it is necessary to mention that the Seikh Noor Uddin Museum of Heritage of the University of Jammu is taking a lead role.

H) **Training Programme:** Training programmes for artists, especially for the new generations who are not interested in the profession, should be conducted regularly. The Seikh Noor Uddin Museum of Heritage of the University of Jammu often conduct training programme and workshop to aware the people of local heritage.

I) **Awareness Programme:** Museums of the Union Territory should recognise and exhibit the Tangible Cultural Heritage so that the culture of the community gets respect in society. Through educational, awareness-raising and information programmes, aimed at the general public, in particular young people of the local community; specific educational and vocational training programmes within the communities and groups concerned make the public aware of the dangers threatening such heritage

J) **Capacity Building:** Capacity-building activities should be taken to make the local people able for safeguarding the tangible cultural heritage. Particular management and scientific research are an integral part of safeguarding tangible heritage and non-formal way of transmitting knowledge.

K) **Community Participation:** Museum alone cannot protect the Tangible Cultural Heritage without the help of local communities. Communities, groups and individuals of the local area must be engaged in conservation activities. Each museum shall endeavour to ensure the broadest possible participation of communities, groups and, where appropriate, individuals who produce, maintain and communicate such heritage, and involve them actively in the management of Tangible Heritage.

24.10 Heritage in the Context of Sustainable Development

Heritage is being placed under increasing stress throughout the world as globalization accelerates, urbanization expands, development pressure increases, cultural tourism diversifies and climate changes. The UNESCO's Convention on World Heritage and Sustainable development in 2015 aimed to extract the potentiality of World Heritage properties and heritage in general, to contribute to sustainable development. Preservation and development of culture matter not only for the holistic development of a country but also for re-energising the neglected areas and communities by providing opportunities to review, re-engage and reinterpret their heritage. Jammu and Kashmir's rich heritage, consisting of art, architecture, monuments, oral traditions, physiography, flora, fauna, if properly preserved, can contribute to sustainable development. The economic aspect of the culture can be recognised in some areas such as culture-oriented jobs, enhancing craftsmanship skills, the revitalization of heritage tourism, adaptive reusing of abandoned heritage buildings *etc.* These can be in the form of socioeconomic factors related to wealth and job creation, urban regeneration and even an enhancement in cultural participation that can lead to social solidarity and community development.

There are many employment opportunity projects in cultural sectors throughout the world. For example, "Employment opportunities for cultural heritage safeguarding in Jordan" project aims to promote sustainable socio-economic development through employment creation among young vulnerable Jordanians and Syrian refugees. Employment opportunities are also available at the World Heritage Centre UNESCO. Many museums and art centres tend to hire skilled and experienced people for the curation and conservation of their collections. Museums and heritage sites also provide jobs as guide lecturers. Mostly local people, who know their heritage well are taken for guiding the tourists. Thus, local employment generation can be done.

World Tourism Organisation estimated that the cultural tourism makes up almost 37 per cent of global tourism and is increasing each year. It helps local communities to hold their own heritage exhibits to visitors, uplifting economic growth. Cultural tours conducted in Jammu and Kashmir are specifically geared toward cultural experiences that motivate local people to celebrate and promote what distinguishes their culture from others. Some local cultural areas can also be brought to attention by cultural tourism, and this can result in a positive change such as community development.

Participation in cultural activities can interpret to an increase in civic awareness, the knowledge of heritage and history, the awareness of own identity, as well as influencing the development of other tourism-related activities, such as restaurants and hotel businesses. However, there is a close relationship between tourism, heritage and the environment. Climate Change has a full impact on cultural tourism and vice versa. Climate change represents a threat to cultural heritage and on the other hand, the increase in global CO_2 emissions and global warming is a result of tourism.

24.11 Conclusion

There are diverse challenges in achieving a consistent and coherent approach to integrating sustainable development with cultural heritage. The findings allow concluding that it could be possible to mainstream across different databases those indicators, which could lead to depicting the overall level of attainment of the UNESCO Culture Agenda 2030 targets on heritage. However, more research is required in developing a strong correlation between national datasets and international targets.

Therefore, many research, initiatives and projects need to be undertaken to build evidence-based indicators that will help to establish a multidimensional, comprehensible and strong narrative on the preservation of cultural heritage and sustainable development.

After analysing the cultural sectors of Jammu and Kashmir, it has been found that traditional economic models fail in important ways to analyse heritage and conservation. The traditional models are framed to express all values in terms of prices. However, it will be kept in mind that all heritage values cannot be put into a single traditional framework, nor can they all be measured in monetary terms. High creative work can also be considered to strengthen socio-economic conditions of a community. Greater engagement between economic and cultural sectors, as well as mutual understanding, is necessary for enabling conservation to play a greater role in society.

REFERENCES

Anita (1991) Ancient temples of Kirmachi, Jay Kay Book House, New Delhi pp. 52

Avruch K (1998) Culture and conflict resolution. US Institute of Peace Press.

Balouria K S (1913) Tawarikh-i-Rajputan Mulk-i-Punjab, 1913 p. 322.

Chaudhary P, Katoch J (2008) Mubarak Mandi: Inheritance of an Ailing Heritage,Saksham Book publishers

Eurostat. Culture Statistics (2019) European Union: Luxembourg, Publications Office of the European Union; Available online: https: //ec.europa. eu/eurostat/documents/3217494/10177894/KS-01-19-712-EN-N. pdf/915f828b-daae-1cca-ba54-a87e90d6b68b (accessed on 17 January 2020).

Hosagrahar J (2019)UNESCO Thematic Indicators for Culture in the 2030 Agenda for Sustainable Development, in Analytical Report of the Consultation with the Member States;

http: //jammutourism.gov.in/MonumentHeritage-Ambaran.html (assessed on 19th June 2022) https: //whc.unesco.org/en/tentativelists/5492/ (assessed on 19th June 2022).

https: //www.astroved.com/astropedia/en/temples/north-india/ shankaracharya-temple (assessed on 23rd June 2022).

Kak R C (2002) Ancient monuments of Kashmir, Gulshan publishers pp 50-65.

Kaul PK (2001) Antiquities of the Chenab Valley in Jammu, Eastern Book linkers.

Matsumoto D, Kouznetsova N, Ray R, Ratzlaff C, Biehl M,Raroque J (1999) Psychological culture, physical health, and subjective well-being. Journal of Gender, Culture and Health. 4(1): 1-18.

Nocca, F (2017) The Role of Cultural Heritage in Sustainable Development: Multidimensional Indicators as Decision-Making Tool. Sustainability 2017, 9, 1882.

Petti L, Trillo C, Makore B N (2020) Cultural heritage and sustainable development targets: a possible harmonisation? Insights from the European Perspective. Sustainability. 12(3): 926.

Rabbani G M (1981) Ancient Kashmir: A historical perspective, Gulshan Booksp.38

Ray S C (1970) Early History and Culture of Kashmir, Munshi Ram Manoharlal pp 215-220.

Sharma A.K. (1982-83) GU'fkraI1981: An Aceramic Neolithic Site in the Kashmir Valley. Asian Perspectives, xxv (2), 1982-1983 (https: //scholarspace.manoa. hawaii.edu/server/api/core/bitstreams/

c46b5213-4451-4cd3-9fb4-9e84c2603636/content)

Spencer-Oatey H, Franklin P (2012) What is culture? A compilation of quotations. Global PAD Core Concepts, 1: 22.

UNESCO (2022) Sustainable Development- World Heritage and Sustainable Development. Available at: https: //whc.unesco.org/en/ sustainabledevelopment/ (Accessed: 07 March 2022).

UNESCO (2021) Employment opportunities for cultural heritage safeguarding in Jordan, Available at https: //en.unesco.org/filedoffice/amman/ Employment-opportunities-for-cultural-heritage-safeguarding-in-Jordan (accessed on 4th May 2022)

UNESCO World Heritage Centre: Paris, France.

UNESCO (1954) Convention for the Protection of Cultural Property in the Event of Armed Conflict with Regulations for the Execution of the Convention 1954; United Nations Educational Scientific and Cultural Organisation: Paris, France, 1954.

Vileniske I G (2008) Influence of Built Heritage on Sustainable Development of Landscape, Landscape Research, 33: 4, 425-437, DOI: 10.1080/01426390801946491

Part – III

Hindu Festivals: Promoting Harmony and Heritage

हमारी संस्कृति हमारे पर्व

पत्तों के झुरमुट में

किसी कली ने मुस्कराया है

फूलों ने खिल-खिला कर

अभिनंदन गीत गाया है

शत शत वंदन है

लो आज फिर नव वर्ष आया है।

वसंत के बाद गूंज रहा फाग का संगीत

गांव में जगह-जगह रंगों की टोलियां हैं

जोश भरा नौजवानों में

बच्चों के हाथों में पिचकारियाँ हैं

देखो होली का रंग-बिरंगा त्यौहार आज आया है।

प्यार भाई -बहना का अद्भुत

आज भाई बहन का रक्षाबंधन का त्यौहार आया है।

बड़ा दिन आज फिर आया है

सेंटा क्लाज ने बच्चों को लुभाया है

प्रभु यीशु ने प्रेम की महिमा को पुनः बताया है।

जल रहे दिए फिर जगह-जगह

है ऋषि दयानंद का निर्वाण दिवस

और महावीर की माया है

शोभा है दीपों की घर- घर

राम का घर आगमन, तो लक्ष्मी का पूजन है

लो आज खुशियों का पर्व दिवाली

जीवन में एक नया प्रकाश लाया है ॥

Chapter 25

Festival of Cultural Harmony and Colours-Holi: Celebrations, Environmental Issues and Eco-friendly Alternatives

Madhu Praskash Srivastava

Department of Botany, Maharishi University of Information Technology, Lucknow, Uttar Pradesh, India
e-mail: madhusrivastava2010@gmail.com

ABSTRACT

Celebrations of festivals help to bring joy, social harmony and feeling of united entities among communities. The use of new cloths, variety of foods and freedom from routine job work creates a feeling of joy and happiness. Holi the celebration of variety, tomfoolery, happiness and congruity and is an old celebration of India and was initially known us 'Holiku'. It is celebrated not only in different parts of the country but at all those nations where Hindu community resides.

The celebration of 'Holi' is ending up with an ecological gamble because of the harmful variety of colours utilized during this festival. Limitless and uncontrolled utilization of colours can prompt grave outcomes concerning human wellbeing and environmental equilibrium. These varieties are profoundly organized polymers and are undeniably challenging to naturally decay. A study was undertaken with a drive to decide the degree of impacts that the 'Holi' colours have on water and soil, individually. The conventional 'Holika Dahan' huge fire produced in this festival is expected to add to deforestation and air contamination. There is likewise worry about the enormous scope wastage of endlessly water contamination because of engineered colours during 'Holi' festivity. Their presence in water, even at extremely low fixations, is exceptionally noticeable and unwanted. At the point when these hued effluents enter streams or any surface water framework they upset natural movement. Ground-water frameworks are additionally impacted by these toxins in light of filtering from the dirt. The effect appraisal and harm control prompted more secure techniques for disposing of the 'Holi' colours, to save our current circumstance. Subsequently this article means to recuperate the solid perspective for commending Holi by utilizing protected, regular and eco-accommodating Holi made by normal colours obtained from plant material.

Keywords: *Environmental, Cultural harmony, Colours, Health issues, Holi, Natural dye.*

25.1 Introduction: Hindu festival of Cultural harmony

Climate has become the dominant focal point in 2015 and with the world gathering COP21 at Paris in September underlining the need for points and objectives for a large group of issues that take steps to rush the speed of corruption in the personal satisfaction, to be tended to determinedly and in union by all nations of the world. Truth be told, an update has been settled upon and an understanding for consistence marked earnestly by most countries. As the need might arise to contribute towards this work by doing however much as could be expected via executing, proliferating and teaching, plans and plans that would quickly resolve the issue of stopping corruption to the climate in their particular areas.

Holi is a festival celebrated by Hindus from religious point of view, however for youth it is spring festival of joy and happiness. Not only in India but also in Bangladesh, Pakistan, Nepal, and other countries with large Indic diaspora populations following Hinduism, in countries such as Suriname, Malaysia, Guyana, South Africa, Trinidad, United Kingdom, United States, Mauritius, and Fiji it is welcomed. In India itself, this festival is known Festival of Colours, or Dol Purnima as in Orissa and Dol Jatra as in Bengali and Basantotsav as in West Bengal. Dhulhandi or Dulhendi is the main day of Holi celebration (Sharma 2013). At the end of the winter season, on the last full moon day of the lunar month of Phalguna (March), (Phalgun Purnima) Holi is celebrated; originally to festive the memorable good harvest and the fertile land. It also has a religious purpose, to honour divine events in Hindu mythology. To fill the gaps between age, gender, status, and caste Holi allows avoiding social custom by throwing colours on each other. No expectation of customs fill the atmosphere with excitement, fun and joy all around.

The Indian celebration of Holi is one extremely large chance to show the way. Everyone, paying little mind to progress in years or abundance, partake in this celebration of variety appearing "solidarity in variety". Holi is additionally called celebration of affection and celebrated by sprinkle of scented colours over one another, wearing new garments ideally white and banquet of delightful dishes as desserts and squeeze particularly 'lassi' and so forth are ceremonies of the celebration. Dr. N.P. Singh narrates in Hindi about the festival of colours – Holi as:

जाने क्यों मन का गुलाल आले में कहीं पड़ा है ।

कितनी देर हुई है कब से द्वारे पर्व खड़ा है ।

भोर किरन रँग रही धरा है ,

किसने ऊषा माँग भरा है ।

कुशल चितेरा कहीं न दीखे,

लेकिन सारा चमन हरा है ।

इंद्र धनुष झरने के नीचे लिए अनंग खड़ा है ।

जाने क्यों मन का गुलाल आले में कहीं पड़ा है ।

होली के उमंग भरा पावन पर्व आज आया है ।

25.1.1 Time and the Day of Celebration

There are two ways of reckoning a lunar month- 'purnimanta' and 'amanta'. In the former, the first day starts after the full moon; and in the latter, after the new moon. Though the amanta reckoning is more common now, the purnimanta was very much in vogue in the earlier days. According to this purnimanta reckoning, Phalguna purnima was the last day of the year and the New Year heralding the Vasanta-ritu (with spring season starting from next day). Thus the full moon festival of Holika gradually became a festival of merrymaking, announcing the commencement of the spring season. This perhaps explains the other names of this festival - Vasanta-Mahotsava and Kama-Mahotsava.

25.1.2 Celebration in Different Regions with Different Names

There are many ways of Holi celebrated in India and the folklore expresses the start of the Holi celebration originally noticed and began in the Barsana district of India including Mathura, Nandgaon, and Vrindavan. In the town of Barsana, Uttar Pradesh, the name given to the celebration is "Lathmar Holi." The custom continued in the town is about ladies pursuing away the men with lathis, yet it can't be said as a beating meeting. In the locale of Uttarakhand, the Kumaon district, where individuals call it "khadi Holi" where individuals move in works singing Khari melodies and wearing conventional garments. In Punjab, it is chiefly called "Hola Mohalla" or champion Holi, seen by Nihang Sikhs and display hand to hand fighting. Individuals of Odisha and West Bengal ordinarily call Dol Jatra and Basant Utsav. Individuals invite the spring season during this time. While youth set up saffron-hued garments and celebrate with colossal bliss in Shantiniketan, simultaneously, on Dol Purnima icons of Radha and Krishna move in a parade. Ranchers of Goa dance like conventional people in the roads and call the celebration as "Shigmo." Table Chongba, a dance of Manipur is seen from the day of the full moon, joining Hindu and native practices. In Kerala, Manjul Kuli is praised in Konkani sanctuary of Gorripuram Thirumala. Bihar locale observes Holi for the sake of Phaghwa while Assam celebrates for the sake of Phakuwah.

25.1.3 Celebrations of Festival in different Parts of the World

This festival is popularly known as the "festival of colours" and signifies the end of winter and arrival of spring as well as the blossoming of love and the chance to laugh, forgive others, and repair broken relationships. The festival of colours is a three-day event full of fun and joy in Kathmandu. Nepali people are mostly Hindu and Holi is celebrated in major cities throughout the country. The people here celebrate the first day by decorating wooden poles called chir with colourful cloth and then burn them through the night to symbolize the end of the old year. The chir represents the tree on which the Hindu god, Krishna, is said to have hung milkmaids' clothes while they were bathing in the Jamuna River in India. Holi is in large part a celebration of Krishna and the many tales about him. Women also commonly wear red saris and walk around the chirs while praying on this day.

On the second day, bonfires are lit to symbolize the death of the demoness Holika who, in Hindu mythology, died in a fire trying to kill Prahlad, a follower of Lord Vishnu. The third day is the day of the largest celebrations when people gather in masses on the streets to join in a large colour fight. Colourful water balloons, paint, and dyed powders are thrown on everyone outside and the city takes on a joyous atmosphere. In addition to throwing coloured powder, the festival also includes feasts, dancing, and singing.

Hindus who have settled in the United Kingdom do not miss out on the excitement of Holi celebrations. Leicester has a large population of people of Indian descent and is particularly known for its love of celebrating Indian festivals, the most loved of which is Holi. Outside of India, Leicester has the largest Holi celebrations in the world. Holi is celebrated at different parks across the city each year. Everyone in the city comes together to watch the lighting of the Holika bonfire on the eve of the festival. During the bonfire, many people in the crowd hold coconuts which will be blessed and cracked open to symbolize hope for new beginnings and the growth of a new harvest. The throwing of colours takes place in Abbey Park with a free-for-all zone and a special family-friendly zone to ensure that even the youngest participants can enjoy the celebrations. During the festival, friends and family apply the tilak on each other which is a red mark worn on the forehead.

New Yorkers of all backgrounds come together at events across the city to celebrate Holi with joy and positive vibes. Attendees throw coloured powder and often, with permission, rub it on each other's faces. To make the colours pop more and get great photos, it is customary to wear white clothing during the festival. The most popular Holi event in New York City takes place on the Play Lawn on Governors Island. Every May, the "Happiest Festival of New

York" returns with free yoga, Sufi and Punjabi entertainment, food vendors, and environmentally safe coloured powders.

Brooklyn celebrates Holi with an annual spring art and music event in Bushwick. The event showcases art from local painters and graphic designers and also includes live DJ performances and a colour fight with traditional Indian powders to make life even more colourful.

Every year at the end of March, the Holi celebrations are held at the Sri Radha Krishna Temple in the town of Spanish Fork, Utah, USA. This Hindu temple is one of the only ones in Utah and is very welcoming towards visitors. No religious affiliation is required to attend the Holi celebrations. The Holi Festival in Utah is a modern adaptation of the traditional Indian celebrations and only coloured powders are used instead of liquid dyes. The colours provided by the temple are made from dyed corn-starch that do not stain your skin. Besides the colour fight, the festival activities include live music, yoga, dancing, food, a bonfire, and the burning of an idol of the demoness

Several events are organized all over in Australia to celebrate Holi, but the Melbourne Holi festival is one of the fun celebrations in the country. This is an alcohol-free, family-friendly festival with non-stop music and multicultural performances. The event also includes Bollywood dancing, drumming performances, kid's activities, food vendors *etc.* The coloured powders are stain-free and non-toxic which makes for easy clean-up and safe use on the skin. It is not advised to paint anyone not participating in the festival.

There are many Indian and non-Indian residents of Singapore who love to take part in the Holi celebration here. The celebration that takes place in Singapore is quite different from those in India. In India, celebrations are more local while Singapore celebrates with extravagant events and fun parties. Rang Barsay is the most famous Holi event in Singapore and people from all over the country come to enjoy the festivities by dancing to party and throwing coloured powder at each other. Rang De Holi is another big event. It is celebrated at the Wave House Sentosa and hosts a large crowd who come together to network and have lots of fun. The event has a DJ who plays party music and serves many kinds of foods and drinks.

The Festival of colours is celebrated with complete fervour in Madrid, Spain. Every year, this city sees a surge of people who come together dressed in white for the celebrations. The festival takes place in the Plaza de Agustín de Lara which becomes packed with hundreds of participants during the colour fight. Participants buy small pouches of coloured powder. When the event begins, handfuls of powder are launched in the air making for excellent photos and turning the sky into a kaleidoscope of colours.

In Sao Paulo of Brazil also the celebrations of Holi are held. Holi Dhamaka is the biggest Holi celebration in Toronto, Canada. In Mauritius, the Holi is celebrated for two days. A week before the actual festival, markets and religious shops start decorating for the holiday and cover everything in a rainbow of colours. Holi here kicks off with a bonfire on the eve of Holi. This is celebrated similarly to how it is in India with the burning of a statue of Holika, the evil demoness, as well as with dancing and folk songs. Later in the day after the colour fight, you'll find people dancing, singing traditional songs, visiting family, and exchanging gifts.

25.2 Other Festivals similar to Holi in different Parts of the World

The festival of colours, with thandai, bhang, phag and ultimate masti is Holi celebrated by Hindu community. Holi – the brightest Indian festival of colours signifies the supremacy of good over evil. During this festival, people smear colours on each other as a sign of love, peace, brotherhood, and unity. But do you know that there are many such Holi-like festivals across the world. Below are listed such vivacious festivals across the globe, who believe in countless hours of mud rolling sessions. If you visit one of these countries and see people throwing tomatoes, eggs and even flour at each other, do not be surprised my friend.

Figure 25.1: Mud Festival of South Korea.

a) Mud Festival: South Korea and Switzerland

This takes place in Boryeong of South Korea during July (summer season) (Figure 25.1). It is exactly 200 km south of Seoul. Millions of people gather here to get crazy and dirty. This interesting Holi-like festival was started in the year 1988 and lasts for 10 days. The reason is to promote the health benefits of local mud that is rich in Germanium and other minerals. It is interesting for those who would like to apply the clay mask to look beautiful. It is the largest messy festival in the entire world, where as many as 1 million people mark their presence.

The largest open air music festival in Switzerland, Paleo, attracts a large crowd. Volunteers along with students have a blast at this musical event which takes place in the last six days of July. Mud adds as an added dose of fun, making the event even more exciting.

b) La Tomatina: Spain

It is the Holi of Spain. It is held in the Valencian town of Buñol where people celebrate throwing tomatoes at each other. But before throwing, they squeeze it well so that it doesn't hurt someone. La Tomatina is celebrated on the last Wednesday of August (Figure 25.2).

Figure 25.2: La Tomatina Festival of Spain.

c) The Orange Battle: Italy

The biggest food fight in Italy, the battle of oranges is similar to the La Tomatina in Spain. Taking into account the costumes involved, it is more unique visually. Participants pelt oranges at one another and make the best of the day. Participants use goggles and helmets.

This one takes place in February every year and it is a wonderful sight. 400 tons of oranges are used in this festival. People make two groups and pelt oranges at each other. There is a history behind this festival. This tradition is alive since the 12th century.

d) Wasserschlacht: Germany

Germans celebrate the onset of summer in an entirely different manner.

During the season, two districts in Berlin gear up and stand up against each other in a massive water fight. Fruits, eggs and flour soon adds up, taking the spirits even higher. This fun event takes place on the Oberbaumbrucke Bridge which for long has been feared to collapse because a large number of participants turn up on this day.

e) Lastonbury Music Festival: Britain

One of the greatest music festivals in Britain, the crowd, the music and the atmosphere simply makes you fall in love with the event. What's the best part about the event, it's mud. The messiness along with good music is simply an icing on the cake. You don't need to buy a cool outfit to attend this festival, at the end of the day, it is going to be covered by mud.

f) Songkran: Thailand

Thailand celebrates New Year on 13th April and this festival lasts from 14th to 15th April. The Thai people believe in celebrating their New Year's Day in a unique manner. The festival which is called Songkran is a huge event in the country. What adds to the unique factor is that people throw ice cold water on each other around the streets and put beige coloured paste on anyone's face they come across. Similar to our Holi festival.

g) Watermelon Festival: Australia

In the Darling town regions of Queensland, this Holi-like unique festival takes place. It is because this region is known as the "Melon Capital of Australia". This melon festival occurs every second year in February where various games are organized. Some of the names are- Melon Bulls Eye, Melon Chariot, Melon Ironman, Melon Skiing, *etc.*

h) Clean Monday Flour War Carnival: Greece

Here in Galaxidi of Greece people celebrate by smearing coloured flour on each other. Some 1500 kg of flour is used. This carnival occurs in early March and marks the beginning of the 40-day lent period.

i) La Merengada: Spain

Spaniards know how to create hot mess, don't they? From oranges to tomatoes, they leave nothing behind. How could sweet nothing be left far behind? The Batalla de Caramelos also known as Candy Fight takes place every year on Fat Thursday in a small town near Barcelona. People usually eat cod fish with red pepper sauce and after that throw their meringue dessert at each other. Once the meringue is finished, the candy starts to fly.

j) Els Enfarinats: Spain

The Spanish version of April fool's Day is celebrated on December 28. The annual Els Enfarinats festival in Ibi is where the citizens throw flour bomb as well as eggs at each other. A little less colourful but even messier than Holi. The cloud of thick flour makes it look like a blizzard. Go ahead, make your childhood fantasies come true.

k) Galaxidi flour festival: Greece

Every year, this small town in Greece, Galaxidi is covered with over 1.5 ton of coloured flour. Over hundreds of people gather around to attend this insane battle of throwing flour at one another. This event, which takes place during February, is also ironically known as Clean Monday. Don't be surprised if you see people dancing on the streets covered in coloured flour.

25.3 Celebrations from Mythology to Present Days

The festival finds mention in Narad and Bhavishya Purana, Jaimini Mimansa and also by Muslim scholars. Some records from paintings, murals and ancient texts are listed below:

25.3.1 Mythology: Reference of Prahlad and Holika

The literal meaning of the word 'Holi' is 'burning'. There are various legends to explain the term, most prominent of all these is associated with demon king Hiranyakashyap. He wanted everybody in his kingdom to worship him only. But to his great disappointment, his own son, Prahlad became an ardent devotee of Lord Narayana. Hiaranyakashyap commanded his sister, Holika to enter a blazing fire with Prahlad in her lap. She had a boon whereby she could enter fire without any damage on herself. However, she was not aware that the boon worked only when she enters the fire alone. As a result she paid a price for her sinister desires, while Prahlad was saved by the grace of the god for his extreme devotion. The festival, therefore, celebrates the victory of good over evil and also the triumph of devotion.

Legend of Lord Krishna is also associated with playing with colours as the Lord himself started the tradition of applying the colour on his beloved Radha and her friends *i.e.* other Gopis. Gradually, the play gained popularity among the people and became a tradition. In some parts of India, especially in Bengal and Orissa, Holi Purnima is also celebrated as the birthday of Shri Chaitanya Mahaprabhu (A.D. 1486-1533). There are also a few other legends associated with the festival - like the legend of Shiva and Kamadeva and those of female monster Ogress Dhundhi in the kingdom of Raghu. All depicting triumph of good over evil- lending a philosophy to the festival.

25.3.2 Reference in Ancient Texts

Besides having a detailed description in the Puranas such as Narad Purana and Bhavishya Purana, the festival of Holi finds a mention in Jaimini Mimansa. A stone inscription belonging to 300 BC found at Ramgarh in the province of Vindhya, which has mention of Holikotsav on it. King Harsha, too has mentioned about holikotsav in his work Ratnavali, written during the 7th century. The famous Muslim tourist - Ulbaruni too has mentioned about holikotsav in his historical memoirs. Other Muslim writers of that period have mentioned, that holikotsav was not only celebrated by the Hindus but also by the Muslims.

25.3.3 Reference in Paintings and Murals

The festival of Holi also finds a reference in the sculptures or murals present on walls of old temples. A 16th century panel sculpted in a temple at Hampi, capital of Vijayanagar, shows a joyous scene of Holi. The painting depicts a Prince and his Princess standing amidst maids waiting with pichkaris to drench the royal couple in coloured water. A 16th century Ahmednagar painting is on the theme of Vasanta Ragini - spring song or music. It shows a royal couple sitting on a grand swing, while maidens are playing music and spraying colours with water sprinklers.

There are a lot of other paintings and murals in the temples of medieval India which provide a pictorial description of Holi. For instance, a Mewar painting (1755) shows the Maharana with his courtiers. While the ruler is bestowing gifts on some people, a merry dance is on, and in the centre is a tank filled with coloured water. A Bundi miniature painting shows a king seated on a tusker and from a balcony above some damsels are showering gulal or coloured powders on him.

25.4. Traditional Celebration of the Festival

Holi festival encompasses three events celebrated every year in March or Falgun month of Hindu calendar. People rejoice by singing local songs called Phag for days much ahead the actual festival. The bonfire is lit on the Holi day and colours both wet and dry (knowns as gulal) is sprinkled among each other.

In many rural pockets of U.P. Chhatisgarh and Madhya Pradesh, the days before Holi are filled with the beats of the nagadiya and the phag, songs that are special and considered auspicious to celebrate the advent of spring and the festival of colours. However, this tradition is fading now only Bollywood songs are played in streets and villages. The sounds of the nagadiya start filling up the pathways in any village of this region. Along with the beats of the nagadiya, are heard snatches of songs called phag, sung especially in this season. "The phag is as old as our ancestors. We have learnt them by word of mouth as they have been handed down over generations," "People would gather in the chaupals (open spaces) in the village with their nagadiyas and dholaks, and sing. It still happens, but not as much as it would before," The singing of phag is closely associated with farming communities. During Holi, the farmers and their families would gather together to sing and enjoy the music. It was their source of entertainment, in days gone by. It is the time for them to take a break from all the toil and celebrate Holi.

All castes and communities living in the village participate in the phag singing. These includes the potters, carpenters, the leather tanner, the ironsmith, the herdsmen, farmers, *etc.* A village was considered complete in itself when all these different communities lived in it in harmony. It is still the case, but the tradition is giving way to modernity. Few lines recited in Hindi are:

फाग खेलन बरसाने आये हैं, नटवर नंद किशोर ।

घेर लई सब गली रंगीली, छाय रही छबि छटा छबीली,

जिन ढोल मृदंग बजाये हैं, बंसी की घनघोर ।

फाग खेलन बरसाने आये हैं, नटवर नंद किशोर ।

"Phag and the nagadiya represent unity and harmony," said the farmer-poet. A week before Holi, preparations are made for the Holi bonfire. Outside the villages, a big wooden pole is fixed on the ground, surrounding which are arranged dry leaves, grass, *etc.* On the eve of Holi, villagers gather at the spot singing phag and the celebrations culminate with the bonfire, which is believed to burn or drive away all the evil. "The leftover ashes from the fire are used to play Holi along with the colours."

"Sadly, the tradition of phag is fading. The younger generation does not have the patience and time for it. And, they have so many other means of entertainment," "I enjoy the sounds of the nagadiyas, but do not understand what the phag songs say," According to a village boy, no one from the older generation has bothered to tell him what the songs were all about. "Maybe it is the generation gap. Chaupals have been replaced by cafes." See few lines:

ka sang khelu faag re

ka sang khelu faag

gaye din holi khelan ke

piya milan ke

ka sang khelu faag re

roop sang mrig anjan dho gaye

sise taal agharam ke

muulhe ke bindiya achanak gir gayi

karke sor sangun ki guye din

holi khelan ke, piya milan ke

ka sang khelu faag re

People bake (offer) wheat ears from new crop, coconut, and sugarcane, in the auspicious Holi fire, which are later distributed as prasad to the friends and family members. The next day after Holi fire and offering prayers the festival of colours start. People visit their friend's houses and smear gulal in the faces of known friends or sprinkle colours on them as a gesture of joy and happiness. The enthusiasm of youth to celebrate this festival can be seen everywhere. If colours are not available ash of Holi or mud is sued to celebrate the festival with joy. People are offered sweets, gujhia, and date palm to eat besides a number of other sweets as a nice welcome gesture. In Gujarat usually dry Holi is preferred for one day but at other places wet Holi can continue for a number of days.

25.5. Natural and Synthetic Colours Used in Holi Festival

Traditionally Holi colours were prepared from the various parts of plants such as flowers, leaves, fruits *etc.* Some of them are Indian coral tree (Parijat), flame of forest (Kesu), marigold, turmeric (Haldi), henna (Mehandi), neem, sandalwood, beetroot *etc.* Turmeric, neem, sandalwood, Kesudo or tesu flowers were 100 per cent natural and have beneficial healing properties as per Ayurveda. But now a days the chemical colours have been replaced by this natural instinct of the fest and many safety issues have been found in all three forms of synthetic colours produced *i.e.* pastes, dry colours and water colours which are prepared from harmful substances like metal oxide, acids, mica, glass powder, alkalis, sand, soil, silica, asbestos, chalk *etc.* and are quite capable of causing serious skin complications and allergies. Nowadays often referred as unholy colours of 'Holi'. Artificial colours can lead to skin irritation, redness, rashes, itching, bumps and also proving a risk to an environment and humane health. This makes the festival less charming in comparison with Holi celebrated in earlier days with natural colours.

In modern times, with the arrival of dry chemical colours, people hve stopped taking pains for extracting the flowers colours and the use of synthetic chemicals is increasing day by day. Holi colours are available in three different forms *i.e.* paste, dry and wet colours. To get darker, long lasting colours and thick texture industrial dyes and metal oxides are mixed with low quality engine oil (for Paste) and are used in Holi. The synthetic dye based Holi powder can cause dermatitis, respiratory problems and allergies and prolong application can cause cancer (Das *et al.*, 2015). Moreover, scientific test have verified that this can cause skin abrasions, skin and eye irritation, allergy and can even triggers asthma (Rawat 2008; Sarmah and Sexena 2013; Sarmah 2013). Similarly, dry synthetic colours are deadly combination of toxic heavy metals which disrupts metabolic functions of body and build up in the body's vital organs such as the kidneys, lung and bones, by the presence of filler base asbestos, chalk or silica. The colorant used in the dry colours, was found to be toxic, with heavy metals causing asthma, skin diseases and temporary blindness (Toxics link, 2000). These chemical colours are not only extremely dangerous to our skin but also are hazardous for the environment (Sarmah and Saxena 2013).

One of the common and easily available water colorant is Gentian violet, it is highly toxic and hazardous to human health. It leads to severe skin ailments and eye problem (Table 25.1). A recent study conducted under the National Biodiversity Strategy and Action Plan revealed that chemical colours have all but wiped out India's wonderful vegetable dyes (Chaube *et al.*, 2010). Chemical colours give delightful pleasure to eyesight but at the same time may act as serious pollutants. In India, these colours are prepared on a small scale and lack any quality checks (Gardner and Lal 2012). Whereas, natural dyes/ colorants derived from flora and fauna are believed to be safe because of their nontoxic, non-carcinogenic and biodegradable nature (Cristea and Vilarem 2003; Ahlstrom *et al.*, 2005). The process is economically viable as the raw materials are available at low cost and so cost of production is also very low. Similar findings were reported in case of marigold, china rose and *Bixa* flowers (Ibrahim *et al.*, 1997). Many tribes of Arunachal Pradesh have been using these plant species traditionally in combination with other plants for extraction and preparation of dyes utilizing indigenous processes (Mahanta and Tiwari 2005). Natural dyes are now a days in demand not only in textile industry but in cosmetics, leather, food and pharmaceuticals. The rich biodiversity of our country has provided us plenty of raw materials, yet sustainable linkage must be developed between cultivation, collection and their use (Gokhale *et al.*, 2004). Hence, the present study was conducted to evaluate the potentiality of eco-friendly Holi colours prepared by using natural dyes extracted from plants. These are not only safe to human beings but will also have zero risk to the environment, with low cost filler base such as Tapioca (*Manihot esculenta*) flour as a substitute for costly starch extracted from corn and other cereals, asbestos or silica which are associated with health issues.

The discharge of the toxic colours in the soil and water has a deleterious effect on the water resources, soil fertility, as well as microorganisms living in these habitats and the ecosystem integrity on the whole. These colours are not readily degradable under natural conditions and are typically not removed from waste water by conventional waste water treatments. Thus, several bacteria have been found to decolorize, transform and completely mineralize coloured soil and water in both aerobic and anaerobic conditions (Papen and Berg 1998). During the last decade, environmental issues associated with the dyestuff production and applications have grown significantly. Although widely recognized and discussed as an annually recurring problem involving many people, lack of any formal study on it has prompted us to undertake the present work. In this chapter, we have studied the side effects of 'Holi' on the environment and utilization dye-degrading microorganisms to help in the removal of toxic colours from polluted soil and water samples. The metabolites produced after biodegradation are mostly non-toxic in nature. We have also suggested healthier practices for a safe and environmental friendly 'Holi.'

25.6 Holi and Use of Water

Recent years have seen water shortage in urban and semi-urban areas due to the excessive growth in population due to increased mortality, migration from villages and a host of other factors that contribute to the excessively inconvenient situations of drinking water available once or twice a week in large sections of cities and towns. This has pre-empted the idea of rain water conservation Holi has always been played with colours and over time wet colours have been in use a plenty. Not only colours dissolved in water, but splashing plain water with large buckets on people around has also become a major source of entertainment. This is again followed by use of a lot of water to get rid of the colours that have stuck on faces and bodies of the Holi revellers. Truly this is a major source concern especially in period of acute shortage and a worsening situation of clean drinking water.

Alternatives suggested include use of herbal colours, use of dry gulal powder *etc.* People could play with mud or wheat flour in rural areas. These are co eco-friendly substances which would not harm the skin, nor require much water to wash them away. Quite a reduction it lead to the consumption of water even for a day. Awareness is needed and people should refrain from harmful chemicals.

25.7 The Ill Effects of the Holi Bonfire

The traditional 'Holika Dahan' bonfire is believed to contribute to deforestation and air pollution (Figure 25.3). There is also concern about the large scale wastage of water and water-pollution due to synthetic colours used during 'Holi' celebrations. Their presence in water, even at very low concentrations, is highly visible and undesirable (Figure 25.4). When these coloured effluents enter rivers or any surface water system they upset biological activity. Ground–water systems are also affected by these pollutants because of leaching from the soil (Alau *et al.*, 2010)

Figure 25.3: Ill Effects of Bone Fire on the Occasion of Holi.

Figure 25.4: Need for Water Conservation in such Trying Times.

The burning of fuel wood to create the bonfire for Holika Dahan presents another serious environmental problem. There was a time when wood the fast depleting resource was collected just to light the Holi fire for Holika Dahan or the burning of the waste in our lives symbolically. So old things were heaped and thrown into the fire as a gesture of having done so. But factually wood was being consumed each year in large quantities for this ritual without grasping the tremendous loss both in terms of wastage of precious fuel and also polluting the environment with the smoke generated. The system of bonfires is still prevalent but reduced to an extent. A solution for this is the lighting of small symbolic fires using branches and twigs and waste material other than wood. Collective celebrations of the Holika Dahan is suggested to reduce the burning of wood and also foster a better festive spirit.

25.8 Preparation of Natural Colours of Holi

Generally Indians are aware of the use of natural and eco-friendly colours but due to changing attitude and changing life style, our priority are totally mismatched with the concept of natural and eco-friendly colours. Yet environmentally it is again the demand of time that we have to take U-turn for these methods. Ancient India was fully aware of the benefits of the fragrant natural and eco-friendly colours for our skin and health and also there therapeutic value. The ingredients of Gulal were purposely chosen for their emollient qualities. In Vrindavan, Holi is still played with actual flower petals chosen for their fragrance and colours such as rajnigandha, (*Polianthes tuberosa*) rose, marigold, jasmine *etc*. By using these safe, natural colours we can help to save our environment and conserve our biodiversity. It is to be noted that Holi can become more soothing after playing with eco-friendly colours. Natural colours are obtained from skin friendly sources such as turmeric (*Curcuma longa*), flower extracts, sandal wood powder, mehndi (*Lawsonia inermis*) *etc*.

25.9 Harmful Chemicals Used as Colours

Chemical colours available in the market are oxidized metals and harmful dyes. Table 25.1 is given to analyse the chemicals present in colours and their effects on health.

Table 25.1: Harmful Chemicals Present in Colours and their Ill Effects

Sl.No.	Colours	Chemical Compound	Health Effects
1	Black	Lead oxide (PbO)	Renal failure and learning disability
2	Blue	Prussian blue [$Fe_7(CN)18$]	Contact dermatitis
3	Green	Copper sulphate ($CuSO_4$)	Eye allergy, temporary blindness
4	Purple	Chromium iodide (CrI2)	Bronchial asthma and other forms of allergy
5	Red	Mercury sulphide (HgS)	Skin cancer and Minamata disease
6	Shiny	Powdered Glass	Skin problems, eye infections and allergy
7	Silver	Aluminum bromide (Al_2Br_6)	Carcinogenic effects

25.10 Common Tips for Celebrating a Safe Holi

Due to worldwide awareness of side effects generated from chemical colours, some researchers proposed following tips to enjoy a joyous Holi. Here it is must to define it in this article. By using natural and eco-friendly colours, one has to oil hair well; it will make it easy to get rid of the colour stuck in the hairs. Also it will limit the effect of chemicals to hairs and skin. During travel, keep the window of your car/bus/train tightly closed. Avoid taking bhang (*Cannabis indica*) and alcohol during festival. Take a bath with luke warm water after the Holi celebrations.

25.11 Removal of Holi Colours

Before people colour you red and green, take a look at these simple, natural tips to get rid of post-Holi colour. Hot water makes the colours faster and tougher to get rid of. The first thing you should do after playing Holi is splash your face with cold water and rinse hair. Make a paste especially with honey and add Multani mitti, scrub on the part of the body or use fruit scrub available in the market. Colours may not come off in one go, but the paste will really lighten the effect.

Mix lemon juice with equal parts of honey and apply it all over the face and body. Apply some coconut oil on a cotton ball and scrub off the colours using it. For face, make a mixture of calamine lotion and rose water and gently rub it into your skin till all of it is absorbed.

Rinse hair with water to allow the extra colours to wash off. Apply egg yolk or curd and leave it for about 30 minutes before using a mild shampoo. Make a hair pack by soaking a few fenugreek (*Trigonella foenum graecum*) seeds in curd. Wash your hair with a shampoo after 30 minutes.

25.12 Removal of Dyes using Aqueous Solution of Neem (*Azadirachta indica*)

In the past a number of conventional biological treatment processes have been used, which were not effective, some of which include coagulation and chemical oxidation, membrane separation process, electrochemical, reverse osmosis and aerobic and anaerobic microbial degradation but all these methods faced one or more limitations and none of them were successful for the complete removal of dye. It has been discovered that dyes can be effectively removed by adsorption process in which dissolved dye compounds were attached to the surface adsorbents. Researchers have exploited many low cost and biodegradable effective adsorbents. They were obtained from natural sources which successfully removed dye from aqueous solutions.

Neem (*Azadirachta indica*) husk activated with $ZnCl_2$, H_3PO_4 and KOH can be effectively used as a raw material for the removal of xylenol orange, procion red and remazolturqoise blue. Neem activated carbon competed favourably with the commercial adsorbent from coconut shell for the adsorption of xylenol orange, procion red and remazolturqoise blue. The three dyes used were subjected to both Langmuir and Freundlich isotherm model; Xylenol orange fit better into the Freundlich isotherm model than the Langmuir isotherm model, procion red fit better to the Langmuir isotherm model than the Freundlich model while remazolturqoise blue adsorption model fit to the two isotherms but fit better in the Langmuir model (Alau *et al.*, 2010).

25.13 Conclusion

In India with the advent of spring season just after winter, it is the season of festivals and festivities for different communities, where a spectrum of culturally very different rituals and modes of festival celebrations are witnessed. With the festival season comes a disaster. Pollution of various types is generated in large amounts all across the country, thereby adding an even greater load of pollutants and contaminants to our already over polluted environment, overburdened rivers, lakes, and seas. The festival of 'Holi' is proving to be an environmental risk due to the toxic colours. Unlimited and uncontrolled use of such dyes can lead to grave consequences in terms of human health and ecological balance. These colours are highly structured polymers and are very difficult to decompose biologically. The impact assessment and damage control led to suggest safer methods of getting rid of the 'Holi' dyes, in order to save our environment.

'Play safe, play smart and enjoy this colourful celebration to the core.'

REFERENCES

Alau K.K, Gimba CE and Kagbu J (2010) Removal of dyes from aqueous solution using Neem (*Azadirachta indica*) husk as activator. Archives of Applied Science Research, 2 (5): 456-461

Ahlström L, Eskilsson C S, Björklund E (2005) Determination of banned azo dyes in consumer goods. Trends Anal. Chem., 24 (1): 49-56.

Chaube, P., Indurkar, H and S, Moghe. S. (2010). Biodegradation and decolorisation of Direct Violet 51 and Tatrazine dye from isolated fungus (TYPE I) Asiatic Journal of Biotechnology Resources. 1: 45-56.

Cristea, G. Y and Vilarem, S. J. (2003). Ultrasound assisted enhancement in natural dye extraction from beetroot for industrial applications and natural dyeing of leather. Ultrason. Sonochem. 16 (6): 782-789

Das P, Goswami N, Borah P (2015) Development of Low cost eco-friendly 'Holi' Powder. International journal of Agriculture Innovation and research. 4(3)2319-1473.

Gardner J J, Lal D (2012). Impact of 'Holi' on the environment: A scientific study. Archives of Applied Science Research. 4 (3): 1403-1410.

Gokhale, S. B., Tatiya, A. U., Bakliwal, S. R and Fursule, R. A. (2004). Natural dye yielding plants in India. Nat Prod Rad. 3(4): 228- 234.

Ibrahim, S. F., Michael, M N, Tera, F. M and Samaha, S. H. (1997). Optimisation of the Dyeing process for chemically modified cotton fabrics. Colourage. pp-27

Mahanta, D, Tiwari, S. C. (2005). Natural dyes yielding plants and indigenous knowledge on dye preparation in Arunachal Pradesh, North East India. Curr Sci., 88: 1474-1480.

Papen H, von Berg. R (1998) Most Probable Number method (MPN) for the estimation of cell numbers of heterotrophic nitrifying bacteria in soil Plant and Soil, 199: 123-130

Rawat V.S. (2008). Unsafe, yet holi colour industry growing at 15 per cent. Business Standard, dated 5, Feb Lucknow India (2013).

Sharma A. (2013). Analysis of conscious awareness for natural colors. Arch Appl. Sci. Res., 5: 273-277.

Sharma A and Saxena R. (2013). Moderation of eco-friendly trends in Indian festival: Holi. Archives of Applied Science Research. 5 (3): 129-133.

Toxics Link (February 2000). The Ugly Truth behind the Colourful World Fact sheet. 8(1)

1. https: //mocomi.com/wpcontent/uploads/2012/03/Mocomi_TheZone_ GoingGreen_EcoFriendlyHoliColours_01.pdf

2. https: //www.unnatisilks.com/blog/holi-celebrate-it-differently-this-time/

3. https: //www.lih.travel/history-of-holi/

4. https: //www.holifestival.org/history-of-holi.html

Hindu Celebrations of Harchhath (Halashashti): An Old Tradition, Promoting Family Bonding and Conserving Diversity of Plants

Shashi Katiyar[1] and Arun Arya[2]

[1]*Pretty Petals School B 2474 Indira Nagar, Lucknow – 226 016, Uttar Pradesh, India*
[2]*Department of Environmental Studies, Faculty of Science, The Maharaja Sayajirao University of Baroda, Vadodara – 390 002, Gujarat, India*
e-mail: sarojarun10arya@rediffmail.com

ABSTRACT

The festivals are important from the cultural point of view. In India some or the other art is made on the festive occasions drawn on the floor or on the walls. Many kind of these things are made on cow dung coating using chalk and vegetable colours. In Janmashtami or festival of Ganesh puja their statues or replica are made in different forms and are displayed prominently for the benefit of devotees in different places. Ramleela is dramatized the by local actors of the village or a society for ten days culminating into death of demon king Ravana depicting victory of truth over injustice. Singing dancing and music promotes the harmony in community and maintains cultural environment. It is obvious from the preceding importance of the festivals in rural life there value may not be only religious but also have other points of view in life. It may be helpful in protecting wildlife increasing plant biodiversity and thus rejuvenating life in mother Earth. A festival is celebrated on the sixth day of Krishna Paksha of Bhadrapada (August- September) month as the birth anniversary of Shri Balramji, the elder brother of Lord Krishna. This fast is observed by the mothers for the long life of the children. The fast of Hal Shashthi is considered very special for the longevity and prosperity of sons. The Plow and Palash are worshiped on this festival. The paintings of family members are made on clean wall and puffed grains of different varieties are used in prayer. The chapter details conservation of plant diversity in certain hindu festivals and particularly in Harchhath.

Keywords: *Hindu celebration, Harchhath (Halashashti), Ritual, Plant diversity, Conservation.*

26.1 Introduction

The term 'cultural heritage' has changed content considerably in recent decades, partially owing to the instruments developed by UNESCO. Cultural heritage does not end at monuments and collections of objects. It also includes traditions or living expressions inherited from our ancestors and passed on to our descendants, such as oral traditions, performing arts, social practice, rituals, festive events, knowledge and practices concerning nature and the universe or the knowledge and skills to produce traditional crafts.

While fragile, intangible cultural heritage is an important factor in maintaining cultural diversity in the face of growing globalization. An understanding of the intangible cultural heritage of different communities helps with intercultural dialogue, and encourages mutual respect for other ways of life. The importance of intangible cultural heritage is not the cultural manifestation itself but rather the wealth of knowledge and skills that is transmitted through it from one generation to the next. The social and economic value of this transmission of knowledge is relevant for minority groups and for mainstream social groups within a State, and is as important for developing States as for developed ones. Romain Rolland, Nobel Laureate and French Philosopher has praised India by saying:

" If there is one place on the face of this Earth where all the dreams of living men have found a home from the very earliest days when man began the dream of existence, it is India."

India is known for its culture and unique way of life style. Its traditions and a number of fairs and festivals. Durga Puja is an annual festival celebrated in September or October, most notably in Kolkata, in West Bengal of India, but also in other parts of India and amongst the Bengali diaspora. It marks the ten-day worship of the Hindu mother-goddess Durga. The worship of the goddess then begins on the inaugural day of Mahalaya, when eyes are painted onto the clay images to bring the goddess to life. It ends on the tenth day, when the images are immersed in the river from where the clay came. Thus, the festival has also come to signify 'home-coming' or a seasonal return to one's roots. Durga Puja is seen as the best instance of the public performance of religion and art, and as a thriving ground for collaborative artists and designers. It was recognized by UNESCO as world heritage in 2021. The festival is characterized by large-scale installations and pavilions in urban areas, as well as by traditional Bengali drumming and veneration of the goddess. During the event, the divides of class, religion and ethnicities collapse as crowds of spectators walk around to admire the installations.

Similar to other festivals is Garba dance performed in large groups in different parts of the country. Particularly in Gujarat it is 9 days grand festival celebrated during Navratri. Details of intangible heritage and particularly Hindu celebration of Harchhath (Halashashti) and rituals performed are described and possible scientific basis of conserving plant diversity is discussed.

26.2 Intangible Cultural Heritage

Intangible cultural heritage refers to the practice representations, expressions, knowledge and skill handed down from generation to generation. This Heritage provides communities with a sense of identity and is continuously recreated in response to their environment search, as the detail turn off the environment destruction of cultural heritage and resources which are the problems worldwide and just demands a combined and holistic effort from the international community.

Until the very last decades of the 20th century this holistic perception of culture had not been adequately perceived by the international community. The main legal instruments adopted with the purpose of protecting cultural heritage, either applicable exclusively in the event of armed conflict (Anonymous 1954) or also in time of peace (Anonymous 1972), were solely devoted to tangible cultural expressions, the significance of which was to be evaluated on the basis of an objective and standardized perception of their artistic, aesthetic, architectural, visual, scientific, and economic value. Thanks to these instruments, this perspective, developed in the Western world, became the globalized evaluation method used by the international community as a whole in order to establish the value of cultural heritage. In meta-juridical terms, this lack of perception of the need to provide adequate safeguarding for immaterial cultural heritage was presumably the result – consciously or not – of the confidence that this heritage, being an essential part of the cultural and social identity of human communities, was automatically and appropriately preserved and developed at the local level, in the context of the social evolution of the communities concerned. In other words, the depositaries of intangible cultural heritage (IHC) were considered to accomplish spontaneously and appropriately the mission of transmitting to future generations the necessary knowledge to preserve and perpetuate their own immaterial heritage, with no need of any international action in that respect. Although this spontaneous process could be considered as having worked out fine for many centuries, its dynamics were abruptly broken by the advancement of the process of globalization which has marked the most recent decades (Lenzerini 2011).

26.2.1 Characters of Intangible Cultural Heritage

☆ **Traditional, contemporary and living at the same time**: intangible cultural heritage does not only represent inherited traditions from the past but also contemporary rural and urban practices in which diverse cultural groups take part;

☆ **Inclusive:** we may share expressions of intangible cultural heritage that are similar to those practiced by others. Whether they are from the neighboring village, from a city on the opposite side of the world, or have been adapted by peoples who have migrated and settled in a different region, they all are intangible cultural heritage: they have been passed from one generation to another, have evolved in response to their environments and they contribute to giving us a sense of identity and continuity, providing a link from our past, through the present, and into our future. Intangible cultural heritage does not give rise to questions of whether or not certain practices are specific to a culture. It contributes to social cohesion, encouraging a sense of identity and responsibility which helps individuals to feel part of one or different communities and to feel part of society at large;

☆ **Representative:** intangible cultural heritage is not merely valued as a cultural good, on a comparative basis, for its exclusivity or its exceptional value. It thrives on its basis in communities and depends on those whose knowledge of traditions, skills and customs are passed on to the rest of the community, from generation to generation, or to other communities;

☆ **Community-based**: intangible cultural heritage can only be heritage when it is recognized as such by the communities, groups or individuals that create, maintain and transmit it – without their recognition, nobody else can decide for them that a given expression or practice is their heritage.

The Japanese government designates Important Intangible Cultural Properties and individuals or groups who perform those skills or techniques in an outstanding manner are recognized as holders or group holders of them, and measures are taken for their protection. The second group, Intangible Folk Cultural Properties, consists of folk culture, in other words, intangible culture that has been passed down in the form of manners or customs, *etc.* It includes festivals and ceremonies, folk performing arts and a variety of techniques used in the production of associated objects. Among these, the traditions that are indispensable for understanding the fundamental basis of Japanese culture are designated as Important Intangible Folk Cultural Properties, and measures are taken for their protection. In line with the Meiji policy, put in place in 1868, of favoring westernization in Japan to the detriment of Japanese traditions that were regarded as impediments to the country's modernization, and the privileging of Shintoism to the detriment of Buddhism in order to enhance Japanese nationalism, many elements of Japanese cultural heritage were destroyed or sold abroad. The Meiji government then realized that there was an urgent need to protect Japanese cultural properties by statutory means, and in 1871 a decree was issued entitled the Preservation of Antique and Ancient Objects which was the first Japanese legal measure to protect cultural property. According to this decree, 'Antique and Ancient Objects' included not only objects that had aesthetic value, but also other objects relating to folklore, such as agricultural tools and children's toys. Antique and ancient objects preserved under this decree were considered as documentation that showed the development of the historical living conditions of the Japanese people (Kawamura 2002; Wada 2002).

The protection of Intangible Cultural Properties must take into account these kinds of qualities. However, the challenge is how to ensure that there are successors to take up and practice the techniques and skills. The cultivation of successors is indispensable; however, an even larger problem is that without the support of related techniques and skills, the heritage itself cannot subsist. The performing arts, for the most part, consist of group performances by a number of people. It follows, therefore, that the performing arts cannot be preserved and carried on simply by ensuring the existence of a leading actor. The system of collective recognition for the entire group of performers is one effort to resolve this problem; however, there remains the matter of those who are not yet at the level of recognition, in other words, future successors. Although there is some government support for their training, the present situation is that we count a great deal on the personal drive and effort of those who carry on the tradition. In addition, make up, costume, and other backstage elements are essential for a performance. Although these should be protected as Conservation Techniques, very few examples are selected from the performing arts, and the reality is that protection does not go beyond awarding a certificate to the holder of the Technique, and as a result, the problem is indeed grave.

26.3 Conserving Culture of the Region through Festivals

Social importance of the festivals perhaps surpasses their religious importance. On the occasion of festivals people of different castes unite, and this helps to reduce the tension in between them. People belonging to the same classes visit each other's house. On the occasion of Holi they embrace each other like in Eid. This serves to strengthen the organization of the society. According to Atharva Veda:

Yena deva na viyanti no ca vidvisate mithah I

Tatkrnmo brahma vo grhe samijnanam porusebhyah II Av.III 30.4

"Blessed by the elders and without hatred against each other, may you be firmly knit in your family and do not fall apart from sweet and lovable life"(Das 2011).

On the occasion of major festivals like Holi and Dussehra, the ceremonies are directed along communal lines. This community consciousness among the people strengthen the organization of the society. And in new versions of festivity girls and boys besides other near relatives therein one house increases harmony among one another. Festivals serves as an excuse for the parents to call their daughters home again in these married women also get an excuse for meeting their relatives once in a while. In this way festivals of old people whether they are of different religions members of different castes in belong to different classes in society bring an opportunity for mutual meeting on these occasions.

Festivals have merrymaking and lot of excitement and enthusiasm. During festival of Diwali people wear new clothes they decorate their houses with many colorful lights and explode crackers. On the other major festivals people take the day off from their work. In a month of Shravan use of swing helps to swing two hearts. On the occasion of Holi there could hardly be more recreation then when colors are flying in the air. People dance with pleasure from the psychological point of view location strike Holi have the importance because they are an opportunity for the people to give expression to their very attendance is all of which make it easier for them to keep their mental balance to cooperate with each other.

26.4 Social Education Enriches Our culture

Maulana Abul Kalam Azad, Education Minister, while inaugurating the village education seminar of UNESCO said social education means education for entire man. It will make him literate so that the knowledge of the world may become available to him. It will tell him exactly how the world become available and how he has to adapt himself to this environment. And how he has to derive the maximum benefit with available modern methods, and this will affect his financial improvement. He will know the science of humanity so that his domestic life may be prosperous and happy. Finally this educated citizen may be able to assist the government to the decision for making the country great and establishing peace after he has acquired some knowledge of the real things originally. The meaning attached with social education was nearly literacy but now it has also included education for the adults. Social education is the stimulation in each citizen of a greater awareness of the advantage of the pattern of society is conceived in the constitution and persistent desire of progress toward such pattern socially economically and culturally. Pt. Jawaharlal Nehru said education will elevate all the various kinds of standards of life be the physical, mental, or intellectual in order that people may be in a position to make its contribution in all its share of national progress (Sharma 1975).

26.5 Conserving Biodiversity through Festivals

India is known to be a land of vast diversities, interesting cultures and different traditions. Some of these festivals and traditions have lasted for generations and continue to exist even during present times. Praying different Gods- Goddesses and observing fast is common ritual. During Karva chauth and Chhat puja even taking water is not allowed. Sweets, flowers and fruits are used routinely. Different Gods have their favorite plants/trees. Nag panchmi is a unique festival where prayer of snakes is performed. Kumbh celebrations are observed after every 12 years at four places Prayagraj, Hardwar, Nasik and Ujjain and lakhs of Hindus visit there and offer prayers. It is said that when Lord Brahma ji was carrying the auspicious urn few drops of elixir or amrit fell at these four places. Chhath puja is an ancient Hindu Vedic festival dedicated to the Hindu Sun God, and Chhathi Maiya. The Surya, is considered as the god of energy and of the life-force. It is worshiped during the Chhath festival in order to thank Surya for sustaining life on earth and to promote well-being, prosperity and progress. The festival's origin could perhaps be traced to the Harappan Civilization and to the matriarchal/matrilineal days of pre-historic India. This fascinating four-day festival is among the few all-women ceremonies that is still in practice today, which is conducted by the women folks of a family, without the supervision of a male priest. It is believed that the ritual of Chhath puja may even predate the ancient Vedas texts, as the Rigveda contains hymns worshiping the Sun god and describes similar rituals. The rituals also find reference in the Sanskrit epic poem Mahabharata in which Draupadi is depicted as observing similar rites. It is said that in the times of the Mahabharata, Chhath Puja was performed by Draupadi, the wife of Pandava Kings. Once during the long exile from their kingdom, thousands of wandering hermits visited their hut. Being devout Hindus, the Pandavas were obliged to feed the monks. But as exiles, the Pandavas were not in a position to offer food to so many hungry hermits. Seeking a quick solution, Draupadi approached Saint Dhaumya, who advised her to worship Surya and observe the rituals of the Chhath for prosperity and abundance.

A) **Teej celebrations:** The word Hartalika comes from 'Harat', which means abduction and 'Aalika' which means female friend. Hartalika Teej Vrat is observed during the Shukla Paksha Tritiya of Bhadrapada month. On this day, statues of Lord Shiva and Goddess Parvati are carved out of sand, and worshiped for marital bliss and progeny. On the occasion of Teej Vrat, married and unmarried women observe fast for a peaceful married life and for a loving husband respectively. Some women even observe *Nirjala* Vrata (fasting without water). While women celebrate Haryali Teej and Kajari Teej by visiting their parent's house, they come back to their in-laws for Hartalika Teej celebration. Women wear new clothes, adorn the best jewellery and use henna. This festival is celebrated in different states of India and Nepal. Hartalika Teej festival is celebrated as Gowri Habba in Tamil Nadu, Andhra Pradesh and Karnataka. On the other hand, in Maharashtra, women conduct Hartalika Puja and observe fast on Teej as well.

B) Kajari Teej is also called Badi Teej. The people in Southern and Northern parts of the country celebrate this festival. The festival is observed in Rajasthan, U P, Gujarat, Bihar and M P. People celebrate the festival in order to welcome monsoon. On this day, women will wear new dresses and apply henna. In a few parts of the country, holy margosa (*Azadirachta indica*) tree is worshipped.

Table 26.1: Some Important Hindu Festivals and Plants Used for Worship and as Prasad

Sl.No.	Festivals	Hindu Month	English Month	Plants used
1.	Sakath	Magh	January	Sweet potato and Sesame or Til
2	Sathtila Ekadshi	Magh	January	Six types of uses of Til (*Sesamum indicum*)/Prasada
3.	Makar sankranti	Magh	January	Til (*Sesamum indicum*) laddos/pulses and rice khichdi
4	Lohri	Magh	January	Wood is used for fire/ground nuts are eaten
5.	Basant Panchmi	Magh	Jan-February	Yellow rice/wheat/Barley/yellow flowers
6.	Holi	Phagun	Mar- April	Wood is used for fire/Palas (*Butea monosperma*) and flower colors are used to celebrate
7.	Amla Ekadashi	Phagun	March	Amla (*Emblica officinalis*)
8.	Shivratri			Leaves of *Aegle marmelos*/*Datura stramonium* fruit *Bhang* (*Cannabis*) is used in drinks
9.	Jaganath rath yatra	Chaitra		Statue of Margosa (*Azadirachta indica*) tree woodCoconut/Puffed rice
10	Nav Sanvatsar	Chaitra	March	Mango leaves for toran/Margosa leaves are eaten/Prayer with flowers and rice
11	Gangor	Chaitra	March	Gori tritiya vrat/Bixa/Henna (*Lawsonia inermis*)
12.	Ram Navami	Chaitra	March	Perform Kanya Puja,kheer is prepared as an offering to God
13	Akshaya Tritiya	Baisakh	May	Malti/Champa/lotus/Basil and pink flowers/Donate barley/eat germinated gram and sweet sattu
14.	Jaith Dushehra	Jaith	May- June	In the evening, perform Ganga Aarti and offer flowers, betel leaves, fruits, in Ganga
15	Bhimseni ekadashi	Jaith	June	Decoration with henna and prayer of Peepal *Ficus religiosa*

Sl.No.	Festivals	Hindu Month	English Month	Plants used
16	Vat Savitri	JaithPurnima	June	Ladies tie thread around *Ficus/* Vad and Pray for the long life of husband/turmeric/*Santalum*/roli used on thread
17.	Assari	Assar	June- July	Santalum/Red rose garland/
18.	Guru Purnima	Assar		Coconut/Food and offerings to teachers and saints
19.	Kokila vrat	Assar	June- July	On Assar Purnima/Amla oil used on body
20.	Raksha Bandhan	Savan	July- Aug.	Festival of brothers and sisters/ Roli/rice/coconut/Mustard/Tying Rakhi to plants
21	JanmaAshtmi	Bhadoon	August- Sept	Charnamrit (Curd/Butter/sugar/ Honey/*Ocimum* leaves) Panjiri (wheat flour/butter, sugar and dry fruits)
22.	Rishi Panchmi	Bhadoon		Snake gourd (*Trichosanthes cucumerina*) other fruits
23.	Anant Chaturdashi	Bhadoon Chaturdashi		Rice/Durva (*Cynodon dactylon*) Cotton (*Gossypium*) Turmeric (*Curcuma longa*)
24.	Karma festival	Bhadoon	Sept.	Karam" (*Mitragyna parvifolia*). Rice beer/flowers
25.	Hartalika Teej	Bhadoon	July- Aug.	Ladies observe fast and prayer with flowers
26.	Kajri Teej			In a few parts of the country, holy neem trees are also worshipped
27.	Kevra Teej			Kevada (*Pandanus*) plant is popularly known as fragrant screw pine. Thazhampoo (Tamil), Ketaki/Dhuli Pushpam (Sanskrit).
28.	Durga Puja			Chandan/mango/durva/wheat/ barley/cotton sut
29.	Onam		Aug.- Sept.	The floral Rangoli, known as *Pookkalam*, The Kurichians tribe of Parambikkulam celebrate it as a festival for eating new grains. The feast include fried pieces of banana coated with Jaggery, Pappadam, Injipuli,Thoran, Mezhukkupuratti, Kaalan, Olan, Avial, Sambhar, Dal served along with ghee, Erisheri, Molosyam, Rasam, Puliseri (also referred to as Velutha curry), Kichadi (not to be confused with Khichdi) and Pachadi (its sweet variant), and Payasam.

Sl.No.	Festivals	Hindu Month	English Month	Plants used
30	Vijaya Dashmi	Kwaar	Sept- Oct	Shami (*Prosopis cineraria*)/ Merrigold (*Tagetes*)/Mango (*Mangifera*) leaves
31.	Karva chauth	Kartik	Oct- Nov.	Rice/udad pulse
32.	Ahoi Ashtami	Kartik	October	Alpana is drawn. After spreading wheat on the floor or on the wooden stool, one water-filled Kalash, nozzle is blocked with the shoots of the grass. Sarai Seenka which is a type of willow. The seven shoots of the grass is also offered. If shoot is not available cotton buds are used.
33.	Kartik Purnima	Kartik	"	Rice kheer/dry fruits
34.	Deepawali	Kartik	"	Sweets/puffed rice
35.	Annakut	Kartik	Oct	Sweets and 56 different types of food items
36.	Tulsi vivah	Kartik ekadshi	Oct.	Basil is worshipped
37.	Indira Ekadashi	Aswin/ Ekadashi		Basil (*Ocimum sanctum*)/Fruits consumed after fast
38.	Chhath	Kartik		Vegetable of gram leaves and rice Milk, rice and jaggery preparation - Kharna Bamboo basket or Supe used containing Coconut, sugarcane, rice, banana leaves, banana, apple, water chestnut, turmeric, radish and ginger plant Sweet potato, betel leaves and betel nut
39.	Harchhath Puja	Bhadon	Sept.	Use baked grains/pulses of 7 types. Eat *Madhuca* (dried flowers) on *Butea monosperma* leaves
40.	Vishu	Chaitra	April	Cassia fistula/Indian laburnum (*Kani Konna*), rice, coconut, cucumber, fruits and other harvest products. golden lemon, *Artocarpus heterophyllus* or jackfruit, Kanmashi, betel leaves, areca nut.

26.5.1 Harchhath Puja (Halashashti) for the Longevity and Prosperity of Sons

This festival is celebrated on the Chhath of Bhadon Krishna Paksha (September). To mark the birth of elder brother of Shri Krishna, known as Balram. This fast is performed by both the sons and daughters, women in Chhattisgarh worship it in a big way. Only women perform this puja in Madhya Pradesh and Uttar Pradesh. In this fast, there is a tradition to eat fruits of trees without sown grains, like pashar chawal *etc.* Only milk of buffalo having a calf is used. The ceremony is performed by all daughter-in-law women (Saubhagyavati and widow). This fast is observed for the long life and prosperity of the sons. In this fast, women fill five or seven roasted grains or dry fruits in seven small earthen pots, depending on the son. One branch of Jari (small thorny bush of *Acacia* or *Zizyphus*), one branch of Palash (*Butea monosperma*) and one branch of Nari (a type of creeper *Ipomoea aquatica*) also called as water spinach is worshiped by burying it in the ground or an earthen pot. The women end the fast by consuming curd made from the milk of a buffalo with calf and dried flowers of mahuva or butter tree (*Madhuca indica*) on palash (*Butea monosperma*) leaves. In Uttar Pradesh, worship is done by making beautiful drawing of family members and Chhat Mata on the clean cow dung plastered wall.

This fast is observed by the mothers for the long life of their child. In the Halashthi fast, women take a vow to observe the Halashashti fast from the early morning after taking a bath *etc.* Plow is worshiped on this festival, which is held on Balaram Jayanti. The idols of Lord Shiva, Parvati, Swami Kartikeya and Ganesha are installed under Kans and Kusha grass (*Desmostachya bipinnata*) and worship is done with devotion with incense sticks, lamps, flowers *etc.* Mahua, Pasai rice, gram, maize, jowar, soybean and paddy lye (puffed grains) and buffalo milk and dung have special importance in the worship of Harchath. In the afternoon, worship is done with reverence by establishing Harchath with Mahua, plum branch, Kansas flower in the house and courtyards. Satgajra oil, bangle, kajal, wooden kakai, bamboo tukaniya, mirror, small doublets, coconut, banana, cucumber are offered. Worship of Satanja (gram, barley, wheat, paddy, tur, maize and moong), barley ears/keep any ornaments and turmeric colored cloth near Harchhath. After worshiping, ladies perform Havan with butter made from buffalo milk.

The importance of intangible cultural heritage is not the cultural manifestation itself but rather the wealth of knowledge and skills that is transmitted. Social and economic values of this transmission of knowledge relevant to minority groups and mainstream social groups, and this is equally important for developing states in saving heritage at risk. Many culture and traditions help in conservation of plant diversity. ICH refers to the practice representations, expressions, knowledge and skill handed down from generation to generation.

Figures 26.1a-f

(a and f) The ladies performing the Harchhath puja in a village; (b) Offering of seven baked grains to the God in the ceremony; (d) A printed paintings is now used by some ladies in the ceremony; (c and e) Hand drawn paintings made on the wall by ladies in the village.

Climate change, continued deforestation and the ongoing spread of deserts inevitably threaten many endangered species and results in the decline of traditional craftsmanship and herbal medicine as raw materials and plant species disappear. The paper discusses how traditions were followed and numerous plants were used in different festivals. These festivals need preserving the traditional culture and plant diversity. The root of the crisis in intangible folk cultural heritage lies in that change. Festivals and such events that lie at the basis of local society have shown deep-rooted staying power; the dates of such festivals, however, tend to differ from the past. Faced with the loss of young people to the cities, small local communities have reached the point where festivals are timed to match the visits of the younger generation in order to ensure continuity through their participation.

Preservation of cultural diversity, as emphasized by Article 1 of the UNESCO Universal Declaration on Cultural Diversity (UNESCO 2001), is embodied in the uniqueness and plurality of the identities of the groups and societies making up humankind. Being a 'source of exchange, innovation and creativity', cultural diversity is vital to humanity and is inextricably linked to the safeguarding of ICH. Safeguarding the ICH means neither collecting samples, as can be done with animal species in a zoo or with vegetable matter in a botanical garden, nor classifying the alleged best manifestations according to the views of external observers. By safeguarding ICH means preserving its link with living cultures and its role in the identity of its holders, as well as allowing the transmission of its different shades and colors to future generations. Protecting the natural environment is often closely linked to safeguarding a community's cosmology, as well as other examples of preserving its intangible cultural heritage observed in form of traditional festivals like Holi, Vatsavitri, Teej and Harchhath Puja or Halashashti *etc.*

REFERENCES

Anonymous (1954) Convention for the Protection of Cultural Property in the Event of Armed Conflict, 249 UNTS 240, and its two Protocols, respectively of 1954 (249 UNTS 358) and 1999 (38 ILM 769).

Anonymous (1990) India: Prehistoric and protohistoric periods. Pub. Div. New Delhi 52pp.

Anonymous (1972) Convention concerning the protection of the World Cultural and Natural Heritage (World Heritage Convention), 1037 UNTS 151.

Anonymous (2013) "The feel of Onam". *Team MetroPlus (15 September 2013). The Hindu.* Retrieved 17 September 2013.

Arya A. (2020) Dastaan: Hamari Tumhari. (in Hindi) Shabd Publication. Vadodara 260 pp

Das P (2011) Glimpses of Atharva Veda. Vaidika Anushandhan Pratisthan, Bhubaneswar. 54 pp

Gangrade P C (2009) Hinduyon ke vrat parve aur Teej teuhar (in Hindi) Pustak Mahal, Delhi 216pp.

Kawamura K (2002) 21 seikai to Bunkazai no Hogo. 21st century society and the protection of cultural property. In: Kawamura K (ed) Essays on policies for the protection of cultural properties, Tokaidaigaku, Shuppankai, Tokyo.

Lenzerini F (2011) Intangible Cultural Heritage: The Living Culture of Peoples. The European Journal of International Law. 22(1): 101-120. doi: 10.1093/ejil/chr006

Sharma R N (1975) Society and culture in India. Rajhans Prakashan, Meerut. 210pp.

UNESCO (2001) General Conference on 2nd Nov. 2001. The text of the Declaration is available at: http: //portal.unesco.org/en/ev.phpURL_ID=13179 and URL_DO=DO_TOPIC and URL_SECTION=201.html

Wada K (2002) Evolution of policies on the protection of cultural properties. In: Kawamura K (ed) Essays on policies for the protection of cultural properties, Tokaidaigaku, Shuppankai, Tokyo.

Index

Sufferings 336

Sulfuryl fluoride 118

Sun temple 136, 250

Sustainable Development 446

Sustainable management 149

Sweet melon 9

Synthetic color 480

Syzygium armoaticum 220

T

Tagore house 435

Taj Mahal 173, 199, 234, 381

Tangible Cultural Heritage 446

Teej Celebrations 493

Temperature 25

Termitary 362

Termite bait 375

Termite mound 365

Termite nest 367

Termites 361

Termopsidae 370

Terrorist attack 173

Textiles 142

Thunderstorm 175

Thymol 220

Thymus vulgaris 220

Tinea pellionella 117

TiO_2 223

Tirthankara 185

Tolypothrix 348

Tolypothrix byssoidea 242

Torula 137

Traditional biocides 239

Trichoderma 137

Tringa guttifer 109

Turin Museum 37

Turmeric 129

Tutankhamun 44

U

UNESCO world heritage 246

Urban spaces 310

Ushkar 451

Usnea

Uttaranchal 109

UV radiation 131

UV-C irradiation 239

V

Vacha 77

Vakatakas period 243

Vanellus gregarious 109

Varied Carpet beetle 113

Vegetable 11

Venice Charter 431

Verrucaria 351

Veshtan 74

W

Wall paintings 159

Wasserschlacht 475

Watermelon Festival 476

Wax

Weathering 282

Wheat 11

Wood 142

Wood decay 366

Wood pulp 81

Wooden monuments 260

Wooden objects 361

World Heritage site 295

Z

Zebtane 290